Graduate Texts in Mathematics 213

Springer
New York
Berlin
Heidelberg
Barcelona
Hong Kong
London
Milan
Paris
Singapore
Tokyo

Graduate Texts in Mathematics

(continued after index)

Klaus Fritzsche
Hans Grauert

From Holomorphic Functions to Complex Manifolds

With 27 Illustrations

Springer

Klaus Fritzsche
Bergische Universität Wuppertal
Gaußstraße 20
D-42119 Wuppertal
Germany
Klaus.Fritzsche@math.uni-wuppertal.de

Hans Grauert
Mathematisches Institut
Georg-August-Universität Göttingen
Bunsenstraße 3-5
D-37073 Göttingen
Germany

Mathematics Subject Classification (2000): 32-01, 32Axx, 32D05, 32Bxx, 32Qxx, 32E35

Library of Congress Cataloging-in-Publication Data
Fritzsche, Klaus.
From holomorphic functions to complex manifolds / Klaus Fritzsche, Hans Grauert.
p. cm. — (Graduate texts in mathematics ; 213)
Includes bibliographical references and indexes.
ISBN 0-387-95395-7 (alk. paper)
1. Complex manifolds. 2. Holomorphic functions. I. Grauert, Hans, 1930– II. Title.
III. Series.
QA331.7 .F75 2002
515′.98—dc21 2001057673

Printed on acid-free paper.

Production managed by Michael Koy; manufacturing supervised by Erica Bresler.
Photocomposed from the authors' LaTeX files.
Printed and bound by Maple-Vail Book Manufacturing Group, York, PA.
Printed in the United States of America.

9 8 7 6 5 4 3 2 1

ISBN 0-387-95395-7 SPIN 10857970

Springer-Verlag New York Berlin Heidelberg
A member of BertelsmannSpringer Science+Business Media GmbH

Preface

The aim of this book is to give an understandable introduction to the theory of complex manifolds. With very few exceptions we give complete proofs. Many examples and figures along with quite a few exercises are included. Our intent is to familiarize the reader with the most important branches and methods in complex analysis of several variables and to do this as simply as possible. Therefore, the abstract concepts involved with sheaves, coherence, and higher-dimensional cohomology are avoided. Only elementary methods such as power series, holomorphic vector bundles, and one-dimensional cocycles are used. Nevertheless, deep results can be proved, for example the Remmert–Stein theorem for analytic sets, finiteness theorems for spaces of cross sections in holomorphic vector bundles, and the solution of the Levi problem.

The first chapter deals with holomorphic functions defined in open subsets of the space $\mathbb{C}^n$. Many of the well-known properties of holomorphic functions of one variable, such as the Cauchy integral formula or the maximum principle, can be applied directly to obtain corresponding properties of holomorphic functions of several variables. Furthermore, certain properties of differentiable functions of several variables, such as the implicit and inverse function theorems, extend easily to holomorphic functions.

In Chapter II the following phenomenon is considered: For $n \geq 2$, there are pairs of open subsets $H \subset P \subset \mathbb{C}^n$ such that every function holomorphic in H extends to a holomorphic function in P. Special emphasis is put on domains $G \subset \mathbb{C}^n$ for which there is no such extension to a bigger domain. They are called domains of holomorphy and have a number of interesting convexity properties. These are described using plurisubharmonic functions. If G is not a domain of holomorphy, one asks for a maximal set E to which all holomorphic functions in G extend. Such an "envelope of holomorphy" exists in the category of Riemann domains, i.e., unbranched domains over $\mathbb{C}^n$.

The common zero locus of a system of holomorphic functions is called an analytic set. In Chapter III we use Weierstrass's division theorem for power series to investigate the local and global structure of analytic sets. Two of the main results are the decomposition of analytic sets into irreducible components and the extension theorem of Remmert and Stein. This is the only place in the book where singularities play an essential role.

Chapter IV establishes the theory of complex manifolds and holomorphic fiber bundles. Numerous examples are given, in particular branched and unbranched coverings of $\mathbb{C}^n$, quotient manifolds such as tori and Hopf manifolds, projective spaces and Grassmannians, algebraic manifolds, modifications, and toric varieties. We do not present the abstract theory of complex spaces, but do provide an elementary introduction to complex algebraic geometry. For example, we prove the theorem of Chow and we cover the theory of divisors and hyperplane sections as well as the process of blowing up points and submanifolds.

The present book grew out of the old book of the authors with the title *Several Complex Variables*, Graduate Texts in Mathematics 38, Springer Heidelberg, 1976. Some of the results in Chapters I, II, III, and V of the old book can be found in the first four chapters of the new one. However, these chapters have been substantially rewritten. Sections on pseudoconvexity and on the structure of analytic sets; the entire theory of bundles, divisors, and meromorphic functions; and a number of examples of complex manifolds have been added.

Our exposition of Stein theory in Chapter V is completely new. Using only power series, some geometry, and the solution of Cousin problems, we prove finiteness and vanishing theorems for certain one-dimensional cohomology groups. Neither sheaf theory nor $\overline{\partial}$ methods are required. As an application Levi's problem is solved. In particular, we show that every pseudoconvex domain in $\mathbb{C}^n$ is a domain of holomorphy.

Through Chapter V we develop everything in full detail. In the last two chapters we deviate a bit from this principle. Toward the end, a number of the results are only sketched. We do carefully define differential forms, higher-dimensional Dolbeault and de Rham cohomology, and Kähler metrics. Using results of the previous sections we show that every compact complex manifold with a positive line bundle has a natural projective algebraic structure. A consequence is the algebraicity of Hodge manifolds, from which the classical period relations are derived. We give a short introduction to elliptic operators, Serre duality, and Hodge and Kodaira decomposition of the Dolbeault cohomology. In such a way we present much of the material from complex differential geometry. This is thought as a preparation for studying the work of Kobayashi and the papers of Ohsawa on pseudoconvex manifolds.

In the last chapter real methods and recent developments in complex analysis that use the techniques of real analysis are considered. Kähler theory is carried over to strongly pseudoconvex subdomains of complex manifolds. We give an introduction to Sobolev space theory, report on results obtained by J.J. Kohn, Diederich, Fornæss, Catlin, and Fefferman ($\overline{\partial}$-Neumann, subelliptic estimates), and sketch an application of harmonic forms to pseudoconvex domains containing nontrivial compact analytic subsets. The Kobayashi metric and the Bergman metric are introduced, and theorems on the boundary behavior of biholomorphic maps are added.

Prerequisites for reading this book are only a basic knowledge of calculus, analytic geometry, and the theory of functions of one complex variable, as well as a few elements from algebra and general topology. Some knowledge about Riemann surfaces would be useful, but is not really necessary. The book is written as an introduction and should be of interest to the specialist and the nonspecialist alike.

We are indebted to many colleagues for valuable suggestions, in particular to K. Diederich, who gave us his view of the state of the art in $\overline{\partial}$-Neumann theory. Special thanks go to A. Huckleberry, who read the manuscript with great care and corrected many inaccuracies. He made numerous helpful suggestions concerning the mathematical content as well as our use of the English language. Finally, we are very grateful to the staff of Springer-Verlag for their help during the preparation of our manuscript.

Wuppertal, Göttingen, Germany
Summer 2001

Klaus Fritzsche
Hans Grauert

Contents

Chapter I

Holomorphic Functions

1. Complex Geometry

Real and Complex Structures. Let V be an n-dimensional complex vector space. Then V can also be regarded as a $2n$-dimensional real vector space, and multiplication by $\mathsf{i} := \sqrt{-1}$ gives a real endomorphism $\mathsf{J} : V \to V$ with $\mathsf{J}^2 = -\mathrm{id}_V$. If $\{a_1, \ldots, a_n\}$ is a complex basis of V, then $\{a_1, \ldots, a_n, \mathsf{i}a_1, \ldots, \mathsf{i}a_n\}$ is a real basis of V.

On the other hand, given a $2n$-dimensional real vector space V, every real endomorphism $\mathsf{J} : V \to V$ with $\mathsf{J}^2 = -\mathrm{id}_V$ induces a *complex structure* on V by

$$(a + \mathsf{i}b) \cdot v := a \cdot v + b \cdot \mathsf{J}(v).$$

We denote this complex vector space also by V, or by (V, J), if we want to emphasize the complex structure.

If a complex structure J is given on V, then $-\mathsf{J}$ is also a complex structure. It is called the *conjugate complex structure*, and the space (V, J) is sometimes denoted by $\overline{V}$. A vector $v \in V$ is also a vector in $\overline{V}$. If z is a complex number, then the product $z \cdot v$, formed in $\overline{V}$, gives the same vector as the product $\overline{z} \cdot v$ in V.

Our most important example is the *complex n-space*

$$\mathbb{C}^n := \{\mathbf{z} := (z_1, \ldots, z_n) \,:\, z_i \in \mathbb{C} \text{ for } i = 1, \ldots, n\},$$

with the *standard basis*

$$\mathbf{e}_1 := (1, 0, \ldots, 0), \ \ldots\ , \mathbf{e}_n := (0, \ldots, 0, 1).$$

We can interpret $\mathbb{C}^n$ as the *real $2n$-space*

$$\mathbb{R}^{2n} = \{(\mathbf{x}, \mathbf{y}) = (x_1, \ldots, x_n, y_1, \ldots, y_n) \,:\, x_i, y_i \in \mathbb{R} \text{ for } i = 1, \ldots, n\}^1,$$

together with the complex structure $\mathsf{J} : \mathbb{R}^{2n} \to \mathbb{R}^{2n}$, given by

$$\mathsf{J}(x_1, \ldots, x_n, y_1, \ldots, y_n) := (-y_1, \ldots, -y_n, x_1, \ldots, x_n).$$

These considerations lead naturally to the idea of "complexification."

---

[1] A row vector is described by a bold symbol, for instance $\mathbf{v}$, whereas the corresponding column vector is written as a transposed vector: $\mathbf{v}^t$.

Definition. Let E be an n-dimensional real vector space. The *complexification* of E is the real vector space $E_c := E \oplus E$, together with the complex structure $\mathsf{J} : E_c \to E_c$, given by

$$\mathsf{J}(v, w) := (-w, v).$$

Furthermore, *conjugation* C in E_c is defined by

$$\mathsf{C}(v, w) := (v, -w).$$

Since $\mathsf{C} \circ \mathsf{J} = -\mathsf{J} \circ \mathsf{C}$, it is clear that C defines a complex isomorphism between E_c and $\overline{E_c}$.

The complexification of $\mathbb{R}^n$ is the complex n-space $\mathbb{C}^n$ identified with $\mathbb{R}^{2n}$ in the way shown above. In this case the conjugation C is given by

$$\mathsf{C} : (z_1, \ldots, z_n) \mapsto (\overline{z}_1, \ldots, \overline{z}_n)$$

and will also be denoted by $\mathbf{z} \mapsto \overline{\mathbf{z}}$.

If $V = E_c$ is the complexification of a real vector space E, then the subspace

$$\mathrm{Re}(V) := \{(v, 0) \, : \, v \in E\} \subset V$$

is called the *real part* of V. Since it is isomorphic to E in a natural way, we can write $V \cong E \oplus \mathrm{i}E$. If V is an arbitrary complex vector space, then V is the complexification of some real vector space as well, but this real part is not uniquely defined. It is given by the real span of any complex basis of V.

Example

Let E be an n-dimensional real vector space and $E^* := \mathrm{Hom}_{\mathbb{R}}(E, \mathbb{R})$ the real dual space of linear forms on E. Then the complexification $(E^*)_c$ can be identified with the space $\mathrm{Hom}_{\mathbb{R}}(E, \mathbb{C})$ of complex-valued linear forms on E.

In the case $E = \mathbb{R}^n$, a linear form $\lambda \in E^*$ is always given by

$$\lambda : \mathbf{v} \mapsto \mathbf{v} \cdot \mathbf{a}^t,$$

with some fixed vector $\mathbf{a} \in \mathbb{R}^n$. An element of the complexification $(E^*)_c$ is then given by $\mathbf{v} \mapsto \mathbf{v} \cdot \mathbf{z}^t$ with $\mathbf{z} = \mathbf{a} + \mathrm{i}\mathbf{b} \in (\mathbb{R}^n)_c = \mathbb{C}^n$.

Now let T be an n-dimensional *complex* vector space and $F(T) := \mathrm{Hom}_{\mathbb{R}}(T, \mathbb{C})$ the space of complex-valued *real* linear forms on T. It contains the subspaces $T' := \mathrm{Hom}_{\mathbb{C}}(T, \mathbb{C})$ of complex linear forms and $\overline{T}' := \mathrm{Hom}_{\mathbb{C}}(\overline{T}, \mathbb{C})$ of complex antilinear forms [2].

[2] A real linear map $\lambda : T \to \mathbb{C}$ is called *complex antilinear* if $\lambda(c \cdot v) = \overline{c} \cdot \lambda(v)$ for $c \in \mathbb{C}$. Therefore, $\overline{T}'$ can be viewed as the set of complex antilinear forms on T.

Let $\{a_1, \dots, a_n\}$ be a complex basis of T, and $b_i := \mathrm{i} a_i$, for $i = 1, \dots, n$. Let $\{\alpha_1, \dots, \alpha_n, \beta_1, \dots, \beta_n\}$ be the basis of $T^* = \mathrm{Hom}_{\mathbb{R}}(T, \mathbb{R})$ that is dual to $\{a_1, \dots, a_n, b_1, \dots, b_n\}$. Then we obtain elements

$$\lambda_i := \alpha_i + \mathrm{i}\beta_i \in F(T), \quad i = 1, \dots, n.$$

Claim. *The forms λ_i are complex-linear.*

PROOF: Consider an element $z = z_1 a_1 + \cdots + z_n a_n \in T$ with $z_i = x_i + \mathrm{i} y_i \in \mathbb{C}$. Then

$$\begin{aligned} \lambda_k(z) &= \lambda_k\Big(\sum_{i=1}^n x_i a_i + \sum_{i=1}^n y_i b_i\Big) \\ &= \sum_{i=1}^n x_i \lambda_k(a_i) + \sum_{i=1}^n y_i \lambda_k(b_i) \\ &= x_k + \mathrm{i} y_k \;=\; z_k. \end{aligned}$$

Now the claim follows. ∎

It is obvious that the λ_i are linearly independent. Therefore, $\{\lambda_1, \dots, \lambda_n\}$ is a basis of T', and $\{\overline{\lambda}_1, \dots, \overline{\lambda}_n\}$ is a basis of $\overline{T}'$.

Since it is also obvious that $T' \cap \overline{T}' = \{0\}$, we see that every element $\lambda \in F(T)$ has a unique representation

$$\lambda = \sum_{i=1}^n c_i \lambda_i + \sum_{i=1}^n d_i \overline{\lambda}_i, \text{ with } c_i, d_i \in \mathbb{C}.$$

Briefly,

$$\lambda = \lambda' + \lambda'', \text{ with } \lambda' \in T' \text{ and } \lambda'' \in \overline{T}'.$$

Here λ is *real*; i.e., $\lambda \in \mathrm{Hom}_{\mathbb{R}}(T, \mathbb{R})$ if and only if $\lambda'' = \overline{\lambda'}$.

Hermitian Forms and Inner Products

Definition. Let T be an n-dimensional complex vector space. A *Hermitian form* on T is a function $H : T \times T \to \mathbb{C}$ with the following properties:

1. $v \mapsto H(v, w)$ is $\mathbb{C}$-linear for every $w \in T$.
2. $H(w, v) = \overline{H(v, w)}$ for $v, w \in T$.

It follows at once that $w \mapsto H(v, w)$ is $\mathbb{C}$-antilinear for every $v \in T$, and $H(v, v)$ is real for every $v \in T$. If $H(v, v) > 0$ for every $v \neq 0$, H is called an *inner product* or *scalar product*.

There is a natural decomposition

$$H(v, w) = S(v, w) + \mathrm{i}A(v, w),$$

with real-valued functions S and A. Since

$$S(w, v) + \mathrm{i}A(w, v) = H(w, v) = \overline{H(v, w)} = S(v, w) - \mathrm{i}A(v, w),$$

it follows that S is symmetric and A antisymmetric.

Example

If k is a field, the set of all matrices with p rows and q columns whose elements lie in k will be denoted by $M_{p,q}(k)$ and the set of square matrices of order n by $M_n(k)$. Here we are interested only in the cases $k = \mathbb{R}$ and $k = \mathbb{C}$.

A Hermitian form on $\mathbb{C}^n$ is given by

$$H : (\mathbf{z}, \mathbf{w}) \mapsto \mathbf{z}\,\mathbf{H}\,\overline{\mathbf{w}}^{\,t},$$

where $\mathbf{H} \in M_n(\mathbb{C})$ is a *Hermitian matrix*, i.e., $\overline{\mathbf{H}}^{\,t} = \mathbf{H}$.

The associated symmetric and antisymmetric real bilinear forms S and A are given by

$$S(\mathbf{z}, \mathbf{w}) = \mathrm{Re}\left(\mathbf{z}\,\mathbf{H}\,\overline{\mathbf{w}}^{\,t}\right) = \frac{1}{2}\left(\mathbf{z}\,\mathbf{H}\,\overline{\mathbf{w}}^{\,t} + \mathbf{w}\,\mathbf{H}\,\overline{\mathbf{z}}^{\,t}\right)$$

and

$$A(\mathbf{z}, \mathbf{w}) = \mathrm{Im}\left(\mathbf{z}\,\mathbf{H}\,\overline{\mathbf{w}}^{\,t}\right) = \frac{1}{2\mathrm{i}}\left(\mathbf{z}\,\mathbf{H}\,\overline{\mathbf{w}}^{\,t} - \mathbf{w}\,\mathbf{H}\,\overline{\mathbf{z}}^{\,t}\right).$$

If H is an inner product, then S is called the associated *Euclidean inner product.*

The identity matrix $\mathbf{E}_n$ yields the *standard Hermitian scalar product*

$$\langle \mathbf{z} \,|\, \mathbf{w} \rangle = \mathbf{z} \cdot \overline{\mathbf{w}}^{\,t} = \sum_{\nu=1}^{n} z_\nu \overline{w}_\nu .$$

Its symmetric part $\left(\mathbf{z} \,|\, \mathbf{w}\right)_{2n} := \mathrm{Re}(\langle \mathbf{z} \,|\, \mathbf{w} \rangle)$ is the *standard Euclidean scalar product.* In fact, if we write $\mathbf{z} = \mathbf{x} + \mathrm{i}\mathbf{y}$ and $\mathbf{w} = \mathbf{u} + \mathrm{i}\mathbf{v}$, with $\mathbf{x}, \mathbf{y}, \mathbf{u}, \mathbf{v} \in \mathbb{R}^n$, then

$$\begin{aligned} \left(\mathbf{z} \,|\, \mathbf{w}\right)_{2n} &= \frac{1}{2}\left(\mathbf{z} \cdot \overline{\mathbf{w}}^{\,t} + \mathbf{w} \cdot \overline{\mathbf{z}}^{\,t}\right) \\ &= \sum_{\nu=1}^{n} \frac{1}{2}\left(v_\nu \overline{w}_\nu + w_\nu \overline{z}_\nu\right) \\ &= \sum_{\nu=1}^{n} (x_\nu u_\nu + y_\nu v_\nu). \end{aligned}$$

If the standard Euclidean scalar product on $\mathbb{R}^n$ is denoted by $(\cdots|\cdots)_n$, we obtain the equation

$$(\mathbf{z}\,|\,\mathbf{w})_{2n} = (\mathbf{x}\,|\,\mathbf{u})_n + (\mathbf{y}\,|\,\mathbf{v})_n.$$

Balls and Polydisks

Definition. The *Euclidean norm* of a vector $\mathbf{z} \in \mathbb{C}^n$ is given by

$$\|\mathbf{z}\| := \sqrt{\langle \mathbf{z}\,|\,\mathbf{z}\rangle} = \sqrt{(\mathbf{z}\,|\,\mathbf{z})_n},$$

the *Euclidean distance* between two vectors $\mathbf{z}, \mathbf{w}$ by

$$\operatorname{dist}(\mathbf{z}, \mathbf{w}) := \|\mathbf{z} - \mathbf{w}\|.$$

An equivalent norm is the *sup-norm* or *modulus* of a vector:

$$|\mathbf{z}| := \max_{\nu=1,\ldots,n} |z_\nu|.$$

This norm is not derived from an inner product, but it defines the same topology on $\mathbb{C}^n$ as the Euclidean norm. This topology coincides with the usual topology on $\mathbb{R}^{2n}$. We assume that the reader is familiar with it and mention only that it has the Hausdorff property.

Definition. $\mathsf{B}_r(\mathbf{z}_0) := \{\mathbf{z} \in \mathbb{C}^n : \operatorname{dist}(\mathbf{z}, \mathbf{z}_0) < r\}$ is called the (*open*) *ball* of radius r with center $\mathbf{z}_0$.

A ball in $\mathbb{C}^n$ is also a ball in $\mathbb{R}^{2n}$, and its topological boundary

$$\partial \mathsf{B}_r(\mathbf{z}_0) = \{\mathbf{z} \in \mathbb{C}^n : \operatorname{dist}(\mathbf{z}, \mathbf{z}_0) = r\}$$

is a $(2n-1)$-dimensional sphere.

Definition. Let $\mathbf{r} = (r_1, \ldots, r_n) \in \mathbb{R}^n$, all $r_\nu > 0$, $\mathbf{z}_0 = (z_1^{(0)}, \ldots, z_n^{(0)}) \in \mathbb{C}^n$. Then

$$\mathsf{P}^n(\mathbf{z}_0, \mathbf{r}) := \{\mathbf{z} \in \mathbb{C}^n : |z_\nu - z_\nu^{(0)}| < r_\nu \text{ for } \nu = 1, \ldots, n\}$$

is called the (*open*) *polydisk* (or *polycylinder*) with *polyradius* $\mathbf{r}$ and center $\mathbf{z}_0$. If $r \in \mathbb{R}_+$ and $\mathbf{r} := (r, \ldots, r)$, we write $\mathsf{P}^n_r(\mathbf{z}_0)$ instead of $\mathsf{P}^n(\mathbf{z}_0, \mathbf{r})$. Then $\mathsf{P}^n_r(\mathbf{z}_0) = \{\mathbf{z} \in \mathbb{C}^n : |\mathbf{z} - \mathbf{z}_0| < r\}$.

If D denotes the open unit disk in $\mathbb{C}$, then $\mathsf{P}^n := \mathsf{P}^n_1(\mathbf{0}) = \underbrace{\mathsf{D} \times \cdots \times \mathsf{D}}_{n \text{ times}}$ is called the *unit polydisk* around $\mathbf{0}$.

We are not interested in the topological boundary of a polydisk. The following part of the boundary is much more important:

Definition. The *distinguished boundary* of the polydisk $\mathsf{P}^n(\mathbf{z}_0, \mathbf{r})$ is the set

$$\mathsf{T}^n(\mathbf{z}_0, \mathbf{r}) = \{\mathbf{z} \in \mathbb{C}^n \,:\, |z_\nu - z_\nu^{(0)}| = r_\nu \text{ for } \nu = 1, \ldots, n\}.$$

The distinguished boundary of a polydisk is the Cartesian product of n circles. It is well known that such a set is diffeomorphic to an n-dimensional *torus*. In the case $n = 1$ a polydisk reduces to a simple disk and its distinguished boundary is equal to its topological boundary.

Connectedness

Connectedness. Both the Euclidean balls and the polydisks form a base of the topology of $\mathbb{C}^n$. By a *region* we mean an ordinary open set in $\mathbb{C}^n$. A region G is *connected* if each two points of G can be joined by a continuous path in G. A connected region is called a *domain.*

If a real hyperplane in $\mathbb{R}^n$ meets a domain, then it cuts the domain into two or more disjoint open pieces. For complex hyperplanes in the complex number space (which have real codimension 2) this is not the case:

1.1 Proposition. *Let $G \subset \mathbb{C}^n$ be a domain and*

$$E := \{\mathbf{z} = (z_1, \ldots, z_n) \in \mathbb{C}^n \,:\, z_1 = 0\}.$$

Then $G' := G - E$ is again a domain.

PROOF: Of course, E is a closed set, without interior points, and $G' = G - E$ is open. Write points of $\mathbb{C}^n$ in the form $\mathbf{z} = (z_1, \mathbf{z}^*)$, with $\mathbf{z}^* \in \mathbb{C}^{n-1}$. Given two points $\mathbf{v} = (v_1, \mathbf{v}^*)$ and $\mathbf{w} = (w_1, \mathbf{w}^*)$ in G', it must be shown that $\mathbf{v}$ and $\mathbf{w}$ can be joined in G' by a continuous path. We do this in two steps.

Step 1: Let $G = \mathsf{P}^n(\mathbf{z}_0, \varepsilon)$ be a small polydisk. Then G' is the product of a punctured disk and a polydisk in $n-1$ variables. Define $\widetilde{\mathbf{z}} := (w_1, \mathbf{v}^*)$. Clearly, $\widetilde{\mathbf{z}} \in G'$, and we can join v_1 and w_1 within the punctured disk, and $\mathbf{v}^*$ and $\mathbf{w}^*$ within the polydisk. Therefore, $\mathbf{v}$ and $\mathbf{w}$ can be joined within G'.

Step 2: Now let G be an arbitrary domain. There is a path $\varphi : I \to G$ joining $\mathbf{v}$ and $\mathbf{w}$. Since $\varphi(I)$ is compact, it can be covered by finitely many polydisks $U_1, \ldots, U_l$ such that $U_\lambda \subset G$ for $\lambda = 1, \ldots, l$.

It is easy to show that there is a $\delta > 0$ such that for all $t', t'' \in I$ with $|t' - t''| < \delta$, $\varphi(t')$ and $\varphi(t'')$ lie in the same polydisk U_k. Then let $a = t_0 < t_1 < \cdots < t_N = b$ be a partition of I with $|t_j - t_{j-1}| < \delta$ for $j = 1, \ldots, N$. Let $\mathbf{z}_j := \varphi(t_j)$ and $\lambda(j) \in \{1, \ldots, l\}$ be chosen such that $U_{\lambda(j)}$ contains $\mathbf{z}_j$ and $\mathbf{z}_{j-1}$ (it can happen that $\lambda(j_1) = \lambda(j_2)$ for $j_1 \neq j_2$). By construction $\mathbf{z}_{j-1}$ lies in $U_{\lambda(j)} \cap U_{\lambda(j-1)}$, and thus $U_{\lambda(j)} \cap U_{\lambda(j-1)} - E$ is always a nonempty open set.

We join $\mathbf{v} = \mathbf{z}_0 \in U_{\lambda(1)}$ and some point $\widehat{\mathbf{z}}_1 \in U_{\lambda(1)} \cap U_{\lambda(2)} - E$ by a path φ_1 interior to $U_{\lambda(1)} - E$. By (1) this is possible. Next we join $\widehat{\mathbf{z}}_1$ and a point $\widehat{\mathbf{z}}_2 \in U_{\lambda(2)} \cap U_{\lambda(3)} - E$ by a path φ_2 interior to $U_{\lambda(2)} - E$, and so on. Finally, φ_N joins $\widehat{\mathbf{z}}_{N-1}$ and $\mathbf{w} = \mathbf{z}_N$ within $U_{\lambda(N)} - E$. The composition of $\varphi_1, \ldots, \varphi_N$ connects $\mathbf{v}$ and $\mathbf{w}$ in G'. ∎

Reinhardt Domains

Definition. The point set

$$\mathscr{V} := \{\mathbf{r} = (r_1, \ldots, r_n) \in \mathbb{R}^n \, : \, r_\nu \geq 0 \text{ for } \nu = 1, \ldots, n\}$$

will be called *absolute space*, the map $\tau : \mathbb{C}^n \to \mathscr{V}$ with $\tau(z_1, \ldots, z_n) := \big(|z_1|, \ldots, |z_n|\big)$ the *natural projection.*

The map τ is continuous and surjective. For any $\mathbf{r} \in \mathscr{V}$, the preimage $\tau^{-1}(\mathbf{r})$ is the torus $\mathsf{T}^n(\mathbf{0}, \mathbf{r})$. For $\mathbf{z} \in \mathbb{C}^n$, we set $\mathsf{P}_\mathbf{z} := \mathsf{P}^n(\mathbf{0}, \tau(\mathbf{z}))$ and $\mathsf{T}_\mathbf{z} := \mathsf{T}^n(\mathbf{0}, \tau(\mathbf{z})) = \tau^{-1}(\tau(\mathbf{z}))$ (see Figure I.1).

Definition. A domain $G \subset \mathbb{C}^n$ is called a *Reinhardt domain* if for every $\mathbf{z} \in G$ the torus $\mathsf{T}_\mathbf{z}$ is also contained in G.

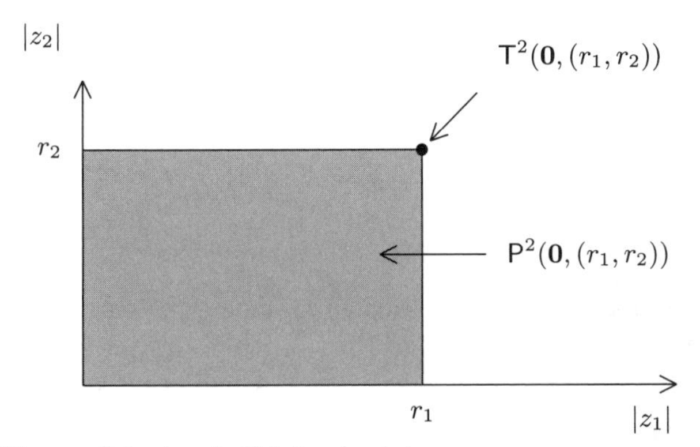

Figure I.1. A polydisk in absolute space

Reinhardt domains G are characterized by their images in absolute space: $\tau^{-1}\tau(G) = G$. Therefore, they can be visualized as domains in $\mathscr{V}$. For example, both balls and polydisks around the origin are Reinhardt domains.

Example

Let $\mathbf{z}_0 \in \mathbb{C}^n$, with $|z_\nu^{(0)}| > 1$ for $\nu = 1, \ldots, n$. Then $\tau(e^{\mathrm{i}\theta} \cdot \mathbf{z}_0) = \tau(\mathbf{z}_0)$, but $|e^{\mathrm{i}\theta} \cdot \mathbf{z}_0 - \mathbf{z}_0| = |e^{\mathrm{i}\theta} - 1| \cdot |\mathbf{z}_0| > |e^{\mathrm{i}\theta} - 1|$, and for suitable θ this expression may be greater than ε. So $\mathsf{P}^n(\mathbf{z}_0, \varepsilon)$ is **not** a Reinhardt domain.

Definition. Let $G \subset \mathbb{C}^n$ be a Reinhardt domain.

1. G is called *proper* if $\mathbf{0} \in G$.
2. G is called *complete* if $\forall \mathbf{z} \in G \cap (\mathbb{C}^*)^n : \mathsf{P}_{\mathbf{z}} \subset G$ (see Figure I.2).

Later on we shall see that for any proper Reinhardt domain G there is a smallest complete Reinhardt domain $\widehat{G}$ containing G.

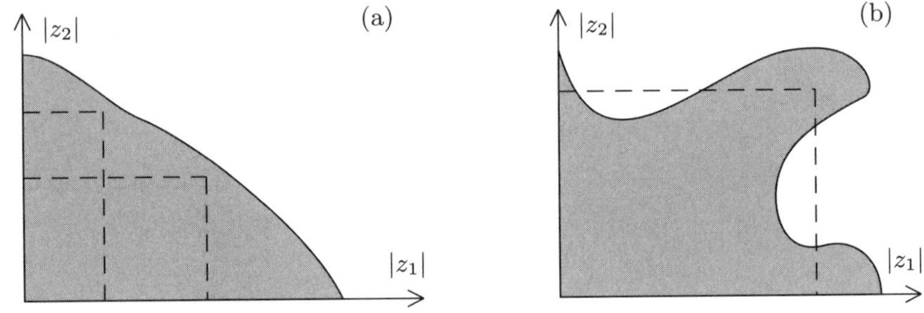

Figure I.2. (a) Complete and (b) noncomplete Reinhardt domain

Exercises

1. Show that there is an open set $B \subset \mathbb{C}^2$ that is not connected but whose image $\tau(B)$ is a domain in absolute space.
2. Which of the following domains is Reinhardt, proper Reinhardt, complete Reinhardt?
   (a) $G_1 := \{\mathbf{z} \in \mathbb{C}^2 \,:\, 1 > |z_1| > |z_2|\}$,
   (b) $G_2 := \{\mathbf{z} \in \mathbb{C}^2 \,:\, |z_1| < 1 \text{ and } |z_2| < 1 - |z_1|\}$,
   (c) G_3 is a domain in $\mathbb{C}^2$ with the property
   $\mathbf{z} \in G \implies e^{\mathrm{i}t} \cdot \mathbf{z} \in G$ for $t \in \mathbb{R}$.
3. Let $G \subset \mathbb{C}^n$ be an arbitrary set. Show that G is a Reinhardt domain $\iff \exists \mathcal{G} \subset \mathscr{V}$ open and connected such that $G = \tau^{-1}(\mathcal{G})$.
4. A domain $G \subset \mathbb{C}^n$ is called *convex*, if for each pair of points $\mathbf{z}, \mathbf{w} \in G$ the line segment from $\mathbf{z}$ to $\mathbf{w}$ is also contained in G. Show that an arbitrary domain G is convex if and only if for every $\mathbf{z} \in \partial G$ there is an affine linear function $\lambda : \mathbb{C}^n \to \mathbb{R}$ with $\lambda(\mathbf{z}) = 0$ and $\lambda|_G < 0$.

2. Power Series

Polynomials. In order to simplify notation, we introduce multi-indices. For $\nu = (\nu_1, \dots, \nu_n) \in \mathbb{Z}^n$ and $\mathbf{z} \in \mathbb{C}^n$ define

$$|\nu| := \sum_{i=1}^{n} \nu_i \quad \text{and} \quad \mathbf{z}^\nu := z_1^{\nu_1} \cdots z_n^{\nu_n}.$$

The notation $\nu \geq 0$ (respectively $\nu > 0$) means that $\nu_i \geq 0$ for each i (respectively $\nu \geq 0$ and $\nu_i > 0$ for at least one i).

A function of the form

$$\mathbf{z} \mapsto p(\mathbf{z}) = \sum_{|\nu| \leq m} a_\nu \mathbf{z}^\nu, \text{ with } a_\nu \in \mathbb{C} \text{ for } |\nu| \leq m,$$

is called a *polynomial* (of *degree* less than or equal to m). If there is a ν with $|\nu| = m$ and $a_\nu \neq 0$, then $p(\mathbf{z})$ is said to have *degree* m. For the zero polynomial no degree is defined. An expression of the form $a_\nu \mathbf{z}^\nu$ with $a_\nu \neq 0$ is called a *monomial* of degree $m := |\nu|$. A polynomial $p(\mathbf{z})$ is called *homogeneous* of degree m if it consists only of monomials of degree m.

2.1 Proposition. *A polynomial $p(\mathbf{z}) \neq 0$ of degree m is homogeneous if and only if*

$$p(\lambda \mathbf{z}) = \lambda^m \cdot p(\mathbf{z}), \qquad \textit{for all } \lambda \in \mathbb{C}.$$

PROOF: Let $p(\mathbf{z}) = a_\nu \mathbf{z}^\nu$ be a monomial of degree m. Then

$$p(\lambda \mathbf{z}) = a_\nu (\lambda \mathbf{z})^\nu = \lambda^m \cdot a_\nu \mathbf{z}^\nu = \lambda^m \cdot p(\mathbf{z}).$$

The same is true for finite sums of monomials.

On the other hand, let $p(\mathbf{z}) = \sum_{|\nu| \leq N} a_\nu \mathbf{z}^\nu$ be an arbitrary polynomial with $p(\lambda \mathbf{z}) = \lambda^m \cdot p(\mathbf{z})$. Gathering monomials of degree i, we obtain a polynomial $p_i(\mathbf{z}) = \sum_{|\nu| = i} a_\nu \mathbf{z}^\nu$ with $p_i(\lambda \mathbf{z}) = \lambda^i \cdot p_i(\mathbf{z})$. Then for fixed $\mathbf{z}$ the two polynomials

$$\lambda \mapsto p(\lambda \mathbf{z}) = \sum_{i=0}^{N} p_i(\mathbf{z}) \cdot \lambda^i \quad \text{and} \quad \lambda \mapsto \lambda^m \cdot p(\mathbf{z})$$

are equal. This is possible only if the coefficients are equal, i.e., $p_m(\mathbf{z}) = p(\mathbf{z})$ and $p_i(\mathbf{z}) = 0$ for $i \neq m$. So $p = p_m$ is homogeneous. ■

Convergence. If for every $\nu \in \mathbb{N}_0^n$ a complex number c_ν is given, one can consider the series $\sum_{\nu \geq 0} c_\nu$ and discuss the matter of convergence. The trouble is that there is no canonical order on $\mathbb{N}_0^n$.

Definition. The series $\sum_{\nu \geq 0} c_\nu$ is called *convergent* if there is a bijective map $\varphi : \mathbb{N} \to \mathbb{N}_0^n$ such that $\sum_{i=1}^\infty |c_{\varphi(i)}| < \infty$. Then the complex number $\sum_{i=1}^\infty c_{\varphi(i)}$ is called the *limit* of the series.

It is clear that this notion of convergence is independent of the chosen map φ, and that it means absolute convergence.

2.2 Proposition. *$\sum_{\nu \geq 0} c_\nu$ is convergent if and only if*

$$\Big\{ \sum_{\nu \in I} |c_\nu| \,:\, I \subset \mathbb{N}_0^n \text{ finite} \Big\}$$

is a bounded set.

The proof is trivial.

2.3 Proposition. *If the series $\sum_{\nu \geq 0} c_\nu$ converges to the complex number c, then for each $\varepsilon > 0$ there exists a finite set $I_0 \subset \mathbb{N}_0^n$ such that:*

1. $\sum_{\nu \in K} |c_\nu| < \varepsilon$, *for any finite set $K \subset \mathbb{N}_0^n$ with $K \cap I_0 = \varnothing$.*
2. $\Big|\sum_{\nu \in I} c_\nu - c\Big| < \varepsilon$, *for any finite set I with $I_0 \subset I \subset \mathbb{N}_0^n$.*

PROOF: We choose a bijective map $\varphi : \mathbb{N} \to \mathbb{N}_0^n$. Then $\sum_{i=1}^\infty c_{\varphi(i)} = c$, and the series is absolutely convergent. For a given $\varepsilon > 0$ there exists an $i_0 \in \mathbb{N}$ such that $\sum_{i=i_0}^\infty |c_{\varphi(i)}| < \varepsilon$ and $\big|\sum_{i=1}^{i_0} c_{\varphi(i)} - c\big| < \varepsilon$.

Setting $I_0 := \varphi(\{1, 2, \ldots, i_0\})$, it follows that $\sum_{\nu \in K} |c_\nu| < \varepsilon$ for any finite set K with $K \cap I_0 = \varnothing$, and $\big|\sum_{\nu \in I_0} c_\nu - c\big| < \varepsilon$.

Then for any finite set I with $I_0 \subset I \subset \mathbb{N}_0^n$,

$$\Big|\sum_{\nu \in I} c_\nu - c\Big| = \Big|\Big(\sum_{\nu \in I_0} c_\nu - c\Big) + \sum_{\nu \in I - I_0} c_\nu\Big| \leq \Big|\sum_{\nu \in I_0} c_\nu - c\Big| + \sum_{\nu \in I - I_0} |c_\nu| < 2\varepsilon.$$

■

Example

Let $q_1, \ldots, q_n$ be real numbers with $0 < q_i < 1$ for $i = 1, \ldots, n$, and $\mathbf{q} := (q_1, \ldots, q_n)$. Then for any $\nu \in \mathbb{N}_0^n$, $\mathbf{q}^\nu = q_1^{\nu_1} \cdots q_n^{\nu_n}$ is a positive real number.

If $I \subset \mathbb{N}_0^n$ is a finite set, then there is a number N such that $I \subset \{0, 1, \ldots, N\}^n$, and therefore

$$\Big|\sum_{\nu \in I} \mathbf{q}^\nu\Big| = \sum_{\nu \in I} \mathbf{q}^\nu \leq \prod_{i=1}^n \sum_{\nu_i = 0}^N q_i^{\nu_i} \leq \prod_{i=1}^n \frac{1}{1 - q_i}.$$

Since the partial sums are bounded, the series is convergent. It is absolutely convergent in any order, and the limit is

$$\sum_{\nu \geq 0} \mathbf{q}^\nu = \prod_{i=1}^{n} \frac{1}{1 - q_i}.$$

We call this series the *generalized geometric series.*

Now let $M \subset \mathbb{C}^n$ be an arbitrary subset, and $\{f_\nu : \nu \in \mathbb{N}_0^n\}$ a family of complex-valued functions on M. We denote by $\|f_\nu\|_M$ the supremum of $|f_\nu|$ on M.

Definition. The series $\sum_{\nu \geq 0} f_\nu$ is called *normally convergent* on M if the series of positive real numbers $\sum_{\nu \geq 0} \|f_\nu\|_M$ is convergent.

2.4 Proposition. *Let the series $\sum_{\nu \geq 0} f_\nu$ be normally convergent on M. Then it is convergent for any $\mathbf{z} \in M$, and for any bijective map $\varphi : \mathbb{N} \to \mathbb{N}_0^n$ the series $\sum_{i=1}^{\infty} f_{\varphi(i)}$ is uniformly convergent on M.*

PROOF: If the series is normally convergent, then $\sum_{\nu \geq 0} |f_\nu(\mathbf{z})|$ is convergent for any $\mathbf{z} \in M$. But then $\sum_{\nu \geq 0} f_\nu(\mathbf{z})$ is also convergent, and there is a complex number $f(\mathbf{z})$ such that $\sum_{i=1}^{\infty} f_{\varphi(i)}(\mathbf{z})$ converges to $f(\mathbf{z})$, for every bijective map $\varphi : \mathbb{N} \to \mathbb{N}_0^n$.

If an $\varepsilon > 0$ is given, there is an i_0 such that $\sum_{i=i_0}^{\infty} \|f_{\varphi(i)}\|_M < \varepsilon$. Then

$$\Big| \sum_{i=k}^{m} f_{\varphi(i)}(\mathbf{z}) \Big| \leq \sum_{i=k}^{m} \|f_{\varphi(i)}\|_M < \varepsilon, \text{ for } m > k \geq i_0 \text{ and } \mathbf{z} \in M.$$

Therefore,

$$\Big| \sum_{i=1}^{k} f_{\varphi(i)}(\mathbf{z}) - f(\mathbf{z}) \Big| = \Big| \sum_{i=k+1}^{\infty} f_{\varphi(i)}(\mathbf{z}) \Big| \leq \varepsilon, \text{ for } k \geq i_0.$$

This proves the uniform convergence. ∎

Power Series. Let $\{a_\nu : \nu \in \mathbb{N}_0^n\}$ be a family of complex numbers, and $\mathbf{z}_0 \in \mathbb{C}^n$ a point. Then the expression

$$\sum_{\nu \geq 0} a_\nu (\mathbf{z} - \mathbf{z}_0)^\nu$$

is called a *(formal) power series* about $\mathbf{z}_0$. It is a series of polynomials. If this series converges normally on a set M to a complex function f, then as a uniform limit of continuous functions f is continuous on M.

2.5 Abel's lemma. *Let $P' \subset\subset P \subset \mathbb{C}^n$ be polydisks around the origin.*[3] *If the power series $\sum_{\nu \geq 0} a_\nu \mathbf{z}^\nu$ converges at some point of the distinguished boundary of P, then it converges normally on P'.*

PROOF: Let $\mathbf{w} \in \partial_0 P$ be a point where $\sum_{\nu \geq 0} a_\nu \mathbf{w}^\nu$ is convergent. Then there is a constant c such that $|a_\nu \mathbf{w}^\nu| \leq c$ for all $\nu \in \mathbb{N}_0^n$.

We choose real numbers q_i with $0 < q_i < 1$ such that $|z_i| \leq q_i |w_i|$ for any $\mathbf{z} = (z_1, \ldots, z_n) \in P'$ and $i = 1, \ldots, n$. It follows that

$$|a_\nu \mathbf{z}^\nu| \leq \mathbf{q}^\nu \cdot c, \text{ for } \mathbf{q} = (q_1, \ldots, q_n),\ \mathbf{z} \in P', \text{ and } \nu \in \mathbb{N}_0^n.$$

Then $\|a_\nu \mathbf{z}^\nu\|_{P'} \leq \mathbf{q}^\nu \cdot c$ as well, and from the convergence of the generalized geometric series it follows that $\sum_{\nu \geq 0} a_\nu \mathbf{z}^\nu$ is normally convergent on P'. ■

Definition. We say that a power series $\sum_{\nu \geq 0} a_\nu (\mathbf{z} - \mathbf{z}_0)^\nu$ *converges compactly* in a domain G if it converges normally on every compact subset $K \subset G$.

2.6 Corollary. *Let $P \subset \mathbb{C}^n$ be a polydisk around the origin and $\mathbf{w}$ be a point of the distinguished boundary of P. If the power series $\sum_{\nu \geq 0} a_\nu \mathbf{z}^\nu$ converges at $\mathbf{w}$, then it converges compactly on P.*

PROOF: Let $K \subset P$ be a compact set. Then there is a q with $0 < q < 1$ such that $K \subset q \cdot P \subset\subset P$. Therefore, the series is normally convergent on K. ■

Let $S(\mathbf{z}) = \sum_{\nu \geq 0} a_\nu \mathbf{z}^\nu$ be a formal power series about the origin, and

$$B := \{\mathbf{z} \in \mathbb{C}^n \,:\, S(\mathbf{z}) \text{ convergent}\}.$$

2.7 Proposition. *The interior B° is a complete Reinhardt domain, and $S(\mathbf{z})$ converges compactly in B°.*

PROOF: Let $\mathbf{w}$ be a point of B°. There is a polydisk $\mathsf{P}^n(\mathbf{w}, \varepsilon) \subset B^\circ$ and a point $\mathbf{v} \in \mathsf{P}^n(\mathbf{w}, \varepsilon) \cap (\mathbb{C}^*)^n$ such that $\mathbf{w} \in P_\mathbf{v}(\mathbf{0})$. Then $T_\mathbf{w} \subset B^\circ$, and if $\mathbf{w} \in (\mathbb{C}^*)^n$, then also $P_\mathbf{w}(\mathbf{0}) \subset B^\circ$.

To see that B° is a complete Reinhardt domain, it remains to show that it is connected. But this is very simple. Every point of B° can be connected to a point in $B^\circ \cap (\mathbb{C}^*)^n$, and then within a suitable polydisk to the origin.

[3] The notation $U \subset\subset V$ means that U lies relatively compact in V; i.e., $\overline{U}$ is a compact set which is contained in V.

From these considerations it follows that B° is the union of relatively compact polydisks around the origin. Therefore, $S(\mathbf{z})$ converges compactly on B°. ∎

The set B° is called the *domain of convergence* of $S(\mathbf{z})$.

2.8 Proposition. *Let G be the domain of convergence of the power series $S(\mathbf{z}) = \sum_{\nu \geq 0} a_\nu \mathbf{z}^\nu$. Then*

$$S_{z_j}(\mathbf{z}) := \sum_{\substack{\nu \geq 0 \\ \nu_j > 0}} a_\nu \cdot \nu_j z_1^{\nu_1} \cdots z_j^{\nu_j - 1} \cdots z_n^{\nu_n}$$

also converges compactly on G.

PROOF: Let $\mathbf{w}$ be any point of $(\mathbb{C}^*)^n \cap G$, and $|a_\nu \mathbf{w}^\nu| \leq c$ for every $\nu \in \mathbb{N}_0^n$. If $0 < q < 1$ and $\mathbf{z} = q \cdot \mathbf{w}$, then

$$|a_\nu \cdot \nu_j z_1^{\nu_1} \cdots z_j^{\nu_j} \cdots z_n^{\nu_n}| = \frac{\nu_j}{|z_j|} \cdot |a_\nu \mathbf{z}^\nu| \leq \frac{c}{|z_j|} \cdot \nu_j \cdot q^{|\nu|}.$$

Now,

$$\sum_{\substack{\nu \geq 0 \\ \nu_j > 0}} \nu_j q^{|\nu|} = \left(\sum_{\nu_1 = 0}^{\infty} q^{\nu_1} \right) \cdots \left(\sum_{\nu_j = 1}^{\infty} \nu_j q^{\nu_j} \right) \cdots \left(\sum_{\nu_n = 0}^{\infty} q^{\nu_n} \right)$$

is convergent. Therefore, $S_{z_j}(\mathbf{z})$ is convergent, and it follows that S_{z_j} is normally convergent on $P_{\mathbf{z}}(\mathbf{0})$. Since every compact set $K \subset G$ can be covered by finitely many polydisks of this kind, S_{z_j} is compactly convergent on P. ∎

Definition. Let $B \subset \mathbb{C}^n$ be an open set. A function $f : B \to \mathbb{C}$ is called *holomorphic* if for every $\mathbf{z}_0 \in B$ there is a neighborhood $U = U(\mathbf{z}_0) \subset B$ and a power series $S(\mathbf{z}) := \sum_{\nu \geq 0} a_\nu (\mathbf{z} - \mathbf{z}_0)^\nu$ that converges on U to $f(\mathbf{z})$.

The set of holomorphic functions on B is denoted by $\mathcal{O}(B)$.

It is immediately clear that every holomorphic function is continuous.

Exercises

1. Let f, g be two nonzero polynomials. Prove that

$$\deg(f \cdot g) = \deg(f) + \deg(g).$$

2. Let $f = f_1 \cdots f_k$ be a homogeneous nonzero polynomial. Show that f_i is homogeneous, for $i = 1, \ldots, k$.

3. Find the domain of convergence for the following power series:

$$f(z,w) = \sum_{k\geq 0} zw^k, \quad g(z,w) = \sum_{k=0}^{\infty}(zw)^k, \quad h(z,w) = \sum_{\nu,\mu\geq 0} \frac{\nu}{\mu!} z^\nu w^\mu .$$

4. Determine the limit and the domain of convergence of the series

$$F(z,w) = \sum_{\nu\geq 0}\Big((2z)^\nu + \sum_{\mu\geq 0} z^\mu\Big) w^\nu .$$

5. A polyradius $\mathbf{r} = (r_1, \ldots, r_n) \in \mathscr{V}$ is called a *radius of convergence* for the power series $f(\mathbf{z}) = \sum_{\nu\geq 0} a_\nu \mathbf{z}^\nu$ if $f(\mathbf{z})$ is convergent in $P = \mathsf{P}^n(\mathbf{0}, \mathbf{r})$, but not convergent in any polydisk $P' = \mathsf{P}^n(\mathbf{0}, \mathbf{r}')$ with $P \subset\subset P'$.

   Prove the following generalization of the root test:

   $\mathbf{r}$ is a radius of convergence for $f(\mathbf{z})$ if and only if $\overline{\lim_{\nu\geq 0}} \sqrt[|\nu|]{|a_\nu|\mathbf{r}^\nu} = 1$.

3. Complex Differentiable Functions

The Complex Gradient

Definition. Let $B \subset \mathbb{C}^n$ be an open set, $\mathbf{z}_0 \in B$ a point. A function $f : B \to \mathbb{C}$ is called *complex differentiable* at $\mathbf{z}_0$ if there exists a map $\Delta : B \to \mathbb{C}^n$ such that the following hold:

1. Δ is continuous at $\mathbf{z}_0$.
2. $f(\mathbf{z}) = f(\mathbf{z}_0) + (\mathbf{z} - \mathbf{z}_0) \cdot \Delta(\mathbf{z})^t$ for $\mathbf{z} \in B$.

Complex differentiability is a local property: For f to be complex differentiable at $\mathbf{z}_0$ it is sufficient that there is a small neighborhood $U = U(\mathbf{z}_0) \subset B$ such that the restriction $f|_U$ is complex differentiable at $\mathbf{z}_0$.

3.1 Proposition. *If f is complex differentiable at $\mathbf{z}_0$, then the value of the function Δ at $\mathbf{z}_0$ is uniquely determined.*

PROOF: Assume that there are two maps Δ_1 and Δ_2 satisfying the conditions of the definition. Then

$$(\mathbf{z} - \mathbf{z}_0) \cdot (\Delta_1(\mathbf{z}) - \Delta_2(\mathbf{z}))^t = 0 \text{ for every } \mathbf{z} \in B.$$

In particular, there is an $\varepsilon > 0$ such that the equation holds for $\mathbf{z} = \mathbf{z}_0 + t\mathbf{e}_i$, with $t \in \mathbb{C}$, $|t| < \varepsilon$, and $i = 1, \ldots, n$. If $\Delta_\lambda = (\Delta_1^{(\lambda)}, \ldots, \Delta_n^{(\lambda)})$, then

$$t \cdot (\Delta_i^{(1)}(\mathbf{z}) - \Delta_i^{(2)}(\mathbf{z})) = 0, \text{ for } |t| < \varepsilon, \quad \mathbf{z} = \mathbf{z}_0 + t\mathbf{e}_i, \text{ and } i = 1, \ldots, n.$$

Because the $\Delta_i^{(\lambda)}$ are continuous at $\mathbf{z}_0$, it follows that $\Delta_1(\mathbf{z}_0) = \Delta_2(\mathbf{z}_0)$. ∎

Definition. Let $f : B \to \mathbb{C}$ be complex differentiable at $\mathbf{z}_0$. If there exists a representation

$$f(\mathbf{z}) = f(\mathbf{z}_0) + (\mathbf{z} - \mathbf{z}_0) \cdot \Delta(\mathbf{z})^t,$$

with Δ continuous at $\mathbf{z}_0$, then the uniquely determined numbers

$$\frac{\partial f}{\partial z_\nu}(\mathbf{z}_0) = f_{z_\nu}(\mathbf{z}_0) := \mathbf{e}_\nu \cdot \Delta(\mathbf{z}_0)^t$$

are called the *partial derivatives* of f at $\mathbf{z}_0$. The vector

$$\nabla f(\mathbf{z}_0) := (f_{z_1}(\mathbf{z}_0), \dots, f_{z_n}(\mathbf{z}_0)) = \Delta(\mathbf{z}_0)$$

is called the *complex gradient* of f at $\mathbf{z}_0$.

Remarks

1. If f is complex differentiable at $\mathbf{z}_0$, then f is continuous there as well.
2. A function f is called *complex differentiable* in an open set B if it is complex differentiable at each point of B. Then the partial derivatives of f define functions f_{z_ν} on B. If each of these partial derivatives is again complex differentiable at $\mathbf{z}_0$, then f is called *twice complex differentiable* at $\mathbf{z}_0$, and one obtains *second derivatives*

$$\frac{\partial^2 f}{\partial z_\nu \partial z_\mu}(\mathbf{z}_0) = f_{z_\nu z_\mu}(\mathbf{z}_0).$$

   By induction, partial derivatives of arbitrary order may be defined.
3. Sums, products, and quotients (with nonvanishing denominators) of complex differentiable functions are again complex differentiable.

Weakly Holomorphic Functions. Let $B \subset \mathbb{C}^n$ be an open set, $\mathbf{z}_0 \in B$ a point, and f a complex-valued function on B. For $\mathbf{w} \neq \mathbf{0}$ let $\varphi_\mathbf{w} : \mathbb{C} \to \mathbb{C}^n$ be defined by

$$\varphi_\mathbf{w}(\zeta) := \mathbf{z}_0 + \zeta\mathbf{w}.$$

Then $f \circ \varphi_\mathbf{w}(\zeta)$ is defined for sufficiently small ζ. If f is complex differentiable at $\mathbf{z}_0$, then we have a representation $f(\mathbf{z}) = f(\mathbf{z}_0) + (z - z_0) \cdot \Delta(\mathbf{z})^t$, with Δ continuous at $\mathbf{z}_0$. It follows that

$$f(\varphi_\mathbf{w}(\zeta)) - f(\varphi_\mathbf{w}(0)) = \zeta\mathbf{w} \cdot \Delta(\varphi_\mathbf{w}(\zeta))^t,$$

and $f \circ \varphi_\mathbf{w}$ is complex differentiable at $\zeta = 0$, with

$$(f \circ \varphi_\mathbf{w})'(0) = \mathbf{w} \cdot \Delta(\mathbf{z}_0)^t = \lim_{\zeta \to 0} \frac{1}{\zeta}[f(\varphi_\mathbf{w}(\zeta)) - f(\varphi_\mathbf{w}(0))].$$

This is the *complex directional derivative* of f at $\mathbf{z}_0$ in the direction $\mathbf{w}$. We denote it by $D_{\mathbf{w}}f(\mathbf{z}_0)$. In particular, $f_{z_\nu}(\mathbf{z}_0) = D_{\mathbf{e}_\nu}f(\mathbf{z}_0)$ for $\nu = 1, \ldots, n$.

An arbitrary function f is called *partially differentiable* at $\mathbf{z}_0$ if all partial derivatives $D_{\mathbf{e}_\nu}f(\mathbf{z}_0)$ exist for $\nu = 1, \ldots, n$.

A function f is called *weakly holomorphic* on B if it is continuous and partially differentiable on B. Then for $\mathbf{z} = (z_1, \ldots, z_n) \in B$ and $\nu = 1, \ldots, n$ the functions

$$\zeta \mapsto f(z_1, \ldots, z_{\nu-1}, \zeta, z_{\nu+1}, \ldots, z_n)$$

are holomorphic functions of one variable.

If f is complex differentiable on B, then f is also weakly holomorphic on B. Later on we shall see that weakly holomorphic functions are always complex differentiable, in contrast to the behavior of real differentiable functions.

Holomorphic Functions

3.2 Proposition. *Let $P \subset \mathbb{C}^n$ be a polydisk around the origin, and $S(z) = \sum_{\nu \geq 0} a_\nu \mathbf{z}^\nu$ a power series that converges compactly on P to a function f. Then f is complex differentiable at $\mathbf{0}$, with*

$$f_{z_1}(\mathbf{0}) = a_{1,0,\ldots,0}\,, \ldots, f_{z_n}(\mathbf{0}) = a_{0,\ldots,0,1}.$$

PROOF: We choose a small polydisk $P_\varepsilon \subset\subset P$ around the origin such that $S(\mathbf{z})$ is normally convergent on P_ε. But then the series obtained by any rearrangement of the terms is also normally convergent, and it converges to the same limit. We write

$$\begin{aligned} f(\mathbf{z}) &= \sum_{\nu \geq 0} a_\nu \mathbf{z}^\nu \\ &= a_{0,0,\ldots,0} + z_1 \cdot \sum_{\substack{\nu_1 > 0 \\ \nu_2,\ldots,\nu_n \geq 0}} a_\nu z_1^{\nu_1 - 1} z_2^{\nu_2} \cdots z_n^{\nu_n} \\ &\quad + z_2 \cdot \sum_{\substack{\nu_1 = 0,\, \nu_2 > 0 \\ \nu_3,\ldots,\nu_n \geq 0}} a_\nu z_2^{\nu_2 - 1} \cdots z_n^{\nu_n} + \cdots + z_n \cdot \sum_{\substack{\nu_1 = \ldots = \nu_{n-1} = 0 \\ \nu_n > 0}} a_\nu z_n^{\nu_n - 1} \\ &= f(\mathbf{0}) + z_1 \cdot \Delta_1(\mathbf{z}) + \cdots + z_n \cdot \Delta_n(\mathbf{z}). \end{aligned}$$

Since the series $\Delta_1(\mathbf{z}), \ldots, \Delta_n(\mathbf{z})$ converge normally on P_ε to continuous functions, f is complex differentiable at $\mathbf{0}$, with $f_{z_\nu}(\mathbf{0}) = \Delta_\nu(\mathbf{0})$. ∎

3.3 Corollary. *If $B \subset \mathbb{C}^n$ is an open set, and $f : B \to \mathbb{C}$ a holomorphic function, then f is complex differentiable on B.*

PROOF: Let $\mathbf{z}_0 \in B$ be an arbitrary point. There is a power series $S(\mathbf{w})$ converging compactly near $\mathbf{0}$ to a holomorphic function g such that

$$f(\mathbf{z}_0 + \mathbf{w}) = g(\mathbf{w}) = g(\mathbf{0}) + w_1 \cdot \Delta_1(\mathbf{w}) + \cdots + w_n \cdot \Delta_n(\mathbf{w}),$$

with continuous functions $\Delta_1, \ldots, \Delta_n$. It follows that f is complex differentiable at $\mathbf{z}_0$. ∎

Exercises

1. Show that there is a function $f : \mathbb{C}^n \to \mathbb{C}$ that is complex differentiable at every point $\mathbf{z} = (z_1, \ldots, z_n)$ with $z_n = 0$, but is nowhere holomorphic.
2. Prove the following chain rule: If $G \subset \mathbb{C}^n$ is a domain, $f : G \to \mathbb{C}$ a complex differentiable function, and $\varphi = (\varphi_1, \ldots, \varphi_n) : \Delta \to G$ a map with holomorphic components φ_i, then $f \circ \varphi : \Delta \to \mathbb{C}$ is a holomorphic function, with $(f \circ \varphi)'(\zeta) = \nabla f(\varphi(\zeta)) \cdot \varphi'(\zeta)^t$.

4. The Cauchy Integral

The Integral Formula. Let $\mathbf{r} = (r_1, \ldots, r_n)$ be an element of $\mathbb{R}^n_+$, $P = \mathsf{P}^n(\mathbf{0}, \mathbf{r})$, $T = \mathsf{T}^n(\mathbf{0}, \mathbf{r})$, and f a continuous function on T. Then

$$k_f(\mathbf{z}, \boldsymbol{\zeta}) := \frac{f(\boldsymbol{\zeta})}{(\boldsymbol{\zeta} - \mathbf{z})^{(1,\ldots,1)}} = \frac{f(\zeta_1, \ldots, \zeta_n)}{(\zeta_1 - z_1) \cdots (\zeta_n - z_n)}$$

defines a continuous function $k_f : P \times T \to \mathbb{C}$.

Definition.

$$\begin{aligned} C_f(\mathbf{z}) &:= \left(\frac{1}{2\pi \mathrm{i}}\right)^n \int_T k_f(\mathbf{z}, \boldsymbol{\zeta})\, d\boldsymbol{\zeta} \\ &:= \left(\frac{1}{2\pi \mathrm{i}}\right)^n \int_{|\zeta_1| = r_1} \cdots \int_{|\zeta_n| = r_n} f(\boldsymbol{\zeta}) \frac{d\zeta_1}{(\zeta_1 - z_1)} \cdots \frac{d\zeta_n}{(\zeta_n - z_n)} \end{aligned}$$

is called the *Cauchy integral* of f over T.

Obviously, C_f is a continuous function on P.

4.1 Theorem (Cauchy integral formula). *Let P, T be as above, and $U = U(\overline{P})$ be an open neighborhood of the closure of P. If f is weakly holomorphic on U, then $C_{f|T}(\mathbf{z}) = f(\mathbf{z})$ for any $\mathbf{z} \in P$.*

PROOF: If $P = \mathsf{D}_{r_1}(0) \times \cdots \times \mathsf{D}_{r_n}(0)$, we may assume that $U = U_1 \times \cdots \times U_n$, with open neighborhoods $U_i = U_i\left(\overline{\mathsf{D}_{r_i}(0)}\right)$, for $i = 1, \ldots, n$.

Since f is weakly holomorphic, we can fix $\mathbf{z}' = (z_1, \ldots, z_{n-1}) \in U_1 \times \cdots \times U_{n-1}$ and apply the Cauchy integral formula in one variable to $\zeta_n \mapsto f(\mathbf{z}', \zeta_n)$. For $z_n \in \mathsf{D}_{r_n}(0)$ it follows that

$$f(\mathbf{z}', z_n) = \frac{1}{2\pi \mathsf{i}} \int_{|\zeta_n|=r_n} \frac{f(\mathbf{z}', \zeta_n)}{\zeta_n - z_n}\, d\zeta_n.$$

Similarly, for the penultimate variable z_{n-1} and $\mathbf{z}'' = (z_1, \ldots, z_{n-2}) \in U_1 \times \cdots \times U_{n-2}$ we obtain

$$\begin{aligned} f(\mathbf{z}'', z_{n-1}, z_n) &= \frac{1}{2\pi \mathsf{i}} \int_{|\zeta_{n-1}|=r_{n-1}} \frac{f(\mathbf{z}'', \zeta_{n-1}, z_n)}{\zeta_{n-1} - z_{n-1}}\, d\zeta_{n-1} \\ &= \left(\frac{1}{2\pi \mathsf{i}}\right)^2 \int_{|\zeta_{n-1}|=r_{n-1}} \int_{|z_n|=r_n} \frac{f(\mathbf{z}'', \zeta_{n-1}, \zeta_n)}{(\zeta_{n-1} - z_{n-1})(\zeta_n - z_n)}\, d\zeta_n\, d\zeta_{n-1}, \end{aligned}$$

and after n steps, $f(\mathbf{z}) = C_{f|T}(\mathbf{z})$, for $\mathbf{z} \in P$. ∎

4.2 Theorem (power series expansion). *Let $P = \mathsf{P}^n(\mathbf{0}, \mathbf{r}) \subset \mathbb{C}^n$ be a polydisk and T its distinguished boundary. If $f : T \to \mathbb{C}$ is a continuous function, then there is a power series $\sum_{\nu \geq 0} a_\nu \mathbf{z}^\nu$ that converges to $C_f(\mathbf{z})$ in all of P.*

The coefficients a_ν of this series are given by

$$a_{\nu_1 \cdots \nu_n} = \left(\frac{1}{2\pi \mathsf{i}}\right)^n \int_T \frac{f(\zeta_1, \ldots, \zeta_n)}{\zeta_1^{\nu_1+1} \cdots \zeta_n^{\nu_n+1}}\, d\zeta_1 \cdots d\zeta_n.$$

PROOF: Setting $\mathbf{1} := (1, \ldots, 1) \in \mathbb{N}_0^n$, for $\mathbf{z} \in P$ and $\boldsymbol{\zeta} \in T$ it follows that

$$\begin{aligned} \frac{1}{(\boldsymbol{\zeta} - \mathbf{z})^{\mathbf{1}}} &= \frac{1}{(\zeta_1 - z_1) \cdots (\zeta_n - z_n)} = \frac{1}{\zeta_1 \cdots \zeta_n \cdot \left(1 - \dfrac{z_1}{\zeta_1}\right) \cdots \left(1 - \dfrac{z_n}{\zeta_n}\right)} \\ &= \frac{1}{\boldsymbol{\zeta}^{\mathbf{1}}} \cdot \sum_{\nu_1=0}^{\infty} \left(\frac{z_1}{\zeta_1}\right)^{\nu_1} \cdots \sum_{\nu_n=0}^{\infty} \left(\frac{z_n}{\zeta_n}\right)^{\nu_n}. \end{aligned}$$

If $\mathbf{r} = (r_1, \ldots, r_n)$, then for fixed $\mathbf{z} \in P$ and arbitrary $\boldsymbol{\zeta} \in T$ we have

$$\left| \frac{z_j}{\zeta_j} \right| = q_j := \frac{|z_j|}{r_j} < 1, \text{ for } j = 1, \ldots, n.$$

Since T is compact and f continuous on T, there is a constant M with $|f(\boldsymbol{\zeta})| \leq M$ on T. Then $\sum_{\nu \geq 0} \left(f(\boldsymbol{\zeta})/\boldsymbol{\zeta}^{\nu+\mathbf{1}}\right)\mathbf{z}^\nu$ is dominated on T by the convergent series $(M/\mathbf{r}^{\mathbf{1}})\sum_{\nu\geq 0}\mathbf{q}^{\nu}$, where $\mathbf{q}=(q_1,\dots,q_n)$, and therefore it is normally convergent as a series of functions on T with limit $f(\boldsymbol{\zeta})/(\boldsymbol{\zeta}-\mathbf{z})^{\mathbf{1}}$. We can interchange summation and integration:

$$C_f(\mathbf{z}) = \left(\frac{1}{2\pi \mathrm{i}}\right)^n \int_T \frac{f(\boldsymbol{\zeta})}{(\boldsymbol{\zeta}-\mathbf{z})^{\mathbf{1}}}\,d\boldsymbol{\zeta} = \sum_{\nu\geq 0} a_\nu \mathbf{z}^\nu,$$

with

$$a_\nu := \left(\frac{1}{2\pi \mathrm{i}}\right)^n \int_T \frac{f(\boldsymbol{\zeta})}{\boldsymbol{\zeta}^{\nu+\mathbf{1}}}\,d\boldsymbol{\zeta}.$$

The series converges for each $\mathbf{z}\in P$. ∎

4.3 Osgood's theorem. *Let $B\subset\mathbb{C}^n$ be an open set. The following statements about a function $f: B\to\mathbb{C}$ are equivalent:*

1. *f is holomorphic.*
2. *f is complex differentiable.*
3. *f is weakly holomorphic.*

PROOF: We already know that a holomorphic function f is complex differentiable, and it is trivial that then f is weakly holomorphic.

On the other hand, let $f: B\to\mathbb{C}$ be weakly holomorphic, and $\mathbf{z}_0\in B$ an arbitrary point. There is a small polydisk P around $\mathbf{z}_0$ that is relatively compact in B. If T is its distinguished boundary, then $f|_P = C_{f|T}$, and the Cauchy integral is the limit of a power series. So f is holomorphic. ∎

In addition, if f is weakly holomorphic on B, $\mathbf{z}_0\in B$ a point, and $P\subset\subset B$ a polydisk around $\mathbf{z}_0$, then there is a power series $S(\mathbf{z})=\sum_{\nu\geq 0}a_\nu(\mathbf{z}-\mathbf{z}_0)^\nu$ that converges to f on all of P.

Holomorphy of the Derivatives

4.4 Weierstrass's convergence theorem. *Let $G\subset\mathbb{C}^n$ be a domain, and (f_k) a sequence of holomorphic functions on G that converges uniformly to a function f. Then f is holomorphic.*

PROOF: The limit function is continuous. Let $\mathbf{z}_0\in G$ be a point, $P\subset\subset G$ a polydisk around $\mathbf{z}_0$, and T its distinguished boundary. Then

$$f|_P = \lim_{k\to\infty} f_k|_P = \lim_{k\to\infty} C_{f_k|T}.$$

Since T is compact, we can interchange integral and limit. Thus, for any fixed $\mathbf{z}\in P$,

$$\lim_{k\to\infty} C_{f_k|T}(\mathbf{z}) = C_{\lim\limits_{k\to\infty} f_k|T}(\mathbf{z}) = C_{f|T}(\mathbf{z}).$$

Since f is continuous on T, the Cauchy integral $C_{f|T}$ has a power series expansion in P. Therefore, f is holomorphic at $\mathbf{z}_0$. ∎

4.5 Proposition. *Let $S(\mathbf{z}) = \sum_{\nu \geq 0} a_\nu \mathbf{z}^\nu$ be a power series and G its domain of convergence. Then the limit function f of $S(\mathbf{z})$ is holomorphic on G, and the formal derivative*

$$S_{z_j}(\mathbf{z}) = \sum_{\substack{\nu \geq 0 \\ \nu_j > 0}} a_\nu \cdot \nu_j z_1^{\nu_1} \cdots z_j^{\nu_j - 1} \cdots z_n^{\nu_n}$$

converges to f_{z_j}. In particular, all partial derivatives of f are likewise holomorphic.

PROOF: Since $S(\mathbf{z})$ converges compactly on G, f is locally the uniform limit of a sequence of polynomials. Then it follows from Weierstrass' theorem that f is holomorphic. But also $S_{z_j}(\mathbf{z})$ converges compactly on G, and its limit function g_j must be holomorphic on G.

Now let $\mathbf{z}_0$ be an arbitrary point of G. Since G is a complete Reinhardt domain, there is a polydisk P around the origin with $\mathbf{z}_0 \in P \subset\subset G$. We define

$$f^*(\mathbf{z}) := \int_0^{z_j} g_j(z_1, \ldots, z_{j-1}, \zeta, z_{j+1}, \ldots, z_n)\, d\zeta + f(z_1, \ldots, 0, \ldots, z_n).$$

For the path of integration we take the connecting segment between 0 and z_j in the z_j-plane. Then f^* is defined on P.

Let $S(\mathbf{z}) = \sum_{i=0}^\infty p_i(\mathbf{z})$ be the expansion into a series of homogeneous polynomials. Then $S_{z_j}(\mathbf{z}) = \sum_{i=0}^\infty (p_i)_{z_j}(\mathbf{z})$, and this series converges uniformly on the compact path of integration we used above. Therefore, we can interchange summation and integration, and consequently,

$$\begin{aligned} f^*(\mathbf{z}) &= \sum_{i=0}^\infty \Big(\int_0^{z_j} (p_i)_{z_j}(z_1, \ldots, \zeta, \ldots, z_n)\, d\zeta + p_j(z_1, \ldots, 0, \ldots, z_n) \Big) \\ &= \sum_{i=0}^\infty p_i(\mathbf{z}) = f(\mathbf{z}), \end{aligned}$$

for $\mathbf{z} \in P$. Hence $f_{z_j}(\mathbf{z}_0) = f^*_{z_j}(\mathbf{z}_0) = g_j(\mathbf{z}_0)$. ∎

4.6 Corollary. *Let $G \subset \mathbb{C}^n$ be a domain and $f : B \to \mathbb{C}$ a holomorphic function. Then f is infinitely complex differentiable in G.*

Let $\nu = (\nu_1, \ldots, \nu_n)$ be a multi-index. Then we use the following abbreviations:

1. $\nu! := \nu_1! \cdots \nu_n!$.
2. If f is sufficiently often complex differentiable at $\mathbf{z}_0$, then

$$D^\nu f(\mathbf{z}_0) := \frac{\partial^{|\nu|} f}{\partial z_1^{\nu_1} \cdots \partial z_n^{\nu_n}}(\mathbf{z}_0).$$

4.7 Identity theorem (for power series).

Let $f(\mathbf{z}) = \sum_{\nu \geq 0} a_\nu \mathbf{z}^\nu$ and $g(\mathbf{z}) = \sum_{\nu \geq 0} b_\nu \mathbf{z}^\nu$ be two convergent power series in a neighborhood $U = U(\mathbf{0}) \subset \mathbb{C}^n$. If there is a neigborhood $V(\mathbf{0}) \subset U$ with $f|_V = g|_V$, then $a_\nu = b_\nu$ for all ν.

PROOF: We know that f and g are holomorphic. Then $D^\nu f(\mathbf{0}) = D^\nu g(\mathbf{0})$ for all ν, and successive differentiation gives

$$D^\nu f(\mathbf{0}) = \nu! \cdot a_\nu \quad \text{and} \quad D^\nu g(\mathbf{0}) = \nu! \cdot b_\nu.$$

∎

4.8 Corollary. *Let $G \subset \mathbb{C}^n$ be a domain, $\mathbf{z}_0 \in G$ a point, and $f : B \to \mathbb{C}$ a holomorphic function. If $f(\mathbf{z}) = \sum_{\nu \geq 0} a_\nu (\mathbf{z} - \mathbf{z}_0)^\nu$ is the (uniquely determined) power series expansion near $\mathbf{z}_0 \in G$, then*

$$a_\nu = \frac{1}{\nu!} \cdot D^\nu f(\mathbf{z}_0), \text{ for each } \nu \in \mathbb{N}_0^n.$$

4.9 Corollary (Cauchy's inequalities). *Let $G \subset \mathbb{C}^n$ be a domain, $f : G \to \mathbb{C}$ holomorphic, $\mathbf{z}_0 \in G$ a point, and $P = \mathsf{P}^n(\mathbf{z}_0, \mathbf{r}) \subset\subset G$ a polydisk with distinguished boundary T. Then*

$$|D^\nu f(\mathbf{z}_0)| \leq \frac{\nu!}{\mathbf{r}^\nu} \cdot \sup_T |f|.$$

PROOF: Let $f(\mathbf{z}) = \sum_{\nu \geq 0} a_\nu (\mathbf{z} - \mathbf{z}_0)^\nu$ be the power series expansion of f at $\mathbf{z}_0$. Then $D^\nu f(\mathbf{z}_0) = \nu! a_\nu$ and

$$a_\nu = \left(\frac{1}{2\pi \mathrm{i}}\right)^n \int_T \frac{f(\boldsymbol{\zeta})}{(\boldsymbol{\zeta} - \mathbf{z}_0)^{\nu + \mathbf{1}}} \, d\boldsymbol{\zeta},$$

and therefore

$$\begin{aligned}
|D^\nu f(\mathbf{z}_0)| &\leq \frac{\nu!}{(2\pi)^n} \int_T \frac{|f(\boldsymbol{\zeta})|}{\mathbf{r}^{\nu + \mathbf{1}}} \, d\boldsymbol{\zeta} \\
&= \frac{\nu!}{(2\pi)^n} \int\limits_{[0,2\pi]^n} \frac{|f(z_1^{(0)} + r_1 e^{\mathrm{i} t_1}, \ldots, z_n^{(0)} + r_n e^{\mathrm{i} t_n})|}{\mathbf{r}^\nu} \, dt_1 \cdots dt_n \\
&\leq \frac{\nu!}{\mathbf{r}^\nu} \cdot \sup_T |f|.
\end{aligned}$$

∎

The Identity Theorem. Let $G \subset \mathbb{C}^n$ be always a domain. The connectedness of G will be decisive in the following.

4.10 Identity theorem (for holomorphic functions).

Let f_1, f_2 be two holomorphic functions on G. If there is a nonempty open subset $U \subset G$ with $f_1|_U = f_2|_U$, then $f_1 = f_2$.

PROOF: We consider $f := f_1 - f_2$ and the set

$$N := \{\mathbf{z} \in G : D^{\nu} f(\mathbf{z}) = 0 \text{ for all } \boldsymbol{\nu}\}.$$

Then $N \neq \varnothing$, since $U \subset N$. Let $\mathbf{z}_0 \in G$ be an arbitrary point, and

$$f(\mathbf{z}) = \sum_{\nu \geq 0} \frac{1}{\nu!} D^{\nu} f(\mathbf{z}_0)(\mathbf{z} - \mathbf{z}_0)^{\nu}$$

the power series expansion of f in a neighborhood $V = V(\mathbf{z}_0) \subset G$. If $\mathbf{z}_0$ belongs to N, then $f|_V \equiv 0$, and also $V \subset N$. This shows that N is open. Because all derivatives $D^{\nu} f$ are continuous, N is closed. Since G is a domain, we get $N = G$ and $f_1 = f_2$. ∎

Remark. In contrast to the theory of one complex variable, it is not sufficient that f_1 and f_2 coincide on a set M that has a cluster point in G. Consider for example, $G = \mathbb{C}^2$ and $M = \{(z_1, z_2) : z_2 = 0\}$. The holomorphic functions $f_1(z_1, z_2) := z_2(z_1 - z_2)$ and $f_2(z_1, z_2) := z_2(z_1 + z_2)$ are equal on M, but $f_1(0, 1) = -1$ and $f_2(0, 1) = 1$.

4.11 Theorem (maximum principle).

Let $f : G \to \mathbb{C}$ be a holomorphic function. If there is a point $\mathbf{z}_0 \in G$ such that $|f|$ has a local maximum at $\mathbf{z}_0$, then f is constant.

PROOF: We consider the map $\varphi_{\mathbf{w}} : \mathbb{C} \to \mathbb{C}^n$ with $\varphi_{\mathbf{w}}(\zeta) = \mathbf{z}_0 + \zeta\mathbf{w}$, for an arbitrary $\mathbf{w} \neq \mathbf{0}$. Then $f \circ \varphi_{\mathbf{w}}$ is a holomorphic function of one complex variable, defined near $\zeta = 0$. Now, since $|f \circ \varphi_{\mathbf{w}}|$ has a local maximum at the origin, this function must be constant in a neighborhood of the origin. But the direction $\mathbf{w}$ was chosen arbitrarily, so f also has to be constant in a neighborhood of $\mathbf{0} \in \mathbb{C}^n$. The identity theorem implies that f is constant on G. ∎

Exercises

1. Prove Liouville's theorem: Every bounded holomorphic function on $\mathbb{C}^n$ is constant.
2. Prove that if $f \in \mathcal{O}(\mathbb{C}^n)$ and $|f(\mathbf{z})| \leq C \cdot \|\mathbf{z}^{\nu}\|$ for some $C > 0$ and some $\nu \in \mathbb{N}_0^n$, then f is a polynomial of degree at most $|\nu|$.

3. Let $G \subset \mathbb{C}^n$ be a domain and $f \in \mathcal{O}(G)$ not constant. Prove that then $f(U) \subset \mathbb{C}$ is open for any open subset $U \subset G$.
4. Let $G \subset \mathbb{C}^n$ be a domain. A set $\mathcal{F}$ of holomorphic functions on G is called *locally bounded*, if for every $\mathbf{z} \in G$ there is a neighborhood $U(\mathbf{z}) \subset G$ such that $\{ \|f\|_U : f \in \mathcal{F}\}$ is bounded. Prove the following:
   (a) (Lemma of Ascoli) If $A \subset G$ is a dense subset and (f_n) is a locally bounded sequence of holomorphic functions in G which converges pointwise on A, then (f_n) is compactly convergent on G.
   (b) (Theorem of Montel) Every locally bounded sequence of holomorphic functions in G has a compactly convergent subsequence.

   Hint: More or less, you can use the well-known proof from the 1-dimensional theory.

5. The Hartogs Figure

Expansion in Reinhardt Domains. Let r'_ν, r''_ν be real numbers with $0 < r'_\nu < r''_\nu$ for $1 \le \nu \le n$. We define

$$\begin{aligned} P &:= \{\mathbf{z} \in \mathbb{C}^n : |z_\nu| < r'_\nu \text{ for all } \nu\}, \\ Q &:= \{\mathbf{z} \in \mathbb{C}^n : r'_\nu < |z_\nu| < r''_\nu \text{ for all } \nu\}. \end{aligned}$$

Clearly, P and Q are Reinhardt domains. Let f be a holomorphic function in Q. Then for all $\mathbf{r} \in \tau(Q)$, the Cauchy integral $C_{f|\mathsf{T}_\mathbf{r}}$ is a holomorphic function in $\mathsf{P}_\mathbf{r}$, and therefore a fortiori in P (see Figure I.3).

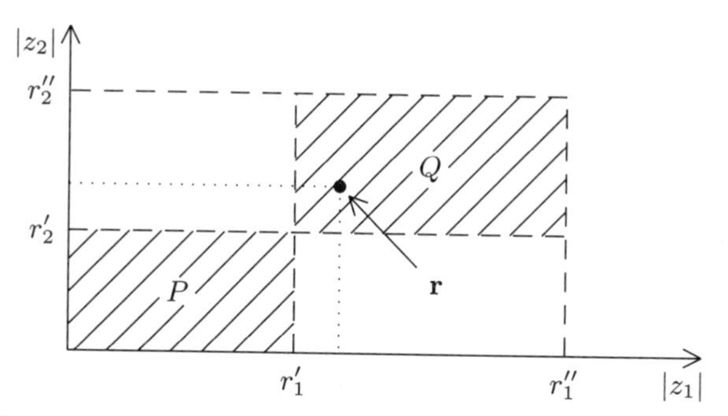

Figure I.3. Expansion in the polydisk

5.1 Proposition. *The function $f_\mathbf{r} : P \to \mathbb{C}$ given by $f_\mathbf{r}(\mathbf{z}) := C_{f|\mathsf{T}_\mathbf{r}}(\mathbf{z})$ is independent of $\mathbf{r}$.*

PROOF: We have

$$f_{\mathbf{r}}(\mathbf{z}) = \left(\frac{1}{2\pi \mathrm{i}}\right)^n \int\limits_{|\zeta_1|=r_1} \cdots \int\limits_{|\zeta_n|=r_n} f(\boldsymbol{\zeta}) \frac{d\zeta_1}{\zeta_1 - z_1} \cdots \frac{d\zeta_n}{\zeta_n - z_n}.$$

In each variable ζ_ν the integrand $f(\boldsymbol{\zeta})/(\zeta_\nu - z_\nu)$ is holomorphic on the annulus $\{\zeta_\nu : r'_\nu < |\zeta_\nu| < r''_\nu\}$. From the Cauchy integral formula for one variable it follows that

$$\int_{|\zeta_\nu|=r_\nu} \frac{f(\boldsymbol{\zeta})}{\zeta_\nu - z_\nu} \, d\zeta_\nu = \int_{|\zeta_\nu|=r^*_\nu} \frac{f(\boldsymbol{\zeta})}{\zeta_\nu - z_\nu} \, d\zeta_\nu$$

if $r'_\nu < r_\nu \le r^*_\nu < r''_\nu$. This yields the proposition. ■

5.2 Proposition. *Let $G \subset \mathbb{C}^n$ be a proper Reinhardt domain, f holomorphic on G. Then for every $\mathbf{z} \in G \cap (\mathbb{C}^*)^n$ the Cauchy integral $C_{f|\mathsf{T}_{\mathbf{z}}}$ coincides with f in a neighborhood of the origin.*

PROOF: $G \cap (\mathbb{C}^*)^n$ is a Reinhardt domain. Therefore, $G_0 := \tau(G \cap (\mathbb{C}^*)^n)$ is a domain in the absolute space.

Let $B := \{\mathbf{r} \in G_0 : C_{f|\mathsf{T}_{\mathbf{r}}}$ coincides with f in the vicinity of $\mathbf{0}\}$. Then $B \ne \varnothing$, because there is a small $\mathbf{r} \in G_0$ such that $\overline{P_{\mathbf{r}}(\mathbf{0})} \subset G$.

B is open: If $\mathbf{r}_0 \in B$, we can find sets P, Q as we did at the beginning of this section such that $\mathbf{r}_0 \in Q \subset G_0$. Then for $\mathbf{r} \in Q$, $f_{\mathbf{r}} = C_{f|\mathsf{T}_{\mathbf{r}}}$ is a holomorphic function on P, and independent of $\mathbf{r}$. But $f_{\mathbf{r}_0}$ coincides with f near the origin. Therefore, $Q \subset B$.

Also, $G_0 - B$ is open. The proof goes as above. Since G_0 is connected, that implies that $B = G_0$. ■

5.3 Corollary. *Let G be a proper Reinhardt domain, f holomorphic in G. Then there is a power series $S(\mathbf{z})$ which converges in G to f.*

PROOF: Let $\mathbf{z}_0 \in G$ be arbitrarily chosen. Then there is a point $\mathbf{w} \in G \cap (\mathbb{C}^*)^n$ with $\mathbf{z}_0 \in P_{\mathbf{w}}$. The holomorphic function $g := C_{f|\mathsf{T}_{\mathbf{w}}}$ has a power series expansion $g(\mathbf{z}) = \sum_{\nu \ge 0} a_\nu \mathbf{z}^\nu$ in $P_{\mathbf{w}}$. Since g coincides with f in a small neighborhood of the origin, the coefficients a_ν are those of the Taylor series of f about $\mathbf{0}$. Since $\mathbf{z}_0$ was arbitrary, it follows that the series converges in all of G. By the identity theorem its limit is equal to f. ■

Definition. If G is a proper Reinhardt domain, then

$$\widehat{G} := \bigcup_{\mathbf{z} \in G \cap (\mathbb{C}^*)^n} P_{\mathbf{z}}$$

is called the *complete hull* of G.

Remarks

1. Every complete Reinhardt domain is proper, but the opposite is in general false. For $n = 1$, Reinhardt domains are open disks around 0, and there is no difference between proper and complete domains.
2. The complete hull $\widehat{G}$ of a proper Reinhardt domain G is again a domain containing G. And it is Reinhardt: For $\mathbf{z} \in \widehat{G}$ there is some $\mathbf{z}_1$ with $\mathbf{z} \in \mathsf{P}_{\mathbf{z}_1} \subset \widehat{G}$. But then also $\mathsf{T}_{\mathbf{z}} \subset \mathsf{P}_{\mathbf{z}_1} \subset \widehat{G}$. The same argument shows that $\widehat{G}$ is complete.
3. Let G_1 be another complete Reinhardt domain with $G \subset G_1$. For $\mathbf{z} \in G \cap (\mathbb{C}^*)^n$, $\mathbf{z}$ also lies in G_1, and by the completeness of G_1 it follows that $\mathsf{P}_{\mathbf{z}} \subset G_1$. So $\widehat{G} \subset G_1$, and we see that $\widehat{G}$ is the smallest complete Reinhardt domain containing G.

An immediate consequence is the following:

5.4 Theorem. *Let G be a proper Reinhardt domain and f be holomorphic in G. Then there is exactly one holomorphic function $\widehat{f}$ in $\widehat{G}$ with $\widehat{f}|_G = f$.*

Hartogs Figures. In the case $n = 1$ the situation above cannot appear. For $n \geq 2$ we can choose sets G and $\widehat{G}$ in $\mathbb{C}^n$ such that $G \neq \widehat{G}$. This reflects an essential difference between the theories of one and several complex variables.

Now let $n \geq 2$, P^n the unit polydisk, $q_1, \ldots, q_n$ real numbers with $0 < q_\nu < 1$ for $\nu = 1, \ldots, n$, and

$$\mathsf{H} = \mathsf{H}(\mathbf{q}) := \{\mathbf{z} \in \mathsf{P}^n \ : \ |z_1| > q_1 \text{ or } |z_\mu| < q_\mu \text{ for } \mu = 2, \ldots, n\}.$$

Then $(\mathsf{P}^n, \mathsf{H})$ is called a *Euclidean Hartogs figure* (see Figure I.4). H is a proper Reinhardt domain and P^n its complete hull.

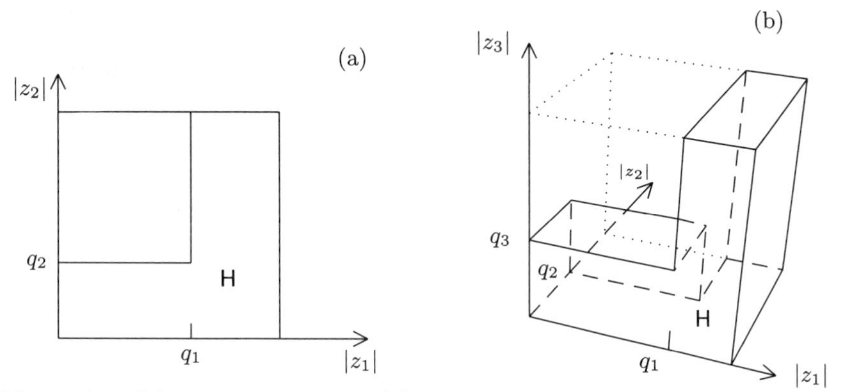

Figure I.4. (a) 2-dimensional, and (b) 3-dimensional Hartogs figure

5.5 Hartogs' theorem. *Let $(\mathsf{P}^n, \mathsf{H})$ be a Euclidean Hartogs figure. Then any holomorphic function f on H has a holomorphic extension $\widehat{f}$ on P^n.*

The theorem follows immediately from our considerations above.

Exercises

1. For $0 < q < 1$ let $G_1, G_2 \subset \mathbb{C}^2$ be defined by

$$\begin{aligned} G_1 &:= \{(z,w) : q < |z| < 1 \text{ and } |w| < 1\}, \\ G_2 &:= \{(z,w) : |z| < 1 \text{ and } |w| < q\}. \end{aligned}$$

   (a) Prove that every holomorphic function f on G_1 has a unique representation

$$f(z,w) = \sum_{n=-\infty}^{\infty} a_n(w) z^n, \text{ with } a_n \in \mathcal{O}(\mathsf{D}).$$

   (b) Prove that every holomorphic function g on G_2 has a unique representation

$$g(z,w) = \sum_{n=0}^{\infty} b_n(w) z^n, \text{ with } b_n \in \mathcal{O}(\mathsf{D}_q(0)).$$

   (c) Use (a) and (b) to prove that every holomorphic function f on $G_1 \cup G_2$ has a unique holomorphic extension to the unit polydisk.

2. Let $G \subset \mathbb{C}^n$ be an arbitrary Reinhardt domain, $f \in \mathcal{O}(G)$. Show that there exists a uniquely determined "Laurent series" $\sum_{\nu \in \mathbb{Z}^n} a_\nu \mathbf{z}^\nu$ converging compactly in G to f.

6. The Cauchy–Riemann Equations

Real Differentiable Functions. Recall the following from real analysis:

Let $B \subset \mathbb{C}^n$ be an open set and $\mathbf{z}_0$ a point of B. A function $f : B \to \mathbb{R}$ is called *differentiable (in the real sense)* if there is a real linear form $L : \mathbb{C}^n \to \mathbb{R}$ and a real-valued function r with:

1. $f(\mathbf{z}) = f(\mathbf{z}_0) + L(\mathbf{z} - \mathbf{z}_0) + r(\mathbf{z} - \mathbf{z}_0)$.
2. $\displaystyle\lim_{\mathbf{w} \to \mathbf{0}} \frac{r(\mathbf{w})}{\|\mathbf{w}\|} = 0$.

The real linear form $Df(\mathbf{z}_0) := L$ is called the *(total) derivative* of f at $\mathbf{z}_0$. It can be given in the form

$$L : \mathbf{u} + \mathrm{i}\mathbf{v} \mapsto \mathbf{u} \cdot \nabla_{\mathbf{x}} f(\mathbf{z}_0)^{\,t} + \mathbf{v} \cdot \nabla_{\mathbf{y}} f(\mathbf{z}_0)^{\,t},$$

with $\nabla_{\mathbf{x}} f(\mathbf{z}_0) = (f_{x_1}(\mathbf{z}_0), \ldots, f_{x_n}(\mathbf{z}_0))$ and $\nabla_{\mathbf{y}} f(\mathbf{z}_0) = (f_{y_1}(\mathbf{z}_0), \ldots, f_{y_n}(\mathbf{z}_0))$. We call $(\nabla_{\mathbf{x}} f(\mathbf{z}_0), \nabla_{\mathbf{y}} f(\mathbf{z}_0))$ the *real gradient* of f at $\mathbf{z}_0$.

If $f = g + \mathrm{i}h : B \to \mathbb{C}$ is a complex-valued function, then f is called *differentiable (in the real sense)*, if g and h are differentiable. The *(real) derivative* of f at $\mathbf{z}_0$ is defined to be the complex-valued real linear form

$$Df(\mathbf{z}_0) := Dg(\mathbf{z}_0) + \mathrm{i}Dh(\mathbf{z}_0).$$

6.1 Proposition. *A function $f : B \to \mathbb{C}$ is (real) differentiable at $\mathbf{z}_0$ if and only if there are maps $\Delta', \Delta'' : B \to \mathbb{C}^n$ such that:*

1. *Δ' and Δ'' are continuous at $\mathbf{z}_0$.*
2. *$f(\mathbf{z}) = f(\mathbf{z}_0) + (\mathbf{z} - \mathbf{z}_0) \cdot \Delta'(\mathbf{z})^{\,t} + (\overline{\mathbf{z}} - \overline{\mathbf{z}}_0) \cdot \Delta''(\mathbf{z})^{\,t}$ for $\mathbf{z} \in B$.*

The values $\Delta'(\mathbf{z}_0)$ and $\Delta''(\mathbf{z}_0)$ are uniquely determined.

PROOF: (1) Let f be differentiable at $\mathbf{z}_0$. Then there is a complex linear form Λ' and a complex antilinear form Λ'' such that

$$Df(\mathbf{z}_0) = \Lambda' + \Lambda''.$$

The decomposition is uniquely determined, and there are vectors $\Delta'(\mathbf{z}_0)$ and $\Delta''(\mathbf{z}_0)$ such that

$$\Lambda'(\mathbf{w}) = \mathbf{w} \cdot \Delta'(\mathbf{z}_0)^{\,t} \quad \text{and} \quad \Lambda''(\mathbf{w}) = \overline{\mathbf{w}} \cdot \Delta''(\mathbf{z}_0)^{\,t}.$$

Now we define

$$\begin{aligned} \Delta'(\mathbf{z}) &:= \Delta'(\mathbf{z}_0) + \frac{r(\mathbf{z} - \mathbf{z}_0)}{2\|\mathbf{z} - \mathbf{z}_0\|^2} \cdot (\overline{\mathbf{z}} - \overline{\mathbf{z}}_0), \\ \Delta''(\mathbf{z}) &:= \Delta''(\mathbf{z}_0) + \frac{r(\mathbf{z} - \mathbf{z}_0)}{2\|\mathbf{z} - \mathbf{z}_0\|^2} \cdot (\mathbf{z} - \mathbf{z}_0). \end{aligned}$$

It is easy to see that

(a) Δ' and Δ'' are continuous at $\mathbf{z}_0$,
(b) $f(\mathbf{z}) = f(\mathbf{z}_0) + (\mathbf{z} - \mathbf{z}_0) \cdot \Delta'(\mathbf{z})^{\,t} + (\overline{\mathbf{z}} - \overline{\mathbf{z}}_0) \cdot \Delta''(\mathbf{z})^{\,t}$.

(2) Now let the decomposition be given, and define

$$\begin{aligned} L(\mathbf{w}) &:= \mathbf{w} \cdot \Delta'(\mathbf{z}_0)^{\,t} + \overline{\mathbf{w}} \cdot \Delta''(\mathbf{z}_0)^{\,t}, \\ r(\mathbf{w}) &:= \mathbf{w} \cdot (\Delta'(\mathbf{z}) - \Delta'(\mathbf{z}_0))^{\,t} + \overline{\mathbf{w}} \cdot (\Delta''(\mathbf{z}) - \Delta''(\mathbf{z}_0))^{\,t}. \end{aligned}$$

Since

$$\frac{|r(\mathbf{w})|}{\|\mathbf{w}\|} \le \|\Delta'(\mathbf{z}) - \Delta'(\mathbf{z}_0)\| + \|\Delta''(\mathbf{z}) - \Delta''(\mathbf{z}_0)\|,$$

it follows that f is differentiable at $\mathbf{z}_0$ with derivative L. ∎

Wirtinger's Calculus

Definition. Let $f : B \to \mathbb{C}$ be real differentiable at $\mathbf{z}_0$. If we have a representation

$$f(\mathbf{z}) = f(\mathbf{z}_0) + (\mathbf{z} - \mathbf{z}_0) \cdot \Delta'(\mathbf{z})^t + (\overline{\mathbf{z}} - \overline{\mathbf{z}}_0) \cdot \Delta''(\mathbf{z})^t,$$

with Δ' and Δ'' continuous at $\mathbf{z}_0$, then the uniquely determined numbers

$$\frac{\partial f}{\partial z_\nu}(\mathbf{z}_0) = f_{z_\nu}(\mathbf{z}_0) := \mathbf{e}_\nu \cdot \Delta'(\mathbf{z}_0)^t$$

and

$$\frac{\partial f}{\partial \overline{z}_\nu}(\mathbf{z}_0) = f_{\overline{z}_\nu}(\mathbf{z}_0) := \mathbf{e}_\nu \cdot \Delta''(\mathbf{z}_0)^t$$

are called the *Wirtinger derivatives* of f at $\mathbf{z}_0$.

The complex linear (respectively antilinear) forms $(\partial f)_{\mathbf{z}_0} : \mathbb{C}^n \to \mathbb{C}$ and $(\overline{\partial} f)_{\mathbf{z}_0} : \mathbb{C}^n \to \mathbb{C}$ are defined by

$$(\partial f)_{\mathbf{z}_0}(\mathbf{w}) := \sum_{\nu=1}^{n} f_{z_\nu}(\mathbf{z}_0) w_\nu \quad \text{and} \quad (\overline{\partial} f)_{\mathbf{z}_0}(\mathbf{w}) := \sum_{\nu=1}^{n} f_{\overline{z}_\nu}(\mathbf{z}_0) \overline{w}_\nu,$$

and the *differential* of f at $\mathbf{z}_0$ by $(df)_{\mathbf{z}_0} := (\partial f)_{\mathbf{z}_0} + (\overline{\partial} f)_{\mathbf{z}_0}$.

Obviously, $Df(\mathbf{z}_0) = (df)_{\mathbf{z}_0}$.

If we introduce the *holomorphic* (respectively *antiholomorphic*) *gradient*

$$\nabla f := (f_{z_1}, \ldots, f_{z_n}) \quad \text{and} \quad \overline{\nabla} f := (f_{\overline{z}_1}, \ldots, f_{\overline{z}_n}),$$

then $(\partial f)_{\mathbf{z}_0}(\mathbf{w}) = \mathbf{w} \cdot \nabla f(\mathbf{z}_0)^t$ and $(\overline{\partial} f)_{\mathbf{z}_0}(\mathbf{w}) = \overline{\mathbf{w}} \cdot \overline{\nabla} f(\mathbf{z}_0)^t$.

6.2 Proposition. *Let f be a (complex-valued) function that is real differentiable at $\mathbf{z}_0$. Then*

$$\begin{aligned} f_{z_\nu}(\mathbf{z}_0) &= \frac{1}{2}(f_{x_\nu}(\mathbf{z}_0) - \mathrm{i} f_{y_\nu}(\mathbf{z}_0)), \\ f_{\overline{z}_\nu}(\mathbf{z}_0) &= \frac{1}{2}(f_{x_\nu}(\mathbf{z}_0) + \mathrm{i} f_{y_\nu}(\mathbf{z}_0)). \end{aligned}$$

PROOF: Let be $L := Df(\mathbf{z}_0)$. Then

$$f_{x_\nu}(\mathbf{z}_0) = L(\mathbf{e}_\nu) = (\partial f)_{\mathbf{z}_0}(\mathbf{e}_\nu) + (\overline{\partial} f)_{\mathbf{z}_0}(\mathbf{e}_\nu) = f_{z_\nu}(\mathbf{z}_0) + f_{\overline{z}_\nu}(\mathbf{z}_0)$$

and

$$f_{y_\nu}(\mathbf{z}_0) = L(\mathrm{i}\mathbf{e}_\nu) = (\partial f)_{\mathbf{z}_0}(\mathrm{i}\mathbf{e}_\nu) + (\overline{\partial} f)_{\mathbf{z}_0}(\mathrm{i}\mathbf{e}_\nu) = \mathrm{i}(f_{z_\nu}(\mathbf{z}_0) - f_{\overline{z}_\nu}(\mathbf{z}_0)).$$

Putting things together we obtain

$$f_{x_\nu}(\mathbf{z}_0) - \mathrm{i} f_{y_\nu}(\mathbf{z}_0) = 2 f_{z_\nu}(\mathbf{z}_0) \quad \text{and} \quad f_{x_\nu}(\mathbf{z}_0) + \mathrm{i} f_{y_\nu}(\mathbf{z}_0) = 2 f_{\overline{z}_\nu}(\mathbf{z}_0).$$

■

Remark. Use these formulas with care! The derivatives f_{x_ν} and f_{y_ν} in general are complex-valued. So the equations do not give the decomposition of f_{z_ν} and $f_{\overline{z}_\nu}$ into real and imaginary parts, respectively!

The Cauchy–Riemann Equations

6.3 Theorem. *Let $f : B \to \mathbb{C}$ be a continuously real differentiable function. Then f is holomorphic if and only if $f_{\overline{z}_\nu}(\mathbf{z}) \equiv 0$ on B, for $\nu = 1, \ldots, n$.*

PROOF: (a) If f is holomorphic, then f is complex differentiable at every point $\mathbf{z}_0 \in B$. Comparing the two decompositions

$$f(\mathbf{z}) = f(\mathbf{z}_0) + (\mathbf{z} - \mathbf{z}_0) \cdot \Delta(\mathbf{z})^t$$

and

$$f(\mathbf{z}) = f(\mathbf{z}_0) + (\mathbf{z} - \mathbf{z}_0) \cdot \Delta'(\mathbf{z})^t + (\overline{\mathbf{z}} - \overline{\mathbf{z}}_0) \cdot \Delta''(\mathbf{z})^t$$

we see that $\Delta'(\mathbf{z}_0) = \Delta(\mathbf{z}_0)$ and $\Delta''(\mathbf{z}_0) = 0$. The latter equation means that $f_{\overline{z}_\nu}(\mathbf{z}_0) \equiv 0$ for $\nu = 1, \ldots, n$.

(b) If $f_{\overline{z}_\nu}(\mathbf{z}) \equiv 0$, then f is holomorphic in each variable and is consequently holomorphic. ■

Remark. Now the following is clear: If f is holomorphic near $\mathbf{z}_0$, then

$$(\overline{\partial} f)_{\mathbf{z}_0} = 0 \quad \text{and} \quad Df(\mathbf{z}_0)(\mathbf{w}) = (df)_{\mathbf{z}_0}(\mathbf{w}) = (\partial f)_{\mathbf{z}_0}(\mathbf{w}) = \sum_{\nu=1}^{n} f_{z_\nu}(\mathbf{z}_0) w_\nu.$$

The equation $(\overline{\partial} f)_{\mathbf{z}} = 0$ is the shortest version of the *Cauchy–Riemann differential equations*. In greater detail, these are the equations

$$f_{\overline{z}_\nu}(\mathbf{z}) \equiv 0, \text{ for } \nu = 1, \ldots, n.$$

Finally, if $f = g + \mathrm{i} h$, then we can write the Cauchy–Riemann equations in their classical form:

$$g_{x_\nu} = h_{y_\nu} \text{ and } h_{x_\nu} = -g_{y_\nu}, \text{ for } \nu = 1, \ldots, n.$$

Exercises

1. Derive the Cauchy–Riemann equations in their classical form.
2. Let $f : G \to \mathbb{C}$ be real differentiable. Prove the formulas

$$\overline{(f_{z_\nu})} = (\overline{f})_{\overline{z}_\nu} \quad \text{and} \quad f_{z_\nu \overline{z}_\mu} = f_{\overline{z}_\mu z_\nu} \text{ for } \nu, \mu = 1, \dots, n.$$

3. Let $G \subset \mathbb{C}^n$ be a domain and $f_1, \dots, f_k : G \to \mathbb{C}$ holomorphic functions. Show that if $\sum_{j=1}^k f_j \overline{f}_j$ is constant, then all f_j are constant.

   Hint: If h is holomorphic, then $\dfrac{\partial^2 |h|^2}{\partial z_j \partial \overline{z}_j} = \left| \dfrac{\partial h}{\partial z_j} \right|^2.$

7. Holomorphic Maps

The Jacobian. Let $B \subset \mathbb{C}^n$ be an open set. A map

$$\mathbf{f} = (f_1, \dots, f_m) : B \to \mathbb{C}^m$$

is called *holomorphic* (respectively *real differentiable*) if all components f_i are holomorphic (respectively real differentiable).

7.1 Proposition. *The map $\mathbf{f} : B \to \mathbb{C}^m$ is holomorphic if and only if for any $\mathbf{z}_0 \in B$ there exists a map $\Delta : B \to M_{m,n}(\mathbb{C})$ with the following properties:*

1. *Δ is continuous at $\mathbf{z}_0$.*
2. *$\mathbf{f}(\mathbf{z}) = \mathbf{f}(\mathbf{z}_0) + (\mathbf{z} - \mathbf{z}_0) \cdot \Delta(\mathbf{z})^t$, for $\mathbf{z} \in B$.*

The value $\Delta(\mathbf{z}_0)$ is uniquely defined.

PROOF: The map $\mathbf{f}$ is holomorphic if there are decompositions

$$f_\mu(\mathbf{z}) = f_\mu(\mathbf{z}_0) + (\mathbf{z} - \mathbf{z}_0) \cdot \Delta_\mu(\mathbf{z})^t,$$

with Δ_μ continuous at $\mathbf{z}_0$, for $\mu = 1, \dots, m$.

Then Δ is given by $\Delta(\mathbf{z})^t = (\Delta_1(\mathbf{z})^t, \dots, \Delta_m(\mathbf{z})^t)$. We leave the further details to the reader. ∎

Definition. If $\mathbf{f} : B \to \mathbb{C}^m$ is holomorphic, then $J_\mathbf{f}(\mathbf{z}_0) := \Delta(\mathbf{z}_0)$ is called the *complex Jacobian (matrix)* of $\mathbf{f}$ at $\mathbf{z}_0$. The associated linear map $\mathbf{f}'(\mathbf{z}_0) : \mathbb{C}^n \to \mathbb{C}^m$ is called the *(complex) derivative* of $\mathbf{f}$ at $\mathbf{z}_0$. It is given by

$$\mathbf{f}'(\mathbf{z}_0)(\mathbf{w}) = \mathbf{w} \cdot J_\mathbf{f}(\mathbf{z}_0)^t.$$

Explicitly, we have

$$J_{\mathbf{f}}(\mathbf{z}) = \begin{pmatrix} (f_1)_{z_1}(\mathbf{z}) & \cdots & (f_1)_{z_n}(\mathbf{z}) \\ \vdots & & \vdots \\ (f_m)_{z_1}(\mathbf{z}) & \cdots & (f_m)_{z_n}(\mathbf{z}) \end{pmatrix}.$$

This matrix is also defined for differentiable maps.

Definition. If $\mathbf{f} = \mathbf{g} + \mathrm{i}\mathbf{h} : B \to \mathbb{C}^m$ is a differentiable map, then the *real Jacobian matrix* $J_{\mathbb{R},\mathbf{f}}(\mathbf{z}_0) \in M_{2m,2n}(\mathbb{R})$ is the real matrix associated to the real linear map

$$(D\mathbf{g}(\mathbf{z}_0), D\mathbf{h}(\mathbf{z}_0)) : \mathbb{C}^n = \mathbb{R}^{2n} \to \mathbb{R}^{2m}.$$

The real Jacobian of $\mathbf{f} = \mathbf{g} + \mathrm{i}\mathbf{h}$ is given by

$$J_{\mathbb{R},\mathbf{f}} = \left(\begin{array}{ccc|ccc} (g_1)_{x_1} & \cdots & (g_1)_{x_n} & (g_1)_{y_1} & \cdots & (g_1)_{y_n} \\ \vdots & & \vdots & \vdots & & \vdots \\ (g_m)_{x_1} & \cdots & (g_m)_{x_n} & (g_m)_{y_1} & \cdots & (g_m)_{y_n} \\ \hline (h_1)_{x_1} & \cdots & (h_1)_{x_n} & (h_1)_{y_1} & \cdots & (h_1)_{y_n} \\ \vdots & & \vdots & \vdots & & \vdots \\ (h_m)_{x_1} & \cdots & (h_m)_{x_n} & (h_m)_{y_1} & \cdots & (h_m)_{y_n} \end{array}\right).$$

The $\mathbb{R}$-linear map $D\mathbf{f}(\mathbf{z}) : \mathbb{C}^n \to \mathbb{C}^n$ is defined by $D\mathbf{f}(\mathbf{z}) := D\mathbf{g}(\mathbf{z}) + \mathrm{i}\, D\mathbf{h}(\mathbf{z})$. Setting $(\partial\mathbf{f})_{\mathbf{z}} := ((\partial f_1)_{\mathbf{z}}, \ldots, (\partial f_m)_{\mathbf{z}})$ and $(\overline{\partial}\mathbf{f})_{\mathbf{z}} := ((\overline{\partial} f_1)_{\mathbf{z}}, \ldots, (\overline{\partial} f_m)_{\mathbf{z}})$, we obtain

$$D\mathbf{f}(\mathbf{z}) = (\partial\mathbf{f})_{\mathbf{z}} + (\overline{\partial}\mathbf{f})_{\mathbf{z}}.$$

7.2 Theorem. *A differentiable map $\mathbf{f} = \mathbf{g} + \mathrm{i}\mathbf{h} : B \to \mathbb{C}^m$ is holomorphic if and only if $D\mathbf{f}(\mathbf{z})$ is $\mathbb{C}$-linear for every $\mathbf{z} \in B$.*

If $\mathbf{f}$ is holomorphic and $n = m$, then $\det(J_{\mathbb{R},\mathbf{f}}(\mathbf{z})) = |\det J_{\mathbf{f}}(\mathbf{z})|^2$.

PROOF: The map $\mathbf{f}$ is holomorphic if and only if $\left(\overline{\partial}\mathbf{f}\right)_{\mathbf{z}} = 0$ for every $\mathbf{z}$. Then $D\mathbf{f}(\mathbf{z}) = (\partial\mathbf{f})_{\mathbf{z}}$, which is complex linear. In this case we have the Cauchy–Riemann equations

$$(g_\mu)_{x_\nu} = (h_\mu)_{y_\nu} \text{ and } (h_\mu)_{x_\nu} = -(g_\mu)_{x_\nu},$$

and therefore

$$(f_\mu)_{z_\nu} = (f_\mu)_{x_\nu} = (g_\mu)_{x_\nu} + \mathrm{i}(h_\mu)_{x_\nu}, \text{ for } \mu = 1, \ldots, m \text{ and } \nu = 1, \ldots, n.$$

If $n = m$, then $J_{\mathbb{R},\mathbf{f}} = \begin{pmatrix} A & B \\ C & D \end{pmatrix}$ with $B = -C$ and $A = D$, and $J_{\mathbf{f}} = A + \mathrm{i}C$.

By elementary transformations,

$$\begin{aligned}\det\begin{pmatrix} A & -C \\ C & A \end{pmatrix} &= \det\begin{pmatrix} A+\mathrm{i}C & -C+\mathrm{i}A \\ C & A \end{pmatrix} \\ &= \det\begin{pmatrix} A+\mathrm{i}C & 0 \\ C & A-\mathrm{i}C \end{pmatrix} \\ &= |\det(A+\mathrm{i}C)|^2.\end{aligned}$$

∎

It follows that holomorphic maps are orientation preserving!

Chain Rules. Let $B \subset \mathbb{C}^n$ be an open set, $\mathbf{f} : B \to \mathbb{C}^m$ a differentiable map, and g a complex-valued differentiable function that is defined on the image of $\mathbf{f}$. Then $g \circ \mathbf{f} : B \to \mathbb{C}$ is differentiable, and the following holds:

7.3 Proposition (complex chain rule).

$$\begin{aligned}(g\circ\mathbf{f})_{z_\nu} &= \sum_{\mu=1}^{m}(g_{w_\mu}\circ\mathbf{f})\cdot(f_\mu)_{z_\nu} + \sum_{\mu=1}^{m}(g_{\overline{w}_\mu}\circ\mathbf{f})\cdot\left(\overline{f}_\mu\right)_{z_\nu},\\ (g\circ\mathbf{f})_{\overline{z}_\nu} &= \sum_{\mu=1}^{m}(g_{w_\mu}\circ\mathbf{f})\cdot(f_\mu)_{\overline{z}_\nu} + \sum_{\mu=1}^{m}(g_{\overline{w}_\mu}\circ\mathbf{f})\cdot\left(\overline{f}_\mu\right)_{\overline{z}_\nu}.\end{aligned}$$

One can use the well-known proof for the chain rule in real analysis, considering z_ν and $\overline{z}_\nu$ as independent variables.

7.4 Corollary. *If* $\mathbf{f}$ *and* g *are holomorphic, then*

$$\begin{aligned}(g\circ\mathbf{f})_{\overline{z}_\nu}(\mathbf{z}) &\equiv 0 \quad \textit{(i.e., } g\circ\mathbf{f} \textit{ is holomorphic)},\\ (g\circ\mathbf{f})_{z_\nu}(\mathbf{z}) &= \sum_{\mu=1}^{m} g_{w_\mu}(\mathbf{f}(\mathbf{z}))\cdot(f_\mu)_{z_\nu}(\mathbf{z}).\end{aligned}$$

The second equation can be abbreviated as

$$\nabla(g\circ\mathbf{f})(\mathbf{z}) = \nabla g(\mathbf{f}(\mathbf{z}))\cdot J_{\mathbf{f}}(\mathbf{z}).$$

Tangent Vectors. In this paragraph we use the term *differentiable* for *infinitely differentiable.*

> **Definition.** A *tangent vector* at a point $\mathbf{z} \in \mathbb{C}^n$ is a pair $\mathsf{t} = (\mathbf{z}, \mathbf{w})$, where the *direction* $\mathbf{w}$ of t is an arbitrary vector of $\mathbb{C}^n$. If the *base point* $\mathbf{z}$ is fixed, we simply write $\mathbf{w}$ instead of t or $(\mathbf{z}, \mathbf{w})$.
>
> The set $T_{\mathbf{z}}$ of all tangent vectors at $\mathbf{z}$ is called the *tangent space* (of $\mathbb{C}^n$) at $\mathbf{z}$.

The notation "tangent vector" is motivated by the following:

Let $B \subset \mathbb{C}^n$ be an open set and $I \subset \mathbb{R}$ an interval containing 0 as an interior point. If $\alpha = (\alpha_1, \ldots, \alpha_n) : I \to B$ is a differentiable path, then $\alpha'(0)$ is the direction of the tangent to the curve α at the point $\alpha(0)$. Therefore,

$$\dot{\alpha}(0) := (\alpha(0), \alpha'(0))$$

is called the *tangent vector of* α at $\mathbf{z} = \alpha(0)$. Each tangent vector $(\mathbf{z}, \mathbf{w}) \in T_{\mathbf{z}}$ can be written in the form $\dot{\alpha}(0)$, e.g., $\alpha(t) := \mathbf{z} + t\mathbf{w}$.

The tangent space $T_{\mathbf{z}}$ carries in a natural way the structure of a complex vector space:

$$\begin{aligned} (\mathbf{z}, \mathbf{w}_1) + (\mathbf{z}, \mathbf{w}_2) &:= (\mathbf{z}, \mathbf{w}_1 + \mathbf{w}_2), \\ \lambda \cdot (\mathbf{z}, \mathbf{w}) &:= (\mathbf{z}, \lambda \cdot \mathbf{w}), \text{ for } \lambda \in \mathbb{C}. \end{aligned}$$

Every tangent vector $\mathfrak{t} = (\mathbf{z}, \mathbf{w})$ operates linearly on the algebra $\mathscr{E}(B)$ of differentiable functions on B by

$$\mathfrak{t}[f] = Df(\mathbf{z})(\mathbf{w}).$$

This is the *directional derivative*, also denoted by $D_{\mathbf{w}}f(\mathbf{z})$. If $\mathfrak{t} = \dot{\alpha}(0)$, for some differentiable path α, then $\mathfrak{t}[f] = (f \circ \alpha)'(0)$, due to the chain rule.

The operator $\mathfrak{t} : \mathscr{E}(B) \to \mathbb{R}$ satisfies the *product rule*:

$$\mathfrak{t}[f \cdot g] = \mathfrak{t}[f] \cdot g(\mathbf{z}) + f(\mathbf{z}) \cdot \mathfrak{t}[g].$$

In general, a linear operator satisfying the product rule is called a *derivation*. In Chapter IV we will show that the tangent space is isomorphic to the vector space of derivations.

The Inverse Mapping.

Let $B_1, B_2 \subset \mathbb{C}^n$ be open sets, and $\mathbf{f} : B_1 \to B_2$ a holomorphic map.

Definition. The map $\mathbf{f}$ is called *biholomorphic* (or an *invertible holomorphic map*) if $\mathbf{f}$ is bijective and $\mathbf{f}^{-1}$ holomorphic.

7.5 Inverse mapping theorem. *Consider a point $\mathbf{z}_0 \in B_1$ and its image $\mathbf{w}_0 = \mathbf{f}(\mathbf{z}_0)$. Then the following are equivalent:*

1. *There are open neighborhoods $U = U(\mathbf{z}_0) \subset B_1$ and $V = V(\mathbf{w}_0) \subset B_2$ such that $\mathbf{f} : U \to V$ is biholomorphic.*
2. $\det J_{\mathbf{f}}(\mathbf{z}_0) \neq 0$.

PROOF: If $\mathbf{f}|_U : U \to V$ is biholomorphic, then $(\mathbf{f}|_U)^{-1} \circ \mathbf{f} = \mathrm{id}_U$ and $1 = \det(\mathbf{E}_n) = \det(J_{(\mathbf{f}|_U)^{-1}}(\mathbf{w}_0) \cdot J_{\mathbf{f}}(\mathbf{z}_0)) = \det(J_{(\mathbf{f}|_U)^{-1}}(\mathbf{w}_0)) \cdot \det(J_{\mathbf{f}}(\mathbf{z}_0))$, and therefore $\det(J_{\mathbf{f}}(\mathbf{z}_0)) \neq 0$.

If $\det(J_{\mathbf{f}}(\mathbf{z}_0)) \neq 0$, then also $\det(J_{\mathbb{R},\mathbf{f}}(\mathbf{z}_0) = |\det J_{\mathbf{f}}(\mathbf{z}_0)|^2 \neq 0$. It follows from real analysis that there are open neighborhoods $U = U(\mathbf{z}_0) \subset B_1$ and $V = V(\mathbf{w}_0) \subset B_2$ such that $\mathbf{f}|_U : U \to V$ is bijective and $\mathbf{g} := (\mathbf{f}|_U)^{-1} : V \to U$ a continuously differentiable map (in the real sense). Then $\mathbf{f} \circ \mathbf{g} = \mathrm{id}_V$ is a holomorphic map, and if we write $\mathbf{f} = (f_1, \ldots, f_n)$ and $\mathbf{g} = (g_1, \ldots, g_n)$, then

$$0 = (f_\nu \circ \mathbf{g})_{\overline{w}_\mu} = \sum_{\lambda=1}^{n} ((f_\nu)_{z_\lambda} \circ \mathbf{g}) \cdot (g_\lambda)_{\overline{w}_\mu}, \text{ for } \nu, \mu = 1, \ldots, n.$$

In the language of matrices this means that

$$\mathbf{0} = J_{\mathbf{f}} \cdot \begin{pmatrix} \overline{\nabla} g_1 \\ \vdots \\ \overline{\nabla} g_n \end{pmatrix}.$$

Since $J_{\mathbf{f}}$ is invertible, it follows that $\overline{\nabla} g_\lambda = 0$ for each λ. Therefore, the map $\mathbf{g}$ is holomorphic. ∎

7.6 Implicit function theorem. *Let $B \subset \mathbb{C}^n \times \mathbb{C}^m$ be an open set, $\mathbf{f} = (f_1, \ldots, f_m) : B \to \mathbb{C}^m$ a holomorphic mapping, and $(\mathbf{z}_0, \mathbf{w}_0) \in B$ a point with $\mathbf{f}(\mathbf{z}_0, \mathbf{w}_0) = 0$ and*

$$\det \left(\frac{\partial f_\mu}{\partial z_\nu}(\mathbf{z}_0, \mathbf{w}_0) \;\middle|\; \begin{array}{l} \mu = 1, \ldots, m \\ \nu = n+1, \ldots, n+m \end{array} \right) \neq 0.$$

Then there is an open neighborhood $U = U' \times U'' \subset B$ and a holomorphic map $\mathbf{g} : U' \to U''$ such that

$$\{(\mathbf{z}, \mathbf{w}) \in U' \times U'' \,:\, \mathbf{f}(\mathbf{z}, \mathbf{w}) = \mathbf{0}\} = \{(\mathbf{z}, \mathbf{g}(\mathbf{z})) \,:\, \mathbf{z} \in U'\}.$$

PROOF: We write $J_{\mathbf{f}}(\mathbf{z}_0, \mathbf{w}_0) = (J' \,|\, J'')$, with $J' \in M_{m,n}(\mathbb{C})$ and $J'' \in M_n(\mathbb{C})$, and define $\mathbf{F} : B \to \mathbb{C}^n \times \mathbb{C}^m$ by $\mathbf{F}(\mathbf{z}, \mathbf{w}) := (\mathbf{z}, \mathbf{f}(\mathbf{z}, \mathbf{w}))$. Then

$$\det J_{\mathbf{F}}(\mathbf{z}_0, \mathbf{w}_0) = \det \begin{pmatrix} \mathbf{E}_n & \mathbf{0} \\ J' & J'' \end{pmatrix} \neq 0.$$

Therefore, there are open neighborhoods $U = U(\mathbf{z}_0, \mathbf{w}_0) \subset B$ and $V = V(\mathbf{z}_0, \mathbf{0}) \subset \mathbb{C}^{n+m}$ such that $\mathbf{F}|_U : U \to V$ is biholomorphic. Obviously, $\mathbf{F}^{-1}(\mathbf{u}, \mathbf{v}) = (\mathbf{u}, \mathbf{h}(\mathbf{u}, \mathbf{v}))$. We may assume that $U = U' \times U'' \subset \mathbb{C}^n \times \mathbb{C}^m$ and $V = U' \times W$, with some open neighborhood $W = W(\mathbf{0}) \subset \mathbb{C}^m$. Defining $\mathbf{g} : U' \to U''$ by $\mathbf{g}(\mathbf{z}) := \mathbf{h}(\mathbf{z}, \mathbf{0})$, it follows that

$$\begin{aligned}\mathbf{f}(\mathbf{z}, \mathbf{w}) = \mathbf{0} \quad &\Longleftrightarrow \quad \mathbf{F}(\mathbf{z}, \mathbf{w}) = (\mathbf{z}, \mathbf{0}) \\ &\Longleftrightarrow \quad (\mathbf{z}, \mathbf{w}) = \mathbf{F}^{-1}(\mathbf{z}, \mathbf{0}) \\ &\Longleftrightarrow \quad \mathbf{w} = \mathbf{h}(\mathbf{z}, \mathbf{0}) = \mathbf{g}(\mathbf{z}).\end{aligned}$$

This completes the proof. ∎

Remark. We can exchange the coordinates in the theorem. If $\operatorname{rk} J_{\mathbf{f}}(\mathbf{z}_0, \mathbf{w}_0) = m$, then there are coordinates $z_{i_1}, \dots, z_{i_n}$ such that $\mathbf{f}^{-1}(\mathbf{0})$ is the graph of a map $\mathbf{g} = \mathbf{g}(z_{i_1}, \dots, z_{i_n})$ near $(\mathbf{z}_0, \mathbf{w}_0)$.

Exercises

1. Let $G = \mathsf{P}^n \subset \mathbb{C}^2$ be the unit polydisk and $\mathbf{f} = (f_1, f_2) : G \to G$ a holomorphic map with $\mathbf{f}(\mathbf{0}) = \mathbf{0}$.
   (a) Show that if $\mathbf{f}(\mathbf{z}) = \mathbf{z} + \sum_{n \geq 2} \mathbf{p}_n(\mathbf{z})$ with pairs $\mathbf{p}_n(\mathbf{z}) = \big(p_1^{(n)}(\mathbf{z}), p_2^{(n)}(\mathbf{z})\big)$ of homogeneous polynomials of degree n, then $\mathbf{f}(\mathbf{z}) \equiv \mathbf{z}$. Hint: Use Cauchy's inequalities and consider the iterated maps $\mathbf{f}^k = \mathbf{f} \circ \cdots \circ \mathbf{f}$ (k times).
   (b) Show that if $\mathbf{f}$ is biholomorphic, then f_1, f_2 are linear.
2. Let $G_1, G_2 \subset \mathbb{C}^n$ be two domains. A continuous map $\mathbf{f} : G_1 \to G_2$ is called *proper* if for every compact subset $K \subset G_2$ the preimage $f^{-1}(K)$ is a compact subset of G_1.
   (a) Show that every biholomorphic map is proper. Give an example of a proper holomorphic map that is not biholomorphic.
   (b) Let G_1 and G_2 be bounded. Show that a continuous map $\mathbf{f} : G_1 \to G_2$ is proper if and only if for every sequence $(\mathbf{z}_k)$ in G_1 tending to ∂G_1, the sequence $(\mathbf{f}(\mathbf{z}_k))$ tends to ∂G_2.
   (c) Let $G', G'' \subset \mathbb{C}$ be bounded domains and $\mathbf{f} : G' \times G'' \to G_2$ a proper holomorphic map onto a bounded domain $G_2 \subset \mathbb{C}^2$. Show that $z \mapsto \mathbf{f}_w(z, w)$ cannot vanish identically on G'. Let $z_0 \in \partial G'$ be an arbitrary point and (z_k) a sequence in G' tending to z_0. Show that the sequence of holomorphic maps $\varphi_k : G'' \to G_2$ with $\varphi_k(w) := \mathbf{f}(z_k, w)$ has a subsequence converging compactly on G'' to a holomorphic map $\varphi_0 : G'' \to \mathbb{C}^2$ with $\varphi_0(G'') \subset \partial G_2$. Show that there must exist at least one point $z_0 \in \partial G'$ such that the corresponding map φ_0 is not constant.
3. Use the results of the last exercise to prove that there is no proper mapping from the unit polydisk to the unit ball in $\mathbb{C}^2$.
4. Let $G_1 \subset \mathbb{C}^n$ and $G_2 \subset \mathbb{C}^m$ be domains and $\mathbf{f} : G_1 \to G_2$ a biholomorphic map. Show that $m = n$.
5. Let $G \subset \mathbb{C}^n$ be a domain and $D : \mathscr{E}(G) \to \mathbb{R}$ a derivation, i.e., an $\mathbb{R}$-linear map satisfying the product rule at $\mathbf{z}_0 \in G$. Show that $D[f]$ depends only on $f|_U$, U an arbitrary small neighborhood of $\mathbf{z}_0$.

6. Let $G \subset \mathbb{C}^n$ be a domain, $\mathbf{f} = (f_1, \ldots, f_m) : G \to \mathbb{C}^m$ a holomorphic mapping, and $M := \{(\mathbf{z}, \mathbf{w}) \in G \times \mathbb{C}^m : \mathbf{w} = \mathbf{f}(\mathbf{z})\}$. Prove the following:

   If $g : G \times \mathbb{C}^m \to \mathbb{C}$ is a holomorphic function with $g|_M = 0$, then for every point $(\mathbf{z}_0, \mathbf{w}_0) \in M$ there is a neighborhood U and an m-tuple $(a_1, \ldots, a_m)$ of holomorphic functions in U such that

$$g(\mathbf{z}, \mathbf{w}) = \sum_{\mu=1}^{m} a_\mu(\mathbf{z}, \mathbf{w}) \cdot (w_\mu - f_\mu(\mathbf{z})) \text{ for } (\mathbf{z}, \mathbf{w}) \in U.$$

8. Analytic Sets

Analytic Subsets. Let $B \subset \mathbb{C}^n$ be an arbitrary region. If $U \subset B$ is an open subset, and $f_1, \ldots, f_q$ are holomorphic functions on U, then their common zero set is denoted by

$$N(f_1, \ldots, f_q) = \{\mathbf{z} \in U : f_1(\mathbf{z}) = \cdots = f_q(\mathbf{z}) = 0\}.$$

Definition. A subset $A \subset B$ is called *analytic* if for every point $\mathbf{z}_0 \in B$ there exists an open neighborhood $U = U(\mathbf{z}_0) \subset B$ and holomorphic functions $f_1, \ldots, f_q$ on U such that $U \cap A = N(f_1, \ldots, f_q)$.

If $\mathbf{z}_0$ is a point of $B - A$, then we can choose an open neighborhood $U = U(\mathbf{z}_0)$ and holomorphic functions $f_1, \ldots, f_q$ on U such that

$$\mathbf{z}_0 \in U' := U - N(f_1, \ldots, f_q) \subset U \subset B.$$

Since the zero set $N(f_1, \ldots, f_q)$ is closed in U, it follows that $B - A$ is open and A closed in B. Therefore, an analytic set in B could have been defined as a **closed** subset $A \subset B$ such that for any $\mathbf{z}_0 \in A$ there exists a neighborhood U and functions $f_1, \ldots, f_q \in \mathcal{O}(U)$ with $A \cap U = N(f_1, \ldots, f_q)$.

Example

In general, analytic sets cannot be given by global equations. We consider the domain $G := G_1 \cup G_2$ with

$$\begin{aligned} G_1 &:= \left\{\mathbf{z} = (z_1, z_2) \in \mathbb{C}^2 : |z_1| < \frac{1}{2} \text{ and } |z_2| < 1\right\}, \\ G_2 &:= \left\{\mathbf{z} = (z_1, z_2) \in \mathbb{C}^2 : |z_1| < 1 \text{ and } \frac{1}{2} < |z_2| < 1\right\}. \end{aligned}$$

For the analytic set we take $A := \{(z_1, z_2) \in G_2 : z_1 = z_2\}$ (see Figure I.5).

The sets G_1, G_2 give an open covering of G with $A \cap G_1 = \varnothing$ and $A \cap G_2 = \{(z_1, z_2) : z_1 - z_2 = 0\}$. So A is an analytic subset of G.

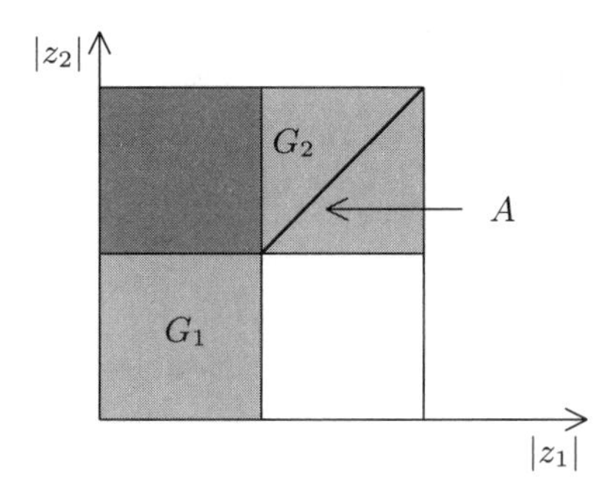

Figure I.5. A not globally defined analytic set

If f is a holomorphic function in G that vanishes on A, then f can be analytically continued to the unit polydisk P^2, since (P^2, G) is a Euclidean Hartogs figure (up to the order of the coordinates). Let $\widehat{f}$ be the continuation. Since $g(z) := \widehat{f}(z, z)$ vanishes for $\frac{1}{2} < |z| < 1$, it also vanishes for $0 \le |z| \le \frac{1}{2}$. This means that f vanishes on $\widehat{A} = \{(z_1, z_2) \in G : z_1 = z_2\}$. Any zero set of finitely many holomorphic functions in G that vanish on A must contain $\widehat{A}$. So A itself cannot be given by global holomorphic functions. In the next chapter we define special domains in $\mathbb{C}^n$ each of which possesses a holomorphic function that cannot be analytically extended to a larger domain. Those domains are called *domains of holomorphy*. On such domains the global representation of analytic sets is possible. The proof of this fact is not contained in this book, because it requires sheaf theory. One has to show that the sheaf of germs of holomorphic functions that vanish on A is "coherent" (cf. [GrRe84], Section 4.2). Then every stalk of this sheaf is generated by global sections (Cartan's theorem A, cf. Chapter V in this book, and [GrRe79], Section IV.5). From that it can be proved that A is the zero set of finitely many global holomorphic functions.

> **Definition.** A subset M of a domain G is called *nowhere dense* in G if the closure of M in G has no interior points.

Since an analytic set $A \subset G$ is always closed in G, it is nowhere dense if in every neighborhood of every point $\mathbf{z} \in G$ there are points outside of A.

8.1 Proposition. *Assume that A is an analytic set in a domain $G \subset \mathbb{C}^n$. If A has an interior point, then $A = G$. If A is nowhere dense in G, then $G - A$ is connected.*

PROOF: To start with we assume that $G = B$ is a ball and that there are holomorphic functions $f_1, \dots, f_q$ on B with $A = N(f_1, \dots, f_q)$.

If $\mathbf{z}_0 \in B$ is an interior point of A, we consider an arbitrary complex line L through $\mathbf{z}_0$. By the identity theorem the functions f_i all vanish on $L \cap B$ and therefore in B.

If A is nowhere dense in B and L an arbitrary complex line, then either $L \cap B \subset A$ or A has only isolated points on $L \cap B$. So any two points of $L \cap B$ outside of A can be connected in $L \cap (B - A)$.

Now let G be an arbitrary domain. If $\mathbf{z}_0 \in G$ is an interior point of A, and $\mathbf{w}_0 \in G$ an arbitrary point, then we can join these points by a continuous path $\alpha : [0,1] \to G$. The compact image of this path can be covered by finitely many balls $B \subset G$ such that $B \cap A$ is the zero set of holomorphic functions on B. Successively it follows that every ball is contained in A. So $A = G$.

If A is nowhere dense in G, then we consider $\mathbf{z}_0, \mathbf{w}_0 \in G - A$ and use the same continuous path. It is clear from above that any point $\mathbf{z}$ in the first ball B that is not an element of A can be joined in $B - A$ to $\mathbf{z}_0$. Applying this successively we obtain a curve between $\mathbf{z}_0$ and $\mathbf{w}_0$ in $B - A$. ■

If $n = 1$, then a nowhere dense analytic set consists only of isolated points.

Bounded Holomorphic Functions. Assume that $G \subset \mathbb{C}^n$ is a domain and $A \subset G$ a proper analytic subset.

8.2 Riemann extension theorem. *If f is a holomorphic function in $G - A$ that is bounded in a neighborhood of every point of A, then f can be holomorphically extended to G.*

PROOF: Since $A \neq G$, A is nowhere dense in G. Let $\mathbf{z}_0 \in A$ be an arbitrary point. Then there is a complex line L through $\mathbf{z}_0$ that in a neighborhood of $\mathbf{z}_0$ intersects A only in $\mathbf{z}_0$.

After a linear change of coordinates we may assume that $\mathbf{z}_0 = \mathbf{0}$ and that $L = \mathbb{C}\mathbf{e}_1$ is the z_1-axis. We can find a polydisk

$$P = \{\mathbf{z} = (z_1, \mathbf{z}') \in \mathbb{C} \times \mathbb{C}^{n-1} \,:\, |z_1| < r_1, |\mathbf{z}'| < r\} \subset\subset G$$

such that $A \cap \{\mathbf{z} \,:\, |z_1| = r_1, |\mathbf{z}'| < r\}$ is empty. For any $\mathbf{c}' \in \mathbb{C}^{n-1}$ with $|\mathbf{c}'| < r$, the set $D = \{\mathbf{z} \,:\, |z_1| \le r_1 \text{ and } \mathbf{z}' = \mathbf{c}'\}$ is a 1-dimensional disc such that $D \cap A$ contains only isolated points, since otherwise $D \subset A$ (see Figure I.6). By the classical Riemann extension theorem in one variable f can be extended to a function $\widehat{f}(z_1, \mathbf{z}')$ that is holomorphic in z_1. By the classical Cauchy integral formula we have

$$\widehat{f}(z_1, \mathbf{z}') = \frac{1}{2\pi \mathrm{i}} \int_{|\zeta| = r_1} \frac{f(\zeta, \mathbf{z}')}{\zeta - z_1} \, d\zeta, \text{ for } |z_1| < r_1 \text{ and } |\mathbf{z}'| < r.$$

The integrand on the right side is holomorphic on P. Consequently, the left side is differentiable (in the real sense), and since integration and differentiation by $\overline{z}_i$ can be exchanged, $\widehat{f}$ is holomorphic on P. If we carry this out

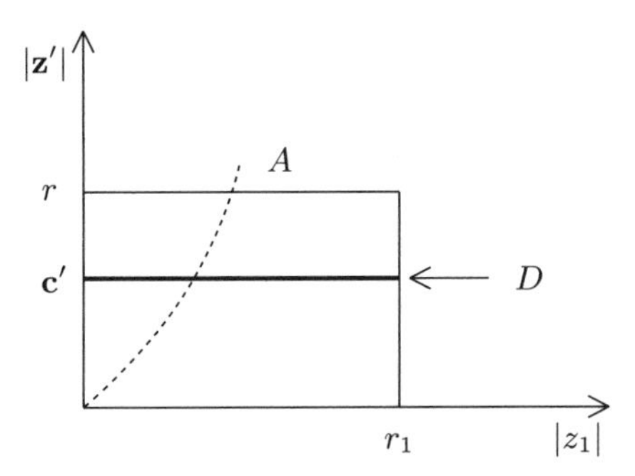

Figure I.6. Riemann extension theorem

at every point $\mathbf{z}_0 \in A$, by the identity theorem we obtain the desired global extension of f to G. ∎

Regular Points. Let $G \subset \mathbb{C}^n$ be a domain, and $\mathbf{z} \in G$ a point. If $f_1, \ldots, f_q$ are holomorphic functions in a neighborhood of $\mathbf{z}$, then we define

$$\operatorname{rk}_{\mathbf{z}}(f_1, \ldots, f_q) := \operatorname{rk} J_{(f_1,\ldots,f_q)}(\mathbf{z}).$$

Definition. An analytic set $A \subset G$ is called *regular of codimension* q *at* $\mathbf{z} \in A$ if there is a neighborhood $U = U(\mathbf{z}) \subset G$ and holomorphic functions $f_1, \ldots, f_q$ on U such that:

1. $A \cap U = N(f_1, \ldots, f_q)$.
2. $\operatorname{rk}_{\mathbf{z}}(f_1, \ldots, f_q) = q$.

The number $n - q$ is called the *dimension* of A at $\mathbf{z}$.

The set A is called *singular* at $\mathbf{z}$ if it is not regular at that point. The set of regular points of A is denoted by $\operatorname{Reg}(A)$ or $\dot{A}$, the set of singular points by $\operatorname{Sing}(A)$.

It is clear that $\dot{A}$ is open in A, and therefore $\operatorname{Sing}(A) \subset A$ closed.

8.3 Theorem (local parametrization of regular points). *Let $A \subset G$ be analytic, $\mathbf{z}_0 \in A$ a point. A is regular of codimension q at $\mathbf{z}_0$ if and only if there are open neighborhoods $U = U(\mathbf{z}_0) \subset G$ and $W = W(\mathbf{0}) \subset \mathbb{C}^n$ and a biholomorphic map $\mathbf{F} : U \to W$ such that $\mathbf{F}(\mathbf{z}_0) = \mathbf{0}$ and*

$$\mathbf{F}(U \cap A) = \{\mathbf{w} = (w_1, \ldots, w_n) \in W \, : \, w_{n-q+1} = \cdots = w_n = 0\}.$$

PROOF: Let A be regular at $\mathbf{z}_0$. There is an open neighborhood $U = U(\mathbf{z}_0)$ such that $A \cap U = N(f_1, \ldots, f_q)$ and $\operatorname{rk}_{\mathbf{z}_0}(f_1, \ldots, f_q) = q$. By renumbering the coordinates we can achieve that

$$J_{(f_1,\ldots,f_q)}(\mathbf{z}_0) = (J' \mid J''),$$

with $J' \in M_{q,n-q}(\mathbb{C})$, $J'' \in M_q(\mathbb{C})$, and $\det J'' \neq 0$. Then define $\mathbf{F} : U \to \mathbb{C}^n$ by

$$\mathbf{F}(z_1,\ldots,z_n) := \big(z_1 - z_1^{(0)},\ldots,z_{n-q} - z_{n-q}^{(0)}, f_1(z_1,\ldots,z_n),\ldots,f_q(z_1,\ldots,z_n)\big).$$

Consequently, the Jacobian has the form

$$J_{\mathbf{F}}(\mathbf{z}_0) = \begin{pmatrix} \mathbf{E}_{n-q} & \mathbf{0} \\ J' & J'' \end{pmatrix},$$

and therefore $\det J_{\mathbf{F}} \neq 0$. Shrinking U if necessary, we have our biholomorphic map $\mathbf{F} : U \to W$, with $\mathbf{F}(\mathbf{z}_0) = \mathbf{0}$ and

$$\mathbf{w} = \mathbf{F}(\mathbf{z}) \text{ for some } \mathbf{z} \in U \cap A \iff w_{n-q+1} = \cdots = w_n = 0.$$

The other direction of the proof is trivial. ∎

Up to this point it is not clear whether or not there exist regular points. In Chapter III we will show that the set of singular points of an analytic set A is a nowhere dense analytic subset of A. At the moment we want only to demonstrate that the zero set of a single holomorphic function contains at least one regular point (and then, of course, a nonempty open set of regular points).

8.4 Proposition. *Let $G \subset \mathbb{C}^n$ be a domain, and f a nonconstant holomorphic function on G. Then the analytic set $N(f)$ contains a regular point.*

PROOF: The case $n = 1$ is trivial. Therefore, we assume $n > 1$.

If every point of $A := N(f)$ is singular, then $\nabla f(\mathbf{z}) \equiv \mathbf{0}$ on A. Since f is not constant, it is impossible that there is a point $\mathbf{z}$ such that $D^\nu f(\mathbf{z}) = 0$ for every multi-index $\boldsymbol{\nu}$. Therefore, we can find a point $\mathbf{z}_0 \in A$, an integer n_0, a multi-index ν_0, and some $\lambda \in \{1,\ldots,n\}$ such that

1. $|\nu_0| = n_0$ and $(D^{\nu_0} f)_{z_\lambda}(\mathbf{z}_0) \neq 0$,
2. $D^\nu f(\mathbf{z}) = 0$ for every $\mathbf{z} \in A$ and every ν with $|\nu| \le n_0$.

The set $M := \{\mathbf{z} \in G : D^{\nu_0} f(\mathbf{z}) = 0\}$ is analytic in G and regular of codimension 1 at $\mathbf{z}_0$. We may assume that $\mathbf{z}_0 = \mathbf{0}$ and $M = \{\mathbf{z} = (z_1, \mathbf{z}') \in G : z_1 = 0\}$, making G sufficiently small.

We have $A \subset M$, and we want to show equality near $\mathbf{z}_0$. It is clear that the function $\zeta \mapsto f(\zeta, \mathbf{0}')$ has exactly one zero at $\zeta = 0$, and it follows easily from Rouché's theorem that for $\mathbf{z}'$ sufficiently close to $\mathbf{0}'$ the functions $\zeta \mapsto f(\zeta, \mathbf{z}')$ also have exactly one zero. This means that there is a neighborhood $V = V(\mathbf{0}) \subset U$ such that $V \cap A = V \cap M$. In particular, $\mathbf{z}_0$ is a regular point of A. ∎

Definition. A *k-dimensional complex submanifold* of a domain $G \subset \mathbb{C}^n$ is an analytic set $A \subset G$ such that A is regular of codimension $n-k$ at every point.

If $A \subset G$ is a k-dimensional complex submanifold, then for every point $\mathbf{z} \in A$ there is an open neighborhood $U = U(\mathbf{z}) \subset G$, an open set $W \subset \mathbb{C}^k$, and a holomorphic map $\varphi : W \to U$ such that:

1. $\operatorname{rk} J_\varphi(\mathbf{w}) = k$ for $\mathbf{w} \in W$.
2. $\varphi(W) = U \cap A$.
3. $\varphi : W \to U \cap A$ is a topological map.[4]

The proof follows immediately from the local parametrization theorem. The map φ is called a *local parametrization.*

Injective Holomorphic Mappings. Let $G \subset \mathbb{C}^n$ be a domain, and $\mathbf{f} = (f_1, \ldots, f_n) : G \to \mathbb{C}^n$ a holomorphic map.

8.5 Theorem. *If $\mathbf{f}$ is injective, then* $\det J_\mathbf{f}(\mathbf{z}) \neq 0$ *everywhere.*

PROOF: We use induction on n. The case $n = 1$ is well known. We consider the case $n > 1$ and define $h := \det J_\mathbf{f}$.

Assume that $N(h) \neq \varnothing$. Then there exists an open subset $U \subset G$ such that $M := U \cap N(h)$ is a nonempty $(n-1)$-dimensional complex submanifold of U.

We claim that $J_\mathbf{f}|_M \equiv 0$. To prove this, we assume that there is a point $\mathbf{z}_0 \in M$ with $J_\mathbf{f}(\mathbf{z}_0) \neq \mathbf{0}$. Without loss of generality, we may assume that

$$\frac{\partial f_n}{\partial z_n}(\mathbf{z}_0) \neq 0.$$

Let $\mathbf{F} : G \to \mathbb{C}^n$ be defined by $\mathbf{F}(\mathbf{z}', z_n) := (\mathbf{z}', f_n(\mathbf{z}', z_n))$. Then $\det J_\mathbf{F}(\mathbf{z}_0) \neq 0$, and there are connected open neighborhoods U of $\mathbf{z}_0$ and V of $\mathbf{w}_0 := \mathbf{F}(\mathbf{z}_0)$ such that $\mathbf{F} : U \to V$ is biholomorphic. There is a holomorphic map $\widetilde{\mathbf{f}} : V \to \mathbb{C}^{n-1}$ such that

$$\mathbf{f} \circ \mathbf{F}^{-1}(\mathbf{w}', w_n) = \bigl(\widetilde{\mathbf{f}}(\mathbf{w}', w_n), w_n\bigr),$$

and we define

$$\mathbf{g} = (g_1, \ldots, g_{n-1}) : W := \bigl\{\mathbf{w}' \in \mathbb{C}^{n-1} \,:\, \bigl(\mathbf{w}', w_n^{(0)}\bigr) \in V\bigr\} \to \mathbb{C}^{n-1}$$

by $\mathbf{g}(\mathbf{w}') := \widetilde{\mathbf{f}}\bigl(\mathbf{w}', w_n^{(0)}\bigr)$.

[4] A map $\varphi : X \to Y$ between topological spaces is called *topological* or a *homeomorphism* if it is continuous and bijective and the inverse mapping $\varphi^{-1} : Y \to X$ is also continuous.

Since $\mathbf{g}$ is injective, we can apply the induction hypothesis and conclude that $\det J_{\mathbf{g}}(\mathbf{w}_0') \neq 0$. Now

$$J_{\mathbf{f}}(\mathbf{z}_0) \cdot J_{\mathbf{F}^{-1}}(\mathbf{w}_0) = \begin{pmatrix} J_{\widetilde{\mathbf{f}}}(\mathbf{w}_0) \\ \mathbf{e}_n \end{pmatrix} = \begin{pmatrix} J_{\mathbf{g}}(\mathbf{w}_0') & \# \\ \mathbf{0}' & 1 \end{pmatrix}.$$

Therefore, $h(\mathbf{z}_0) = \det J_{\mathbf{F}}(\mathbf{z}_0) \cdot \det J_{\mathbf{g}}(\mathbf{w}_0') \neq 0$ as well. This is a contradiction.

We have demonstrated that $J_{\mathbf{f}}(\mathbf{z}) = 0$ for every $\mathbf{z} \in M$. Since $\mathbf{f}$ is holomorphic, also $D\mathbf{f}(\mathbf{z}) \equiv 0$ on M, and using a local parametrization of M we obtain that $\mathbf{f}|_M$ is locally constant. But this is impossible, since $\mathbf{f}$ is injective. The set $N(h)$ must be empty. ■

8.6 Corollary. *If $G \subset \mathbb{C}^n$ is a domain, and $\mathbf{f} : G \to \mathbb{C}^n$ an injective holomorphic mapping, then also $\mathbf{f}(G)$ is a domain, and $\mathbf{f} : G \to \mathbf{f}(G)$ is biholomorphic.*

PROOF: Let $\mathbf{w}_0 := \mathbf{f}(\mathbf{z}_0)$ be a point of $G' := \mathbf{f}(G)$. Then $\det J_{\mathbf{f}}(\mathbf{z}_0) \neq 0$, and there are open neighborhoods $U = U(\mathbf{z}_0) \subset G$ and $V = V(\mathbf{w}_0) \subset \mathbb{C}^n$ such that $\mathbf{f} : U \to V$ is biholomorphic. It follows that $\mathbf{w}_0$ is an interior point of G' and that $\mathbf{f}^{-1}$ is holomorphic at $\mathbf{w}_0$. ■

Exercises

1. Prove the following properties:
   (a) Finite intersections and unions of analytic sets are analytic.
   (b) If $\mathbf{f} : G_1 \to G_2$ is a holomorphic map between domains and $A \subset G_2$ an analytic set, then $\mathbf{f}^{-1}(A) \subset G_1$ is analytic as well.
   (c) If $A_1 \subset G_1$ and $A_2 \subset G_2$ are analytic sets, then $A_1 \times A_2$ is an analytic subset of $G_1 \times G_2$.
2. Let $U \subset \mathbb{C}^n$ be an open neighborhood of the origin and $A \subset U$ be an analytic subset containing the origin. For $1 \leq k \leq n-1$ and $I = \{i_1, \ldots, i_k\} \subset \{1, \ldots, n\}$ let $p_I : \mathbb{C}^n \to \mathbb{C}^k$ be defined by

   $$p_I(z_1, \ldots, z_n) := (z_{i_1}, \ldots, z_{i_k}).$$

   Prove: If A is regular of codimension $n-k$ at the origin, then there exists an I and open neighborhoods $V = V(\mathbf{0}) \subset U$, $W = W(\mathbf{0}) \subset \mathbb{C}^k$ such that $p_I : A \cap V \to W$ is bijective.
3. Show that $A := \{(w, z_1, z_2) \in \mathbb{C}^3 : w^2 = z_1 z_2\}$ is an analytic set that is regular of codimension 1 outside the origin and singular at $\mathbf{0}$.
4. Let A_1, A_2 be two analytic sets in a neighborhood of the origin in $\mathbb{C}^n$ such that $\mathbf{0} \in A := A_1 \cap A_2$. Suppose that $U \cap A_1 \neq U \cap A_2$ for every neighborhood U of $\mathbf{0}$. Show that A is singular at $\mathbf{0}$.

Chapter II

Domains of Holomorphy

1. The Continuity Theorem

General Hartogs Figures. The subject of this chapter is the continuation of holomorphic functions. We consider domains in $\mathbb{C}^n$, for $n \geq 2$. A typical example is the *Euclidean Hartogs figure* $(\mathsf{P}^n, \mathsf{H})$, where $\mathsf{P}^n = \mathsf{P}^n(\mathbf{0}, 1)$ is the unit polydisk, and

$$\mathsf{H} = \{\mathbf{z} \in \mathsf{P}^n \,:\, |z_1| > q_1 \text{ or } |z_\nu| < q_\nu \text{ for } \nu = 2, \dots, n\}.$$

Here $q_1, \dots, q_n$ are real numbers with $0 < q_\nu < 1$ for $\nu = 1, \dots, n$. Every holomorphic function f on H has a holomorphic extension $\widehat{f}$ on P^n.

Definition. Let $\mathbf{g} = (g_1, \dots, g_n) : \mathsf{P}^n \to \mathbb{C}^n$ be an injective holomorphic mapping, $\widetilde{P} := \mathbf{g}(\mathsf{P}^n)$ and $\widetilde{H} := \mathbf{g}(\mathsf{H})$. Then $\big(\widetilde{P}, \widetilde{H}\big)$ is called a *general Hartogs figure.*

We use the symbolic picture that appears as Figure II.1

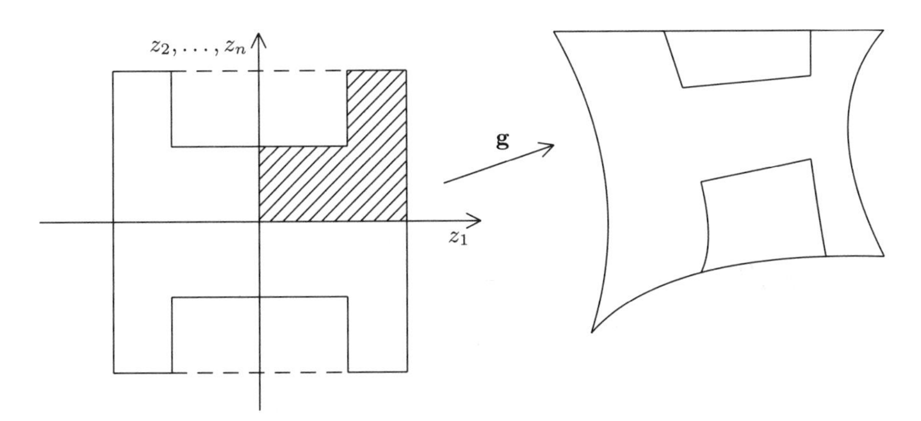

Figure II.1. General Hartogs figure

1.1 Continuity theorem. *Let $G \subset \mathbb{C}^n$ be domain, $\big(\widetilde{P}, \widetilde{H}\big)$ a general Hartogs figure with $\widetilde{H} \subset G$, f a holomorphic function on G. If $G \cap \widetilde{P}$ is connected, then f can be continued uniquely to $G \cup \widetilde{P}$.*

PROOF: Let $\mathbf{g} : \mathsf{P}^n \to \mathbb{C}^n$ be an injective holomorphic mapping such that $\widetilde{P} := \mathbf{g}(\mathsf{P}^n)$ and $\widetilde{H} := \mathbf{g}(\mathsf{H})$. The function $h := f \circ \mathbf{g}$ is holomorphic in H. Therefore, there exists exactly one holomorphic function $\widehat{h}$ on P^n with $\widehat{h}|_{\mathsf{H}} = h$. Since $\mathbf{g} : \mathsf{P}^n \to \widetilde{P}$ is biholomorphic, the function $f_0 := \widehat{h} \circ \mathbf{g}^{-1}$ is defined on $\widetilde{P}$, and it is a holomorphic extension of $f|_{\widetilde{H}}$. We define

$$\widehat{f}(\mathbf{z}) := \begin{cases} f(\mathbf{z}) & \text{for } \mathbf{z} \in G, \\ f_0(\mathbf{z}) & \text{for } \mathbf{z} \in \widetilde{P}. \end{cases}$$

Since $G \cap \widetilde{P}$ is connected and $f = f_0$ on $\widetilde{H}$, it follows from the identity theorem that $\widehat{f}$ is a well-defined holomorphic function on $G \cup \widetilde{P}$. This is the desired extension of f. ∎

Example

Let $n \geq 2$ and $P' \subset\subset P$ be polydiscs around the origin in $\mathbb{C}^n$. Then every holomorphic function f on $P - \overline{P'}$ can be extended uniquely to a holomorphic function on P.

For a proof we may assume that $P = \mathsf{P}^n$ is the unit polydisk, and $P' = \mathsf{P}^n(\mathbf{0}, \mathbf{r})$, with $\mathbf{r} = (r_1, \ldots, r_n)$ and $0 < r_\nu < 1$ for $\nu = 1, \ldots, n$. It is clear that $G := P - \overline{P'}$ is a domain.

Given a point $\mathbf{z}_0 = \left(z_1^{(0)}, \ldots, z_n^{(0)}\right) \in G$ with $|z_n^{(0)}| > r_n$, we choose real numbers $q_1, \ldots, q_n$ as follows: For $\nu = 1, \ldots, n-1$, let q_ν be arbitrary numbers, with $r_\nu < q_\nu < 1$. To obtain a suitable q_n, we define an automorphism T of the unit disk D by

$$T(\zeta) := \frac{\zeta - z_n^{(0)}}{\overline{z}_n^{(0)}\zeta - 1}.$$

This automorphism maps $z_n^{(0)}$ onto 0 and a small disk $D \subset \{\zeta \in \mathbb{C} : r_n < |\zeta| < 1\}$ around $z_n^{(0)}$ onto a disk $K \subset \mathsf{D}$ with $0 \in K$. Notice that 0 need not be the center of K. We choose $q_n > 0$ such that $\mathsf{D}_{q_n}(0) \subset K$.

If we define $\mathsf{H} := \{\mathbf{z} \in \mathsf{P}^n : |z_1| > q_1 \text{ or } |z_\nu| < q_\nu \text{ for } \nu = 2, \ldots, n\}$, then $(\mathsf{P}^n, \mathsf{H})$ is a Euclidean Hartogs figure. The mapping $\mathbf{g} : \mathsf{P}^n \to \mathsf{P}^n$ defined by

$$\mathbf{g}(z_1, \ldots, z_n) := (z_1, \ldots, z_{n-1}, T^{-1}(z_n))$$

is biholomorphic, and $\left(\widetilde{P}, \widetilde{H}\right) = (\mathsf{P}^n, \mathbf{g}(\mathsf{H}))$ is a general Hartogs figure, with

$$\widetilde{H} \subset \{\mathbf{z} \in \mathsf{P}^n : |z_1| > r_1 \text{ or } |z_n| > r_n\} \subset G.$$

Since $\widetilde{P} \cap G = G$ is connected, the continuity theorem may be applied. The preceding example is a special case of the so-called *Kugelsatz* which we shall prove in Chapter VI.

Since G satisfies the continuity principle, we obtain that $g(D_{\mathbf{w}}) = S_1(\mathbf{w})$ is contained in G. This is valid for every $\mathbf{w} \in P''$. Therefore, $\widetilde{P} \subset G$, and G is Hartogs convex. ∎

1.6 Corollary. *The unit polydisk* P^n *is Hartogs convex.*

Domains of Holomorphy

Definition. Let $G \subset \mathbb{C}^n$ be a domain, f holomorphic in G, and $\mathbf{z}_0 \in \partial G$ a point. The function f is called *completely singular* at $\mathbf{z}_0$ if for every connected neighborhood $U = U(\mathbf{z}_0) \subset \mathbb{C}^n$ and every connected component C of $U \cap G$ there is no holomorphic function g on U for which $g|_C = f|_C$.

Example

Let $G := \mathbb{C} - \{x \in \mathbb{R} : x \leq 0\}$ and let f be a branch of the logarithm on G. Then f is completely singular at $z = 0$ but not at any point $x \in \mathbb{R}$ with $x < 0$.

Definition. A domain $G \subset \mathbb{C}^n$ is called a *weak domain of holomorphy* if for every point $\mathbf{z} \in \partial G$ there is a function $f \in \mathcal{O}(G)$ that is completely singular at $\mathbf{z}$.

The domain G is called a *domain of holomorphy* if there is a function $f \in \mathcal{O}(G)$ that is completely singular at **every** point $\mathbf{z} \in \partial G$.

Examples

1. Since $\mathbb{C}^n$ has no boundary point, it trivially satisfies the requirements of a domain of holomorphy.
2. It is easy to see that every domain $G \subset \mathbb{C}$ is a weak domain of holomorphy: If z_0 is a point in ∂G, then $f(z) := 1/(z - z_0)$ is holomorphic in G and completely singular at z_0.

   For $G = \mathsf{D}$ we can show even more! The function $f(z) := \sum_{\nu=0}^{\infty} z^{\nu!}$ is holomorphic in the unit disk and becomes completely singular at any boundary point. Therefore, D is a domain of holomorphy. At the end of this chapter we will see that every domain in $\mathbb{C}$ is a domain of holomorphy.
3. If $f : \mathsf{D} \to \mathbb{C}$ is a holomorphic function that becomes completely singular at every boundary point, then the same is true for $\widehat{f} : \mathsf{P}^n = \mathsf{D} \times \cdots \times \mathsf{D} \to \mathbb{C}$, defined by $\widehat{f}(z_1, \ldots, z_n) := f(z_1) + \cdots + f(z_n)$. In fact, if $\mathbf{z}_0$ is a boundary point of P^n, then there exists an i such that the ith component $z_i^{(0)}$ is a boundary point of D. If $\widehat{f}$ could be extended holomorphically across $\mathbf{z}_0$,

then $\widehat{f}_i(\zeta) := \widehat{f}(z_1^{(0)}, \ldots, \zeta, \ldots, z_n^{(0)})$ would also have a holomorphic extension. But then f could not be completely singular at $z_i^{(0)}$. Therefore, the unit polydisk is a domain of holomorphy.

4. If $(\mathsf{P}^n, \mathsf{H})$ is a Euclidean Hartogs figure, then H is not a domain of holomorphy.

1.7 Proposition. *Let $G \subset \mathbb{C}^n$ be a domain. If for every point $\mathbf{z}_0 \in \partial G$ there is an open neighborhood $U = U(\mathbf{z}_0) \subset \mathbb{C}^n$ and a holomorphic function $f : G \cup U \to \mathbb{C}$ with $f(\mathbf{z}_0) = 0$ and $f(\mathbf{z}) \neq 0$ for $\mathbf{z} \in G$, then G is a weak domain of holomorphy.*

PROOF: We show that $1/f$ is completely singular at $\mathbf{z}_0$. For this assume that there is a connected open neighborhood $V = V(\mathbf{z}_0)$, a connected component $C \subset V \cap G$, and a holomorphic function F on V with $F|_C = (1/f)\,|_C$. The set $V' := V - N(f)$ is still connected and contains C. By the identity theorem the functions F and $1/f$ must coincide in V'. Then F is clearly not holomorphic at $\mathbf{z}_0$. This is a contradiction. ∎

1.8 Corollary. *Every convex domain in $\mathbb{C}^n$ is a weak domain of holomorphy.*

PROOF: If $\mathbf{z}_0 \in \partial G$, then because of the convexity there is a real linear form λ on $\mathbb{C}^n$ with $\lambda(\mathbf{z}) < \lambda(\mathbf{z}_0)$ for $\mathbf{z} \in G$. We can write λ in the form

$$\lambda(\mathbf{z}) = \sum_{\nu=1}^{n} \alpha_\nu z_\nu + \sum_{\nu=1}^{n} \overline{\alpha}_\nu \overline{z}_\nu, \quad \text{with } \boldsymbol{\alpha} := (\alpha_1, \ldots, \alpha_n) \neq \mathbf{0}.$$

So $\lambda = \operatorname{Re} h(\mathbf{z})$, where $h(\mathbf{z}) := 2 \cdot \sum_{\nu=1}^{n} \alpha_\nu z_\nu$ is holomorphic on $\mathbb{C}^n$.

Since the function $f(\mathbf{z}) := h(\mathbf{z}) - h(\mathbf{z}_0)$ is holomorphic on $\mathbb{C}^n$, $f(\mathbf{z}_0) = 0$, and $f(\mathbf{z}) \neq 0$ on G, the proposition may be applied. ∎

We will show that every weak domain of holomorphy is Hartogs convex. As a tool we need the following simple geometric lemma, which will be useful in other situations as well.

1.9 Lemma (on boundary components). *Let $G \subset \mathbb{C}^n$ be a domain, $U \subset \mathbb{C}^n$ an open set with $U \cap G \neq \varnothing$ and $(\mathbb{C}^n - U) \cap G \neq \varnothing$.*

Then $G \cap \partial C \cap \partial U \neq \varnothing$ for any connected component C of $U \cap G$.

PROOF: We choose points $\mathbf{z}_1 \in C \subset U \cap G$ and $\mathbf{z}_2 \in (\mathbb{C}^n - U) \cap G$. There is a continuous path $\gamma : [0,1] \to G$ with $\gamma(0) = \mathbf{z}_1$ and $\gamma(1) = \mathbf{z}_2$. Let $t_0 := \sup\{t \in [0,1] : \gamma(t) \in C\}$ and $\mathbf{z}_0 := \gamma(t_0)$. Clearly, $\mathbf{z}_0 \in \partial C \cap G$, but $\mathbf{z}_0 \notin C$. Since C is a connected component of $U \cap G$, $\mathbf{z}_0$ cannot lie in $U \cap G$ and therefore even not in U. Since $\gamma(t) \in U$ for $t < t_0$, it follows that $\mathbf{z}_0 \in \partial U$. ∎

1.10 Theorem. *Let $G \subset \mathbb{C}^n$ be a weak domain of holomorphy. Then G is Hartogs convex.*

PROOF: Assume that G is not Hartogs convex. Then there is a general Hartogs figure (P, H) with $H \subset G$ but $P \cap G \neq P$. We choose an arbitrary $\mathbf{z}_0$ in H and set $C := C_{P \cap G}(\mathbf{z}_0)$.[1] Since H lies in $P \cap G$ and is connected, it follows that $H \subset C$. Furthermore, $C \subsetneqq P$.

Since $P \cap G \neq \varnothing$ and $(\mathbb{C}^n - G) \cap P \neq \varnothing$, by the lemma there is a point $\mathbf{z}_1 \in \partial C \cap \partial G \cap P$ (see Figure II.5).

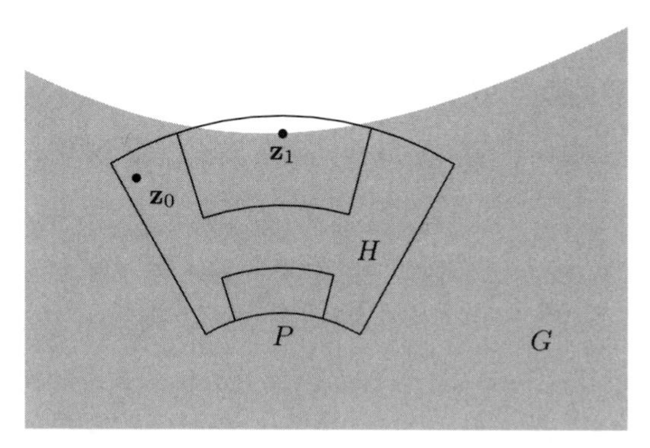

Figure II.5. G is not Hartogs convex

Let f be an arbitrary holomorphic function in G. Then $f|_C$ is also holomorphic, and by the continuity theorem it has a holomorphic extension F on P. Since P is an open connected neighborhood of $\mathbf{z}_1$, we obtain that f is not completely singular at $\mathbf{z}_1$. This completes the proof by contradiction. ∎

It follows, for example, that every convex domain is Hartogs convex. As a consequence, we see that every ball is Hartogs convex.

1.11 Theorem. *Every domain of holomorphy is Hartogs convex.*

The proof is trivial.

For the converse of this theorem one has to construct on any Hartogs convex domain a global holomorphic function that becomes completely singular at every boundary point, something that is rather difficult. It was done in 1910 by E.E. Levi in very special cases. The general case is called *Levi's problem.*

In 1942 K. Oka gave a proof for $n = 2$. At the beginning of the 1950s Oka, Bremermann, and Norguet solved Levi's problem for arbitrary n. It was gen-

[1] We denote by $C_M(\mathbf{z})$ the connected component of M containing $\mathbf{z}$.

eralized for complex manifolds (H. Grauert, 1958) and complex spaces (R. Narasimhan, 1962). Finally, in 1965 L. Hörmander published a proof that used Hilbert space methods and partial differential equations.

Exercises

1. Prove the following statements:
   (a) Finite intersections of Hartogs convex domains are Hartogs convex.
   (b) If $G_1 \subset G_2 \subset G_3 \subset \cdots$ is an ascending chain of Hartogs convex domains, then the union of all G_i is also Hartogs convex.
2. Let $G \subset \mathbb{C}^n$ be a domain, $0 \le r < R$, and $\mathbf{a} \in G$ a point. Let $U = U(\mathbf{a}) \subset G$ be an open neighborhood and define $Q := \{\mathbf{w} \in \mathbb{C}^m : r < |\mathbf{w}| < R\}$. Prove that every holomorphic function on $(G \times Q) \cup (U \times \mathsf{P}^m(\mathbf{0}, R))$ has a unique holomorphic extension to $G \times \mathsf{P}^m(\mathbf{0}, R)$.
3. Let $0 < r < R$ be given. Use Hartogs figures to prove that every holomorphic function on $\mathsf{B}_R(\mathbf{0}) - \overline{\mathsf{B}_r(\mathbf{0})}$ has a unique holomorphic extension to the whole ball $\mathsf{B}_R(\mathbf{0})$.
4. For $\varepsilon \ge 0$, consider the domain
$$G_\varepsilon := \{(z, w) \in \mathsf{P}^2(\mathbf{0}, 1) \,:\, |z| < |w|^2 + \varepsilon\}.$$
   Prove that G_ε is Hartogs convex if and only if $\varepsilon = 0$.
5. Let $G \subset \mathbb{C}^n$ be a domain and $f : G \to \mathsf{D}_R(0) \subset \mathbb{C}$ a function, $\Gamma = \{(\mathbf{z}, w) \in G \times \mathsf{D}_R(0) \,:\, w = f(\mathbf{z})\}$ its graph. Sow that if there is a holomorphic function F in $G \times \mathsf{D}_R(0)$ that is completely singular at every point of Γ, then f is continuous. (With more effort one can show that f is holomorphic.)
6. Show that the "Hartogs triangle" $\{(z, w) \in \mathbb{C}^2 \,:\, |w| < |z| < 1\}$ is a weak domain of holomorphy.

2. Plurisubharmonic Functions

Subharmonic Functions. Recall some facts from complex analysis of one variable. A twice differentiable real-valued function h on a domain $G \subset \mathbb{C}$ is called *harmonic* if $h_{z\overline{z}}(z) \equiv 0$ on G. The real part of a holomorphic function is always harmonic, and on an open disk every harmonic function is the real part of some holomorphic function.

If $D = \mathsf{D}_r(a) \subset \mathbb{C}$ is an open disk and $\beta : \mathbb{R} \to \mathbb{R}$ a continuous periodic function with period 2π, then there is a continuous function $h : \overline{D} \to \mathbb{R}$ that is harmonic on D such that $h(re^{\mathrm{i}t}) = \beta(t)$ for every t (Dirichlet's principle).

An upper semicontinuous function $\varphi : G \to \mathbb{R} \cup \{-\infty\}$ is said to satisfy the *weak mean value property* if the following holds:

For every $a \in G$ there is an $r > 0$ with $\mathsf{D}_r(a) \subset\subset G$ and

$$\varphi(a) \le \frac{1}{2\pi}\int_0^{2\pi} \varphi(a + \varrho e^{\mathrm{i}t})\,dt \qquad \text{for } 0 < \varrho \le r.$$

Remarks

1. If $\varphi : G \to \mathbb{R} \cup \{-\infty\}$ is an upper semicontinuous function, then the sets $U_\nu := \{z \in G : \varphi(z) < \nu\}$ are open, and therefore φ is bounded from above on every compact subset $K \subset G$. It follows that the integral in the definition always exists.
2. Harmonic functions satisfy the weak mean value property (even the stronger *mean value property* with "=" instead of "≤").
3. If $f : G \to \mathbb{C}$ is a nowhere identically vanishing holomorphic function, then $\log|f|$ satisfies the weak mean value property. In fact, the function $\varphi := \log|f|$ is harmonic on $G - N(f)$, because it can be written locally as $\mathrm{Re}(\log f)$, with a suitable branch of the logarithm. And at any point $z_0 \in N(f)$ we have $\varphi(z_0) = -\infty$, so the inequality of the weak mean value property is satisfied.

2.1 Proposition. *Let $\varphi : G \to \mathbb{R}$ satisfy the weak mean value property. If φ has a global maximum in G, then φ is constant.*

PROOF: Let $a \in G$ be any point with $c := \varphi(a) \ge \varphi(z)$ for $z \in G$. We choose an $r > 0$ such that

$$\mathsf{D}_r(a) \subset\subset G \text{ and } \varphi(a) \le \frac{1}{2\pi}\int_0^{2\pi} \varphi(a + \varrho e^{\mathrm{i}t})\,dt \text{ for } 0 < \varrho \le r.$$

Assume that there is a $b \in \mathsf{D}_r(a)$ with $\varphi(b) < \varphi(a)$. We write $b = a + \varrho e^{\mathrm{i}t_0}$ and get

$$\varphi(a) \le \frac{1}{2\pi}\int_0^{2\pi} \varphi(a + \varrho e^{\mathrm{i}t})\,dt < \frac{1}{2\pi}\int_0^{2\pi} \varphi(a)\,dt = \varphi(a).$$

This is a contradiction, so φ must be constant on $\mathsf{D}_r(a)$. Now we define the set $M := \{z \in G : \varphi(z) = c\}$. Obviously, M is closed in G and not empty, and we just showed that M is open. So $M = G$, and φ is constant. ∎

Definition. Let $G \subset \mathbb{C}$ be a domain. A function $s : G \to \mathbb{R} \cup \{-\infty\}$ is called *subharmonic* if the following hold:

1. s is upper semicontinuous on G.
2. If $D \subset\subset G$ is a disk, $h : \overline{D} \to \mathbb{R}$ continuous, $h|_D$ harmonic, and $h \ge s$ on ∂D, then $h \ge s$ on D.

2.2 Proposition. *Let $s_\nu : G \to \mathbb{R} \cup \{-\infty\}$ be a monotonically decreasing sequence of subharmonic functions. Then $s := \lim_{\nu\to\infty} s_\nu$ is subharmonic.*

PROOF: The limit $s = \lim_{\nu\to\infty} s_\nu = \inf\{s_\nu\}$ is upper semicontinuous. Let $D \subset\subset G$ be a disk, $h : \overline{D} \to \mathbb{R}$ continuous and harmonic on D, with $s \le h$ on ∂D. For fixed ε we consider the compact sets

$$K_\nu := \{z \in \partial D \,:\, s_\nu(z) \ge h(z) + \varepsilon\}.$$

Then $K_{\nu+1} \subset K_\nu$ and $\bigcap_{\nu=1}^\infty K_\nu = \varnothing$. Therefore, there is a $\nu_0 \in \mathbb{N}$ with $K_\nu = \varnothing$ for $\nu \ge \nu_0$. This means that for $\nu \ge \nu_0$, $s_\nu < h + \varepsilon$ on ∂D, and therefore the same is true on D. Since the s_ν are decreasing, $s < h + \varepsilon$ on D. This holds for every $\varepsilon > 0$, and consequently $s \le h$ on D. ∎

2.3 Proposition. *Let $(s_\alpha)_{\alpha\in A}$ be a family of subharmonic functions on G. If $s := \sup s_\alpha$ ist upper semicontinuous and finite everywhere, then s is subharmonic.*

PROOF: If $s \le h$ on ∂D, where $D \subset\subset G$ and $h : \overline{D} \to \mathbb{R}$ is continuous and harmonic on D, then $s_\alpha \le h$ on ∂D for every $\alpha \in A$. Since the s_α are subharmonic, it follows that $s_\alpha \le h$ on D for every $\alpha \in A$. But then $s \le h$ on D as well. ∎

Examples

1. Clearly, every harmonic function is subharmonic.
2. Let $s : G \to \mathbb{R}$ be a continuous subharmonic function such that $-s$ is also subharmonic. Then s is harmonic. To show this, we look at an arbitrary point $a \in G$ and choose an $r > 0$ such that $D := \mathsf{D}_r(a) \subset\subset G$. Then there is a continuous function $h : \overline{D} \to \mathbb{R}$ with $h|_{\partial D} = s|_{\partial D}$ that is harmonic on D (Dirichlet's principle). It follows that $s \le h$ on D. But because $-h$ is also harmonic, we have $-s \le -h$ on D as well. Together this gives $s = h$ on D.
3. Let $f : G \to \mathbb{C}$ be a holomorphic function. Then $s := \log|f|$ is subharmonic. In fact, if $f(z) \equiv 0$ on G, then we have $s(z) \equiv -\infty$, and there is nothing to prove. Otherwise, s is harmonic on $G - N(f)$, and we have only to look at an isolated zero a of f. We choose $D = \mathsf{D}_r(a) \subset\subset G$ and a function h that is continuous on $\overline{D}$ and harmonic on D, with $s \le h$ on ∂D. We know that s, and therefore also $s - h$, has the weak mean value property on D, and it is certainly not constant. So it must take its maximum on the boundary ∂D. This means that $s \le h$ on D.
4. Let $G \subset \mathbb{C}$ be an arbitrary domain. The *boundary distance* $\delta_G : G \to \mathbb{R}_+ \cup \{+\infty\}$ is defined by

$$\delta_G(z) := \sup\{r \in \mathbb{R} \,:\, \mathsf{D}_r(z) \subset G\}.$$

Claim: $s := -\log \delta_G$ is subharmonic on G.

PROOF: If $G = \mathbb{C}$, then $s(z) \equiv -\infty$ and there is nothing to prove. If $G \ne \mathbb{C}$, then s is real-valued and continuous. For $w \in \partial G$ we define

$s_w : G \to \mathbb{R}$ by setting $s_w(z) := -\log|z - w|$. Then $s(z) = \sup\{s_w(z) : w \in \partial G\}$. By Proposition 2.3 the claim follows. ∎

The Maximum Principle

2.4 Theorem. *Let $s : G \to \mathbb{R} \cup \{-\infty\}$ be a subharmonic function on a domain $G \subset \mathbb{C}$. If s takes its maximum on G, then it must be constant.*

PROOF: Assume that $c := s(a) \ge s(z)$ for every $z \in G$. As in the case of functions that have the weak mean value property it suffices to show that s is constant in a neighborhood of a. If this is not the case, there is a small disk $D = \mathsf{D}_r(a) \subset\subset G$ and $b \in \partial D$ with $s(a) > s(b)$. Since s is upper semicontinuous, there is a continuous function h on ∂D with $s \le h \le c$ and $h(b) < c$. Solving Dirichlet's problem we can construct a harmonic continuation of h on D. Now

$$h(a) = \frac{1}{2\pi} \int_0^{2\pi} h(a + re^{\mathrm{i}t})\, dt < c = s(a).$$

This is a contradiction. ∎

For later use we give the following criterion for a function to be subharmonic:

2.5 Theorem. *Let $G \subset \mathbb{C}$ be a domain and $s : G \to \mathbb{R} \cup \{-\infty\}$ an upper semicontinuous function. Suppose that for every disk $D \subset\subset G$ and every function $f \in \mathcal{O}(\overline{D})$ with $s < \mathrm{Re}(f)$ on ∂D it follows that $s < \mathrm{Re}(f)$ on D. Then s is subharmonic.*

PROOF: Let $D = \mathsf{D}_r(a) \subset\subset G$, $h : \overline{D} \to \mathbb{R}$ continuous and harmonic on D, and $s \le h$ on ∂D. For simplicity we assume $a = 0$.

For $\nu \in \mathbb{N}$, a harmonic function h_ν on $D_\nu := \mathsf{D}_{(\nu/(\nu-1))r}(0) \supset D$ is given by

$$h_\nu(z) := h\Bigl(\Bigl(1 - \frac{1}{\nu}\Bigr) z\Bigr).$$

Then (h_ν) converges on $\overline{D}$ uniformly, increasing monotonically to h. Furthermore, for every ν there is a holomorphic function f_ν on D_ν with $\mathrm{Re}(f_\nu) = h_\nu$.

Let $\varepsilon > 0$ be given. Then there is a ν_0 such that $|h - h_\nu| < \varepsilon$ on $\overline{D}$ for $\nu \ge \nu_0$. Therefore, $s < h_\nu + \varepsilon = \mathrm{Re}(f_\nu + \varepsilon)$ on ∂D for $\nu \ge \nu_0$. By definition it follows that $s < h_\nu + \varepsilon$ on D. Since (h_ν) is increasing, it follows that $s < h + \varepsilon$ and therefore $s \le h$ on D. ∎

Differentiable Subharmonic Functions

2.6 Lemma. *Let $s : G \to \mathbb{R}$ be a $\mathscr{C}^2$ function such that $s_{z\overline{z}} > 0$ on G. Then s is subharmonic.*

PROOF: Let $D = \mathsf{D}_r(a) \subset\subset G$ and let a continuous function $h : \overline{D} \to \mathbb{R}$ be given such that h is harmonic on D and $s \le h$ on ∂D. We define $\varphi := s - h$.

Assume that φ takes its maximum at some interior point z_0 of D. Then we look at the Taylor expansion of φ at z_0 in a small neighborhood about z_0:

$$\varphi(z_0 + z) = \varphi(z_0) + 2 \operatorname{Re} Q(z) + \varphi_{z\overline{z}}(z_0) z\overline{z} + R(z),$$

where $Q(z) := \varphi_z(z_0)z + \frac{1}{2}\varphi_{zz}(z_0)z^2$ is holomorphic and $R(z)/|z|^2 \to 0$ for $z \to 0$. The function $\psi(z) := 2\operatorname{Re} Q(z)$ is harmonic, with $\psi(0) = 0$. Since it cannot assume a maximum or a minimum, it must have zeros arbitrarily close to but not equal to 0. On the other hand, $\varphi(z_0 + z) - \varphi(z_0) \le 0$ and $\varphi_{z\overline{z}}(z_0)z\overline{z} > 0$ outside $z = 0$. This is a contradiction. Thus φ must assume its maximum on the boundary of D, and $s \le h$ on D. ∎

2.7 Theorem. *Let $s : G \to \mathbb{R}$ be a $\mathscr{C}^2$ function. Then s is subharmonic if and only if $s_{z\overline{z}} \ge 0$ on G.*

PROOF: (a) Let $s_{z\overline{z}}(z) \ge 0$ for every $z \in G$. Then we define s_ν on G by setting $s_\nu := s + (1/\nu)z\overline{z}$. Obviously, $(s_\nu)_{z\overline{z}} = s_{z\overline{z}} + (1/\nu) > 0$. Then s_ν is subharmonic by the above lemma. Since (s_ν) converges, monotonically decreasing, to s, it follows that s is subharmonic.

(b) Let s be subharmonic on G. We assume that $s_{z\overline{z}}(a) < 0$ for some $a \in G$. Then there is a connected open neighborhood $U = U(a) \subset G$ such that $s_{z\overline{z}} < 0$ on U. By the lemma it follows that $-s$ is subharmonic on U. Then s must be harmonic on U. So $s_{z\overline{z}}(a) = 0$, contrary to assumption. ∎

Plurisubharmonic Functions. We return to the study of domains in arbitrary dimensions. Let $G \subset \mathbb{C}^n$ be a domain and $(\mathbf{a}, \mathbf{w})$ a tangent vector at $\mathbf{a} \in G$. We use the holomorphic mapping $\alpha_{\mathbf{a},\mathbf{w}} : \mathbb{C} \to \mathbb{C}^n$ defined by $\alpha_{\mathbf{a},\mathbf{w}}(\zeta) := \mathbf{a} + \zeta\mathbf{w}$.

Definition. Let $G \subset \mathbb{C}^n$ be a domain. An upper semicontinuous function $p : G \to \mathbb{R} \cup \{-\infty\}$ is called *plurisubharmonic* on G if for every tangent vector $(\mathbf{a}, \mathbf{w})$ in G the function

$$p_{\mathbf{a},\mathbf{w}}(\zeta) := p \circ \alpha_{\mathbf{a},\mathbf{w}}(\zeta) = p(\mathbf{a} + \zeta\mathbf{w})$$

is subharmonic on the connected component $G(\mathbf{a}, \mathbf{w})$ of the set $\alpha_{\mathbf{a},\mathbf{w}}^{-1}(G) \subset \mathbb{C}$ containing 0.

Remarks

1. Plurisubharmonicity is a local property.

2. If $f \in \mathcal{O}(G)$, then $\log|f|$ is plurisubharmonic.
3. If p_1, p_2 are plurisubharmonic, then so is $p_1 + p_2$.
4. If p is plurisubharmonic and $c > 0$, then $c \cdot p$ is plurisubharmonic.
5. If (p_ν) is a monotonically decreasing sequence of plurisubharmonic functions, then $p := \lim_{\nu\to\infty} p_\nu$ is also plurisubharmonic.
6. Let $(p_\alpha)_{\alpha\in A}$ be a family of plurisubharmonic functions. If $p := \sup(p_\alpha)$ is upper semicontinuous and finite, then it is also plurisubharmonic.
7. If a plurisubharmonic function p takes its maximum at a point of the domain G, then p is constant on G.

The Levi Form

Definition. Let $U \subset \mathbb{C}^n$ be an open set, $f \in \mathscr{C}^2(U;\mathbb{R})$, and $\mathbf{a} \in U$. The quadratic form[2] $\operatorname{Lev}(f) : T_{\mathbf{a}} \to \mathbb{R}$ with

$$\operatorname{Lev}(f)(\mathbf{a}, \mathbf{w}) := \sum_{\nu,\mu} f_{z_\nu \overline{z}_\mu}(\mathbf{a}) w_\nu \overline{w}_\mu$$

is called the *Levi form* of f at $\mathbf{a}$.

Obviously, $\operatorname{Lev}(f)$ is linear in f.

Examples

1. In the case $n = 1$ we have $\operatorname{Lev}(s)(a, w) = s_{z\overline{z}}(a) w\overline{w}$. So s is subharmonic if and only if $\operatorname{Lev}(s)(a, w) \geq 0$ for every $a \in G$ and $w \in \mathbb{C}$.
2. Let $f(\mathbf{z}) := \|\mathbf{z}\|^2 = \sum_{i=1}^n z_i \overline{z}_i$. Then $\operatorname{Lev}(f)(\mathbf{a}, \mathbf{w}) = \|\mathbf{w}\|^2$ for every $\mathbf{a}$.
3. If $f \in \mathscr{C}^2(U;\mathbb{R})$ and $\varrho : \mathbb{R} \to \mathbb{R}$ is twice continuously differentiable, then

$$\operatorname{Lev}(\varrho \circ f)(\mathbf{a}, \mathbf{w}) = \varrho''(f(\mathbf{a})) \cdot |(\partial f)_{\mathbf{a}}(\mathbf{w})|^2 + \varrho'(f(\mathbf{a})) \cdot \operatorname{Lev}(f)(\mathbf{a}, \mathbf{w}).$$

4. If $\mathbf{F} : U \to V \subset \mathbb{C}^m$ is a holomorphic map and $g \in \mathscr{C}^2(V;\mathbb{R})$, then

$$\operatorname{Lev}(g \circ \mathbf{F})(\mathbf{a}, \mathbf{w}) = \operatorname{Lev}(g)(\mathbf{F}(\mathbf{a}), \mathbf{F}'(\mathbf{a})(\mathbf{w})).$$

5. For $f \in \mathscr{C}^2(U;\mathbb{R})$ the Taylor expansion at $\mathbf{a} \in U$ gives

$$f(\mathbf{z}) = f(\mathbf{a}) + 2\operatorname{Re}(Q_f(\mathbf{z} - \mathbf{a})) + \operatorname{Lev}(f)(\mathbf{a}, \mathbf{z} - \mathbf{a}) + R(\mathbf{z} - \mathbf{a}),$$

where $Q_f(\mathbf{w}) = \sum_{\nu=1}^n f_{z_\nu}(\mathbf{a}) w_\nu + \frac{1}{2}\sum_{\nu,\mu} f_{z_\nu z_\mu}(\mathbf{a}) w_\nu w_\mu$ is a holomorphic quadratic polynomial, and

$$\lim_{\mathbf{z}\to\mathbf{a}} \frac{R(\mathbf{z} - \mathbf{a})}{\|\mathbf{z} - \mathbf{a}\|^2} = 0.$$

[2] If $H : T \times T \to \mathbb{C}$ is a Hermitian form on a complex vextor space, then the associated *quadratic form* $Q : V \to \mathbb{R}$ is given by $Q(v) := H(v, v)$.

2.8 Theorem. *A function $f \in \mathscr{C}^2(U;\mathbb{R})$ is plurisubharmonic if and only if* $\operatorname{Lev}(f)(\mathbf{a},\mathbf{w}) \geq 0$ *for every* $\mathbf{a} \in U$ *and every* $\mathbf{w} \in T_{\mathbf{a}}$.

PROOF: Let $(\mathbf{a},\mathbf{w})$ be a tangent vector in G and $\alpha := \alpha_{\mathbf{a},\mathbf{w}}$. Then $f \circ \alpha(0) = f(\mathbf{a})$ and

$$(f \circ \alpha)_{\zeta\overline{\zeta}}(0) = \operatorname{Lev}(f \circ \alpha)(0,1) = \operatorname{Lev}(f)(\mathbf{a},\mathbf{w}).$$

Now, f is plurisubharmonic if and only if $f \circ \alpha$ is subharmonic near 0 for any $\alpha = \alpha_{\mathbf{a},\mathbf{w}}$. Equivalently, $(f \circ \alpha)_{\zeta\overline{\zeta}}(0) \geq 0$ for any such α. But this is true if and only if $\operatorname{Lev}(f)(\mathbf{a},\mathbf{w}) \geq 0$ for any tangent vector $(\mathbf{a},\mathbf{w})$ in G. ∎

2.9 Corollary. *Let $G_1 \subset \mathbb{C}^n$ and $G_2 \subset \mathbb{C}^m$ be domains, $\mathbf{F} : G_1 \to G_2$ a holomorphic map, and $g \in \mathscr{C}^2(G_1;\mathbb{R})$ plurisubharmonic. Then $g \circ \mathbf{F}$ is plurisubharmonic on G_1.*

PROOF: This is trivial, because of the formula in Example 4 above. ∎

Exhaustion Functions

For every domain $G \subset \mathbb{C}$ the function $-\log \delta_G$ is subharmonic. In higher dimensions it is in general not true that this function is plurisubharmonic for every domain G.

> **Definition.** Let $G \subset \mathbb{C}^n$ be a domain. A nonconstant continuous function $f : G \to \mathbb{R}$ is called an *exhaustion function* for G if for $c < \sup_G(f)$ all sublevel sets
>
> $$G_c(f) := \{\mathbf{z} \in G \,:\, f(\mathbf{z}) < c\}$$
>
> are relatively compact in G.

Example

For $G = \mathbb{C}^n$, the function $f(\mathbf{z}) := \|\mathbf{z}\|^2$ is an exhaustion function. For $G \neq \mathbb{C}^n$, we define the *boundary distance* δ_G by

$$\delta_G(\mathbf{z}) := \operatorname{dist}(\mathbf{z}, \mathbb{C}^n - G).$$

Then $-\delta_G$ is a bounded, and $-\log \delta_G$ an unbounded, exhaustion function. We only have to show that δ_G is continuous:

For every point $\mathbf{z} \in G$ there is a point $\mathbf{r}(\mathbf{z}) \in \mathbb{C}^n - G$ such that

$$\delta_G(\mathbf{z}) = \operatorname{dist}(\mathbf{z}, \mathbf{r}(\mathbf{z})) \leq \operatorname{dist}(\mathbf{z}, \mathbf{w}) \text{ for every } \mathbf{w} \in \mathbb{C}^n - G.$$

Then for two arbitrary points $\mathbf{u}, \mathbf{v} \in G$ we have

$$\begin{aligned} \delta_G(\mathbf{u}) = \|\mathbf{u} - \mathbf{r}(\mathbf{u})\| &\leq \|\mathbf{u} - \mathbf{r}(\mathbf{v})\| \leq \|\mathbf{u} - \mathbf{v}\| + \delta_G(\mathbf{v}), \\ \text{and in the same way } \delta_G(\mathbf{v}) &\leq \|\mathbf{u} - \mathbf{v}\| + \delta_G(\mathbf{u}). \end{aligned}$$

Therefore, $|\delta_G(\mathbf{u}) - \delta_G(\mathbf{v})| \leq \|\mathbf{u} - \mathbf{v}\|$.

Definition. A function $f \in \mathscr{C}^2(G;\mathbb{R})$ is called *strictly plurisubharmonic* if $\mathrm{Lev}(f)(\mathbf{a},\mathbf{w}) > 0$ for $\mathbf{a} \in G$, $\mathbf{w} \in T_{\mathbf{a}}$, and $\mathbf{w} \neq \mathbf{0}$.

For a proof of the following result we refer to [Ra86], Chapter II, Proposition 4.14.

2.10 Smoothing lemma. *Let $G \subset \mathbb{C}^n$ be a domain, $f : G \to \mathbb{R}$ a continuous plurisubharmonic exhaustion function, $K \subset G$ compact, and $\varepsilon > 0$. Then there exists a $\mathscr{C}^\infty$ exhaustion function $g : G \to \mathbb{R}$ such that:*

1. *$g \geq f$ on G.*
2. *g is strictly plurisubharmonic.*
3. *$|g(\mathbf{z}) - f(\mathbf{z})| < \varepsilon$ on K.*

Exercises

1. Let $G \subset \mathbb{C}$ be a domain. Prove the following statements:
   (a) If $f : G \to \mathbb{C}$ is a holomorphic function, then $|f|^\alpha$ is subharmonic for $\alpha > 0$.
   (b) If u is subharmonic on G, then u^p is subharmonic for $p \in \mathbb{N}$.
   (c) Let $u \not\equiv -\infty$ be subharmonic on G. Then $\{z \in G : u(z) = -\infty\}$ does not contain any open subset.
2. Let $G \subset \mathbb{C}$ be a domain, $s \not\equiv -\infty$ a subharmonic function on G, $P := \{z \in G : s(z) = -\infty\}$. Show that if u is a continuous function on G and subharmonic on $G - A$, then u is subharmonic on G.
3. Let $U \subset \mathbb{C}^n$ be open, $\mathbf{f} : U \to \mathbb{C}^k$ a holomorphic map, and $\mathbf{A} \in M_k(\mathbb{R})$ a positive semidefinite matrix. Show that $\varphi(\mathbf{z}) := \mathbf{f}(\mathbf{z}) \cdot \mathbf{A} \cdot \mathbf{f}(\mathbf{z})^{\,t}$ is plurisubharmonic.
4. Let $G = \{(z,w) \in \mathbb{C}^2 : |w| < |z| < 1\}$ be the Hartogs triangle. Prove that there does not exist any bounded plurisubharmonic exhaustion function on G.
5. Are the following functions plurisubharmonic (respectively strictly plurisubharmonic)?
$$\begin{aligned} p_1(\mathbf{z}) &:= \log(1 + \|\mathbf{z}\|^2), \text{ for } \mathbf{z} \in \mathbb{C}^n, \\ p_2(\mathbf{z}) &:= -\log(1 - \|\mathbf{z}\|^2), \text{ for } \|\mathbf{z}\| < 1, \\ p_3(\mathbf{z}) &:= \|\mathbf{z}\|^2 e^{-\mathrm{Re}(z_n)}, \text{ for } \mathbf{z} \in \mathbb{C}^n. \end{aligned}$$
6. Consider a domain $G \subset \mathbb{C}^n$ and a function $f \in \mathscr{C}^2(G)$. Prove that f is strictly plurisubharmonic if and only if for every open set $U \subset\subset G$ there is an $\varepsilon > 0$ such that $f(\mathbf{z}) - \varepsilon\|\mathbf{z}\|^2$ is plurisubharmonic on U.

3. Pseudoconvexity

Pseudoconvexity

Definition. A domain $G \subset \mathbb{C}^n$ is called *pseudoconvex* if there is a strictly plurisubharmonic $\mathscr{C}^\infty$ exhaustion function on G.

Remarks

1. By the smoothing lemma the following is clear: If $-\log \delta_G$ is plurisubharmonic, then G is pseudoconvex.
2. Pseudoconvexity is invariant under biholomorphic transformations.

3.1 Theorem. *If $G \subset \mathbb{C}^n$ is a pseudoconvex domain, then G satisfies the continuity principle.*

PROOF: Let $p : G \to \mathbb{R}$ be a strictly plurisubharmonic exhaustion function. Suppose that there exists a family $\{S_t : 0 \le t \le 1\}$ of analytic disks given by a continuous mapping $\varphi : \overline{\mathsf{D}} \times [0,1] \to \mathbb{C}^n$ such that $S_0 \subset G$ and $bS_t \subset G$ for every $t \in [0,1]$, but not all S_t are contained in G.

The functions $p \circ \varphi_t : \mathsf{D} \to G$ are subharmonic for every t with $S_t \subset G$. It follows by the maximum principle that $p|S_t \le \max_{bS_t} p$ for all those t.

We define $t_0 := \inf\{t \in [0,1] : S_t \not\subset G\}$. Then $t_0 > 0$, $S_{t_0} \subset \overline{G}$, and S_{t_0} meets ∂G in at least one point $\mathbf{z}_0$. We can find an increasing sequence (t_ν) converging to t_0 and a sequence of points $\mathbf{z}_\nu \in S_{t_\nu}$ converging to $\mathbf{z}_0$. So $p(\mathbf{z}_\nu) \to c_0 := \sup_G(p)$, but there is a $c < c_0$ such that $p|_{bS_t} \le c$ for every $t \in [0,1]$. This is a contradiction. ∎

3.2 Corollary. *If G is pseudoconvex, then G is Hartogs convex.*

The Boundary Distance

3.3 Theorem. *If $G \subset \mathbb{C}^n$ is a Hartogs convex domain, then $-\log \delta_G$ is plurisubharmonic on G.*

PROOF: For $\mathbf{z} \in G$ and $\mathbf{u} \in \mathbb{C}^n$ with $\|\mathbf{u}\| = 1$ we define

$$\delta_{G,\mathbf{u}}(\mathbf{z}) := \sup\{t > 0 : \mathbf{z} + \tau\mathbf{u} \in G \text{ for } |\tau| \le t\}.$$

Then $\delta_G(\mathbf{z}) = \inf\{\delta_{G,\mathbf{u}}(\mathbf{z}) : \|\mathbf{u}\| = 1\}$, and it is sufficient to show that $-\log \delta_{G,\mathbf{u}}$ is plurisubharmonic for fixed $\mathbf{u}$.

(a) Unfortunately, $\delta_{G,\mathbf{u}}$ does not need to be continuous, but it is lower semicontinuous:

Let $\mathbf{z}_0 \in G$ be an arbitrary point and $c < \delta_{G,\mathbf{u}}(\mathbf{z}_0)$. Then the compact set $K := \{\mathbf{z} = \mathbf{z}_0 + \tau\mathbf{u} \,:\, |\tau| \le c\}$ is contained in G, and there is a $\delta > 0$ such that $\{\mathbf{z} \,:\, \mathrm{dist}(K, \mathbf{z}) < \delta\} \subset G$.

For $\mathbf{z} \in B_\delta(\mathbf{z}_0)$ and $|\tau| \le c$ we have

$$\|(\mathbf{z} + \tau\mathbf{u}) - (\mathbf{z}_0 + \tau\mathbf{u})\| = \|\mathbf{z} - \mathbf{z}_0\| < \delta, \text{ and therefore } \delta_{G,\mathbf{u}}(\mathbf{z}) \ge c.$$

(b) The function $-\log \delta_{G,\mathbf{u}}$ is upper semicontinuous, and we have to show that

$$s(\zeta) := -\log \delta_{G,\mathbf{u}}(\mathbf{z}_0 + \zeta\mathbf{b})$$

is subharmonic for fixed $\mathbf{u}, \mathbf{z}_0, \mathbf{b}$. First consider the case that $\mathbf{u}$ and $\mathbf{b}$ are linearly dependent: $\mathbf{b} = \lambda\mathbf{u}$, $\lambda \ne 0$.

Let G_0 be the connected component of 0 in $\{\zeta \in \mathbb{C} \,:\, \mathbf{z}_0 + \zeta\mathbf{b} \in G\}$. Then

$$\begin{aligned} \delta_{G,\mathbf{u}}(\mathbf{z}_0 + \zeta\mathbf{b}) &= \sup\{t > 0 \,:\, \mathbf{z}_0 + \zeta\mathbf{b} + \tau\mathbf{u} \in G \text{ for } |\tau| \le t\} \\ &= \sup\{t > 0 \,:\, \zeta + \tau/\lambda \in G_0 \text{ for } |\tau| \le t\} \\ &= |\lambda| \cdot \sup\{r > 0 \,:\, \zeta + \sigma \in G_0 \text{ for } |\sigma| \le r\} \\ &= |\lambda| \cdot \delta_{G_0}(\zeta), \end{aligned}$$

and this function is in fact subharmonic.

(c) Now assume that $\mathbf{u}$ and $\mathbf{b}$ are linearly independent. Since these vectors are fixed, we can restrict ourselves to the following special situation:

$$n = 2, \quad \mathbf{z}_0 = \mathbf{0}, \quad \mathbf{b} = \mathbf{e}_1, \quad \text{and} \quad \mathbf{u} = \mathbf{e}_2.$$

Then $s(\zeta) = -\log \sup\{t > 0 \,:\, (\zeta, \tau) \in G \text{ for } |\tau| \le t\}$. We use holomorphic functions to show that s is subharmonic. Let $R > r > 0$ be real numbers such that $(\zeta, 0) \in G$ for $|\zeta| < R$, and let $f : \mathsf{D}_R(0) \to \mathbb{C}$ be a holomorphic function such that $s < h := \mathrm{Re}\, f$ on $\partial\mathsf{D}_r(0)$. We have to show that $s < h$ on $\mathsf{D}_r(0)$.

We have the following equivalences:

$$\begin{aligned} s(\zeta) < h(\zeta) &\iff \sup\{t > 0 \,:\, (\zeta, \tau) \in G \text{ for } |\tau| \le t\} > e^{-h(\zeta)} \\ &\iff \big(\zeta, c \cdot e^{-f(\zeta)}\big) \in G \text{ for } c \in \overline{\mathsf{D}}. \end{aligned}$$

(d) Define a holomorphic map $\mathbf{F}$ by

$$\mathbf{F}(z_1, z_2) := \big(rz_1, z_2 e^{-f(rz_1)}\big).$$

Then $\mathbf{F}$ is well defined on a neighborhood of the unit polydisk $\mathsf{P}^2 = \mathsf{P}^2(\mathbf{0}, 1)$. It must be shown that $\mathbf{F}(\mathsf{P}^2) \subset G$. We already know the following:

1. $\mathbf{F}(z_1, z_2) \in G$ for $|z_1| = 1$ and $|z_2| \le 1$, because $s(t) < h(t)$ on $\partial\mathsf{D}_r(0)$.
2. $\mathbf{F}(z_1, 0) \in G$ for $|z_1| \le 1$, because $(\zeta, 0) \in G$ for $|\zeta| \le r$.

These facts will be used to construct an appropriate Hartogs figure. First, note that

$$J_{\mathbf{F}}(z_1, z_2) = \begin{pmatrix} r & 0 \\ * & e^{-f(rz_1)} \end{pmatrix}, \qquad \text{so } \det J_{\mathbf{F}}(z_1, z_2) \neq 0.$$

By the inverse function theorem it follows that $\mathbf{F}$ is biholomorphic.

For $0 < \delta < 1$ we define $\mathbf{h}_\delta : \mathbb{C}^2 \to \mathbb{C}^2$ by $\mathbf{h}_\delta(z_1, z_2) := (z_1, \delta z_2)$ and apply $\mathbf{h}_\delta$ to the compact set

$$C := \{(z_1, z_2) \in \mathbb{C}^2 \,:\, (|z_1| \le 1,\ z_2 = 0) \text{ or } (|z_1| = 1,\ |z_2| \le 1)\} \subset \overline{\mathsf{P}^2}.$$

Consequently,

$$C_\delta := \mathbf{h}_\delta(C) = \{(z_1, z_2) \in \mathbb{C}^2 \,:\, (|z_1| \le 1,\ z_2 = 0) \text{ or } (|z_1| = 1,\ |z_2| \le \delta)\}.$$

Then $\mathbf{F}(C_\delta) \subset G$, as we saw above, and therefore $C_\delta \subset \mathbf{F}^{-1}(G)$.

For $0 < \varepsilon < \min(\delta, 1 - \delta)$ we define a neighborhood U_ε of C_δ by $U_\varepsilon :=$

$$\{(z_1, z_2) \in \mathbb{C}^2 \,:\, (|z_1| < 1+\varepsilon,\ |z_2| < \varepsilon) \text{ or } (1-\varepsilon < |z_1| < 1+\varepsilon,\ |z_2| < \delta+\varepsilon)\}.$$

If we choose ε small enough, then $U_\varepsilon \subset \mathbf{F}^{-1}(G)$.

Finally, we define $\mathsf{H}_\varepsilon := \mathbf{h}_\delta^{-1}(U_\varepsilon \cap \mathsf{P}^2) \cap \mathsf{P}^2$ (see Figure II.6). Then

$$\begin{aligned} \mathsf{H}_\varepsilon &= \{(z_1, z_2) \in \mathsf{P}^2 \,:\, (z_1, \delta z_2) \in U_\varepsilon \cap \mathsf{P}^2\} \\ &= \Big\{(z_1, z_2) \in \mathbb{C}^2 \,:\, \big(|z_1| < 1,\ |z_2| < \frac{\varepsilon}{\delta}\big) \text{ or } (1 - \varepsilon < |z_1| < 1,\ |z_2| < 1)\Big\}. \end{aligned}$$

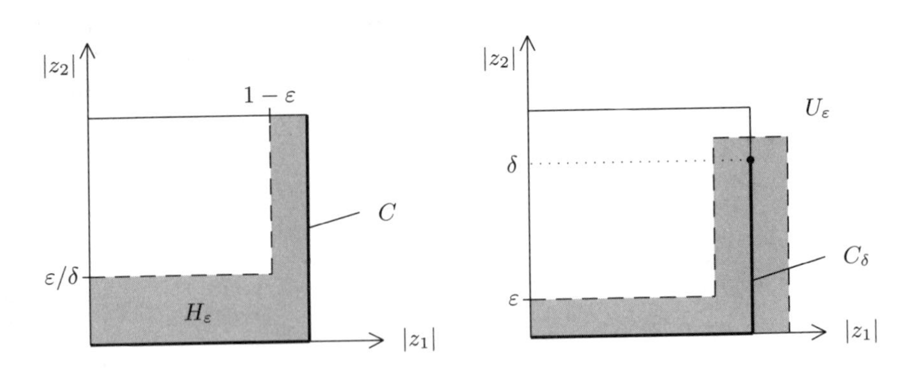

Figure II.6. Construction of the Hartogs figure

Since $(\mathsf{P}^2, \mathsf{H}_\varepsilon)$ is a Euclidean Hartogs figure, $(\mathbf{F} \circ \mathbf{h}_\delta(\mathsf{P}^2), \mathbf{F} \circ \mathbf{h}_\delta(\mathsf{H}_\varepsilon))$ is a general Hartogs figure with $\mathbf{F} \circ \mathbf{h}_\delta(\mathsf{H}_\varepsilon) \subset \mathbf{F}(U_\varepsilon \cap \mathsf{P}^2) \subset G$. Since G is Hartogs pseudoconvex, it follows that $\mathbf{F} \circ h_\delta(\mathsf{P}^2) \subset G$. This is valid for every $\delta < 1$. But $\mathsf{P}^2 = \bigcup_{0<\delta<1} \mathbf{h}_\delta(\mathsf{P}^2)$. Therefore, $\mathbf{F}(\mathsf{P}^2) \subset G$, which was to be shown. ∎

3.4 Theorem. *The following properties of a domain $G \subset \mathbb{C}^n$ are equivalent:*

1. *G satisfies the continuity principle.*
2. *G is Hartogs pseudoconvex.*
3. *$-\log \delta_G$ is plurisubharmonic on G.*
4. *G is pseudoconvex.*

PROOF:

(1) $\Longrightarrow$ (2) is Theorem 1.5,
(2) $\Longrightarrow$ (3) is Theorem 3.3,
(3) $\Longrightarrow$ (4) follows from the smoothing lemma,
(4) $\Longrightarrow$ (1) was proved in Theorem 3.1. ∎

Properties of Pseudoconvex Domains

3.5 Theorem. *If $G_1, G_2 \subset \mathbb{C}^n$ are pseudoconvex domains, then $G_1 \cap G_2$ is pseudoconvex.*

PROOF: The statement is trivial if one uses Hartogs pseudoconvexity. ∎

3.6 Theorem. *Let $G_1 \subset G_2 \subset \ldots \subset \mathbb{C}^n$ be an ascending chain of pseudoconvex domains. Then $G := \bigcup_{\nu=1}^{\infty} G_\nu$ is again pseudoconvex.*

PROOF: This follows immediately from the continuity principle. ∎

3.7 Theorem. *A domain $G \subset \mathbb{C}^n$ is pseudoconvex if and only if there is an open covering $(U_\iota)_{\iota \in I}$ of $\overline{G}$ such that $U_\iota \cap G$ is pseudoconvex for every $\iota \in I$.*

PROOF:

"$\Longrightarrow$" is trivial. The other direction will be proved in two steps. At first, we assume that G is bounded.

For any point $\mathbf{z}_0 \in \partial G$ there is an open set U_ι such that $\mathbf{z}_0 \in U_\iota$ and $G \cap U_\iota$ is pseudoconvex. If we choose a neighborhood $W = W(\mathbf{z}_0) \subset U_\iota$ so small that $\operatorname{dist}(\mathbf{z}, \partial U_\iota) > \operatorname{dist}(\mathbf{z}, \mathbf{z}_0)$ for every $\mathbf{z} \in W \cap G$, then $\delta_G(\mathbf{z}) = \delta_{G \cap U_\iota}(\mathbf{z})$ on $W \cap G$. This shows that there is an open neighborhood $U = U(\partial G)$ such that $-\log \delta_G$ is plurisubharmonic on $U \cap G$ (we use the fact that ∂G is compact). Now, $G - U \subset\subset G$. We define

$$c := \sup\{-\log \delta_G(\mathbf{z}) \,:\, \mathbf{z} \in G - U\},$$

and

$$p(\mathbf{z}) := \max\bigl(-\log \delta_G(\mathbf{z}), \|z\|^2 + c + 1\bigr).$$

Then p is a plurisubharmonic exhaustion function, and by the smoothing lemma, G is pseudoconvex.

If G is unbounded, we write it as an ascending union of the domains $G_\nu := \mathsf{B}_\nu(\mathbf{0}) \cap G$. Each G_ν is bounded and satisfies the hypothesis, so is pseudoconvex. Then G is also a pseudoconvex domain. ∎

Exercises

1. Suppose that $G_1 \subset \mathbb{C}^n$ and $G_2 \subset \mathbb{C}^m$ are domains.
   (a) Show that if G_1 and G_2 are pseudoconvex, then $G_1 \times G_2$ is a pseudoconvex domain in $\mathbb{C}^{n+m}$
   (b) Show that if there is a proper holomorphic map $\mathbf{f} : G_1 \to G_2$ and G_2 is pseudoconvex, then G_1 is also pseudoconvex.
2. Let $G \subset \mathbb{C}^n$ be a domain and $\varrho : G \to \mathbb{R}$ a lower semicontinuous positive function. Prove that

$$\widehat{G} := \{(\mathbf{z}', w) \in G \times \mathbb{C} \,:\, |w| < \varrho(\mathbf{z}')\}$$

   is pseudoconvex if and only if $-\log \varrho$ is plurisubharmonic.
3. A domain $G \subset \mathbb{C}^n$ is pseudoconvex if and only if for every compact set $K \subset G$ the set

$$\widehat{K}_{\text{pl}} := \Bigl\{\mathbf{z} \in G \,:\, p(\mathbf{z}) \le \sup_K p \text{ for all plurisubharmonic functions } p \text{ on } G\Bigr\}$$

   is relatively compact in G.

4. Levi Convex Boundaries

Boundary Functions

Definition. Let $G \subset \mathbb{C}^n$ be a domain. The boundary of G is called *smooth* at $\mathbf{z}_0 \in \partial G$ if there is an open neighborhood $U = U(\mathbf{z}_0) \subset \mathbb{C}^n$ and a function $\varrho \in \mathscr{C}^\infty(U; \mathbb{R})$ such that:
1. $U \cap G = \{\mathbf{z} \in U \,:\, \varrho(\mathbf{z}) < 0\}$.
2. $(d\varrho)_\mathbf{z} \neq 0$ for $\mathbf{z} \in U$.

The function ϱ is called a local *defining function* (or *boundary function*).

Remark. Without loss of generality we may assume that $\varrho_{y_n} \neq 0$. Then by the implicit function theorem there are neighborhoods

$$U' \text{ of } (\mathbf{z}'_0, x_n^{(0)}) = \big(z_1^{(0)}, \ldots, z_{n-1}^{(0)}, x_n^{(0)}\big) \in \mathbb{C}^{n-1} \times \mathbb{R}, \quad U'' \text{ of } y_n^{(0)} \in \mathbb{R},$$

and a $\mathscr{C}^\infty$ function $\gamma : U' \to U''$ such that $\{(\mathbf{z}', x_n, y_n) \in U' \times U'' : \varrho(\mathbf{z}', x_n + \mathrm{i}y_n) = 0\} = \{(\mathbf{z}', x_n, \gamma(\mathbf{z}', x_n)) : (\mathbf{z}', x_n) \in U'\}$.

Making the neighborhood $U := \{(\mathbf{z}', x_n + \mathrm{i}y_n) : (\mathbf{z}', x_n) \in U' \text{ and } y_n \in U''\}$ small enough and correcting the sign if necessary, one can achieve that

$$U \cap G = \{(\mathbf{z}', x_n + \mathrm{i}y_n) \in U : y_n < \gamma(\mathbf{z}', x_n)\}.$$

In particular, $U \cap \partial G = \{\mathbf{z} \in U : \varrho(\mathbf{z}) = 0\}$ is a $(2n-1)$-dimensional differentiable submanifold of U.

4.1 Lemma. *Let ∂G be smooth at $\mathbf{z}_0$, and let ϱ_1, ϱ_2 be two local defining functions on $U = U(\mathbf{z}_0)$. Then there is a $\mathscr{C}^\infty$ function h on U such that:*

1. *$h > 0$ on U.*
2. *$\varrho_1 = h \cdot \varrho_2$ on U.*
3. *$(d\varrho_1)_\mathbf{z} = h(\mathbf{z}) \cdot (d\varrho_2)_\mathbf{z}$ for $\mathbf{z} \in U \cap \partial G$.*

PROOF: Define $h := \varrho_1/\varrho_2$ on $U - \partial G$. After a change of coordinates, we have $\mathbf{z}_0 = \mathbf{0}$ and $\varrho_2 = y_n$. Then $g(t) := \varrho_1(\mathbf{z}', x_n + \mathrm{i}t)$ is a smooth function that vanishes at $t = 0$. Therefore,

$$\begin{aligned} \varrho_1(\mathbf{z}', z_n) &= g(y_n) - g(0) \\ &= \int_0^{y_n} g'(s)\,ds = y_n \cdot \int_0^1 g'(ty_n)\,dt \\ &= \varrho_2(\mathbf{z}', x_n + \mathrm{i}y_n) \cdot h(\mathbf{z}', z_n), \end{aligned}$$

where

$$h(\mathbf{z}', x_n + \mathrm{i}y_n) = \int_0^1 \frac{\partial \varrho_1}{\partial y_n}(\mathbf{z}', x_n + \mathrm{i}ty_n)\,dt$$

is smooth.

For $\mathbf{z} \in \partial G$ we have $(d\varrho_1)_\mathbf{z} = h(\mathbf{z}) \cdot (d\varrho_2)_\mathbf{z}$. Therefore, $h(\mathbf{z}) \neq 0$, and even greater than 0, since $h(\mathbf{z}) \geq 0$ by continuity. ∎

4.2 Theorem. *Let $G \subset\subset \mathbb{C}^n$ be a bounded domain with smooth boundary. Then ∂G is a differentiable submanifold, and there exists a global defining function.*

PROOF: We can find open sets $V_i \subset\subset U_i \subset \mathbb{C}^n$, $i = 1, \ldots, N$, such that:

1. $\{V_1, \ldots, V_N\}$ is an open covering of ∂G.
2. For each i there exists a local defining function ϱ_i for G on U_i.

3. For each i there is a smooth function $\varphi_i : U_i \to \mathbb{R}$ with $\varphi_i|_{V_i} \equiv 1$, $\varphi_i|_{\mathbb{C}^n - U_i} \equiv 0$, and $\varphi_i \geq 0$ in general.

Define $\varphi := \sum_i \varphi_i$ (so $\varphi > 0$ on ∂G) and $\psi_i := \varphi_i/\varphi$. Then $\sum_i \psi_i \equiv 1$ on ∂G. The system of the functions ψ_i is called a *partition of unity* on ∂G.

The function $\varrho := \sum_{i=1}^N \psi_i \varrho_i$ is now a global defining function for G. We leave it to the reader to check the details. ∎

The Levi Condition. For the remainder of this section let $G \subset\subset \mathbb{C}^n$ be a bounded domain with smooth boundary, and $\varrho : U = U(\partial G) \to \mathbb{R}$ a global defining function. Then at any $\mathbf{z}_0 \in \partial G$ the real tangent space of the boundary

$$T_{\mathbf{z}_0}(\partial G) := \{\mathbf{v} \in T_{\mathbf{z}_0} : (d\varrho)_{\mathbf{z}_0}(\mathbf{v}) = 0\}$$

is a $(2n-1)$-dimensional real subspace of $T_{\mathbf{z}_0}$. The space

$$H_{\mathbf{z}_0}(\partial G) := T_{\mathbf{z}_0}(\partial G) \cap \mathrm{i}T_{\mathbf{z}_0}(\partial G) = \{\mathbf{v} \in T_{\mathbf{z}_0} : (\partial\varrho)_{\mathbf{z}_0}(\mathbf{v}) = 0\}$$

is called the *complex* (or *holomorphic*) *tangent space* of the boundary at $\mathbf{z}_0$. It is a $(2n-2)$-dimensional real subspace of $T_{\mathbf{z}_0}$, with a natural complex structure, so an $(n-1)$-dimensional complex subspace[3].

> **Definition.** The domain G is said to satisfy the *Levi condition* (respectively the *strict Levi condition*) at $\mathbf{z}_0 \in \partial G$ if $\mathrm{Lev}(\varrho)$ is positive semidefinite (respectively positive definite) on $H_{\mathbf{z}_0}(\partial G)$. The domain G is called *Levi convex* (respectively *strictly Levi convex*) if G satisfies the Levi condition (respectively the strict Levi condition) at every point $\mathbf{z} \in \partial G$.

Remark. The Levi conditions do not depend on the choice of the boundary function, and they are invariant under biholomorphic transformations.

If $\varrho_1 = h \cdot \varrho_2$, with $h > 0$, then for $\mathbf{z} \in \partial G$,

$$\mathrm{Lev}(\varrho_1)(\mathbf{z}, \mathbf{w}) = h(\mathbf{z}) \cdot \mathrm{Lev}(\varrho_2)(\mathbf{z}, \mathbf{w}) + 2\,\mathrm{Re}\{(\overline{\partial}h)_{\mathbf{z}}(\mathbf{w}) \cdot (\partial\varrho_2)_{\mathbf{z}}(\mathbf{w})\}.$$

So on $H_{\mathbf{z}}(\partial G)$ the Levi forms of ϱ_1 and ϱ_2 differ only by a positive constant.

Affine Convexity. Recall some facts from real analysis:

A set $M \subset \mathbb{R}^n$ is *convex* if for every two points $\mathbf{x}, \mathbf{y} \in M$, the closed line segment from $\mathbf{x}$ to $\mathbf{y}$ is contained in M. In that case, for each point $\mathbf{x}_0 \in \mathbb{R}^n - M$ there is a real hyperplane $H \subset \mathbb{R}^n$ with $\mathbf{x}_0 \in H$ and $M \cap H = \varnothing$. This property was already used in Section 1.

[3] $H_{\mathbf{z}}(\partial G)$ is often denoted by $T^{1,0}_{\mathbf{z}}(\partial G)$.

If $\mathbf{a} \in \mathbb{R}^n$, $U = U(\mathbf{a})$ is an open neighborhood and $\varphi : U \to \mathbb{R}$ is at least $\mathscr{C}^2$, then the quadratic form

$$\operatorname{Hess}(\varphi)(\mathbf{a}, \mathbf{w}) := \sum_{\nu,\mu} \varphi_{x_\nu x_\mu}(\mathbf{a}) w_\nu w_\mu$$

is known as the *Hessian* of φ at $\mathbf{a}$.

4.3 Proposition. *Let $G \subset\subset \mathbb{R}^n$ be a domain with smooth boundary, and ϱ a global defining function with $(d\varrho)_\mathbf{x} \neq 0$ for $\mathbf{x} \in \partial G$. Then G is convex if and only if* $\operatorname{Hess}(\varrho)$ *is positive semidefinite on every tangent space $T_\mathbf{x}(\partial G)$.*

Proof: Let G be convex, and $\mathbf{x}_0 \in \partial G$ an arbitrary point. Then $T := T_{\mathbf{x}_0}(\partial G)$ is a real hyperplane with $T \cap G = \varnothing$. For $\mathbf{w} \in T$ and $\alpha(t) := \mathbf{x}_0 + t\mathbf{w}$ we have

$$(\varrho \circ \alpha)''(0) = \operatorname{Hess}(\varrho)(\mathbf{x}_0, \mathbf{w}).$$

Since $\varrho(\mathbf{x}_0) = 0$ and $\varrho \circ \alpha(t) \geq 0$, it follows that $\varrho \circ \alpha$ has a minimum at $t = 0$. Then $(\varrho \circ \alpha)''(0) \geq 0$, and $\operatorname{Hess}(\varrho)$ is positive semidefinite on T.

Now let the criterion be fulfilled, assume that $\mathbf{0} \in G$, and define ϱ_ε by

$$\varrho_\varepsilon(\mathbf{x}) := \varrho(\mathbf{x}) + \frac{\varepsilon}{N}\|\mathbf{x}\|^N.$$

For small ε and large N the set $G_\varepsilon := \{\mathbf{x} : \varrho_\varepsilon(\mathbf{x}) < 0\}$ is a domain. We have $G_\varepsilon \subset G_{\varepsilon'} \subset G$ for $\varepsilon' < \varepsilon$, and $\bigcup_{\varepsilon > 0} G_\varepsilon = G$. Therefore, it is sufficient to show that G_ε is convex.

The Hessian of ϱ_ε is positive definite on $T_\mathbf{x}(\partial G)$ for every $\mathbf{x} \in \partial G$. Thus this also holds in a neighborhood U of ∂G. If ε is small enough, then $\partial G_\varepsilon \subset U$. We consider

$$S := \{(\mathbf{x}, \mathbf{y}) \in G_\varepsilon \times G_\varepsilon : t\mathbf{x} + (1-t)\mathbf{y} \in G_\varepsilon, \text{ for } 0 < t < 1\}.$$

Then S is an open subset of the connected set $G_\varepsilon \times G_\varepsilon$. Suppose that S is not a closed subset. Then there exist points $\mathbf{x}_0, \mathbf{y}_0 \in G_\varepsilon$ and a $t_0 \in (0,1)$ with $t_0\mathbf{x}_0 + (1-t_0)\mathbf{y}_0 \in \partial G_\varepsilon$. So the function $t \mapsto \varrho_\varepsilon \circ \alpha(t)$, with $\alpha(t) := t\mathbf{x}_0 + (1-t)\mathbf{y}_0$, has a maximum at t_0. Then $(\varrho_\varepsilon \circ \alpha)''(t_0) \leq 0$ and $\operatorname{Hess}(\varrho_\varepsilon)(\alpha(t_0), \mathbf{x}_0 - \mathbf{y}_0) \leq 0$. This is a contradiction. ∎

A domain $G = \{\varrho < 0\}$ is called *strictly convex* at $\mathbf{x}_0 \in \partial G$ if $\operatorname{Hess}(\varrho)$ is positive definite at $\mathbf{x}_0$. This property is independent of ϱ and invariant under affine transformations.

Now we return to Levi convexity.

4.4 Lemma. *Let $U \subset \mathbb{C}^n$ be open and $\varphi \in \mathscr{C}^2(U; \mathbb{R})$. Then*

$$\operatorname{Lev}(\varphi)(\mathbf{z}, \mathbf{w}) = \frac{1}{4}\left(\operatorname{Hess}(\varphi)(\mathbf{z}, \mathbf{w}) + \operatorname{Hess}(\varphi)(\mathbf{z}, \mathrm{i}\mathbf{w})\right).$$

Proof: This is a simple calculation! ∎

4.5 Theorem. *Let $G \subset\subset \mathbb{C}^n$ be a domain with smooth boundary. Then the following statements are equivalent:*

1. *G is strictly Levi convex.*
2. *There is an open neighborhood $U = U(\partial G)$ and a strictly plurisubharmonic function $\varrho \in \mathscr{C}^\infty(U; \mathbb{R})$ such that $U \cap G = \{\mathbf{z} \in U : \varrho(\mathbf{z}) < 0\}$ and $(d\varrho)_\mathbf{z} \neq 0$ for $\mathbf{z} \in U$.*
3. *For every $\mathbf{z} \in \partial G$ there is an open neighborhood $W = W(\mathbf{z}) \subset \mathbb{C}^n$, an open set $V \subset \mathbb{C}^n$, and a biholomorphic map $\mathbf{F} : W \to V$ such that $\mathbf{F}(W \cap G)$ is convex and even strictly convex at every point of $\mathbf{F}(W \cap \partial G)$.*

Proof:

(1) $\Longrightarrow$ (2) : We choose a global defining function ϱ for G, and an open neighborhood $U = U(\partial G)$ such that ϱ is defined on U with $(d\varrho)_\mathbf{z} \neq 0$ for $\mathbf{z} \in U$. Let $A > 0$ be a real constant, and $\varrho_A := e^{A\varrho} - 1$. Then ϱ_A is also a global defining function, and

$$\operatorname{Lev}(\varrho_A)(\mathbf{z}, \mathbf{w}) = Ae^{A\varrho(\mathbf{z})} \left[\operatorname{Lev}(\varrho)(\mathbf{z}, \mathbf{w}) + A|(\partial\varrho)_\mathbf{z}(\mathbf{w})|^2\right].$$

The set $K := \partial G \times S^{2n-1}$ is compact, and

$$K_0 := \{(\mathbf{z}, \mathbf{w}) \in K : \operatorname{Lev}(\varrho)(\mathbf{z}, \mathbf{w}) \leq 0\}$$

is a closed subset. Since $\operatorname{Lev}(\varrho)$ is positive definite on $H_\mathbf{z}(\partial G)$, we have $(\partial\varrho)_\mathbf{z}(\mathbf{w}) \neq 0$ for $(\mathbf{z}, \mathbf{w}) \in K_0$. Therefore,

$$\begin{aligned} M &:= \min_K \operatorname{Lev}(\varrho)(\mathbf{z}, \mathbf{w}) > -\infty, \\ C &:= \min_{K_0} |(\partial\varrho)_\mathbf{z}(\mathbf{w})|^2 > 0. \end{aligned}$$

We choose A so large that $A \cdot C + M > 0$. Then

$$\operatorname{Lev}(\varrho_A)(\mathbf{z}, \mathbf{w}) = A \cdot [\operatorname{Lev}(\varrho)(\mathbf{z}, \mathbf{w}) + A|(\partial\varrho)_\mathbf{z}(\mathbf{w})|^2] \geq A \cdot (M + AC) > 0$$

for $(\mathbf{z}, \mathbf{w}) \in K_0$, and

$$\operatorname{Lev}(\varrho_A)(\mathbf{z}, \mathbf{w}) > A^2 \cdot |(\partial\varrho)_\mathbf{z}(\mathbf{w})|^2 \geq 0$$

for $(\mathbf{z}, \mathbf{w}) \in K - K_0$.

So $\operatorname{Lev}(\varrho_A)(\mathbf{z}, \mathbf{w}) > 0$ for every $\mathbf{z} \in \partial G$ and every $\mathbf{w} \in \mathbb{C}^n - \{\mathbf{0}\}$. By continuity, ϱ_A is strictly plurisubharmonic in a neighborhood of ∂G.

(2) $\Longrightarrow$ (3) : We consider a point $\mathbf{z}_0 \in \partial G$ and make some simple coordinate transformations:

By the translation $\mathbf{z} \mapsto \mathbf{w} = \mathbf{z} - \mathbf{z}_0$ we replace $\mathbf{z}_0$ by the origin, and a permutation of coordinates ensures that $\varrho_{w_1}(\mathbf{0}) \neq 0$.

The linear transformation

$$\mathbf{w} \mapsto \mathbf{u} = \big(\varrho_{w_1}(\mathbf{0})w_1 + \cdots + \varrho_{w_n}(\mathbf{0})w_n, w_2, \ldots, w_n\big)$$

gives $u_1 = \mathbf{w} \cdot \nabla\varrho(\mathbf{0})^{\,t}$, and therefore

$$\begin{aligned}
\varrho(\mathbf{u}) &= 2\,\mathrm{Re}\,\big(\mathbf{u} \cdot \nabla(\varrho \circ \mathbf{w})(\mathbf{0})^{\,t}\big) + \text{ terms of degree } \geq 2 \\
&= 2\,\mathrm{Re}\,\big(\mathbf{u} \cdot J_{\mathbf{w}}(\mathbf{0})^{\,t} \cdot \nabla\varrho(\mathbf{0})^{\,t}\big) + \text{ terms of degree } \geq 2 \\
&= 2\,\mathrm{Re}\,\big(\mathbf{w} \cdot \nabla\varrho(\mathbf{0})^{\,t}\big) + \text{ terms of degree } \geq 2 \\
&= 2\,\mathrm{Re}(u_1) + \text{ terms of degree } \geq 2.
\end{aligned}$$

Finally, we write $\varrho(\mathbf{u}) = 2\,\mathrm{Re}(u_1 + Q(\mathbf{u})) + \mathrm{Lev}(\varrho)(\mathbf{0}, \mathbf{u}) + \cdots$, where Q is a quadratic holomorphic polynomial, and make the biholomorphic transformation

$$\mathbf{u} \mapsto \mathbf{v} = (u_1 + Q(\mathbf{u}), u_2, \ldots, u_n).$$

It follows that

$$\varrho(\mathbf{v}) = 2\,\mathrm{Re}(v_1) + \mathrm{Lev}(\varrho)(\mathbf{0}, \mathbf{v}) + \text{ terms of order } \geq 3.$$

By the uniqueness of the Taylor expansion

$$\varrho(\mathbf{v}) = D\varrho(\mathbf{0})(\mathbf{v}) + \frac{1}{2}\mathrm{Hess}(\varrho)(\mathbf{0}, \mathbf{v}) + \text{ terms of order } \geq 3\,,$$

and therefore $\mathrm{Hess}(\varrho)(\mathbf{0}, \mathbf{v}) = 2 \cdot \mathrm{Lev}(\varrho)(\mathbf{0}, \mathbf{v}) > 0$ for $\mathbf{v} \neq \mathbf{0}$ (in the new coordinates). Everything works in a neighborhood that may be chosen to be convex.

(3) $\Longrightarrow$ (1) : This follows from Lemma 4.4:

$$\mathrm{Hess}(\varrho) > 0 \text{ on } T_{\mathbf{z}}(\partial G) \implies \mathrm{Lev}(\varrho) > 0 \text{ on } H_{\mathbf{z}}(\partial G).$$

The latter property is invariant under biholomorphic transformations. ∎

A Theorem of Levi. Let $G \subset\subset \mathbb{C}^n$ be a domain with smooth boundary. If G is strictly Levi convex, then it is easy to see that G is pseudoconvex. We wish to demonstrate that even the weaker Levi convexity is equivalent to pseudoconvexity. For that purpose we extend the boundary distance to a function on $\mathbb{C}^n$.

$$d_G(\mathbf{z}) := \begin{cases} \delta_G(\mathbf{z}) & \text{for } \mathbf{z} \in G, \\ 0 & \text{for } \mathbf{z} \in \partial G, \\ -\delta_{\mathbb{C}^n - \overline{G}}(\mathbf{z}) & \text{for } \mathbf{z} \notin \overline{G}. \end{cases}$$

4.6 Lemma. *$-d_G$ is a smooth defining function for G.*

PROOF: We use real coordinates $\mathbf{x} = (x_1, \ldots, x_N)$ with $N = 2n$. It is clear that $G = \{\mathbf{x} : -d_G(\mathbf{x}) < 0\}$.

Let $\mathbf{x}_0 \in \partial G$ be an arbitrary point and $\varrho : U(\mathbf{x}_0) \to \mathbb{R}$ a local defining function. We may assume that $\varrho_{x_N}(\mathbf{x}_0) \neq 0$. Then by the implicit function theorem there is a product neighborhood $U' \times U''$ of $\mathbf{x}_0$ in U and a smooth function $h : U' \to \mathbb{R}$ such that

$$\{(\mathbf{x}', x_N) \in U' \times U'' : \varrho(\mathbf{x}', x_N) = 0\} = \{(\mathbf{x}', h(\mathbf{x}')) : \mathbf{x}' \in U'\}.$$

It follows that $\mathbf{0} = \nabla_{\mathbf{x}'}\varrho(\mathbf{x}', h(\mathbf{x}')) + \varrho_{x_N}(\mathbf{x}', h(\mathbf{x}')) \cdot \nabla h(\mathbf{x}')$.

At the point $(\mathbf{x}', h(\mathbf{x}')) \in \partial G$ the gradient $\nabla\varrho(\mathbf{x}', h(\mathbf{x}'))$ is normal to ∂G and directed outward from G. Every point $\mathbf{y}$ in a small neighborhood of the boundary has a unique representation $\mathbf{y} = \mathbf{x} + t \cdot \nabla\varrho(\mathbf{x})$, where $t = -d_G(\mathbf{y})$ and $\mathbf{x}$ is the point where the perpendicular from $\mathbf{y}$ to ∂G meets the boundary. Therefore, we define the smooth map $\mathbf{F} : U' \times \mathbb{R} \to \mathbb{R}^N$ by

$$\mathbf{y} = \mathbf{F}(\mathbf{x}', t) := (\mathbf{x}', h(\mathbf{x}')) + t \cdot \nabla\varrho(\mathbf{x}', h(\mathbf{x}')).$$

Then there are smooth functions $\mathbf{A}$ and $\mathbf{b}$ such that

$$J_{\mathbb{R},\mathbf{F}}(\mathbf{x}', t) = \begin{pmatrix} \mathbf{E}_{N-1} + t \cdot \mathbf{A}(\mathbf{x}') & \nabla_{\mathbf{x}'}\varrho(\mathbf{x}', h(\mathbf{x}'))^t \\ \nabla h(\mathbf{x}') + t \cdot \mathbf{b}(\mathbf{x}') & \varrho_{x_N}(\mathbf{x}', h(\mathbf{x}')) \end{pmatrix},$$

and therefore

$$\begin{aligned} \det J_{\mathbb{R},\mathbf{F}}(\mathbf{x}', 0) &= \det \begin{pmatrix} \mathbf{E}_{N-1} & -\varrho_{x_N}(\mathbf{x}', h(\mathbf{x}')) \cdot \nabla h(\mathbf{x}')^t \\ \nabla h(\mathbf{x}') & \varrho_{x_N}(\mathbf{x}', h(\mathbf{x}')) \end{pmatrix} \\ &= \varrho_{x_N}(\mathbf{x}', h(\mathbf{x}')) \cdot \det \begin{pmatrix} \mathbf{E}_{N-1} & -\nabla h(\mathbf{x}')^t \\ \mathbf{0}' & 1 + \|\nabla h(\mathbf{x}')\|^2 \end{pmatrix} \\ &= \varrho_{x_N}(\mathbf{x}', h(\mathbf{x}'))(1 + \|\nabla h(\mathbf{x}')\|^2) \neq 0. \end{aligned}$$

It follows that there exists an $\varepsilon > 0$ such that $\mathbf{F}$ maps $U' \times (-\varepsilon, \varepsilon)$ diffeomorphically onto a neighborhood $W = W(\mathbf{x}_0)$, and $U' \times \{0\}$ onto $\partial G \cap W$. Moreover, since $d_G(\mathbf{x} + t \cdot \nabla\varrho(\mathbf{x})) = -t$ for $|t| < \varepsilon$ and ε small enough, it follows that $d_G = (-t) \circ \mathbf{F}^{-1}$ is a smooth function near ∂G. If $\mathbf{p}'$ is defined by $\mathbf{p}'(\mathbf{x}', t) := (\mathbf{x}', 0)$, then the projection

$$\mathbf{p} = \mathbf{p}' \circ \mathbf{F}^{-1} : \mathbf{x} + t \cdot \nabla\varrho(\mathbf{x}) \mapsto \mathbf{x}, \text{ for } \mathbf{x} \in \partial G,$$

is a smooth map, and d_G is given by $d_G(\mathbf{y}) = \sigma \cdot \|\mathbf{y} - \mathbf{p}(\mathbf{y})\|$, where $\sigma = 1$ for $\mathbf{y} \in G$ and $\sigma = -1$ elsewhere.

For $\mathbf{y} \notin \partial G$ we have

$$
\begin{aligned}
(d_G)_{y_\nu}(\mathbf{y}) &= \frac{\sigma}{\|\mathbf{y}-\mathbf{p}(\mathbf{y})\|} \cdot \sum_{k=1}^{N} (y_k - p_k(\mathbf{y}))(\delta_{k\nu} - (p_k)_{y_\nu}(\mathbf{y})) \\
&= \frac{\sigma}{\|\mathbf{y}-\mathbf{p}(\mathbf{y})\|} \cdot \left[y_\nu - p_\nu(\mathbf{y}) - \big(\mathbf{y}-\mathbf{p}(\mathbf{y}) \,|\, \mathbf{p}_{y_\nu}(\mathbf{y})\big)_N \right],
\end{aligned}
$$

and therefore

$$
\nabla d_G(\mathbf{y}) = \frac{\sigma}{\|\mathbf{y}-\mathbf{p}(\mathbf{y})\|} \cdot \left[\mathbf{y} - \mathbf{p}(\mathbf{y}) - D\mathbf{p}(\mathbf{y})(\mathbf{y}-\mathbf{p}(\mathbf{y}))\right].
$$

Since $\varrho(\mathbf{p}(\mathbf{y})) \equiv 0$, it follows that $D\mathbf{p}(\mathbf{y})(\nabla\varrho(\mathbf{p}(\mathbf{y}))) = 0$. But $\mathbf{y} - \mathbf{p}(\mathbf{y})$ is a multiple of $\nabla\varrho(\mathbf{p}(\mathbf{y}))$. Together this gives

$$
\nabla d_G(\mathbf{y}) = \sigma \cdot \frac{\mathbf{y}-\mathbf{p}(\mathbf{y})}{\|\mathbf{y}-\mathbf{p}(\mathbf{y})\|} = \pm \frac{\nabla\varrho(\mathbf{p}(\mathbf{y}))}{\|\nabla\varrho(\mathbf{p}(\mathbf{y}))\|}.
$$

If $\mathbf{y}$ tends to ∂G, we obtain that $\nabla d_G(\mathbf{y}) \neq \mathbf{0}$. ∎

E.E. Levi showed that every domain of holomorphy with smooth boundary is Levi convex, and locally the boundary of a strictly Levi convex domain G is the "natural boundary" for some holomorphic function in G. Here we prove the following result, which is sometimes called "Levi's theorem".

4.7 Theorem. *A domain G with smooth boundary is pseudoconvex if and only if it is Levi convex.*

PROOF:

(1) Let G be pseudoconvex. The function $-d_G$ is a smooth boundary function for G, and $-\log d_G = -\log \delta_G$ is plurisubharmonic on G, because of the pseudoconvexity. We calculate

$$
\operatorname{Lev}(-\log d_G)(\mathbf{z},\mathbf{w}) = \frac{1}{d_G(\mathbf{z})} \cdot \operatorname{Lev}(-d_G)(\mathbf{z},\mathbf{w}) + \frac{1}{d_G(\mathbf{z})^2} \cdot |(\partial(d_G))_{\mathbf{z}}(\mathbf{w})|^2.
$$

This is nonnegative in G. If $\mathbf{z} \in G$, $\mathbf{w} \in T_{\mathbf{z}}$, and $(\partial(d_G))_{\mathbf{z}}(\mathbf{w}) = 0$, it follows that $\operatorname{Lev}(-d_G)(\mathbf{z},\mathbf{w}) \geq 0$. This remains true for $\mathbf{z} \to \partial G$, so $-d_G$ satisfies the Levi condition.

(2) Let G be Levi convex, and suppose that G is not pseudoconvex. Then in any neighborhood U of the boundary there exists a point $\mathbf{z}_0$ where the Levi form of $-\log \delta_G$ has a negative eigenvalue. This means that there is a vector $\mathbf{w}_0$ such that

$$
\varphi_{\zeta\overline{\zeta}}(0) = \operatorname{Lev}(\log \delta_G)(\mathbf{z}_0,\mathbf{w}_0) > 0, \text{ for } \varphi(\zeta) := \log \delta_G(\mathbf{z}_0 + \zeta \mathbf{w}_0).
$$

Consider the Taylor expansion

$$
\begin{aligned}
\varphi(\zeta) &= \varphi(0) + 2\operatorname{Re}(\varphi_\zeta(0)\zeta + \frac{1}{2}\varphi_{\zeta\zeta}(0)\zeta^2) + \varphi_{\zeta\overline{\zeta}}(0)|\zeta|^2 + \cdots \\
&= \varphi(0) + \operatorname{Re}(A\zeta + B\zeta^2) + \lambda|\zeta|^2 + \cdots,
\end{aligned}
$$

with complex constants A, B and a real constant $\lambda > 0$.

We choose a point $\mathbf{p}_0 \in \partial G$ with $\delta_G(\mathbf{z}_0) = \|\mathbf{p}_0 - \mathbf{z}_0\|$, and an arbitrary $\varepsilon > 0$. Then an analytic disk $\psi : \mathsf{D}_\varepsilon(0) \to \mathbb{C}^n$ can be defined by

$$\psi(\zeta) := \mathbf{z}_0 + \zeta \mathbf{w}_0 + \exp(A\zeta + B\zeta^2)(\mathbf{p}_0 - \mathbf{z}_0).$$

We have $\psi(0) = \mathbf{p}_0$, and we wish to show that $\psi(\zeta) \in G$, for $0 < |\zeta| < \varepsilon$ and ε sufficiently small.

Since $\varphi(\zeta) \geq \varphi(0) + \operatorname{Re}(A\zeta + B\zeta^2) + (\lambda/2)|\zeta|^2$ near $\zeta = 0$, it follows that

$$\begin{aligned} \delta_G(\mathbf{z}_0 + \zeta \mathbf{w}_0) &= \exp(\varphi(\zeta)) \\ &\geq \exp(\varphi(0)) \cdot \left| \exp\left(A\zeta + B\zeta^2\right) \right| \cdot \exp\left(\frac{\lambda}{2}|\zeta|^2\right) \\ &> \delta_G(\mathbf{z}_0) \cdot \left| \exp\left(A\zeta + B\zeta^2\right) \right| \\ &= \left\| \exp\left(A\zeta + B\zeta^2\right)(\mathbf{p}_0 - \mathbf{z}_0) \right\|, \end{aligned}$$

for ζ small and $\neq 0$. This means that we can choose the ε in such a way that $\psi(\zeta) \in G$, for $0 < |\zeta| < \varepsilon$. The analytic disc is tangent to ∂G from the interior of G.

Now $f(\zeta) = d_G(\psi(\zeta))$ is a smooth function with a local minimum at $\zeta = 0$. Therefore $(\partial d_G)_{\mathbf{p}_0}(\psi'(0)) = (\partial f)_0(1) = 0$, and

$$f(\zeta) = \operatorname{Re}\left(f_{\zeta\zeta}(0)\zeta^2\right) + f_{\zeta\overline{\zeta}}|\zeta|^2 + \text{terms of order } \geq 3.$$

Since $\operatorname{Re}\left(f_{\zeta\zeta}(0)e^{2it}\right) + f_{\zeta\overline{\zeta}} \geq 0$ for every t, it follows that

$$\operatorname{Lev}(d_G)(\mathbf{p}_0, \psi'(0)) = f_{\zeta\overline{\zeta}}(0) > 0.$$

This is a contradiction to the Levi condition at $\mathbf{p}_0$, because $-d_G$ is a defining function for G. ∎

Exercises

1. Prove Lemma 4.4.
2. Assume that $G \subset\subset \mathbb{C}^2$ has a smooth boundary that is Levi convex outside a point $\mathbf{a}$ that is not isolated in ∂G. Show that G is pseudoconvex.
3. Assume that $G \subset \mathbb{C}^2$ is an arbitrary domain and that $S \subset G$ is a smooth real surface with the following property: In every point of S the tangent to S is not a complex line. Prove that for every compact set $K \subset G$ there are arbitrarily small pseudoconvex neighborhoods of $S \cap K$.
4. Assume that $G \subset\subset \mathbb{C}^2$ is a domain with smooth boundary. Then G is strictly Levi convex at a point $\mathbf{z}_0 \in \partial G$ if and only if the following condition is satisfied:

   There is a neighborhood $U = U(\mathbf{z}_0)$, a holomorphic function $\varphi : \mathsf{D} \to U$ with $\varphi(0) = \mathbf{z}_0$ and $\varphi'(0) \neq 0$, and a local defining function ϱ on U such that $(\varrho \circ \varphi)(\zeta) > 0$ on $\mathsf{D} - \{0\}$ and $(\varrho \circ \varphi)_{\zeta\overline{\zeta}}(0) > 0$.

5. Let $G \subset\subset \mathbb{C}^n$ be a domain with smooth boundary. If G satisfies the strict Levi condition at $\mathbf{z}_0 \in \partial G$, then prove that the following hold:
(a) There is no analytic disk $\varphi : \mathsf{D} \to \mathbb{C}^n$ with

$$\varphi(0) = \mathbf{z}_0 \quad \text{and} \quad \lim_{\zeta \to 0} \frac{\delta_G(\varphi(\zeta))}{\|\varphi(\zeta) - \varphi(0)\|^2} = 0\,.$$

(b) There are a neighborhood $U = U(\mathbf{z}_0)$ and a holomorphic function f in U with $\overline{G} \cap \{\mathbf{z} \in U \,:\, f(\mathbf{z}) = 0\} = \{\mathbf{z}_0\}$.

6. A bounded domain $G \subset \mathbb{C}^n$ is called *strongly pseudoconvex* if there are a neighborhood $U = U(\partial G)$ and a strictly plurisubharmonic function $\varrho \in \mathscr{C}^2(U)$ such that $G \cap U = \{\mathbf{z} \in U \,:\, \varrho(\mathbf{z}) < 0\}$. Notice that a strongly pseudoconvex domain does not necessarily have a smooth boundary!

Prove the following results about a strongly pseudoconvex bounded domain G:
(a) G is pseudoconvex.
(b) If G has a smooth boundary, then G is strictly Levi convex.
(c) For every $\mathbf{z} \in \partial G$ there is a neighborhood $U = U(\mathbf{z})$ such that $U \cap G$ is a weak domain of holomorphy.

7. Let $G \subset \mathbb{C}^n$ be a pseudoconvex domain. Then prove that there is a family of domains $G_\nu \subset G$ such that the following hold:
(a) $G_\nu \subset\subset G_{\nu+1}$ for every ν.
(b) $\bigcup_{\nu=1}^{\infty} G_\nu = G$.
(c) For every ν there is a strictly plurisubharmonic function $f_\nu \in \mathscr{C}^\infty(G_{\nu+1})$ such that G_ν is a connected component of the set

$$\{\mathbf{z} \in G_{\nu+1} \,:\, f_\nu(\mathbf{z}) < 0\}\,.$$

5. Holomorphic Convexity

Affine Convexity We will investigate relationships between pseudoconvexity and affine convexity. Let us begin with some observations about convex domains in $\mathbb{R}^N$.

Let $\mathscr{L}$ be the set of affine linear functions $f : \mathbb{R}^N \to \mathbb{R}$ with

$$f(\mathbf{x}) = a_1 x_1 + \cdots + a_N x_N + b, \quad a_1, \ldots, a_N, b \in \mathbb{R}.$$

If M is a convex set and $\mathbf{x}_0$ a point not contained in M, then there exists a function $f \in \mathscr{L}$ with $f(\mathbf{x}_0) = 0$ and $f|_M < 0$. For any $c \in \mathbb{R}$, the set $\{\mathbf{x} \in \mathbb{R}^N \,:\, f(\mathbf{x}) < c\}$ is a convex half-space.

Definition. Let $M \subset \mathbb{R}^N$ be an arbitrary subset. Then the set

$$H(M) := \Big\{\mathbf{x} \in \mathbb{R}^N \,:\, f(\mathbf{x}) \le \sup_M f, \text{ for all } f \in \mathscr{L}\Big\}$$

is called the *affine convex hull* of M.

5.1 Proposition. *Let $M, M_1, M_2 \subset \mathbb{R}^N$ be arbitrary subsets. Then*

1. $M \subset H(M)$.
2. $H(M)$ *is closed and convex.*
3. $H(H(M)) = H(M)$.
4. *If* $M_1 \subset M_2$, *then* $H(M_1) \subset H(M_2)$.
5. *If* M *is closed and convex, then* $H(M) = M$.
6. *If* M *is bounded, then* $H(M)$ *is also bounded.*

PROOF: (1) is trivial.

(2) If $\mathbf{x}_0 \notin H(M)$, then there is an $f \in \mathscr{L}$ with $f(\mathbf{x}_0) > \sup_M f$. By continuity, $f(\mathbf{x}) > \sup_M f$ in a neighborhood of $\mathbf{x}_0$. Therefore, $H(M)$ is closed.

If $\mathbf{x}_0, \mathbf{y}_0$ are two points in $H(M)$, then they are contained in every convex half-space $E = \{\mathbf{x} : f(\mathbf{x}) < \sup_M f\}$, and also the closed line segment from $\mathbf{x}_0$ to $\mathbf{y}_0$ is contained in each of these half-spaces. This shows that $H(M)$ is convex.

(3) We have to show that $H(H(M)) \subset H(M)$. If $\mathbf{x} \in H(H(M))$ is an arbitrary point and f an element of $\mathscr{L}$, then $f(\mathbf{x}) \le \sup_{H(M)} f \le \sup_M f$, by the definition of $H(M)$.

(4) is trivial.

(5) Let M be closed and convex. If $\mathbf{x}_0 \notin M$, then there is a point $\mathbf{y}_0 \in M$ such that $\operatorname{dist}(\mathbf{x}_0, M) = \operatorname{dist}(\mathbf{x}_0, \mathbf{y}_0)$ (because M is closed). Let $\mathbf{z}_0$ be a point in the open line segment from $\mathbf{x}_0$ to $\mathbf{y}_0$. Then $\mathbf{z}_0 \notin M$, and there is a function $f \in \mathscr{L}$ with $f(\mathbf{z}_0) = 0$ and $f|_M < 0$. Since $t \mapsto f(t\mathbf{x}_0 + (1-t)\mathbf{y}_0)$ is a monotone function, it follows that $f(\mathbf{x}_0) > 0$ and therefore $\mathbf{x}_0 \notin H(M)$.

(6) If M is bounded, there is an $R > 0$ such that M is contained in the closed convex set $\overline{B_R(\mathbf{0})}$. Thus $H(M) \subset \overline{B_R(\mathbf{0})}$. ∎

Remark. $H(M)$ is the smallest closed convex set that contains M.

5.2 Theorem. *A domain $G \subset \mathbb{R}^N$ is convex if and only if $K \subset\subset G$ implies that $H(K) \subset\subset G$.*

PROOF: Let G be a convex domain, and $M \subset\subset G$ a subset. Then $H(M)$ is closed and contained in the bounded set $H(\overline{M})$. Therefore, $H(M)$ is compact, and it remains to show that $H(M) \subset G$. If there is a point $\mathbf{x}_0 \in H(M) - G$, then there is a function $f \in \mathscr{L}$ with $f(\mathbf{x}_0) = 0$ and $f|_G < 0$. It follows that $\sup_{\overline{M}} f < 0$, and $f(\mathbf{x}_0) > \sup_M f$. This is a contradiction to $\mathbf{x}_0 \in H(M)$.

On the other hand, let the criterion be fulfilled. If $\mathbf{x}_0, \mathbf{y}_0$ are two points of G, then $K := \{\mathbf{x}_0, \mathbf{y}_0\}$ is a relatively compact subset of G. It follows that $H(K)$ is contained in G. Since $H(K)$ is closed and convex, the closed line segment from $\mathbf{x}_0$ to $\mathbf{y}_0$ is also contained in G. Therefore G is convex. ∎

Holomorphic Convexity. Now we replace affine linear functions by holomorphic functions.

Definition. Let $G \subset \mathbb{C}^n$ be a domain and $K \subset G$ a subset. The set

$$\widehat{K} = \widehat{K}_G := \Big\{\mathbf{z} \in G \,:\, |f(\mathbf{z})| \le \sup_K |f|, \text{ for all } f \in \mathcal{O}(G)\Big\}$$

is called the *holomorphically convex hull* of K in G.

5.3 Proposition. *Let $G \subset \mathbb{C}^n$ be a domain, and K, K_1, K_2 subsets of G. Then*

1. *$K \subset \widehat{K}$.*
2. *$\widehat{K}$ is closed in G.*
3. *$\widehat{\widehat{K}} = \widehat{K}$.*
4. *If $K_1 \subset K_2$, then $\widehat{K_1} \subset \widehat{K_2}$.*
5. *If K is bounded, then $\widehat{K}$ is also bounded.*

PROOF: (1) is trivial.

(2) Let $\mathbf{z}_0$ be a point of $G - \widehat{K}$. Then there exists a holomorphic function f on G with $|f(\mathbf{z}_0)| > \sup_K |f|$. By continuity, this inequality holds on an entire neighborhood $U = U(\mathbf{z}_0) \subset G$. So $G - \widehat{K}$ is open.

(3) $\sup_{\widehat{K}} |f| \le \sup_K |f|$.

(4) is trivial.

(5) If K is bounded, it is contained in a closed polydisk $\overline{\mathsf{P}^n(\mathbf{0}, r)}$. The coordinate functions z_ν are holomorphic in G. For $\mathbf{z} \in \widehat{K}$ we have $|z_\nu| \le \sup_K |z_\nu| \le r$. Hence $\widehat{K}$ is also bounded. ∎

Definition. A domain $G \subset \mathbb{C}^n$ is called *holomorphically convex* if $K \subset\subset G$ implies that $\widehat{K} \subset\subset G$.

Example

In $\mathbb{C}$ every domain is holomorphically convex:

Let $K \subset\subset G$ be an arbitrary subset. Then $\widehat{K}$ is bounded, and it remains to show that the closure of $\widehat{K}$ is contained in G. If there is a point $z_0 \in \overline{\widehat{K}} - G$, then z_0 lies in $\partial\widehat{K} \cap \partial G$. We consider the holomorphic function $f(z) := 1/(z - z_0)$ in G. If (z_ν) is a sequence in $\widehat{K}$ converging to z_0, then $|f(z_\nu)| \le \sup_K|f| \le \sup_{\overline{K}}|f| < \infty$. This is a contradiction. For $n \ge 2$, we will show that there are domains that are not holomorphically convex. But we have the following result.

5.4 Proposition. *If $G \subset \mathbb{C}^n$ is an affine convex domain, then it is holomorphically convex.*

PROOF: Let K be relatively compact in G. Then $H(K) \subset\subset G$. If $\mathbf{z}_0$ is a point of $G - H(K)$, then there exists an affine linear function $\lambda \in \mathcal{L}$ with $\lambda(\mathbf{z}_0) > \sup_K \lambda$. Replacing λ by $\lambda - \lambda(\mathbf{0})$ we may assume that λ is a homogeneous linear function of the form

$$\lambda(\mathbf{z}) = 2\,\mathrm{Re}(\alpha_1 z_1 + \cdots + \alpha_n z_n).$$

Then $f(\mathbf{z}) := \exp(2 \cdot (\alpha_1 z_1 + \cdots + \alpha_n z_n))$ is holomorphic in G, and $|f(\mathbf{z})| = \exp(\lambda(\mathbf{z}))$. Therefore, $|f(\mathbf{z}_0)| > \sup_K|f|$, and $\mathbf{z}_0 \in G - \widehat{K}$. This proves $\widehat{K} \subset\subset G$. ∎

In general, holomorphic convexity is a much weaker property than affine convexity.

The Cartan–Thullen Theorem. Let $G \subset \mathbb{C}^n$ be a domain, and $\varepsilon > 0$ a small real number. We define

$$G_\varepsilon := \{\mathbf{z} \in G \,:\, \delta_G(\mathbf{z}) \ge \varepsilon\}.$$

Here are some properties of the set G_ε:

1. If $\mathbf{z} \in G$, then there is an $\varepsilon > 0$ such that $\delta_G(\mathbf{z}) \ge \varepsilon$. Therefore, $G = \bigcup_{\varepsilon>0} G_\varepsilon$.
2. If $\varepsilon_1 \le \varepsilon_2$, then $G_{\varepsilon_1} \supset G_{\varepsilon_2}$.
3. G_ε is a closed subset of $\mathbb{C}^n$. In fact, if $\mathbf{z}_0 \in \mathbb{C}^n - G_\varepsilon$, then $\delta_G(\mathbf{z}_0) < \varepsilon$ or $\mathbf{z}_0 \notin G$. In the latter case, the ball $\mathsf{B}_\varepsilon(\mathbf{z}_0)$ is contained in $\mathbb{C}^n - G_\varepsilon$. If $\mathbf{z}_0 \in G - G_\varepsilon$ and $\delta := \delta_G(\mathbf{z}_0)$, then $\mathsf{B}_{\varepsilon-\delta}(\mathbf{z}_0) \subset \mathbb{C}^n - G_\varepsilon$. So $\mathbb{C}^n - G_\varepsilon$ is open.

5.5 Lemma. *Let $G \subset \mathbb{C}^n$ be a domain, $K \subset G$ a compact subset, and f a holomorphic function in G. If $K \subset G_\varepsilon$, then for any δ with $0 < \delta < \varepsilon$ there exists a constant $C > 0$ such that the following inequality holds:*

$$\sup_K |D^\alpha f(\mathbf{z})| \le \frac{\alpha!}{\delta^{|\alpha|}} \cdot C.$$

PROOF: For $0 < \delta < \varepsilon$, $G' := \{\mathbf{z} \in G : \operatorname{dist}(K, \mathbf{z}) < \delta\}$ is open and relatively compact in G, and for any $\mathbf{z} \in K$ the closed polydisk $\overline{\mathsf{P}^n(\mathbf{z}, \delta)}$ is contained in $\overline{G'} \subset G$. If T is the distinguished boundary of the polydisk and $|f| \le C$ on $\overline{G'}$, then the Cauchy inequalities yield

$$|D^\alpha f(\mathbf{z})| \le \frac{\alpha!}{\delta^{|\alpha|}} \cdot \sup_T |f| \le \frac{\alpha!}{\delta^{|\alpha|}} \cdot C.$$

■

5.6 Theorem (Cartan–Thullen). *If G is a weak domain of holomorphy, then G is holomorphically convex.*

PROOF: Let $K \subset\subset G$. We want to show that $\widehat{K} \subset\subset G$. Let $\varepsilon := \operatorname{dist}(K, \mathbb{C}^n - G) \ge \operatorname{dist}(\overline{K}, \mathbb{C}^n - G) > 0$. Clearly, K lies in G_ε.

We assert that the holomorphically convex hull $\widehat{K}$ lies even in G_ε. Suppose this is not so. Then there is a $\mathbf{z}_0 \in \widehat{K} - G_\varepsilon$. Now let f be any holomorphic function in G. In a neighborhood $U = U(\mathbf{z}_0) \subset G$, f has a Taylor expansion

$$f(\mathbf{z}) = \sum_{\nu \ge 0} a_\nu (\mathbf{z} - \mathbf{z}_0)^\nu, \text{ with } a_\nu = \frac{1}{\nu!} D^\nu f(\mathbf{z}_0).$$

The function $\mathbf{z} \mapsto a_\nu(\mathbf{z}) := \frac{1}{\nu!} D^\nu f(\mathbf{z})$ is holomorphic in G. Therefore, $|a_\nu(\mathbf{z}_0)| \le \sup_K |a_\nu(\mathbf{z})|$. By the lemma, for any δ with $0 < \delta < \varepsilon$ there exists a $C > 0$ such that $\sup_K |a_\nu(\mathbf{z})| \le C/\delta^{|\nu|}$, and then

$$|a_\nu(\mathbf{z} - \mathbf{z}_0)^\nu| \le C \cdot \left(\frac{|z_1 - z_1^{(0)}|}{\delta} \right)^{\nu_1} \cdots \left(\frac{|z_n - z_n^{(0)}|}{\delta} \right)^{\nu_n}.$$

On any polydisk $\mathsf{P}^n(\mathbf{z}_0, \delta)$ the Taylor series is dominated by a geometric series. Therefore, it converges on $P = \mathsf{P}^n(\mathbf{z}_0, \varepsilon)$ to a holomorphic function $\widehat{f}$. We have $f = \widehat{f}$ near $\mathbf{z}_0$, and then on the connected component Q of $\mathbf{z}_0$ in $P \cap G$. Since P meets G and $\mathbb{C}^n - G$, it follows from Lemma 1..9 that there is a point $\mathbf{z}_1 \in P \cap \partial Q \cap \partial G$. Then f cannot be completely singular at $\mathbf{z}_1$. This is a contradiction, because f is an arbitrary holomorphic function in G, and G is a weak domain of holomorphy. ■

Exercises

1. Let $G_1 \subset G_2 \subset \mathbb{C}^n$ be domains. Assume that for every $f \in \mathcal{O}(G_1)$ there is a sequence of functions $f_\nu \in \mathcal{O}(G_2)$ converging compactly on G_1 to f. Show that for every compact set $K \subset G_1$ it follows that $\widehat{K}_{G_2} \cap G_1 = \widehat{K}_{G_1}$.

2. Let $\mathbf{F} : G_1 \to G_2$ be a proper holomorphic map between domains in $\mathbb{C}^n$, respectively $\mathbb{C}^m$. Show that if G_2 is holomorphically convex, then so is G_1.
3. Let $G \subset \mathbb{C}^n$ be a domain and $S \subset G$ be a closed analytic disk with boundary bS. Show that $S \subset \widehat{(bS)}_G$.
4. Define the domain $G \subset \mathbb{C}^2$ by $G := \mathsf{P}^2(\mathbf{0}, 1) - \overline{\mathsf{P}^2(\mathbf{0}, 1/2)}$. Construct the holomorphically convex hull $\widehat{K}_G$ for $K := \{(z_1, z_2) : z_1 = 0 \text{ and } |z_2| = 3/4\}$. Is $\widehat{K}_G$ a relatively compact subset of G?
5. Let $\mathcal{F}$ be a family of functions in the domain G. For a compact subset $K \subset G$ we define

$$\widehat{K}_{\mathcal{F}} := \left\{ \mathbf{z} \in G : |f(\mathbf{z})| \le \sup_K |f| \text{ for all } f \in \mathcal{F} \right\}.$$

   The domain G is called *convex with respect to* $\mathcal{F}$, provided that $\widehat{K}_{\mathcal{F}}$ is relatively compact in G whenever K is. Prove:
   (a) Every bounded domain is convex with respect to the family $\mathscr{C}^0(G)$ of all continuous functions.
   (b) The unit ball $\mathsf{B} = \mathsf{B}_1(\mathbf{0})$ is convex with respect to the family of holomorphic functions $z_\nu^k \cdot z_\mu^l$ with $\nu, \mu = 1, \ldots, n$ and $k, l \in \mathbb{N}_0$.

6. Singular Functions

Normal Exhaustions. Let $G \subset \mathbb{C}^n$ be a domain. If G is holomorphically convex, we want to construct a holomorphic function in G that is completely singular at every boundary point. For that we use "normal exhaustions."

Definition. A *normal exhaustion* of G is a sequence (K_ν) of compact subsets of G such that:
1. $K_\nu \subset\subset (K_{\nu+1})^\circ$, for every ν.
2. $\bigcup_{\nu=1}^\infty K_\nu = G$.

6.1 Theorem. *Any domain G in $\mathbb{C}^n$ admits a normal exhaustion. If G is holomorphically convex, then there is a normal exhaustion (K_ν) with $\widehat{K}_\nu = K_\nu$ for every ν.*

PROOF: In the general case, $K_\nu := \overline{\mathsf{P}^n(\mathbf{0}, \nu)} \cap G_{1/\nu}$ gives a normal exhaustion. If G is holomorphically convex, $\widehat{K}_\nu \subset\subset G$ for every ν. We construct a new exhaustion by induction.

Let $K_1^* := \widehat{K}_1$. Suppose that compact sets $K_1^*, \ldots, K_{\nu-1}^*$ have been constructed, with $\widehat{K}_j^* = K_j^*$ for $j = 1, \ldots, \nu - 1$, and $K_j^* \subset\subset (K_{j+1}^*)^\circ$. Then there exists a $\lambda(\nu) \in \mathbb{N}$ such that $K_{\nu-1}^* \subset (K_{\lambda(\nu)})^\circ$. Let $K_\nu^* := \widehat{K}_{\lambda(\nu)}$.

It is clear that (K_ν^*) is a normal exhaustion with $\widehat{K}_\nu^* = K_\nu^*$. ∎

Unbounded Holomorphic Functions. Again let $G \subset \mathbb{C}^n$ be a domain.

6.2 Theorem. *Let (K_ν) be a normal exhaustion of G with $\widehat{K}_\nu = K_\nu$, $\lambda(\mu)$ a strictly monotonic increasing sequence of natural numbers, and $(\mathbf{z}_\mu)$ a sequence of points with $\mathbf{z}_\mu \in K_{\lambda(\mu)+1} - K_{\lambda(\mu)}$.*

Then there exists a holomorphic function f in G such that $|f(\mathbf{z}_\mu)|$ is unbounded.

PROOF: The function f is constructed as the limit function of an infinite series $f = \sum_{\mu=1}^\infty f_\mu$. By induction we define holomorphic functions f_μ in G such that:

1. $|f_\mu|_{K_{\lambda(\mu)}} < 2^{-\mu}$ for $\mu \geq 1$.
2. $|f_\mu(\mathbf{z}_\mu)| > \mu + 1 + \sum_{j=1}^{\mu-1} |f_j(\mathbf{z}_\mu)|$ for $\mu \geq 2$.

Let $f_1 := 0$. Now for $\mu \geq 2$ suppose that $f_1, \ldots, f_{\mu-1}$ have been constructed. Since $\mathbf{z}_\mu \in K_{\lambda(\mu)+1} - K_{\lambda(\mu)}$ and $\widehat{K}_{\lambda(\mu)} = K_{\lambda(\mu)}$, there exists a function g holomorphic in G such that $|g(\mathbf{z}_\mu)| > q := \sup_{K_{\lambda(\mu)}} |g|$. By multiplication by a suitable constant we can make

$$|g(\mathbf{z}_\mu)| > 1 > q.$$

If we set $f_\mu := g^k$ with a sufficiently large k, then f_μ has the properties (1) and (2).

We assert that $\sum_\mu f_\mu$ converges compactly in G. To prove this, first note that for $K \subset G$ an arbitrary compact subset, there is a $\mu_0 \in \mathbb{N}$ such that $K \subset K_{\lambda(\mu_0)}$. By construction $\sup_K |f_\mu| < 2^{-\mu}$ for $\mu \geq \mu_0$. Since the geometric series $\sum_\mu 2^{-\mu}$ dominates $\sum_\mu f_\mu$ in K, the series of the f_μ is normally convergent on K. This shows that $f = \sum_\mu f_\mu$ is holomorphic in G. Moreover,

$$\begin{aligned}
|f(\mathbf{z}_\mu)| &\geq |f_\mu(\mathbf{z}_\mu)| - \sum_{\nu \neq \mu} |f_\nu(\mathbf{z}_\mu)| \\
&> \mu + 1 - \sum_{\nu > \mu} |f_\nu(\mathbf{z}_\mu)| \\
&> \mu + 1 - \sum_{\nu > \mu} 2^{-\nu} \quad (\text{since } \mathbf{z}_\mu \in K_{\lambda(\nu)} \text{ for } \nu > \mu) \\
&\geq \mu \quad (\text{since } \sum_{\nu \geq 1} 2^{-\nu} = 1).
\end{aligned}$$

It follows that $|f(\mathbf{z}_\mu)| \to \infty$ for $\mu \to \infty$. ∎

The following is an important consequence:

6.3 Theorem. *A domain G is holomorphically convex if and only if for any infinite set D that is discrete in G there exists a function f holomorphic in G such that $|f|$ is unbounded on D.*

PROOF: (1) Let G be holomorphically convex, $D \subset G$ infinite and discrete. Moreover, let (K_ν) be a normal exhaustion of G with $\widehat{K}_\nu = K_\nu$. Then $K_\nu \cap D$ is finite (or empty) for every $\nu \in \mathbb{N}$. We construct a sequence of points $\mathbf{z}_\mu \in D$ by induction.

Let $\mathbf{z}_1 \in D - K_1$ be arbitrary, and $\lambda(1) \in \mathbb{N}$ minimal with the property that $\mathbf{z}_1$ lies in $K_{\lambda(1)+1}$. Now suppose the points $\mathbf{z}_1, \ldots, \mathbf{z}_{\mu-1}$ and the numbers $\lambda(1), \ldots, \lambda(\mu-1)$ have been constructed such that

$$\mathbf{z}_\nu \in K_{\lambda(\nu)+1} - K_{\lambda(\nu)}, \text{ for } \nu = 1, \ldots, \mu - 1.$$

Then we choose $\mathbf{z}_\mu \in D - K_{\lambda(\mu-1)+1}$ and $\lambda(\mu)$ minimal with the property that $\mathbf{z}_\mu$ lies in $K_{\lambda(\mu)+1}$. By the theorem above there is a holomorphic function f in G such that $|f(\mathbf{z}_\mu)| \to \infty$ for $\mu \to \infty$. Therefore, $|f|$ is unbounded on D.

(2) Now suppose that the criterion is satisfied, and $K \subset\subset G$. Then $\widehat{K} \subset G$, and we have to show that $\widehat{K}$ is compact. Let $(\mathbf{z}_\nu)$ be any sequence of points in $\widehat{K}$. Then

$$\sup\{|f(\mathbf{z}_\nu)| \, : \, \nu \in \mathbb{N}\} \le \sup_K |f| < \infty, \text{ for every } f \in \mathcal{O}(G).$$

Therefore, $\{\mathbf{z}_\nu \, : \, \nu \in \mathbb{N}\}$ cannot be discrete in G. Thus the sequence $(\mathbf{z}_\nu)$ has a cluster point $\mathbf{z}_0$ in G. Since $\widehat{K}$ is closed, $\mathbf{z}_0$ belongs to $\widehat{K}$. So G is holomorphically convex. ∎

Sequences. For a domain $G \subset \mathbb{C}^n$ we wish to construct a sequence that accumulates at every point of its boundary.

6.4 Theorem. *Let (K_ν) be a normal exhaustion of G. Then there exists a strictly monotonic increasing sequence $\lambda(\mu)$ of natural numbers and a sequence $(\mathbf{z}_\mu)$ of points in G such that:*

1. *$\mathbf{z}_\mu \in K_{\lambda(\mu)+1} - K_{\lambda(\mu)}$, for every μ.*
2. *If $\mathbf{z}_0$ is a boundary point of G and $U = U(\mathbf{z}_0)$ an open connected neighborhood, then every connected component of $U \cap G$ contains infinitely many points of the sequence $(\mathbf{z}_\mu)$.*

PROOF: This is a purely topological result, since we make no assumption about G. The proof is carried out in several steps.

(1) Let $\mathcal{B} = \{B_\nu : \nu \in \mathbb{N}\}$ be the countable system of balls with rational center and rational radius meeting ∂G. Every intersection $B_\nu \cap G$ has at most countably many connected components. Thus we obtain a countable family

$$\mathcal{C} = \{C_\mu : \exists \nu \in \mathbb{N} \text{ such that } C_\mu \text{ is a connected component of } B_\nu \in \mathcal{B}\}.$$

(2) By induction, the sequences $\lambda(\mu)$ and $(\mathbf{z}_\mu)$ are constructed. Let $\mathbf{z}_1$ be arbitrary in $C_1 - K_1$. Then there is a unique number $\lambda(1)$ such that $\mathbf{z}_1 \in K_{\lambda(1)+1} - K_{\lambda(1)}$.

Now suppose $\mathbf{z}_1, \dots, \mathbf{z}_{\mu-1}$ have been constructed such that

$$\mathbf{z}_j \in C_j \cap (K_{\lambda(j)+1} - K_{\lambda(j)}), \text{ for } j = 1, \dots, \mu - 1.$$

We choose $\mathbf{z}_\mu \in C_\mu - K_{\lambda(\mu-1)+1}$ and $\lambda(\mu)$ as usual. That is possible, since there is a point $\mathbf{w} \in B_{\nu(\mu)} \cap \partial C_\mu \cap \partial G$ if C_μ is a connected component of $B_{\nu(\mu)} \cap G$. Then $\mathbb{C}^n - K_{\lambda(\mu-1)+1}$ is an open neighborhood of $\mathbf{w}$ and contains points of C_μ.

(3) Now we show that property (2) of the theorem is satisfied. Let $\mathbf{z}_0$ be a point of ∂G, $U = U(\mathbf{z}_0)$ an open connected neighborhood, and Q a connected component of $U \cap G$. We assume that only finitely many $\mathbf{z}_\mu$ lie in Q, say $\mathbf{z}_1, \dots, \mathbf{z}_m$. Then

$$U^* := U - \{\mathbf{z}_1, \dots, \mathbf{z}_m\} \quad \text{and} \quad Q^* := Q - \{\mathbf{z}_1, \dots, \mathbf{z}_m\}$$

are open connected sets that contain no $\mathbf{z}_\mu$. Obviously, Q^* is a connected component of $G \cap U^*$.

There is a point $\mathbf{w}_0$ in $U^* \cap \partial Q^* \cap \partial G$, and a ball $B_\nu \subset U^*$ with $\mathbf{w}_0 \in B_\nu$. Then $B_\nu \cap G \subset U^* \cap G$. Moreover, $B_\nu \cap G$ must contain a point $\mathbf{w}_1 \in Q^*$. The connected component C^* of $\mathbf{w}_1$ in $B_\nu \cap G$ is a subset of the connected component of $\mathbf{w}_1$ in $U^* \cap G$. But C^* is an element C_{μ_0} of $\mathcal{C}$. By construction it contains the point $\mathbf{z}_{\mu_0}$. That is a contradiction. Infinitely many members of the sequence belong to Q. ∎

6.5 Theorem. *If G is holomorphically convex, then it is a domain of holomorphy.*

PROOF: Let (K_ν) be a normal exhaustion of G with $\widehat{K}_\nu = K_\nu$ and choose sequences $\lambda(\mu) \in \mathbb{N}$ and $(\mathbf{z}_\mu)$ in G such that $\mathbf{z}_\mu \in K_{\lambda(\mu)+1} - K_{\lambda(\mu)}$. We may assume that for every point $\mathbf{z}_0 \in \partial G$, every open connected neighborhood $U = U(\mathbf{z}_0)$, and every connected component Q of $U \cap G$ there are infinitely many $\mathbf{z}_\mu$ in Q.

Now let f be holomorphic in G and unbounded on $D := \{\mathbf{z}_\mu : \mu \in \mathbb{N}\}$. It is clear that f is completely singular at every point $\mathbf{z}_0 \in \partial G$. ∎

Remark. It is not necessary that a completely singular holomorphic function is unbounded. In 1978, D. Catlin showed in his dissertation that if $G \subset\subset \mathbb{C}^n$ is a holomorphically convex domain with smooth boundary, then there exists a function holomorphic in G and smooth in a neighborhood of $\overline{G}$ that is completely singular at every point of the boundary of G.

Exercises

1. A domain $G \subset\subset \mathbb{C}^n$ is holomorphically convex if and only if for every $\mathbf{z} \in \partial G$ there is a neighborhood $U(\mathbf{z})$ such that $U \cap G$ is a domain of holomorphy.
2. Let $G_1 \subset \mathbb{C}^n$ and $G_2 \subset \mathbb{C}^m$ be domains of holomorphy. If $\mathbf{f} : G_1 \to \mathbb{C}^m$ is a holomorphic mapping, then $f^{-1}(G_2) \cap G_1$ is a domain of holomorphy.
3. Find a bounded holomorphic function on the unit disk D that is singular at every boundary point.

7. Examples and Applications

Domains of Holomorphy

7.1 Proposition. *Every domain in the complex plane $\mathbb{C}$ is a domain of holomorphy.*

PROOF: We have already shown that every domain in $\mathbb{C}$ is holomorphically convex. Therefore, such a domain is also a domain of holomorphy. ∎

7.2 Theorem. *The following statements about domains $G \in \mathbb{C}^n$ are equivalent:*

1. *G is a weak domain of holomorphy.*
2. *G is holomorphically convex.*
3. *For every infinite discrete subset $D \subset G$ there exists a holomorphic function f in G such that $|f|$ is unbounded on D.*
4. *G is a domain of holomorphy.*

The equivalences have all been proved in the preceding paragraphs. Furthermore, we know that every domain of holomorphy is pseudoconvex. Still missing here is the proof of the Levi problem: Every pseudoconvex domain is holomorphically convex. We say more about this in Chapter V.

Every affine convex open subset of $\mathbb{C}^n$ is a domain of holomorphy. The n-fold Cartesian product of plane domains is a further example.

7.3 Proposition. *If $G_1, \ldots, G_n \subset \mathbb{C}$ are arbitrary domains, then $G := G_1 \times \cdots \times G_n$ is a domain of holomorphy.*

Proof: Let $D = \{\mathbf{z}_\mu = (z_1^\mu, \dots, z_n^\mu) \,:\, \mu \in \mathbb{N}\}$ be an infinite discrete subset of G. Then there is an i such that (z_i^μ) has no cluster point in G_i, and there is a holomorphic function f in G_i with $\lim_{\mu\to\infty}\big|\, f(z_i^\mu)\,\big| = \infty$. The function $\widehat{f}$ in G, defined by $\widehat{f}(z_1, \dots, z_n) := f(z_i)$, is holomorphic in G and unbounded on D. ∎

Remark. The same proof shows that every Cartesian product of domains of holomorphy is again a domain of holomorphy.

Complete Reinhardt Domains. Let $G \subset \mathbb{C}^n$ be a complete Reinhardt domain (see Section I.1). We will give criteria for G to be a domain of holomorphy. For that purpose we define a map log from the absolute value space $\mathscr{V}$ to $\mathbb{R}^n$ by

$$\log(r_1, \dots, r_n) := (\log r_1, \dots, \log r_n).$$

Definition. A Reinhardt domain G is called *logarithmically convex* if $\log \tau(G \cap (\mathbb{C}^*)^n)$ is an affine convex domain in $\mathbb{R}^n$.

Remark. For $\mathbf{z} = (z_1, \dots, z_n) \in G$ we have $\log \tau(\mathbf{z}) = (\log|z_1|, \dots, \log|z_n|)$. If $\mathbf{z} \in (\mathbb{C}^*)^n$, then $|z_i| > 0$ for each i, and $\log \tau(\mathbf{z})$ is in fact an element of $\mathbb{R}^n$.

7.4 Proposition. *The domain of convergence of a power series* $S(\mathbf{z}) = \sum_{\nu\geq 0} a_\nu \mathbf{z}^\nu$ *is logarithmically convex.*

Proof: Let G be the domain of convergence of $S(\mathbf{z})$, and $M := \log \tau(G \cap (\mathbb{C}^*)^n) \subset \mathbb{R}^n$. We consider two points $\mathbf{x}, \mathbf{y} \in M$ and points $\mathbf{z}, \mathbf{w} \in G \cap (\mathbb{C}^*)^n$ with $\log \tau(\mathbf{z}) = \mathbf{x}$ and $\log \tau(\mathbf{w}) = \mathbf{y}$. If $\lambda > 1$ is small enough, $\lambda\mathbf{z}$ and $\lambda\mathbf{w}$ still belong to $G \cap (\mathbb{C}^*)^n$. Since $S(\mathbf{z})$ is convergent in $\lambda\mathbf{z}, \lambda\mathbf{w}$, there is a constant $C > 0$ such that

$$|a_\nu| \cdot \lambda^{|\nu|} \cdot |\mathbf{z}^\nu| \leq C \quad \text{and} \quad |a_\nu| \cdot \lambda^{|\nu|} \cdot |\mathbf{w}^\nu| \leq C, \text{ for every } \nu \in \mathbb{N}_0^n.$$

Thus

$$|a_\nu| \cdot \lambda^{|\nu|} \cdot |\mathbf{z}^\nu|^t \cdot |\mathbf{w}^\nu|^{1-t} \leq C, \text{ for every } \nu \text{ and } 0 \leq t \leq 1.$$

It follows from Abel's lemma that $S(\mathbf{z})$ is convergent in a neighborhood of

$$\mathbf{z}_t := \big(|z_1|^t |w_1|^{1-t}, \dots, |z_n|^t |w_n|^{1-t}\big).$$

This means that $\mathbf{z}_t \in G$ and $t\mathbf{x} + (1-t)\mathbf{y} = \log \tau(\mathbf{z}_t) \in M$, for $0 \leq t \leq 1$. Therefore, M is convex. ∎

7.5 Proposition. *Let G be a complete Reinhardt domain. If G is logarithmically convex, then it is holomorphically convex.*

PROOF: Let K be a relatively compact subset of G. Since G is a complete Reinhardt domain and $\overline{K}$ a compact subset of G, there are points $\mathbf{z}_1, \dots, \mathbf{z}_k \in G \cap (\mathbb{C}^*)^n$ such that

$$K \subset G' := \bigcup_{i=1}^{k} \mathsf{P}^n(\mathbf{0}, \mathbf{q}_i) \subset G, \qquad \text{where } \mathbf{q}_i := \tau(\mathbf{z}_i).$$

We consider the set $\mathscr{M} = \{m(\mathbf{z}) = \mathbf{z}^\nu \,:\, \nu \in \mathbb{N}_0^n\}$ of monomials, which is a subset of $\mathcal{O}(G)$. For $\mathbf{z} \in \mathsf{P}^n(\mathbf{0}, \mathbf{q}_i)$ and $m \in \mathscr{M}$ we have

$$|m(\mathbf{z})| = |\mathbf{z}^\nu| < \mathbf{q}_i^\nu = |m(\mathbf{q}_i)|.$$

Let $Z := \{\mathbf{z}_1, \dots, \mathbf{z}_k\}$. Then for $\mathbf{z} \in \widehat{K}$ it follows that

$$|m(\mathbf{z})| \le \sup_K |m| \le \sup_{G'} |m| \le \sup_Z |m|, \text{ for every } m \in \mathscr{M}.$$

Suppose that $\widehat{K}$ is not relatively compact in G. Then $\widehat{K}$ has a cluster point $\mathbf{z}_0$ in ∂G, and it follows that $|m(\mathbf{z}_0)| \le \sup_Z |m|$, for every $m \in \mathscr{M}$.

Let $h(\mathbf{z}) := \log \tau(\mathbf{z})$, for $\mathbf{z} \in (\mathbb{C}^*)^n$. Since G is logarithmically convex, the domain $G_0 := h(G \cap (\mathbb{C}^*)^n) \subset \mathbb{R}^n$ is affine convex. For the time being we assume that $\mathbf{z}_0 \in (\mathbb{C}^*)^n$. Then $\mathbf{x}_0 := h(\mathbf{z}_0) \in \partial G_0$, and there is a real linear function $\lambda(\mathbf{x}) = a_1 x_1 + \cdots + a_n x_n$ such that $\lambda(\mathbf{x}) < \lambda(\mathbf{x}_0)$ for $\mathbf{x} \in G_0$.

Let $\mathbf{x} = \log \tau(\mathbf{z})$ be a point of G_0, and $\mathbf{u} \in \mathbb{R}^n$ with $u_j \le x_j$ for $j = 1, \dots, n$. Then $e^{u_j} \le e^{x_j} = |z_j|$, and therefore (since G is a complete Reinhardt domain) $\mathbf{w} = (e^{u_1}, \dots, e^{u_n}) \in G \cap (\mathbb{C}^*)^n$ and $\mathbf{u} \in G_0$. In particular,

$$\lambda(\mathbf{x}) - n a_j = \lambda(\mathbf{x} - n\mathbf{e}_j) < \lambda(\mathbf{x}_0), \text{ for every } n \in \mathbb{N}.$$

Therefore, $a_j \ge 0$ for $j = 1, \dots, n$.

Now we choose rational numbers $r_j > a_j$ and define $\widetilde{\lambda}(\mathbf{x}) := r_1 x_1 + \cdots + r_n x_n$. If we choose the r_j sufficiently close to a_j, the inequality $\widetilde{\lambda}(\mathbf{q}_i) < \widetilde{\lambda}(\mathbf{x}_0)$ holds for $i = 1, \dots, k$, and it still holds after multiplying by the common denominator of the r_j. Therefore, we may assume that the r_j are natural numbers, and we can define a special monomial m_0 by $m_0(\mathbf{z}) := z_1^{r_1} \cdots z_n^{r_n}$. Then

$$|m_0(\mathbf{z}_i)| = e^{\widetilde{\lambda}(\mathbf{q}_i)} < e^{\widetilde{\lambda}(\mathbf{x}_0)} = |m_0(\mathbf{z}_0)|, \text{ for } i = 1, \dots, k.$$

So $|m_0(\mathbf{z}_0)| > \sup_Z |m_0|$, and this is a contradiction.

If $\mathbf{z}_0 \notin (\mathbb{C}^*)^n$, then after a permutation of the coordinates we may assume that $z_1^{(0)} \cdots z_l^{(0)} \ne 0$ and $z_{l+1}^{(0)} = \cdots = z_n^{(0)} = 0$. We can project on the space

$\mathbb{C}^l$ and work with monomials in the variables $z_1, \dots, z_l$. Then the proof goes through as above. ∎

Now we get the following result:

7.6 Theorem. *Let $G \subset \mathbb{C}^n$ be a complete Reinhardt domain. Then the following statements are equivalent:*

1. *G is the domain of convergence of a power series.*
2. *G is logarithmically convex.*
3. *G is holomorphically convex.*
4. *G is a domain of holomorphy.*

PROOF: We have only to show that if G is a complete Reinhardt domain and a domain of holomorphy, then it is the domain of convergence of a power series. By hypothesis, there is a function f that is holomorphic in G and completely singular at every boundary point. In Section I.5 we proved that for every holomorphic function in a proper Reinhardt domain there is a power series $S(\mathbf{z})$ that converges in G to f. By the identity theorem it does not converge on any domain strictly larger than G. ∎

Analytic Polyhedra. Let $G \subset \mathbb{C}^n$ be a domain.

Definition. Let $U \subset G$, $V_1, \dots, V_k \subset \mathbb{C}$ open subsets, and $f_1, \dots, f_k$ holomorphic functions in G. The set

$$P := \{\mathbf{z} \in U \, : \, f_i(\mathbf{z}) \in V_i, \text{ for } i = 1, \dots, k\}$$

is called an *analytic polyhedron* in G if $P \subset\subset U$.

If, in addition, $V_1 = \cdots = V_k = \mathsf{D}$, then one speaks of a *special analytic polyhedron* in G.

Remark. An analytic polyhedron P need not be connected. The set U in the definition ensures that each union of connected components of P is also an analytic polyhedron if it has a positive distance from every other connected component of P.

7.7 Theorem. *Every connected analytic polyhedron P in G is a domain of holomorphy.*

PROOF: We have only to show that P is a weak domain of holomorphy. If $\mathbf{z}_0 \in \partial P$, then there is an i such that $f_i(\mathbf{z}_0) \in \partial V_i$. Therefore, $f(\mathbf{z}) := (f_i(\mathbf{z}) - f_i(\mathbf{z}_0))^{-1}$ is holomorphic in P and completely singular at $\mathbf{z}_0$. ∎

Example

Let $q < 1$ be a positive real number, and

$$P := \{\mathbf{z} = (z_1, z_2) \in \mathbb{C}^2 \,:\, |z_1| < 1,\, |z_2| < 1 \text{ and } |z_1 \cdot z_2| < q\}.$$

Then P (see Figure II.7) is clearly an analytic polyhedron, but neither affine

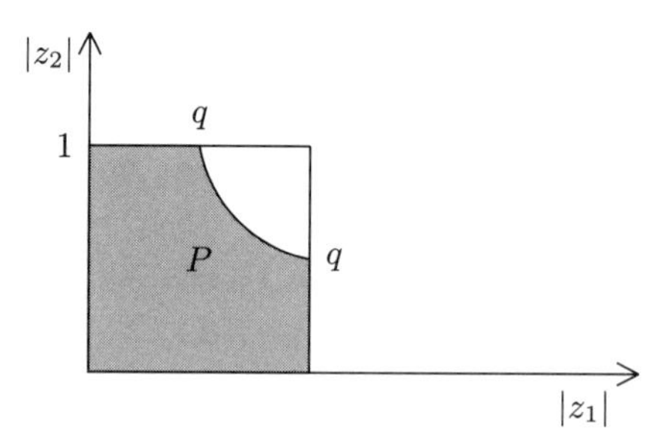

Figure II.7. An analytic polyhedron

convex nor a Cartesian product of domains. So the analytic polyhedra enrich our stock of examples of domains of holomorphy.

We will show that every domain of holomorphy is "almost" an analytic polyhedron.

7.8 Theorem. *If $G \subset \mathbb{C}^n$ is a domain of holomorphy, then there exists a sequence (P_ν) of special analytic polyhedra in G with $P_\nu \subset\subset P_{\nu+1}$ and $\bigcup_{\nu=1}^{\infty} P_\nu = G$.*

PROOF: Let (K_ν) be a normal exhaustion of G with $\widehat{K}_\nu = K_\nu$. If $\mathbf{z} \in \partial K_{\nu+1}$ is an arbitrary point, then $\mathbf{z}$ does not lie in $K_\nu \subset (K_{\nu+1})^\circ$, and therefore not in $\widehat{K}_\nu$. Hence there exists a function f holomorphic in G for which $q := \sup_{K_\nu} |f| < |f(\mathbf{z})|$. By multiplication by a suitable constant we obtain $q < 1 < |f(\mathbf{z})|$, and then there is an entire neighborhood $U = U(\mathbf{z})$ such that $|f| > 1$ on U.

Since the boundary $\partial K_{\nu+1}$ is compact, we can find finitely many open neighborhoods $U_{\nu,j}$ of $\mathbf{z}_{\nu,j} \in \partial K_{\nu+1}$, $j = 1, \ldots, k_\nu$, and corresponding functions $f_{\nu,j}$ holomorphic in G such that $|f_{\nu,j}| > 1$ on $U_{\nu,j}$, and $\partial K_{\nu+1} \subset \bigcup_{j=1}^{k_\nu} U_{\nu,j}$. We define

$$P_\nu := \{\mathbf{z} \in (K_{\nu+1})^\circ \,:\, |f_{\nu,j}(\mathbf{z})| < 1 \text{ for } j = 1, \ldots, k_\nu\}.$$

Clearly, $K_\nu \subset P_\nu \subset (K_{\nu+1})^\circ$. Furthermore, $M := K_{\nu+1} - (U_{\nu,1} \cup \cdots \cup U_{\nu,k_\nu})$ is a compact set with $P_\nu \subset M \subset (K_{\nu+1})^\circ$. Consequently, $P_\nu \subset\subset K_{\nu+1}$. Thus P_ν is a special analytic polyhedron in G. It follows trivially that the sequence (P_ν) exhausts the domain G. ∎

In the theory of Stein manifolds one proves the converse of this theorem.

Exercises

1. If R is a domain in the real number space $\mathbb{R}^n$, then

$$T_R = R + \mathrm{i}\mathbb{R}^n := \{\mathbf{z} \in \mathbb{C}^n : (\mathrm{Re}(z_1), \ldots, \mathrm{Re}(z_n)) \in R\}$$

   is called the *tube domain* associated with R. Prove that the following properties are equivalent:
   (a) R is convex.
   (b) T_R is (affine) convex.
   (c) T_R is holomorphically convex.
   (d) T_R is pseudoconvex.
   Hint: To show $(d) \Longrightarrow (a)$ choose $\mathbf{x}_0, \mathbf{y}_0 \in R$. Then the function $\varphi(\zeta) := -\ln \delta_{T_R}(\mathbf{x}_0 + \zeta(\mathbf{y}_0 - \mathbf{x}_0))$ is subharmonic in D. Since $\delta_{T_R}(\mathbf{x} + \mathrm{i}\mathbf{y}) = \delta_R(\mathbf{x})$, one concludes that $t \mapsto -\ln \delta_R(\mathbf{x}_0 + t(\mathbf{y}_0 - \mathbf{x}_0))$ assumes its maximum at $t = 0$ or $t = 1$.
2. Let $G \subset \mathbb{C}^n$ be a domain. A domain $\widehat{G} \subset \mathbb{C}^n$ is called the *envelope of holomorphy* of G if every holomorphic function f in G has a holomorphic extension to $\widehat{G}$. Prove:
   (a) If $R \subset \mathbb{R}^n$ is a domain and $H(R)$ its affine convex hull, then $\widehat{G} := H(R) + \mathrm{i}\mathbb{R}^n$ is the envelope of holomorphy of the tube domain $G = R + \mathrm{i}\mathbb{R}^n$.
   (b) If $G \subset \mathbb{C}^n$ is a Reinhardt domain and $\widehat{G}$ the smallest logarithmically convex complete Reinhardt domain containing G, then $\widehat{G}$ is the envelope of holomorphy of G. Hint: Use the convex hull of $\log \tau(G)$.
3. Construct the envelope of holomorphy of the domain

$$G_q := \mathsf{P}^2(\mathbf{0}, (1, q)) \cup \mathsf{P}^2(\mathbf{0}, (q, 1)).$$

4. A domain $G \subset \mathbb{C}^n$ is called a *Runge domain* if for every holomorphic function f in G there is a sequence (p_ν) of polynomials converging compactly in G to f.

   Prove that the Cartesian product of n simply connected subdomains of $\mathbb{C}$ is a Runge domain in $\mathbb{C}^n$.
5. A domain $G \subset \mathbb{C}^n$ is called *polynomially convex* if it is convex with respect to the family of all polynomials (cf. Exercise 5.5). Prove that every polynomially convex domain is a holomorphically convex Runge domain.

8. Riemann Domains over $\mathbb{C}^n$

Riemann Domains. It turns out that for general domains in $\mathbb{C}^n$ the envelope of holomorphy (cf. Exercise 7.2) cannot exist in $\mathbb{C}^n$. Therefore, we have to consider domains covering $\mathbb{C}^n$.

Definition. A *(Riemann) domain over* $\mathbb{C}^n$ is a pair (G, π) with the following properties:

1. G is a connected Hausdorff space.[4]
2. $\pi : G \to \mathbb{C}^n$ is a local homeomorphism (that is, for each point $x \in G$ and its "base point" $\mathbf{z} := \pi(x) \in \mathbb{C}^n$ there exist open neighborhoods $U = U(x) \subset X$ and $V = V(\mathbf{z}) \subset \mathbb{C}^n$ such that $\pi : U \to V$ is a homeomorphism).

Remarks

1. Let (G, π) be a Riemann domain. Then G is pathwise connected, and the map $\pi : G \to \mathbb{C}^n$ is continuous and open. The latter means that the images of open sets are again open.
2. If (G_ν, π_ν) are domains over $\mathbb{C}^n$ for $\nu = 1, \ldots, l$, and $x_\nu \in G_\nu$ are points over the same base point $\mathbf{z}_0$, then there are open neighborhoods $U_\nu = U_\nu(x_\nu) \subset G_\nu$ and a connected open neighborhood $V = V(\mathbf{z}_0) \subset \mathbb{C}^n$ such that $\pi_\nu|_{U_\nu} : U_\nu \to V$ is a homeomorphism for $\nu = 1, \ldots, l$.

Examples

1. If G is a domain in $\mathbb{C}^n$, then (G, id_G) is a Riemann domain.
2. The Riemann surface of $\sqrt{z}$ (without the branch point) is the set

$$G := \{(z, w) \in \mathbb{C}^* \times \mathbb{C} \,:\, w^2 = z\}.$$

If G is provided with the topology induced from $\mathbb{C}^* \times \mathbb{C}$, then it is a Hausdorff space. The mapping $\varphi : \mathbb{C}^* \to G$ defined by $\zeta \mapsto (\zeta^2, \zeta)$ is continuous and bijective. Therefore, G is connected. The mapping φ is called a *uniformization* of G.

Now let $\pi : G \to \mathbb{C}$ be defined by $\pi(z, w) := z$. Clearly, π is continuous. If $(z_0, w_0) \in G$ is an arbitrary point, then $z_0 \neq 0$, and we can find a simply connected neighborhood $V(\mathbf{z}_0) \subset \mathbb{C}^*$. Then there exists a holomorphic function f in V with $f^2(z) \equiv z$ and $f(z_0) = w_0$. We denote $f(z)$ by $\sqrt{z}$. The image $W := f(V)$ is open, and the set $\pi^{-1}(V)$ can be written as the union of two disjoint open sets

$$U_\pm := \{(z, \pm f(z)) \,:\, z \in V\} = (V \times (\pm W)) \cap G.$$

Let $\widehat{f}(z) := (z, f(z))$. Then $\widehat{f} : V \to G$ is continuous, and $\pi \circ \widehat{f}(z) \equiv z$. The open set $U := U_+$ is a neighborhood of (z_0, w_0), with $\widehat{f}(V) = U$ and $\widehat{f} \circ \pi(z, w) = (z, w)$ on U; that is, $\pi|_U : U \to V$ is topological. Hence (G, π) is a Riemann domain over $\mathbb{C}$.

---

[4] A general topological space X is said to be *connected* if it is not the union of two disjoint nonempty open sets. A space X is called *pathwise connected* if each two points of X can be joined by a continuous path in X. For open sets in $\mathbb{C}^n$ these two notions are equivalent.

The space G can be visualized in the following manner: We cover $\mathbb{C}$ with two additional copies of $\mathbb{C}$, cut both these "sheets" along the positive real axis, and paste them crosswise to one another (this is not possible in $\mathbb{R}^3$ without self intersection, but in higher dimensions, it is). This is illustrated in Figure II.8.

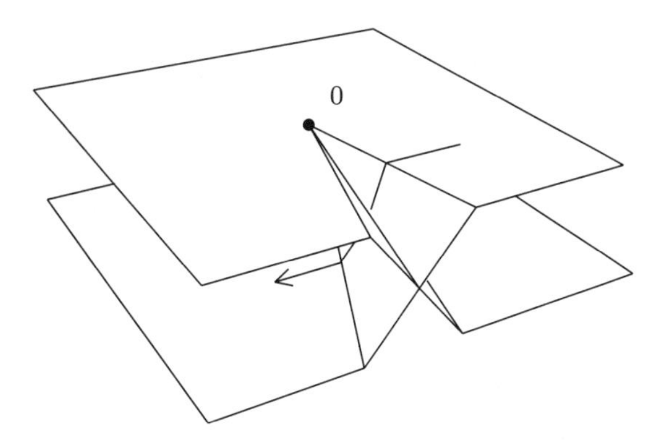

Figure II.8. The Riemann surface of $\sqrt{z}$

8.1 Proposition (on the uniqueness of lifting). *Let (G, π) be a domain over $\mathbb{C}^n$ and Y a connected topological space. Let $y_0 \in Y$ be a point and $\psi_1, \psi_2 : Y \to G$ continuous mappings with $\psi_1(y_0) = \psi_2(y_0)$ and $\pi \circ \psi_1 = \pi \circ \psi_2$. Then $\psi_1 = \psi_2$.*

PROOF: Let $M := \{y \in Y : \psi_1(y) = \psi_2(y)\}$. By assumption, $y_0 \in M$, so $M \neq \varnothing$. Since G is a Hausdorff space, it follows immediately that M is closed. Now let $y \in Y$ be chosen arbitrarily, and set $x := \psi_1(y) = \psi_2(y)$ and $\mathbf{z} := \pi(x)$. There are open neighborhoods $U = U(x) \subset G$ and $V = V(\mathbf{z}) \subset \mathbb{C}^n$ such that $\pi : U \to V$ is topological, and there is an open neigborhood $W = W(y)$ with $\psi_\lambda(W) \subset U$ for $\lambda = 1, 2$. Then

$$\psi_1|_W = (\pi|_U)^{-1} \circ \pi \circ \psi_1|_W = (\pi|_U)^{-1} \circ \pi \circ \psi_2|_W = \psi_2|_W,$$

and therefore $W \subset M$. Hence M is open, and since Y is connected, it follows that $M = Y$.

∎

Definition. Let $\mathbf{z}_0 \in \mathbb{C}^n$ be fixed. A *(Riemann) domain over $\mathbb{C}^n$ with distinguished point* is a triple $\mathcal{G} = (G, \pi, x_0)$ for which:

1. (G, π) is a domain over $\mathbb{C}^n$.
2. x_0 is a point of G with $\pi(x_0) := \mathbf{z}_0$.

Definition. Let $\mathcal{G}_j = (G_j, \pi_j, x_j)$ be domains over $\mathbb{C}^n$ with distinguished point. We say that $\mathcal{G}_1$ *is contained in* $\mathcal{G}_2$ (denoted by $\mathcal{G}_1 \prec \mathcal{G}_2$) if there is a continuous map $\varphi : G_1 \to G_2$ with the following properties:

1. $\pi_2 \circ \varphi = \pi_1$ (called "φ preserves fibers").
2. $\varphi(x_1) = x_2$.

8.2 Proposition. *If $\mathcal{G}_1 \prec \mathcal{G}_2$, then the fiber preserving map $\varphi : G_1 \to G_2$ with $\varphi(x_1) = x_2$ is uniquely determined.*

This follows immediately from the uniqueness of lifting.

8.3 Proposition. *The relation "$\prec$" is a weak ordering; that is:*

1. $\mathcal{G} \prec \mathcal{G}$.
2. $\mathcal{G}_1 \prec \mathcal{G}_2$ *and* $\mathcal{G}_2 \prec \mathcal{G}_3 \implies \mathcal{G}_1 \prec \mathcal{G}_3$.

The proof is trivial.

Definition. Two domains $\mathcal{G}_1, \mathcal{G}_2$ over $\mathbb{C}^n$ with fundamental point are called *isomorphic* or *equivalent* (symbolically $\mathcal{G}_1 \cong \mathcal{G}_2$) if $\mathcal{G}_1 \prec \mathcal{G}_2$ and $\mathcal{G}_2 \prec \mathcal{G}_1$.

8.4 Proposition. *Two domains $\mathcal{G}_j = (G_j, \pi_j, x_j)$, $j = 1, 2$, are isomorphic if and only if there exists a* **topological**[5] *fiber preserving map $\varphi : G_1 \to G_2$ with $\varphi(x_1) = x_2$.*

PROOF: If we have fiber preserving mappings $\varphi_1 : G_1 \to G_2$ and $\varphi_2 : G_2 \to G_1$, with $\varphi_1(x_1) = x_2$ and $\varphi_2(x_2) = x_1$, it follows easily from the uniqueness of fiber preserving maps that $\varphi_2 \circ \varphi_1 = \mathrm{id}_{G_1}$ and $\varphi_1 \circ \varphi_2 = \mathrm{id}_{G_2}$. The other direction of the proof is trivial. ∎

Definition. A domain $\mathcal{G} = (G, \pi, x_0)$ with $\pi(x_0) = \mathbf{z}_0$ is called *schlicht* if it is isomorphic to a domain $\mathcal{G}_0 = (G_0, \mathrm{id}_{G_0}, \mathbf{z}_0)$ with $G_0 \subset \mathbb{C}^n$.

8.5 Proposition. *Let $\mathcal{G}_j = (G_j, \mathrm{id}_{G_j}, x_j)$, $j = 1, 2$, be two schlicht domains with $G_1, G_2 \subset \mathbb{C}^n$. Then $\mathcal{G}_1 \prec \mathcal{G}_2$ if and only if $G_1 \subset G_2$.*

Example

[5] Recall that a "topological map" is a homeomorphism!

Let $G_1 := \{(z,w) \in \mathbb{C}^2 : w^2 = z \text{ and } z \neq 0\}$ and $\pi_1(z,w) := z$. Then $\mathcal{G}_1 = (G_1, \pi_1, (1,1))$ is the Riemann surface of $\sqrt{z}$, with distinguished point $(1,1)$. The domain $\mathcal{G}_1$ is contained in the schlicht domain $\mathcal{G}_2 = (\mathbb{C}, \mathrm{id}_{\mathbb{C}}, 1)$, by $\varphi(z,w) := z$. But the two domains are not isomorphic.

Union of Riemann Domains. We begin with the definition of the union of two Riemann domains. Let $\mathcal{G}_j = (G_j, \pi_j, x_j)$, $j = 1,2$, be two Riemann domains over $\mathbb{C}^n$ with distinguished point, and $\mathbf{z}_0 := \pi_1(x_1) = \pi_2(x_2)$. We want to glue G_1, G_2 in such a way that x_1 and x_2 will also be glued.

To get a rough idea of the construction, assume that we already have a Riemann domain $\mathcal{G} = (G, \pi, x_0)$ that is in some sense the union of $\mathcal{G}_1$ and $\mathcal{G}_2$. Then there should exist continuous fiber preserving maps $\varphi_1 : G_1 \to G$ with $\varphi_1(x_1) = x_0$, and $\varphi_2 : G_2 \to G$ with $\varphi_2(x_2) := x_0$. If $\alpha : [0,1] \to G_1$ and $\beta : [0,1] \to G_2$ are two continuous paths with $\alpha(0) = x_1$, $\beta(0) = x_2$ and $\pi_1 \circ \alpha = \pi_2 \circ \beta$, then $\gamma_1 := \varphi_1 \circ \alpha$ and $\gamma_2 := \varphi_2 \circ \beta$ are continuous paths in G with $\pi \circ \gamma_1 = \pi \circ \gamma_2$ and $\gamma_1(0) = \gamma_2(0) = x_0$. Due to the uniqueness of lifting, it follows that $\gamma_1 = \gamma_2$. This means that $\alpha(t)$ and $\beta(t)$ have to be glued for every $t \in [0,1]$. Unfortunately, this is an ambiguous rule. For example, we could say that $x \in G_1$ and $y \in G_2$ have to be glued if $\pi_1(x) = \pi_2(y)$. Then the desired property is fulfilled, but it may be that there are no paths α from x_1 to x and β from x_2 to y with $\pi_1 \circ \alpha = \pi_2 \circ \beta$.

Therefore, we proceed in the following way: Start with the disjoint union $G_1 \dot\cup G_2$, and take the "finest" equivalence relation $\sim$ on this set with the following property:

1. $x_1 \sim x_2$.
2. If there are continuous paths $\alpha : [0,1] \to G_1$ and $\beta : [0,1] \to G_2$ with $\alpha(0) = x_1$, $\beta(0) = x_2$, and $\pi_1 \circ \alpha = \pi_2 \circ \beta$, then $\alpha(1) \sim \beta(1)$.

One can equip $G := (G_1 \dot\cup G_2)/\sim$ with the structure of a Riemann domain. This will now be carried out in a more general context.

Let X be an arbitrary set. An equivalence relation on X is given by a partition $\mathscr{X} = \{X_\nu : \nu \in N\}$ of X into subsets with:

1. $\bigcup_{\nu \in N} X_\nu = X$.
2. $X_\nu \cap X_\mu = \varnothing$ for $\nu \neq \mu$.

The sets X_ν are the equivalence classes.

Now let a family $(\mathscr{X}_\iota)_{\iota \in I}$ of equivalence relations on X be given with $\mathscr{X}_\iota = \{X^\iota_{\nu_\iota} : \nu_\iota \in N_\iota\}$ for $\iota \in I$. We set $N := \prod_{\iota \in I} N_\iota$, and

$$X_\nu := \bigcap_{\iota \in I} X^\iota_{\nu_\iota}, \qquad \text{for } \nu := (\nu_\iota)_{\iota \in I} \in N.$$

Then $\mathscr{X} = \{X_\nu : \nu \in N\}$ is again an equivalence relation (simple exercise), and it is finer than any $\mathscr{X}_\iota$. This means that for every $\iota \in I$ and every $\nu \in N$, there is a $\nu_\iota \in N_\iota$ with $X_\nu \subset X^\iota_{\nu_\iota}$.

We apply this to the disjoint union $X = \dot{\bigcup}_{\lambda \in L} G_\lambda$, for a given family $(\mathcal{G}_\lambda)_{\lambda \in L}$ of Riemann domains $\mathcal{G}_\lambda = (G_\lambda, \pi_\lambda, x_\lambda)$ over $\mathbb{C}^n$ with distinguished point. An equivalence relation on X is said to have property (P) if the following hold:

1. $x_\lambda \sim x_\varrho$, for $\lambda, \varrho \in L$.
2. If $\alpha : [0,1] \to G_\lambda$ and $\beta : [0,1] \to G_\varrho$ are continuous paths with $\alpha(0) \sim \beta(0)$ and $\pi_\lambda \circ \alpha = \pi_\varrho \circ \beta$, then $\alpha(1) \sim \beta(1)$.

We consider the family of all equivalence relations on X with property (P). It is not empty, as seen above in the case of two domains. Therefore we can construct an equivalence relation (as above) that is finer than any equivalence relation with property (P). We denote it by $\sim_P$. It is clear that $\pi_\lambda(x) = \pi_\varrho(y)$ if $x \in G_\lambda$, $y \in G_\varrho$, and $x \sim_P y$. The relation $\sim_P$ also has property (P), and the elements of an equivalence class X_ν all lie over the same point $\mathbf{z} = \mathbf{z}(X_\nu)$. We define $\widetilde{G} := X/\sim_P$ and $\widetilde{\pi}(X_\nu) := \mathbf{z}(X_\nu)$. The equivalence class of all x_λ will be denoted by $\widetilde{x}$.

8.6 Lemma. *Let $y \in G_\lambda$ and $x \in G_\varrho$ be given with $\pi_\varrho(x) = \pi_\lambda(y) =: \mathbf{z}$. If we choose open neighborhoods $U = U(y) \subset G_\lambda$, $V = V(x) \subset G_\varrho$, and an open connected neighborhood $W = W(\mathbf{z})$ such that $\pi_\lambda : U \to W$ and $\pi_\varrho : V \to W$ are topological mappings, then for $\varphi := (\pi_\varrho|_V)^{-1} \circ \pi_\lambda : U \to V$ the following hold:*

1. *$\varphi(y) = x$.*
2. *If $x \sim_P y$, then $\varphi(y') \sim_P y'$ for every $y' \in U$.*

PROOF: The first statement is trivial. Now let $\alpha : [0,1] \to W$ be a continuous path with $\alpha(0) = \mathbf{z}$ and $\alpha(1) = \pi_\lambda(y')$ for some $y' \in U$. Then $\beta := (\pi_\lambda|_U)^{-1} \circ \alpha$ and $\gamma := \varphi \circ \beta$ are continuous paths in U and V with $\beta(0) = y \sim_P x = \varphi(y) = \gamma(0)$ and $\pi_\lambda \circ \beta = \pi_\varrho \circ \varphi \circ \beta = \pi_\varrho \circ \gamma$. Therefore, $y' = \beta(1) \sim_P \gamma(1) = \varphi(y')$. ■

8.7 Theorem. *There is a topology on $\widetilde{G}$ such that*

$$\widetilde{\mathcal{G}} := (\widetilde{G}, \widetilde{\pi}, \widetilde{x})$$

is a Riemann domain over $\mathbb{C}^n$ with distinguished point $\widetilde{x}$, and all maps $\varphi_\lambda : G_\lambda \to \widetilde{G}$ with

$$\varphi_\lambda(x) := \textit{equivalence class of } x$$

are continuous and fiber preserving.

PROOF: (1) Sets of the form $\varphi_\lambda(M)$ for M open in G_λ together with $\widetilde{G}$ constitute a base of a topology for $\widetilde{G}$. To see this it remains to show that the intersection of two such sets is again of this form.

Let $M \subset G_\lambda$ and $N \subset G_\varrho$ be open subsets. Then

$$\varphi_\lambda(M) \cap \varphi_\varrho(N) = \varphi_\varrho(N \cap \varphi_\varrho^{-1}(\varphi_\lambda(M))).$$

But $\varphi_\varrho^{-1}(\varphi_\lambda(M))$ is open in G_ϱ. In fact, let $x \in \varphi_\varrho^{-1}(\varphi_\lambda(M))$ be given, and $y \in M$ be chosen such that $\varphi_\lambda(y) = \varphi_\varrho(x)$ (and therefore $y \sim_P x$). Let $\mathbf{z} := \pi_\lambda(y) = \pi_\varrho(x)$. Then there exist open neighborhoods $U = U(y)$ and $V = V(x)$ and an open connected neighborhood $W = W(\mathbf{z})$ such that $\pi_\lambda : U \to W$ and $\pi_\varrho : V \to W$ are topological mappings. Let $\varphi := (\pi_\varrho|_V)^{-1} \circ \pi_\lambda : U \to V$. By the lemma, $\varphi(y) = x$ and $\varphi(y') \sim_P y'$ for every $y' \in U$.

So $V' := \varphi(M \cap U)$ is a neighborhood of x in G_ϱ, and since $\varphi_\varrho(\varphi(y')) = \varphi_\lambda(y')$ for every $y' \in U$, it follows that $V' \subset \varphi_\varrho^{-1}(\varphi_\lambda(M))$.

Consequently, every φ_λ is a continuous map.

(2) Remark: Since every $y \in \widetilde{G}$ is an equivalence class $\varphi_\lambda(x)$, we have

$$M = \bigcup_{\lambda \in L} \varphi_\lambda(\varphi_\lambda^{-1}(M)) \text{ for any subset } M \subset \widetilde{G}.$$

(3) $\widetilde{\pi} : \widetilde{G} \to \mathbb{C}^n$ is continuous: Let $V \subset \mathbb{C}^n$ be an arbitrary open set, and $M := \widetilde{\pi}^{-1}(V)$. Then $\varphi_\lambda^{-1}(M) = \pi_\lambda^{-1}(V)$ is open in G_λ, and therefore $M = \bigcup_{\lambda \in L} \varphi_\lambda(\varphi_\lambda^{-1}(M))$ is open in $\widetilde{G}$.

(4) $\widetilde{G}$ is a Hausdorff space: Let $y_1, y_2 \in \widetilde{G}$ with $y_1 \neq y_2$, and $\mathbf{z}_1 := \widetilde{\pi}(y_1)$, $\mathbf{z}_2 := \widetilde{\pi}(y_2)$.

There are two cases. If $\mathbf{z}_1 \neq \mathbf{z}_2$, then there are open neighborhoods $V_1(\mathbf{z}_1)$ and $V_2(\mathbf{z}_2)$ with $V_1 \cap V_2 = \varnothing$. Then $\widetilde{\pi}^{-1}(V_1)$ and $\widetilde{\pi}^{-1}(V_2)$ are disjoint open neighborhoods of y_1 and y_2. If $\mathbf{z}_1 = \mathbf{z}_2$, then we choose elements $x_1 \in G_\lambda$, $x_2 \in G_\varrho$ with $\varphi_\lambda(x_1) = y_1$ and $\varphi_\varrho(x_2) = y_2$, and since x_1 and x_2 are not equivalent, the above lemma implies that there are disjoint neigborhoods of y_1 and y_2.

(5) $\widetilde{G}$ is connected: Let $y = \varphi_\lambda(x)$ be an arbitrary point of $\widetilde{G}$. Then there is a continuous path $\alpha : [0,1] \to G_\lambda$ that connects the distinguished point x_λ to x. Then $\varphi_\lambda \circ \alpha$ connects $\widetilde{x}$ to y.

(6) $\widetilde{\pi}$ is locally topological: Let $y = \varphi_\lambda(x)$ be a point of $\widetilde{G}$, and $\mathbf{z} = \widetilde{\pi}(y) = \pi_\lambda(x)$. Then there exist open neighborhoods $U = U(x) \subset G_\lambda$ and $W = W(\mathbf{z}) \subset \mathbb{C}^n$ such that $\pi_\lambda : U \to W$ is a topological mapping. $\widetilde{U} := \varphi_\lambda(U)$ is an open neighborhood of y, with $\widetilde{\pi}(\widetilde{U}) = \pi_\lambda(U) = W$. In addition, $\widetilde{\pi}|_{\widetilde{U}}$ is injective, since $\widetilde{\pi} \circ \varphi_\lambda = \pi_\lambda$ and $\pi_\lambda|_U$ is injective.

(7) Clearly, the maps $\varphi_\lambda : G_\lambda \to \widetilde{G}$ are fiber preserving, and it was already shown that they are continuous. ■

Now $\widetilde{\mathcal{G}}$ has the following properties:

1. $\mathcal{G}_\lambda \prec \widetilde{\mathcal{G}}$, for every $\lambda \in L$.
2. If $\mathcal{G}^*$ is a domain over $\mathbb{C}^n$ with $\mathcal{G}_\lambda \prec \mathcal{G}^*$ for every λ, then $\widetilde{\mathcal{G}} \prec \mathcal{G}^*$.

PROOF: (of the second statement)

If $\mathcal{G}^*$ is given, then there exist fiber preserving mappings $\varphi^*_\lambda : G_\lambda \to G^*$. We introduce a new equivalence relation $\simeq$ on the disjoint union X of the G_λ by

$$x \simeq x' :\iff x \in G_\lambda,\ x' \in G_\varrho \text{ and } \varphi^*_\lambda(x) = \varphi^*_\varrho(x').$$

It follows from the uniqueness of lifting that $\simeq$ has the property (P). Now we define a map $\varphi : \widetilde{G} \to G^*$ by

$$\varphi(\varphi_\lambda(x)) := \varphi^*_\lambda(x).$$

Since $\sim_P$ is the finest equivalence relation with property (P), φ is well defined. Also it is clear that φ is continuous and fiber preserving. ■

Therefore $\widetilde{\mathcal{G}}$ is the smallest Riemann domain over $\mathbb{C}^n$ that contains all domains $\mathcal{G}_\lambda$.

Definition. The domain $\widetilde{\mathcal{G}}$ constructed as above is called the *union of the domains* $\mathcal{G}_\lambda$, and we write $\widetilde{\mathcal{G}} = \bigcup_{\lambda \in L} \mathcal{G}_\lambda$.

Special cases:

1. From $\mathcal{G}_1 \prec \mathcal{G}$ and $\mathcal{G}_2 \prec \mathcal{G}$ it follows that $\mathcal{G}_1 \cup \mathcal{G}_2 \prec \mathcal{G}$.
2. From $\mathcal{G}_1 \prec \mathcal{G}_2$ it follows that $\mathcal{G}_1 \cup \mathcal{G}_2 \cong \mathcal{G}_2$.
3. $\mathcal{G} \cup \mathcal{G} \cong \mathcal{G}$.
4. $\mathcal{G}_1 \cup \mathcal{G}_2 \cong \mathcal{G}_2 \cup \mathcal{G}_1$.
5. $\mathcal{G}_1 \cup (\mathcal{G}_2 \cup \mathcal{G}_3) \cong (\mathcal{G}_1 \cup \mathcal{G}_2) \cup \mathcal{G}_3$.

Example

Let $\mathcal{G}_1 = (G_1, \pi_1, x_1)$ be the Riemann surface of $\sqrt{z}$ with distinguished point $x_1 = (1, 1)$ and $\mathcal{G}_2 = (G_2, \mathrm{id}, x_2)$ the schlicht domain

$$G_2 = \Big\{ z \in \mathbb{C} : \frac{1}{2} < |z| < 2 \Big\}$$

with distinguished point $x_2 = 1$.

Then $\mathcal{G}_1 \cup \mathcal{G}_2 = (\widetilde{G}, \widetilde{\pi}, \widetilde{x}_0)$, where $\widetilde{G} = (G_1 \dot{\cup} G_2)/ \sim_P$.

Let $y \in \pi_1^{-1}(G_2) \subset G_1$. Then we can connect y to the point x_1 by a path α in $\pi_1^{-1}(G_2)$, and $\pi_1(y)$ to x_2 by the path $\pi_1 \circ \alpha$ in G_2. But $x_1 \sim_P x_2$, so

$y \sim_P \pi_1(y)$ as well. This shows that over each point of G_2 there is exactly one equivalence class.

Now let $z \in \mathbb{C} - \{0\}$ be arbitrary. The line through z and 0 in $\mathbb{C}$ contains a segment $\alpha : [0,1] \to \mathbb{C}^*$ that connects z to a point $z^* \in G_2$. There are two paths α_1, α_2 in G_1 with $\pi_1 \circ \alpha_1 = \pi_1 \circ \alpha_2 = \alpha$. Since $\alpha_1(1) \sim_P \alpha_2(1)$, it follows that $\alpha_1(0) \sim_P \alpha_2(0)$.

Then it follows that $\mathcal{G}_1 \cup \mathcal{G}_2 = (\mathbb{C} - \{0\}, \mathrm{id}, 1)$.

Exercises

1. For $\mathbf{t} = (t_1, \ldots, t_n) \in \mathscr{V}$ define $\Phi_\mathbf{t} : \mathbb{C}^n \to \mathbb{C}^n$ by

$$\Phi_\mathbf{t}(z_1, \ldots, z_n) := \big(e^{\mathrm{i}t_1} z_1, \ldots, e^{\mathrm{i}t_n} z_n\big).$$

   A Riemann domain $\mathcal{G} = (G, \pi, x_0)$ is called a *Reinhardt domain over* $\mathbb{C}^n$ if $\pi(x_0) = \mathbf{0}$ and for every $\mathbf{t} \in \mathscr{V} - (\mathbb{C}^*)^n$ there is an isomorphism $\varphi_\mathbf{t} : \mathcal{G} \to \mathcal{G}$ with $\pi \circ \varphi_\mathbf{t} = \Phi_\mathbf{t} \circ \pi$. Prove:
   (a) If $G \subset \mathbb{C}^n$ is a proper Reinhardt domain, then $\mathcal{G} = (G, \mathrm{id}, \mathbf{0})$ is a Reinhardt domain over $\mathbb{C}^n$.
   (b) Let $G_1, G_2 \subset \mathbb{C}^2$ be defined by

$$\begin{aligned} G_1 &:= \mathsf{P}^2(\mathbf{0}, 1) - \Big\{(z, w) \,:\, |z| = \frac{1}{2} \text{ and } |w| \le \frac{1}{2}\Big\}, \\ G_2 &:= \Big\{(z, w) \in \mathsf{P}^2(\mathbf{0}, 1) \,:\, |w| < \frac{1}{2}\Big\}. \end{aligned}$$

   Gluing G_1 and G_2 along $\big\{(z, w) \,:\, \frac{1}{2} < |z| < 1 \text{ and } |w| < \frac{1}{2}\big\}$ one obtains a Riemann domain over $\mathbb{C}^2$ that is a Reinhardt domain over $\mathbb{C}^2$, but not schlicht. Show that this domain can be obtained as the union of $\mathcal{G}_1 = \big(G_1, \mathrm{id}, (\frac{3}{4}, \frac{1}{4})\big)$ and $\mathcal{G}_2 = \big(G_2, \mathrm{id}, (\frac{3}{4}, \frac{1}{4})\big)$.
2. Let $J = \{0, 1, 2, 3, \ldots\} \subset \mathbb{N}_0$ be a finite or infinite sequence of natural numbers and $P_i = \mathsf{P}^n(\mathbf{z}_i, r_i)$, $i \in J$, a sequence of polydisks in $\mathbb{C}^n$. Assume that for every pair $(i, j) \in J \times J$ an "incidence number" $\varepsilon_{ij} \in \{0, 1\}$ is given such that the following hold:
   (a) $\varepsilon_{ij} = \varepsilon_{ji}$ and $\varepsilon_{ii} = 1$.
   (b) $\varepsilon_{ij} = 0$ if $P_i \cap P_j = \varnothing$.
   (c) For every $i > 0$ in J there is a $j < i$ with $\varepsilon_{ij} = 1$.
   (d) If $P_i \cap P_j \cap P_k \neq \varnothing$ and $\varepsilon_{ij} = 1$, then $\varepsilon_{ik} = \varepsilon_{jk}$.

   Points $\mathbf{z} \in P_i$ and $\mathbf{w} \in P_j$ are called equivalent ($\mathbf{z} \sim \mathbf{w}$) if $\mathbf{z} = \mathbf{w}$ and $\varepsilon_{ij} = 1$. Prove that $G := \dot{\bigcup} P_i / \sim$ carries in a natural way the structure of a Riemann domain over $\mathbb{C}^n$.

   Let $\pi : G \to \mathbb{C}^n$ be the canonical projection and suppose that there is a point $\mathbf{z}_0 \in \bigcap_{i \in J} P_i$. Is there a point $x_0 \in G$ such that (G, π, x_0) can be written as the union of the Riemann domains $(P_j, \mathrm{id}, \mathbf{z}_0)$?

9. The Envelope of Holomorphy

Holomorphy on Riemann Domains

Definition. Let (G, π) be a domain over $\mathbb{C}^n$. A function $f : G \to \mathbb{C}$ is called *holomorphic at a point* $x \in G$ if there are open neighborhoods $U = U(x) \subset G$ and $V = V(\pi(x)) \subset \mathbb{C}^n$ such that $\pi|_U : U \to V$ is topological and $f \circ (\pi|_U)^{-1} : V \to \mathbb{C}$ is holomorphic. The function f is called *holomorphic on* G if f is holomorphic at every point $x \in G$.

Remark. A holomorphic function is always continuous. For schlicht domains in $\mathbb{C}^n$ the new notion of holomorphy agrees with the old one.

Definition. Let $\mathcal{G}_j = (G_j, \pi_j, x_j)$, $j = 1, 2$, be domains over $\mathbb{C}^n$ with distinguished point, and $\mathcal{G}_1 \prec \mathcal{G}_2$ by virtue of a continuous mapping $\varphi : G_1 \to G_2$. For every function f on G_2 we define $f|_{G_1} := f \circ \varphi$.

9.1 Proposition. *If $f : G_2 \to \mathbb{C}$ is holomorphic and $\mathcal{G}_1 \prec \mathcal{G}_2$, then $f|_{G_1}$ is holomorphic on G_1.*

PROOF: Trivial, since φ is a local homeomorphism with $\pi_2 \circ \varphi = \pi_1$. ∎

Definition.

1. Let (G, π) be a domain over $\mathbb{C}^n$ and $x \in G$ a point. If f is a holomorphic function defined near x, then the pair (f, x) is called a *local holomorphic function* at x.
2. Let (G_1, π_1), (G_2, π_2) be domains over $\mathbb{C}^n$, and $x_1 \in G_1$, $x_2 \in G_2$ points with $\pi_1(x_1) = \pi_2(x_2) =: \mathbf{z}$. Two local holomorphic functions (f_1, x_1), (f_2, x_2) are called *equivalent* if there exist open neighborhoods $U_1(x_1) \subset G_1$, $U_2(x_2) \subset G_2$, $V(\mathbf{z})$, and topological mappings $\pi_1 : U_1 \to V$, $\pi_2 : U_2 \to V$ with $f_1 \circ (\pi_1|_{U_1})^{-1} = f_2 \circ (\pi_2|_{U_2})^{-1}$.
3. The equivalence class of a local holomorphic function (f, x) is denoted by f_x.

Remark. If $(f_1)_{x_1} = (f_2)_{x_2}$, then clearly, $f_1(x_1) = f_2(x_2)$. In particular, if $G_1 = G_2$, $\pi_1 = \pi_2$, and $x_1 = x_2$, then it follows that f_1 and f_2 coincide in an open neighborhood of $x_1 = x_2$.

9.2 Proposition. *Let (G_1, π_1), (G_2, π_2) be domains over $\mathbb{C}^n$, $\alpha_\lambda : [0,1] \to G_\lambda$ continuous paths with $\pi_1 \circ \alpha_1 = \pi_2 \circ \alpha_2$. Additionally, let f_λ be holomorphic on G_λ, for $\lambda = 1, 2$. If $(f_1)_{\alpha_1(0)} = (f_2)_{\alpha_2(0)}$, then also $(f_1)_{\alpha_1(1)} = (f_2)_{\alpha_2(1)}$.*

PROOF: Let $M := \{t \in [0,1] : (f_1)_{\alpha_1(t)} = (f_2)_{\alpha_2(t)}\}$. Then $M \neq \varnothing$, since $0 \in M$. It is easy to see that M is open and closed in $[0,1]$, because of the identity theorem for holomorphic functions. So $M = [0,1]$. ■

9.3 Proposition. *Let $\mathcal{G}_j = (G_j, \pi_j, x_j)$, $j = 1, 2$, be domains over $\mathbb{C}^n$ with distinguished point, and $\mathcal{G}_1 \prec \mathcal{G}_2$. Then for every holomorphic function f on G_1 there is at most one holomorphic function F on G_2 with $F|_{G_1} = f$, i.e., a possible holomorphic extension of f is uniquely determined.*

PROOF: Let F_1, F_2 be holomorphic extensions of f to G_2. We choose neighborhoods $U_\lambda(x_\lambda) \subset G_\lambda$ such that the given fiber-preserving map $\varphi : G_1 \to G_2$ maps U_1 topologically onto U_2. We have $F_j \circ \varphi|_{U_1} = f|_{U_1}$, for $j = 1, 2$, and therefore $F_1|_{U_2} = F_2|_{U_2}$. It follows that $(F_1)_{x_2} = (F_2)_{x_2}$. Since each point of G_2 can be joined to x_2, the equality $F_1 = F_2$ follows. ■

Envelopes of Holomorphy

Definition. Let $\mathcal{G} = (G, \pi, x_0)$ be a domain over $\mathbb{C}^n$ with distinguished point and $\mathscr{F}$ a nonempty set of holomorphic functions on G.

Let $(\mathcal{G}_\lambda)_{\lambda \in L}$ be the system of all domains over $\mathbb{C}^n$ with the following properties:

1. $\mathcal{G} \prec \mathcal{G}_\lambda$ for every $\lambda \in L$.
2. For every $f \in \mathscr{F}$ and every $\lambda \in L$ there is a holomorphic function F_λ on G_λ with $F_\lambda|_G = f$.

Then $H_{\mathscr{F}}(\mathcal{G}) := \bigcup_{\lambda \in L} \mathcal{G}_\lambda$ is called the *$\mathscr{F}$-hull* of $\mathcal{G}$.

If $\mathscr{F} = \mathcal{O}(G)$ is the set of all holomorphic functions on G, then $H(\mathcal{G}) := H_{\mathcal{O}(G)}(\mathcal{G})$ is called the *envelope of holomorphy* of $\mathcal{G}$. If $\mathscr{F} = \{f\}$ for some holomorphic function f on G, then $H_f(\mathcal{G}) := H_{\{f\}}(\mathcal{G})$ is called the *domain of existence* of the function f.

9.4 Theorem. *Let $\mathcal{G} = (G, \pi, x_0)$ be a domain over $\mathbb{C}^n$, $\mathscr{F}$ a nonempty set of holomorphic functions on G, and $H_{\mathscr{F}}(\mathcal{G}) = (\widehat{G}, \widehat{\pi}, \widehat{x}_0)$ the $\mathscr{F}$-hull. Then the following hold:*

1. *$\mathcal{G} \prec H_{\mathscr{F}}(\mathcal{G})$.*
2. *For each function $f \in \mathscr{F}$ there exists exactly one holomorphic function F on $\widehat{G}$ with $F|_G = f$.*

3. *If* $\mathcal{G}_1 = (G_1, \pi_1, x_1)$ *is a domain over* $\mathbb{C}^n$ *such that* $\mathcal{G} \prec \mathcal{G}_1$ *and every function* $f \in \mathscr{F}$ *can be holomorphically extended to* G_1*, then* $\mathcal{G}_1 \prec H_{\mathscr{F}}(\mathcal{G})$.

PROOF: $H_{\mathscr{F}}(\mathcal{G})$ is the union of all Riemann domains $\mathcal{G}_\lambda = (G_\lambda, \pi_\lambda, x_\lambda)$ to which each function $f \in \mathscr{F}$ can be extended. We have fiber-preserving maps $\varphi_\lambda : G \to G_\lambda$ and $\widehat{\varphi}_\lambda : G_\lambda \to \widehat{G}$.

Let $\sim_P$ be the finest equivalence relation on $X := \dot{\bigcup}_{\lambda \in L} G_\lambda$ with property (P).[6] Then $\widehat{G}$ is the set of equivalence classes in X relative to $\sim_P$. We define a new equivalence relation $\simeq$ on X by

$$x \simeq x' \quad :\iff \quad x \in G_\lambda,\ x' \in G_\varrho,\ \pi_\lambda(x) = \pi_\varrho(x'), \text{ and for each } f \in \mathcal{F}$$
$$\text{and its holomorphic extensions } F_1, F_2 \text{ on } G_\lambda, \text{ respectively } G_\varrho,$$
$$\text{we have } (F_\lambda)_x = (F_\varrho)_{x'}.$$

Then $\simeq$ has property (P):

(i) For any λ we can find open neighborhoods $U = U(x_0)$, $V = V(x_\lambda)$, and $W = W(\pi(x_0))$ such that all mappings in the following commutative diagram are homeomorphisms:

$$\begin{array}{ccc} U & \xrightarrow{\varphi_\lambda} & V \\ & \pi \searrow \quad \swarrow \pi_\lambda & \\ & W & \end{array}$$

Then for $f \in \mathscr{F}$ and its holomorphic extension F_λ on G_λ we have that $F_\lambda \circ (\pi_\lambda|_V)^{-1} = F_\lambda \circ \varphi_\lambda \circ (\pi|_U)^{-1} = f \circ (\pi|_U)^{-1}$ is independent of λ. Therefore, all distinguished points x_λ are equivalent.

(ii) If $\alpha : [0,1] \to G_\lambda$ and $\beta : [0,1] \to G_\varrho$ are continuous paths with $\alpha(0) \simeq \beta(0)$ and $\pi_\lambda \circ \alpha = \pi_\varrho \circ \beta$, then $(F_\lambda)_{\alpha(0)} = (F_\varrho)_{\beta(0)}$. It follows that $(F_\lambda)_{\alpha(1)} = (F_\varrho)_{\beta(1)}$ as well, and therefore $\alpha(1) \simeq \beta(1)$.

Since $\mathcal{G} \prec \mathcal{G}_\lambda$ and $\mathcal{G}_\lambda \prec H_{\mathscr{F}}(\mathcal{G})$, it follows that $\mathcal{G} \prec H_{\mathcal{F}}(\mathcal{G})$. Furthermore, the fiber preserving map $\widehat{\varphi} := \widehat{\varphi}_\lambda \circ \varphi_\lambda$ does not depend on λ.

Now let a function $f \in \mathscr{F}$ be given. We construct a holomorphic extension F on $\widehat{G}$ as follows:

If $y \in \widehat{G}$ is an arbitrary point, then there is a $\lambda \in L$ and a point $y_\lambda \in G_\lambda$ such that $y = \widehat{\varphi}_\lambda(y_\lambda)$, and we define

$$F(y) := F_\lambda(y_\lambda).$$

If $y = \widehat{\varphi}_\varrho(y_\varrho)$ as well, then $y_\lambda \sim_P y_\varrho$, and therefore $y_\lambda \simeq y_\varrho$ as well. It follows that $F_\lambda(y_\lambda) = F_\varrho(y_\varrho)$, and F is well defined.

[6] For the definition of property (P) see page 92.

We have $F \circ \widehat{\varphi} = F \circ \widehat{\varphi}_\lambda \circ \varphi_\lambda = F_\lambda \circ \varphi_\lambda = f$ on G. This shows that F is an extension of f, and from the equation $F \circ \widehat{\varphi}_\lambda = F_\lambda$ it follows that F is holomorphic (since $\widehat{\varphi}_\lambda$ is locally topological).

The maximality of $H_{\mathscr{F}}(\mathcal{G})$ follows by construction. ∎

The $\mathscr{F}$-hull $H_{\mathscr{F}}(\mathcal{G})$ is therefore the largest domain into which all functions $f \in \mathscr{F}$ can be holomorphically extended.

9.5 Identity theorem. *Let $\mathcal{G}_j = (G_j, \pi_j, x_j)$, $j = 1, 2$, be domains over $\mathbb{C}^n$, and $\widetilde{\mathcal{G}} = (\widetilde{G}, \widetilde{\pi}, \widetilde{x})$ the union of $\mathcal{G}_1$ and $\mathcal{G}_2$. Let $f_j : G_j \to \mathbb{C}$ be holomorphic functions and $\mathcal{G} = (G, \pi, x)$ a domain with $\mathcal{G} \prec \mathcal{G}_j$ for $j = 1, 2$ such that $f_1|_G = f_2|_G$. Then there is a holomorphic function $\widetilde{f}$ on $\widetilde{G}$ with $\widetilde{f}|_{G_j} = f_j$, for $j = 1, 2$.*

PROOF: Let $f := f_1|_G = f_2|_G$, and $\mathscr{F} := \{f\}$. Since $\mathcal{G}_1 \prec H_{\mathscr{F}}(\mathcal{G})$ and $\mathcal{G}_2 \prec H_{\mathscr{F}}(\mathcal{G})$, it follows that $\mathcal{G}_1 \cup \mathcal{G}_2 \prec H_{\mathscr{F}}(\mathcal{G})$.

Let $\widehat{f}$ be a holomorphic extension of f to $\widehat{G}$ (where $H_{\mathscr{F}}(\mathcal{G}) = (\widehat{G}, \widehat{\pi}, \widehat{x})$), and $\widetilde{f} := \widehat{f}|_{\widetilde{G}}$. Then

$$(\widetilde{f}|_{G_j})|_G = \widetilde{f}|_G = (\widehat{f}|_{\widetilde{G}})|_G = \widehat{f}|_G = f.$$

Therefore, $\widetilde{f}|_{G_j}$ is a holomorphic extension of f to G_j. Due to the uniqueness of holomorphic extension, $\widetilde{f}|_{G_j} = f_j$ for $j = 1, 2$. ∎

Pseudoconvexity. Let $\mathsf{P}^n \subset \mathbb{C}^n$ be the unit polydisk, $(\mathsf{P}^n, \mathsf{H})$ a Euclidean Hartogs figure, and $\Phi : \mathsf{P}^n \to \mathbb{C}^n$ an injective holomorphic mapping. Then $(\Phi(\mathsf{P}^n), \Phi(\mathsf{H}))$ is a generalized Hartogs figure. $\mathcal{P} = (\mathsf{P}^n, \Phi, \mathbf{0})$ and $\mathcal{H} = (\mathsf{H}, \Phi, \mathbf{0})$ are Riemann domains with $\mathcal{H} \prec \mathcal{P}$. We regard the pair $(\mathcal{P}, \mathcal{H})$ as a generalized Hartogs figure.

9.6 Proposition. *Let (G, π) be a domain over $\mathbb{C}^n$, $(\mathcal{P}, \mathcal{H})$ a generalized Hartogs figure, and $x_0 \in G$ a point for which $\mathcal{H} \prec \mathcal{G} := (G, \pi, x_0)$.*

Then every holomorphic function f on G can be extended holomorphically to $\mathcal{G} \cup \mathcal{P}$.

The proposition follows immediately from the identity theorem.

Definition. A domain (G, π) over $\mathbb{C}^n$ is called *Hartogs convex* if the fact that $(\mathcal{P}, \mathcal{H})$ is a generalized Hartogs figure and $x_0 \in G$ a point with $\mathcal{H} \prec \mathcal{G} := (G, \pi, x_0)$ implies $\mathcal{G} \cup \mathcal{P} \cong \mathcal{G}$.

A domain $\mathcal{G} = (G, \pi, x_0)$ over $\mathbb{C}^n$ is called a *domain of holomorphy* if there exists a holomorphic function f on G such that its domain of existence is equal to $\mathcal{G}$.

Remark. If $G \subset \mathbb{C}^n$ is a schlicht domain, then the new definition agrees with the old one.

9.7 Theorem.

1. *If $\mathcal{G} = (G, \pi, x_0)$ is a domain over $\mathbb{C}^n$ and $\mathscr{F}$ a nonempty set of holomorphic functions on G, then $H_{\mathscr{F}}(\mathcal{G})$ is Hartogs convex.*
2. *Every domain of holomorphy is Hartogs convex.*

PROOF: Let $(\mathcal{P}, \mathcal{H})$ be a generalized Hartogs figure with $\mathcal{H} \prec H_{\mathscr{F}}(\mathcal{G})$. Then every function $f \in \mathscr{F}$ has a holomorphic continuation to $H_{\mathscr{F}}(\mathcal{G}) \cup \mathcal{P}$. Therefore, $H_{\mathscr{F}}(\mathcal{G}) \cup \mathcal{P} \prec H_{\mathscr{F}}(\mathcal{G})$. On the other hand, we also have $H_{\mathscr{F}}(\mathcal{G}) \prec H_{\mathscr{F}}(\mathcal{G}) \cup \mathcal{P}$. So $H_{\mathscr{F}}(\mathcal{G}) \cup \mathcal{P} \cong H_{\mathscr{F}}(\mathcal{G})$. ∎

A Riemann domain (G, π) is called *holomorphically convex* if for every infinite discrete subset $D \subset G$ there exists a holomorphic function f on G that is unbounded on D.

9.8 Theorem (Oka, 1953). *If a Riemann domain (G, π) is Hartogs pseudoconvex, it is holomorphically convex (and therefore a domain of holomorphy).*

This is the solution of Levi's problem for Riemann domains over $\mathbb{C}^n$. We cannot give the proof here.

It seems possible to construct the holomorphic hull by adjoining Hartogs figures (cf. H. Langmaak, [La60]). It is conceivable that such a construction may be realized with the help of a computer, but until now (spring 2002) no successful attempt is known. We assume that parallel computer methods are necessary.

Boundary Points.

In the literature other notions of pseudoconvexity are used. We want to give a rough idea of these methods.

Definition. Let X be a topological space. A *filter (basis)* on X is a nonempty set $\mathcal{R}$ of subsets of X with the following properties:

1. $\varnothing \notin \mathcal{R}$.
2. The intersection of two elements of $\mathcal{R}$ contains again an element of the set $\mathcal{R}$.

Example

1. If x_0 is a point of X, then every fundamental system of neighborhoods of x_0 in X is a filter, called a *neighborhood filter* of x_0.

2. Let (x_n) be a sequence of points of X. If we define $S_N := \{x_n : n \geq N\}$, then $\mathcal{R} := \{S_N : N \in \mathbb{N}\}$ is the so-called *elementary filter* of the sequence (x_n). A filter is therefore the generalization of a sequence.

Definition. A point $x_0 \in X$ is called a *cluster point* of the filter $\mathcal{R}$ if $x_0 \in \overline{A}$, for every $A \in \mathcal{R}$. The point x_0 is called a *limit* of the filter $\mathcal{R}$ if every element of a fundamental system of neighborhoods of x_0 contains an element of $\mathcal{R}$.

For sequences the new notions agree with the old ones.

If $f : X \to Y$ is a continuous map, then the image of any filter on X is a filter on Y, the so-called *direct image*.

Definition. Let (G, π) be a Riemann domain over $\mathbb{C}^n$. An *accessible boundary point* of (G, π) is a filter $\mathcal{R}$ on G with the following properties:

1. $\mathcal{R}$ has no cluster point in G.
2. The direct image $\pi(\mathcal{R})$ has a limit $\mathbf{z}_0 \in \mathbb{C}^n$.
3. For every connected open neighborhood $V = V(\mathbf{z}_0) \subset \mathbb{C}^n$ there is exactly one connected component of $\pi^{-1}(V)$ that belongs to $\mathcal{R}$.
4. For every element $U \in \mathcal{R}$ there is a neighborhood $V = V(\mathbf{z}_0)$ such that U is a connected component of $\pi^{-1}(V)$.

Remark. For a Hausdorff space X the following hold:

1. A filter in X has at most one limit.
2. If a filter in X has the limit x_0, then x_0 is the only cluster point of this filter.

(for a proof see Bourbaki, [Bou66], § 8.1)

Therefore, the limit $\mathbf{z}_0$ in the definition above is uniquely determined.

There is an equivalent description of accessible boundary points that avoids the filter concept. For this consider sequences (x_ν) of points of G with the following properties:

1. (x_ν) has no cluster point in G.
2. The sequence of the images $\pi(x_\nu)$ has a limit $\mathbf{z}_0 \in \mathbb{C}^n$.
3. For every connected open neighborhood $V = V(\mathbf{z}_0) \subset \mathbb{C}^n$ there is an $n_0 \in \mathbb{N}$ such that for $n, m \geq n_0$ the points x_n and x_m can be joined by a continuous path $\alpha : [0, 1] \to G$ with $\pi \circ \alpha([0, 1]) \subset V$.

Two such sequences $(x_\nu), (y_\nu)$ are called equivalent if:

1. $\lim_{\nu \to \infty} \pi(x_\nu) = \lim_{\nu \to \infty} \pi(y_\nu) = \mathbf{z}_0$.
2. For every connected open neighborhood $V = V(\mathbf{z}_0)$ there is an n_0 such that for $n, m \geq n_0$ the points x_n and y_m can be joined by a continuous path $\alpha : [0, 1] \to G$ with $\pi \circ \alpha([0, 1]) \subset V$.

An accessible boundary point is an equivalence class of such sequences.

Let $\breve{\partial} G$ be the set of all accessible boundary points of G. Even if G is schlicht, this set may be different from the topological boundary ∂G. There may be points in ∂G that are not accessible, and it may be happen that an accessible boundary point is the limit of two inequivalent sequences.

We define $\breve{G} := G \cup \breve{\partial} G$. If $r_0 = [x_n]$ is an accessible boundary point, we define a neighborhood of r_0 in $\breve{G}$ as follows: Take a connected open set $U \subset G$ such that almost all x_n lie in U and $\pi(U)$ is contained in a neighborhood of $\mathbf{z}_0 := \lim_{n\to\infty} \pi(x_n)$. Then add all boundary points $r = [y_n]$ such that almost all y_n lie in U and $\lim_{n\to\infty} \pi(y_n)$ is a cluster point of $\pi(U)$. With this neighborhood definition $\breve{G}$ becomes a Hausdorff space, and $\breve{\pi} : \breve{G} \to \mathbb{C}^n$ with

$$\breve{\pi}(x) := \begin{cases} \pi(x) & \text{if } x \in G, \\ \lim\limits_{n\to\infty} \pi(x_n) & \text{if } x = [x_n] \in \breve{\partial} G, \end{cases}$$

is a continuous mapping.

Definition. A boundary point $r \in \breve{\partial} G$ is called *removable* if there is a connected open neighborhood $U = U(r) \subset \breve{G}$ such that $(U, \breve{\pi})$ is a schlicht Riemann domain over $\mathbb{C}^n$ and $\breve{\partial} G \cap U$ is locally contained in a proper analytic subset of U.

A subset $M \subset \breve{\partial} G$ is called *thin* if for every $r_0 \in M$ there is an open neighborhood $U = U(r_0) \subset \breve{G}$ and a nowhere identically vanishing holomorphic function f on $U \cap G$ such that for every $r \in M \cap U$ there exists a sequence (x_n) in $U \cap G$ converging to r such that $\lim_{n\to\infty} f(x_n) = 0$.

Example

Let $G \subset \mathbb{C}^n$ be a (schlicht) domain and $A \subset G$ a nowhere dense analytic subset. Then every point of A is a removable boundary point of $G' := G - A$.

The points of the boundary of the hyperball $\mathsf{B}_r(\mathbf{0}) \subset \mathbb{C}^n$ are all not removable.

Let B be a ball in the affine hyperplane $H = \{(z_0, \dots, z_n) \in \mathbb{C}^{n+1} : z_0 = 1\}$, and $G \subset \mathbb{C}^{n+1} - \{\mathbf{0}\}$ the cone over B. Then every boundary point of G is not removable, since locally the boundary has real dimension $2n+1$. The set $M := \{\mathbf{0}\}$ is thin in the boundary, as is seen by choosing $f(z_0, \dots, z_n) := z_0$.

Analytic Disks. Let (G, π) be a Riemann domain over $\mathbb{C}^n$. If $\varphi : \overline{\mathsf{D}} \to \breve{G}$ is a continuous mapping, $\breve{\pi} \circ \varphi : \mathsf{D} \to \mathbb{C}^n$ holomorphic, and $(\breve{\pi} \circ \varphi)'(\zeta) \neq 0$ for $\zeta \in \mathsf{D}$, then $S := \varphi(\mathsf{D})$ is called an *analytic disk* in $\breve{G}$. The set $bS := \varphi(\partial \mathsf{D})$ is called its *boundary*.

Let $I := [0,1]$ be the unit interval. A family $(S_t)_{t \in I}$ of analytic disks $\varphi_t(\mathsf{D})$ in $\breve{G}$ is called continuous if the mapping $(z,t) \mapsto \varphi_t(z)$ is continuous. It is called distinguished if $S_t \subset G$ for $0 \le t < 1$ and $bS_t \subset G$ for $0 \le t \le 1$.

Definition. The domain G is called *pseudoconvex* if for every distinguished continuous family $(S_t)_{t \in I}$ of analytic disks in $\breve{G}$ it follows that $S_1 \subset G$.

The domain G is called *pseudoconvex at* $r \in \breve{\partial} G$ if there is a neighborhood $U = U(r) \subset \breve{G}$ and an $\varepsilon > 0$ such that for every distinguished continuous family $(S_t)_{t \in I}$ of analytic disks in $\breve{G}$ with $\breve{\pi}(S_t) \subset \mathsf{B}_\varepsilon(\breve{\pi}(r))$ it follows that $S_t \cap U \subset G$ for $t \in I$.

As in $\mathbb{C}^n$ one can show that a Riemann domain is pseudoconvex if and only if it is Hartogs pseudoconvex.

9.9 Theorem (Oka). *A Riemann domain (G, π) is pseudoconvex if and only if it is pseudoconvex at every point $r \in \breve{\partial} G$.*

9.10 Corollary. *If (G, π) is a domain of holomorphy, then G is pseudoconvex at every accessible boundary point.*

The converse theorem is Oka's solution of Levi's problem.

Finally, we mention the following result:

9.11 Theorem. *Let (G, π) be a Riemann domain over $\mathbb{C}^n$, and $M \subset \breve{\partial} G$ a thin set of nonremovable boundary points. If G is pseudoconvex at every point of $\breve{\partial} G - M$, then G is pseudoconvex.*

PROOF: See [GrRe56], §3, Satz 4. ∎

Exercises

1. Prove that a Reinhardt domain $\mathcal{G}$ over $\mathbb{C}^n$ must be schlicht if it is a domain of holomorphy.
2. Prove that if (G, π) is a Reinhardt domain, then for every $f \in \mathcal{O}(G)$ there is a power series $S(\mathbf{z})$ at the origin such that $f(x) = S(\pi(x))$ for $x \in G$.
3. Prove that the envelope of holomorphy of a Reinhardt domain is again a Reinhardt domain.
4. Prove that the Riemann surface of the function $f(z) = \log(z)$ has just one boundary point over $0 \in \mathbb{C}$.
5. Find a schlicht Riemann domain in $\mathbb{C}^2$ whose envelope of holomorphy is not schlicht.
6. Construct a Riemann domain $\mathcal{G} = (G, \pi, x_0)$ over $\mathbb{C}^2$ such that for all $x, y \in \pi^{-1}(\pi(x_0))$ and every $f \in \mathcal{O}(G)$ the equality $f(x) = f(y)$ holds.

7. Let (G, π) be a Riemann domain and $\big(\widehat{G}, \widehat{\pi}\big)$ its envelope of holomorphy. If f is a holomorphic function on G and F its holomorphic extension to $\widehat{G}$, then $f(G) = F(\widehat{G})$.
8. Consider

$$G := \Big\{(z,w) : \frac{1}{2} < |z| < 1,\ |w| < 1\Big\} - \Big\{\big(re^{\mathrm{i}t}, w\big) : \frac{1}{2} < r < 1,\ t \ge 0 \text{ and } |w| = \frac{t}{1+t}\Big\}.$$

   Determine the envelope of holomorphy of G.
9. Let $G \subset \mathbb{C}^n$ be a domain and $p : G \to \mathbb{R}$ a plurisubharmonic function. If $\mathbf{z}_0$ is an accessible boundary point of $B := \{\mathbf{z} \in G : p(\mathbf{z}) < c\} \subset\subset G$, then B is pseudoconvex at $\mathbf{z}_0$, in the sense of the last paragraph.

Chapter III

Analytic Sets

1. The Algebra of Power Series

The Banach Algebra $B_{\mathbf{t}}$. In this chapter we shall deal more extensively with power series in $\mathbb{C}^n$. Our objective is to find a division algorithm for power series that will facilitate our investigation of analytic sets.

We denote by $\mathbb{C}[\![\mathbf{z}]\!]$ the ring of formal power series $\sum_{\nu \geq 0} a_\nu \mathbf{z}^\nu$ about the origin. Let $\mathbb{R}_+^n$ be the set of n-tuples of positive real numbers.

Definition. Let $\mathbf{t} = (t_1, \ldots, t_n) \in \mathbb{R}_+^n$ and $f = \sum_{\nu \geq 0} a_\nu \mathbf{z}^\nu \in \mathbb{C}[\![\mathbf{z}]\!]$. We define the "number" $\|f\|_{\mathbf{t}}$ by

$$\|f\|_{\mathbf{t}} := \begin{cases} \sum_{\nu \geq 0} |a_\nu| \mathbf{t}^\nu & \text{if this series converges,} \\ \infty & \text{otherwise.} \end{cases}$$

Let $B_{\mathbf{t}} := \{f \in \mathbb{C}[\![\mathbf{z}]\!] \,:\, \|f\|_{\mathbf{t}} < \infty\}$.

Remark. One can introduce a weak ordering on $\mathbb{R}_+^n$ if one defines

$$(t_1, \ldots, t_n) \leq (t_1^*, \ldots, t_n^*) \;:\Longleftrightarrow\; t_i \leq t_i^* \text{ for } i = 1, \ldots, n.$$

For fixed f, the function $\mathbf{t} \mapsto \|f\|_{\mathbf{t}}$ is monotone: If $\mathbf{t} \leq \mathbf{t}^*$, then $\|f\|_{\mathbf{t}} \leq \|f\|_{\mathbf{t}^*}$.

Definition. A set B is called a *complex Banach algebra* if the following conditions are satisfied:

1. There are operations

$$+ : B \times B \to B, \quad \cdot : \mathbb{C} \times B \to B \quad \text{and} \quad \circ : B \times B \to B$$

such that

(a) $(B, +, \cdot)$ is a complex vector space,
(b) $(B, +, \circ)$ is a commutative ring with 1,
(c) $c \cdot (f \circ g) = (c \cdot f) \circ g = f \circ (c \cdot g)$ for all $f, g \in B$ and $c \in \mathbb{C}$.

2. To every $f \in B$ a real number $\|f\| \geq 0$ is assigned that has the properties of a norm:

(a) $\|c \cdot f\| = |c| \cdot \|f\|$, for $c \in \mathbb{C}$ and $f \in B$,
(b) $\|f + g\| \leq \|f\| + \|g\|$, for $f, g \in B$,

(c) $\|f\| = 0 \iff f = 0$.

3. $\|f \circ g\| \le \|f\| \cdot \|g\|$, for $f, g \in B$.
4. B is complete; i.e., every sequence in B that is Cauchy with respect to the norm has a limit in B.

1.1 Theorem. *$B_{\mathbf{t}} = \{f \in \mathbb{C}[\![\mathbf{z}]\!] : \|f\|_{\mathbf{t}} < \infty\}$ is a complex Banach algebra for any $\mathbf{t} \in \mathbb{R}_+^n$.*

PROOF: Clearly, $\mathbb{C}[\![\mathbf{z}]\!]$ is a commutative $\mathbb{C}$-algebra with 1. Straightforward calculations show that $\|\ldots\|_{\mathbf{t}}$ satisfies the properties (2a), (2b), (2c) and (3). It follows that $B_{\mathbf{t}}$ is closed under the algebraic operations, and all that remains to be shown is completeness.

Let (f_λ) be a Cauchy sequence in $B_{\mathbf{t}}$ with $f_\lambda = \sum_{\nu \ge 0} a_\nu^{(\lambda)} \mathbf{z}^\nu$. Then for every $\varepsilon > 0$ there is an $n = n(\varepsilon) \in \mathbb{N}$ such that for all $\lambda, \mu \ge n$,

$$\sum_{\nu \ge 0} |a_\nu^{(\lambda)} - a_\nu^{(\mu)}| \mathbf{t}^\nu = \|f_\lambda - f_\mu\|_{\mathbf{t}} < \varepsilon.$$

Since $\mathbf{t}^\nu = t_1^{\nu_1} \cdots t_n^{\nu_n} \neq 0$, it follows that

$$\left| a_\nu^{(\lambda)} - a_\nu^{(\mu)} \right| < \frac{\varepsilon}{\mathbf{t}^\nu} \quad \text{for every } \nu \in \mathbb{N}_0^n.$$

For fixed ν, $(a_\nu^{(\lambda)})$ is therefore a Cauchy sequence in $\mathbb{C}$ which converges to a complex number a_ν.

Let $f := \sum_{\nu \ge 0} a_\nu \mathbf{z}^\nu$. This is an element of $\mathbb{C}[\![\mathbf{z}]\!]$.

Given $\delta > 0$, it follows that there exists an $n = n(\delta)$ such that

$$\sum_{\nu \ge 0} \left| a_\nu^{(\lambda)} - a_\nu^{(\lambda+\mu)} \right| \mathbf{t}^\nu < \frac{\delta}{2} \quad \text{for } \lambda \ge n \text{ and } \mu \in \mathbb{N}.$$

Let $I \subset \mathbb{N}_0^n$ be an arbitrary finite set. For any $\lambda \ge n$ there exists a $\mu = \mu(\lambda) \in \mathbb{N}$ such that $\sum_{\nu \in I} |a_\nu^{(\lambda+\mu)} - a_\nu| \mathbf{t}^\nu < \delta/2$, and then

$$\sum_{\nu \in I} \left| a_\nu^{(\lambda)} - a_\nu \right| \mathbf{t}^\nu < \delta, \quad \text{for } \lambda \ge n.$$

In particular, $\|f_\lambda - f\|_{\mathbf{t}} \le \delta$. Thus $f_\lambda - f$ (and then also f) belongs to $B_{\mathbf{t}}$, and (f_λ) converges to f. ∎

Expansion with Respect to z_1. For the following we require some additional notation:

If $\nu \in \mathbb{N}_0^n$, $\mathbf{t} \in \mathbb{R}_+^n$, and $\mathbf{z} \in \mathbb{C}^n$, write

$$\nu = (\nu_1, \nu'), \quad \mathbf{t} = (t_1, \mathbf{t}'), \quad \text{and} \quad \mathbf{z} = (z_1, \mathbf{z}').$$

An element $f = \sum_{\nu \geq 0} a_\nu \mathbf{z}^\nu \in \mathbb{C}[\![\mathbf{z}]\!]$ can be written in the form

$$f = \sum_{\lambda=0}^{\infty} f_\lambda z_1^\lambda, \quad \text{with} \quad f_\lambda(\mathbf{z}') = \sum_{\nu' \geq 0} a_{(\lambda,\nu')} \{\mathbf{z}'\}^{\nu'}.$$

The series f_λ are formal power series in the variables $z_2, \ldots, z_n$. We call this representation of f the *expansion of f with respect to z_1*. Now the following assertions hold:

1. $f \in B_\mathbf{t} \iff f_\lambda \in B_{\mathbf{t}'}$ for all λ, and $\sum_{\lambda=0}^{\infty} \|f_\lambda\|_{\mathbf{t}'} t_1^\lambda < \infty$.
2. $\|z_1^s \cdot f\|_\mathbf{t} = t_1^s \cdot \|f\|_\mathbf{t}$, for $s \in \mathbb{N}_0$.

PROOF:

(1) Since we are dealing with absolute convergence, it is clear that

$$\|f\|_\mathbf{t} = \sum_{\lambda=0}^{\infty} \|f_\lambda\|_{\mathbf{t}'} t_1^\lambda.$$

(2) We have $z_1^s \cdot f = \sum_{\lambda=0}^{\infty} f_\lambda z_1^{\lambda+s}$. The right side is the unique expansion of $z_1^s \cdot f$ with respect to z_1. Now the formula can be easily derived. ∎

Convergent Series in Banach Algebras. Let B be a complex Banach algebra and (f_λ) a sequence of elements of B. The series $\sum_{\lambda=1}^{\infty} f_\lambda$ converges to an element $f \in B$ if the sequence $F_n := \sum_{\lambda=1}^{n} f_\lambda$ converges to f with respect to the given norm.

1.2 Proposition. *Every $f \in B$ with $\|1 - f\| < 1$ is a unit in B with*

$$f^{-1} = \sum_{\lambda=0}^{\infty} (1-f)^\lambda \text{ and } \|f^{-1}\| \leq \frac{1}{1 - \|1-f\|}.$$

PROOF: Let $\varepsilon := \|1 - f\|$. Then $0 \leq \varepsilon < 1$, and the convergent geometric series $\sum_{\lambda=0}^{\infty} \varepsilon^\lambda$ dominates the series $\sum_{\lambda=0}^{\infty}(1-f)^\lambda$. As usual, it follows that this series converges to an element $g \in B$. We have

$$\begin{aligned} f \cdot \sum_{\lambda=0}^{n} (1-f)^\lambda &= (1-(1-f)) \cdot \sum_{\lambda=0}^{n} (1-f)^\lambda \\ &= \sum_{\lambda=0}^{n} (1-f)^\lambda - \sum_{\lambda=1}^{n+1} (1-f)^\lambda \\ &= 1 - (1-f)^{n+1}. \end{aligned}$$

As n tends to infinity we obtain $f \cdot g = 1$ and $\|g\| \leq \sum_{\lambda=0}^{\infty} \varepsilon^\lambda = 1/(1-\varepsilon)$. ∎

Convergent Power Series. A formal power series $f = \sum_{\nu \geq 0} a_\nu \mathbf{z}^\nu$ is called *convergent* if it is convergent in some polydisk P around the origin. In that case there exists a point $\mathbf{t} \in P \cap \mathbb{R}^n_+$, and since f converges absolutely at $\mathbf{t}$, it follows that $f \in B_\mathbf{t}$. On the other hand, if $f \in B_\mathbf{t}$, then by Abel's lemma f converges in $P = \mathsf{P}^n(\mathbf{0}, \mathbf{t})$.

1.3 Theorem.

1. *$H_n := \{f \in \mathbb{C}[\![\mathbf{z}]\!] \,:\, \exists \mathbf{t} \in \mathbb{R}^n_+$ with $\|f\|_\mathbf{t} < \infty\}$ is the set of convergent power series.*
2. *H_n is a $\mathbb{C}$-algebra.*
3. *There is no* zero divisor *in H_n: If $f \cdot g = 0$ in H_n, then $f = 0$ or $g = 0$.*

We have already proved the first part, and then the second part follows easily. The last part is trivial, since $\mathbb{C}[\![\mathbf{z}]\!]$ contains no zero divisors.

Remark. If f is convergent and $f(\mathbf{0}) = 0$, then for every $\varepsilon > 0$ there is a $\mathbf{t} \in \mathbb{R}^n_+$ with $\|f\|_\mathbf{t} < \varepsilon$. In fact, since $f(\mathbf{0}) = 0$, we have a representation $f = z_1 f_1 + \cdots + z_n f_n$. If $\|f\|_\mathbf{t} < \infty$, then also $\|f_i\|_\mathbf{t} < \infty$ for $i = 1, \ldots, n$, and

$$\|f\|_t = \sum_{i=1}^n t_i \|f_i\|_\mathbf{t} \leq \max(t_1, \ldots, t_n) \cdot \sum_{i=1}^n \|f_i\|_\mathbf{t}.$$

This expression becomes arbitrarily small as $\mathbf{t} \to \mathbf{0}$.

When we go from $B_\mathbf{t}$ to H_n, we lose the norm and the Banach algebra structure, but we gain new algebraic properties:

1. $f \in H_n$ is a unit $\iff$ $f(\mathbf{0}) \neq 0$.

   PROOF: One direction is trivial. For the other one suppose that $f(\mathbf{0}) \neq 0$. Then $g := f \cdot f(\mathbf{0})^{-1} - 1$ is an element of H_n with $g(\mathbf{0}) = 0$. So there exists a $\mathbf{t}$ with $\|g\|_\mathbf{t} < 1$, and $f \cdot f(\mathbf{0})^{-1}$ is a unit in $B_\mathbf{t}$. Thus f is a unit in H_n. ∎

2. The set $\mathfrak{m} := \{f \in H_n \,:\, f(\mathbf{0}) = 0\}$ of all nonunits in H_n is an *ideal*:
   (a) $f_1, f_2 \in \mathfrak{m} \implies f_1 + f_2 \in \mathfrak{m}$.
   (b) $f \in \mathfrak{m}$ and $h \in H_n \implies h \cdot f \in \mathfrak{m}$.

An ideal $\mathfrak{a}$ in a ring R is called *maximal* if for every ideal $\mathfrak{b}$ with $\mathfrak{a} \subset \mathfrak{b} \subset R$ it follows that $\mathfrak{a} = \mathfrak{b}$ or $\mathfrak{b} = R$.

One can show that in any commutative ring with $1 \neq 0$ there exists a maximal ideal. If $\mathfrak{a} \subset R$ is maximal, then $R/\mathfrak{a}$ is a field.

1.4 Theorem. *The set $\mathfrak{m}$ of nonunits is the unique maximal ideal in H_n, and $H_n/\mathfrak{m} \cong \mathbb{C}$.*

PROOF: If $\mathfrak{a} \subset H_n$ is a proper ideal, then it cannot contain a unit. Therefore, it is contained in $\mathfrak{m}$. The homomorphism $\varphi : H_n \to \mathbb{C}$ given by $f \mapsto f(0)$ is surjective and has $\mathfrak{m}$ as kernel. ∎

Distinguished Directions. An element $f \in H_n$ is called *z_1-regular of order k* if there exists a power series $f_0(z_1)$ in one variable such that:

1. $f(z_1, 0, \dots, 0) = z_1^k \cdot f_0(z_1)$.
2. $f_0(0) \neq 0$.

If f is z_1-regular of some order, f is called *z_1-regular.*

Let $f(\mathbf{z}) = \sum_{\lambda=0}^{\infty} f_\lambda z_1^\lambda$ be the expansion of f with respect to z_1. Then f is z_1-regular of order k if and only if $f_0(\mathbf{0}') = \cdots = f_{k-1}(\mathbf{0}') = 0$ and $f_k(\mathbf{0}') \neq 0$. f is z_1-regular if and only if $f(z_1, 0, \dots, 0) \not\equiv 0$.

We often need the following properties:

1. f is a unit in $H_n \iff f$ is z_1-regular of order 0.
2. If f_λ is z_1-regular of order k_λ, for $\lambda = 1, 2$, then $f_1 \cdot f_2$ is z_1-regular of order $k_1 + k_2$.

There are elements $f \neq 0$ of H_n that $f(\mathbf{0}) = 0$ which are not z_1-regular, even after exchanging the coordinates.

Definition. Let $\mathbf{c} = (c_2, \dots, c_n)$ be an element of $\mathbb{C}^{n-1}$. The linear map $\sigma_\mathbf{c} : \mathbb{C}^n \to \mathbb{C}^n$ with

$$\sigma_\mathbf{c}(z_1, \dots, z_n) := (z_1, z_2 + c_2 z_1, \dots, z_n + c_n z_1)$$

is called a *shear.*

The set Σ of all shears is a subgroup of the group of linear automorphisms of $\mathbb{C}^n$, with $\sigma_\mathbf{0} = \mathrm{id}_{\mathbb{C}^n}$.

We can write $\sigma_\mathbf{c}(\mathbf{z}) := \mathbf{z} + z_1 \cdot (0, \mathbf{c})$. In particular, we have $\sigma_\mathbf{c}(\mathbf{e}_1) = (1, \mathbf{c})$.

1.5 Theorem. *Let $f \in H_n$ be a nonzero element. Then there exists a shear σ such that $f \circ \sigma$ is z_1-regular.*

PROOF: Assume that f converges in the polydisk P. If we had $f(z_1, \mathbf{z}') = 0$ for every point $(z_1, \mathbf{z}') \in P$ with $z_1 \neq 0$, then by continuity we would have $f = 0$, which can be excluded. Therefore, there exists a point $\mathbf{a} = (a_1, \mathbf{a}') \in P$ with $a_1 \neq 0$ and $f(\mathbf{a}) \neq 0$. We define $\mathbf{c} := (a_1)^{-1} \cdot \mathbf{a}'$ and $\sigma := \sigma_\mathbf{c}$. Now,

$$f \circ \sigma(a_1, \mathbf{0}') = f(a_1 \cdot \sigma_\mathbf{c}(\mathbf{e}_1)) = f(a_1 \cdot (1, \mathbf{c})) = f(\mathbf{a}) \neq 0.$$

So $f \circ \sigma(z_1, \mathbf{0}') \not\equiv 0$, and $f \circ \sigma$ is z_1-regular. ∎

Remark. If $f_1, \dots, f_l$ are nonzero elements in H_n, then $f := f_1 \cdots f_l \neq 0$ converges on a polydisc P, and there exists a point $\mathbf{a} \in P$ with $f(\mathbf{a}) \neq 0$ and $a_1 \neq 0$. As in the proof above we obtain a shear σ such that $f_1 \circ \sigma, \dots, f_l \circ \sigma$ are z_1-regular.

Exercises

1. Let $f(\mathbf{z}) = \sum_{\nu \geq 0} a_\nu \mathbf{z}^\nu$ be a formal power series.
   (a) Prove the "Cauchy estimates": $f \in B_\mathbf{t} \implies |a_\nu| \leq \|f\|_\mathbf{t}/\mathbf{t}^\nu$ for almost every ν.
   (b) Prove that if there is a constant C with $|a_\nu|\mathbf{s}^\nu \leq C$, then $f \in B_\mathbf{t}$ for $t \leq \mathbf{s}$.
   (c) Let $f_n(\mathbf{z}) = \sum_{\nu \geq 0} a_{\nu,n}\mathbf{z}^\nu$ be a sequence of power series with $\|f_n\|_\mathbf{s} \leq C$. If every sequence $(a_{\nu,n})$ converges in $\mathbb{C}$ to a number a_ν, then show that (f_n) is a Cauchy sequence in $B_\mathbf{t}$ converging to $f(\mathbf{z}) = \sum_{\nu \geq 0} a_\nu \mathbf{z}^\nu$, for every $\mathbf{t} \leq \mathbf{s}$.
2. The *Krull topology* on H_n is defined as follows: A sequence (f_n) converges in H_n to f if for every $k \in \mathbb{N}$ there is an n_0 with $f - f_n \in \mathfrak{m}^k$ for $n \geq n_0$. What are the open sets in H_n? Is H_n with the Krull topology a Hausdorff space?
3. Let B be a complex Banach algebra with 1. Show that for every $f \in B$ the series $\exp(f) = \sum_{n=0}^\infty f^n/n!$ is convergent, and that $\exp(f)$ is a unit in B.
4. If f is a formal power series and $f = \sum_{\lambda=0}^\infty p_\lambda$ its expansion into homogeneous polynomials, then the *order* of f is defined to be the number

$$\operatorname{ord}(f) := \min\{s \in \mathbb{N}_0 \,:\, p_s \neq 0\}.$$

   Now let (f_n) be a sequence of formal power series such that for every $k \in \mathbb{N}$ there is an n_0 with $\operatorname{ord}(f_n) \geq k$ for $n \geq n_0$. Show that $\sum_{n=1}^\infty f_n$ is a formal power series. Use this technique also for the following:

   If $g_1, \ldots, g_m$ are elements of H_n with $\operatorname{ord}(g_i) \geq 1$, then

$$\sum_{\mu \geq 0} a_\mu \mathbf{w}^\mu \mapsto \sum_{\mu \geq 0} a_\mu (g_1(\mathbf{z}), \ldots, g_m(\mathbf{z}))^\mu$$

   defines a homomorphism $\varphi : H_m \to H_n$ of complex algebras.

2. The Preparation Theorem

Division with Remainder in $B_\mathbf{t}$. Let a fixed element $\mathbf{t} \in \mathbb{R}_+^n$ be chosen. When no confusion is possible we write B in place of $B_\mathbf{t}$, B' in place of $B_{\mathbf{t}'}$, and $\|f\|$ in place of $\|f\|_\mathbf{t}$. The ring of polynomials in z_1 with coefficients in B' is denoted by $B'[z_1]$.

2.1 Weierstrass Formula in $B_\mathbf{t}$. *Let f and $g = \sum_{\lambda=0}^\infty g_\lambda z_1^\lambda$ be two elements of B. Assume that there exists an $s \in \mathbb{N}_0$ and a real number ε with $0 < \varepsilon < 1$ such that g_s is a unit in B' and $\|z_1^s - g \cdot g_s^{-1}\| < \varepsilon \cdot t_1^s$.*

Then there exists exactly one $q \in B$ *and one* $r \in B'[z_1]$ *with* $\deg(r) < s$ *such that*

$$f = q \cdot g + r,$$

with

$$\|g_s \cdot q\| \leq t_1^{-s} \cdot \|f\| \cdot \frac{1}{1-\varepsilon}$$

and

$$\|r\| \leq \|f\| \cdot \frac{1}{1-\varepsilon}.$$

PROOF: Let us first try to explain the idea of the proof. If $h \in B$, then there is a unique decomposition $h = q_h \cdot z_1^s + r_h$, where $r_h \in B'[z_1]$ and $\deg(r_h) < s$. If g is given, we define an operator $T = T_g : B \to B$ by

$$T(h) := g \cdot g_s^{-1} \cdot q_h + r_h.$$

If T were an isomorphism, then $f = T(T^{-1}f) = g \cdot (g_s^{-1} \cdot q_{T^{-1}f}) + r_{T^{-1}f}$ would be the desired decomposition. One knows from Banach space theory that T is an isomorphism if $\mathrm{id}_B - T$ is "small" in some sense. Since $\|(\mathrm{id}_B - T)(h)\| = \|z_1^s - gg_s^{-1}\| \cdot \|q_h\|$, one can, in fact, conclude from the hypothesis of the theorem that $\mathrm{id}_B - T$ is "small." Now $T^{-1} = \sum_{\lambda=0}^{\infty}(\mathrm{id}_B - T)^\lambda$. Since $(\mathrm{id}_B - T)^0 f = f$ and $(\mathrm{id}_B - T)^1 f = (z_1^s - gg_s^{-1})q_f$, we obtain the following algorithm:

Inductively we define sequences f_λ, q_λ, r_λ beginning with $f_0 = f = z_1^s q_0 + r_0$. If $f_\lambda = z_1^s q_\lambda + r_\lambda$ has been constructed for some $\lambda \geq 0$, then we define

$$f_{\lambda+1} := \left(z_1^s - gg_s^{-1}\right) q_\lambda,$$

and obtain $q_{\lambda+1}$ and $r_{\lambda+1}$ by the unique decomposition

$$f_{\lambda+1} = z_1^s q_{\lambda+1} + r_{\lambda+1}, \quad r_{\lambda+1} \in B'[z_1] \quad \text{with } \deg(r_{\lambda+1}) < s.$$

If we define $q := \sum_{\lambda=0}^{\infty} g_s^{-1} q_\lambda$ and $r := \sum_{\lambda=0}^{\infty} r_\lambda$, then

$$\begin{aligned} f = f_0 &= \sum_{\lambda=0}^{\infty} f_\lambda - \sum_{\lambda=0}^{\infty} f_{\lambda+1} \\ &= \sum_{\lambda=0}^{\infty} (f_\lambda - f_{\lambda+1}) \\ &= \sum_{\lambda=0}^{\infty} \left(gg_s^{-1} q_\lambda + r_\lambda\right) = g \cdot q + r. \end{aligned}$$

When using this algorithm we do not need the abstract transformation T and the Banach theory of such transformations. However, it is necessary to prove the convergence of all of the series that were used.

For this let $h := -(z_1^s - gg_s^{-1})$. Then $\|h\| < \varepsilon \cdot t_1^s$ and $gg_s^{-1} = z_1^s + h$.

From $f_\lambda = z_1^s q_\lambda + r_\lambda$ it follows that $\|r_\lambda\| \le \|f_\lambda\|$ and $\|q_\lambda\| \le t_1^{-s} \cdot \|f_\lambda\|$. Furthermore, from $f_{\lambda+1} = -h \cdot q_\lambda$ it follows that

$$\|f_{\lambda+1}\| \le \|h\| \cdot \|q_\lambda\| < \varepsilon \cdot \|f_\lambda\|.$$

Thus $\|f_\lambda\| < \varepsilon^\lambda \cdot \|f\|$ and $\sum_{\lambda=0}^\infty f_\lambda$ converges.

Since

$$\|g_s^{-1} q_\lambda\| < \varepsilon^\lambda t_1^{-s} \|g_s^{-1}\| \cdot \|f\| \quad \text{and} \quad \|r_\lambda\| < \varepsilon^\lambda \|f\|,$$

the series $q = \sum_{\lambda=0}^\infty g_s^{-1} q_\lambda$ and $r = \sum_{\lambda=0}^\infty r_\lambda$ also converge.

The estimates for $\|g_s q\|$ and $\|f\|$ follow readily:

$$\begin{aligned} \|g_s q\| &\le \sum_{\lambda=0}^\infty \|q_\lambda\| \le t_1^{-s} \|f\| \cdot \sum_{\lambda=0}^\infty \varepsilon^\lambda = t_1^{-s} \|f\| \cdot \frac{1}{1-\varepsilon}, \\ \|r\| &\le \sum_{\lambda=0}^\infty \|r_\lambda\| \le \|f\| \cdot \frac{1}{1-\varepsilon}. \end{aligned}$$

It still remains to show uniqueness. Assuming that there are two expressions of the form

$$f = q_1 g + r_1 = q_2 g + r_2,$$

it follows that

$$0 = (q_1 - q_2) \cdot g + (r_1 - r_2) = (q_1 - q_2) g_s z_1^s + (q_1 - q_2) g_s h + (r_1 - r_2)$$

and

$$\begin{aligned} \|(q_1 - q_2) g_s z_1^s\| &\le \|(q_1 - q_2) g_s z_1^s + (r_1 - r_2)\| \\ &= \|(q_1 - q_2) g_s h\| \\ &\le \varepsilon \cdot t_1^s \cdot \|(q_1 - q_2) g_s\| \\ &= \varepsilon \cdot \|(q_1 - q_2) g_s z_1^s\|. \end{aligned}$$

Since $0 < \varepsilon < 1$, $(q_1 - q_2) g_s z_1^s = 0$. Therefore, $q_1 = q_2$ and $r_1 = r_2$. ∎

2.2 Corollary. *If the assumptions of the theorem are satisfied and if in addition $f \in B'[z_1]$, $g \in B'[z_1]$, and $\deg(g) = s$, then $q \in B'[z_1]$ with $q = 0$ or $\deg(q) = \deg(f) - s$.*

PROOF: Let $d := \deg(f)$. For $d < s$ we have the decomposition $f = 0 \cdot g + f$. Hence we have to consider only the case $d \ge s$.

We assume that $\deg(f_\mu) \le d$ for $\mu = 0, \dots, \lambda$. Then $\deg(q_\lambda) \le d - s$, and therefore

$$\deg(f_{\lambda+1}) = \deg\big(f_\lambda - r_\lambda - g g_s^{-1} q_\lambda\big) \le \max(d, s-1, s+(d-s)) = d.$$

Hence $\deg(f_\lambda) \le d$ and $\deg(q_\lambda) \le d-s$ for all λ. It follows that $\deg(q) \le d-s$, and from $f = g \cdot q + r$ we can conclude that $\deg(q) = d - s$. ∎

The Weierstrass Condition. We use the notation from above.

Definition. Let $s \in \mathbb{N}_0$. An element $g = \sum_{\lambda=0}^{\infty} g_\lambda z_1^\lambda \in B$ satisfies the *Weierstrass condition* (or *W-condition*) at position s if:

1. g_s is a unit in B'.
2. $\|z_1^s - g g_s^{-1}\| < \frac{1}{2} t_1^s$.

Let R be an *integral domain.*[1] A polynomial $f(u) = f_s u^s + f_{s-1} u^{s-1} + \cdots + f_1 u + f_0 \in R[u]$ is called *monic* or *normalized* if $f_s = 1$. A polynomial $f \in B'[z_1]$ is normalized if and only if it is z_1-regular of some order $k \le s$.

2.3 Weierstrass preparation theorem in $B_\mathbf{t}$. *If $g \in B$ satisfies the W-condition at position s, then there exists exactly one normalized polynomial $\omega \in B'[z_1]$ of degree s and one unit $e \in B$ such that $g = e \cdot \omega$.*

PROOF: We apply the Weierstrass formula to $f = z_1^s$. There are uniquely determined elements $q \in B$ and $r \in B'[z_1]$ with $z_1^s = q \cdot g + r$ and $\deg(r) < s$ (we choose an $\varepsilon < \frac{1}{2}$ such that $\|z_1^s - g g_s^{-1}\| < \varepsilon t_1^s$).

But then $z_1^s - g g_s^{-1} = (q - g_s^{-1}) g + r$ is a decomposition in the sense of the Weierstrass formula. Therefore, we have the estimate

$$\|g_s q - 1\| \le t_1^{-s} \|z_1^s - g g_s^{-1}\| \cdot \frac{1}{1-\varepsilon} < \frac{\varepsilon}{1-\varepsilon} < 1.$$

That means that $g_s q$ and hence q is a unit in B. Let $e := q^{-1}$ and $\omega := z_1^s - r$. Then ω is a normalized polynomial of degree s, and $e \cdot \omega = q^{-1}(z_1^s - r) = g$.

If there are two decompositions $g = e_1(z_1^s - r_1) = e_2(z_1^s - r_2)$, then

$$z_1^s = e_1^{-1} \cdot g + r_1 = e_2^{-1} \cdot g + r_2.$$

From the uniqueness condition in the Weierstrass formula it follows that $e_1 = e_2$ and $r_1 = r_2$. ∎

2.4 Corollary. *If g is a polynomial in z_1, then e is also a polynomial in z_1.*

[1] An integral domain is a commutative nonzero ring in which the product of two nonzero elements is nonzero.

PROOF: We use the decomposition $z_1^s - gg_s^{-1} = (q - g_s^{-1})g + r$. From the Weierstrass formula it follows that

$$\|r\| \le \|z_1^s - gg_s^{-1}\| \cdot \frac{1}{1-\varepsilon} < t_1^s \cdot \frac{\varepsilon}{1-\varepsilon} < t_1^s.$$

Since $\omega_s = 1$, it is also true that

$$\|z_1^s - \omega\omega_s^{-1}\| = \|z_1^s - \omega\| = \|r\| < t_1^s.$$

Therefore, $g = e \cdot \omega + 0$ is a decomposition in the sense of the Weierstrass formula, and the proposition follows from Corollary 2.2. ∎

The Weierstrass preparation theorem serves as a "preparation for the examination of the zeros of a holomorphic function." If the function is represented by a convergent power series g, and there exists a decomposition $g = e \cdot \omega$ with a unit e and a "pseudopolynomial" $\omega(z_1, \mathbf{z}') = z_1^s + A_1(\mathbf{z}')z_1^{s-1} + \cdots + A_s(\mathbf{z}')$, then g and ω have the same zeros. However, the examination of ω is simpler than that of g.

Weierstrass Polynomials. Now we turn to the proof of the Weierstrass formula and the preparation theorem for convergent power series.

The ring H_n is an integral domain with 1. If $f \in H_n$ and $f(\mathbf{z}) = \sum_{\lambda=0}^{\infty} f_\lambda(\mathbf{z}')z_1^\lambda$ with $f_\lambda = 0$ for $\lambda > s$, then f is an element of the polynomial ring $H_{n-1}[z_1]$. If $f_s \ne 0$, then $\deg(f) = s$. If f is normalized and $f_\lambda(\mathbf{0}') = 0$ for $\lambda < s$, then f is z_1-regular exactly of order s, and $f(z_1, \mathbf{0}') = z_1^s$.

Definition. A normalized polynomial $\omega \in H_{n-1}[z_1]$ with $\deg(\omega) = s$ and $\omega(z_1, \mathbf{0}') = z_1^s$ is called a *Weierstrass polynomial.*

We have seen that a normalized polynomial $\omega \in H_{n-1}[z_1]$ with $\deg(\omega) = s$ is a Weierstrass polynomial if and only if it is z_1-regular of order s. It follows easily that the product of two Weierstrass polynomials is again a Weierstrass polynomial.

If $g = e \cdot \omega$ is the product of a unit and a Weierstrass polynomial of degree s, then we also have that g is z_1-regular of order s, since the unit e is z_1-regular of order 0. We now show that conversely, every z_1-regular convergent power series is the product of a unit and a Weierstrass polynomial.

2.5 Theorem. *Let $g \in H_n$ be z_1-regular of order s. Then for every $\varepsilon > 0$ and every $\mathbf{t}_0 \in \mathbb{R}_+^n$ there exists a $\mathbf{t} \le \mathbf{t}_0$ such that g lies in $B_\mathbf{t}$, g_s is a unit in $B_{\mathbf{t}'}$, and $\|z_1^s - gg_s^{-1}\|_\mathbf{t} \le \varepsilon \cdot t_1^s$.*

PROOF: Let $g = \sum_{\lambda=0}^{\infty} g_\lambda z_1^\lambda$ be the expansion of g with respect to z_1. Then $g_\lambda(\mathbf{0}') = 0$ for $\lambda = 0, 1, \dots, s-1$ and $g_s(\mathbf{0}') \ne 0$.

Since g is convergent, there exists a $\mathbf{t}_1 \le \mathbf{t}_0$ with $\|g\|_{\mathbf{t}_1} < \infty$. Then $g_\lambda \in B_{\mathbf{t}_1'}$ for all λ, and in particular,

$$f(\mathbf{z}') := g_s(\mathbf{0}')^{-1} g_s(\mathbf{z}') - 1 \in B_{\mathbf{t}'_1}.$$

Now, $f(\mathbf{0}') = 0$. Thus there exists a $\mathbf{t}_2 \le \mathbf{t}_1$ such that $\|f\|_{\mathbf{t}'} < 1$ for all $\mathbf{t} \le \mathbf{t}_2$. Therefore, g_s is a unit in $B_{\mathbf{t}'}$, and g an element of $B_{\mathbf{t}}$.

Let $h := z_1^s - g g_s^{-1}$. Then $h \in B_{\mathbf{t}}$ for all $\mathbf{t} \le \mathbf{t}_2$, and we have an expansion $h = \sum_{\lambda=0}^{\infty} h_\lambda z_1^\lambda$ with $h_s = 0$, $h_\lambda = -g_\lambda g_s^{-1}$ for $\lambda \ne s$, and $h_\lambda(\mathbf{0}') = 0$ for $\lambda = 0, 1, \ldots, s-1$.

If $t_1 > 0$ is sufficiently small, then

$$\Big\| \sum_{\lambda=s+1}^{\infty} h_\lambda z_1^\lambda \Big\|_{\mathbf{t}} \le t_1^{s+1} \cdot \Big\| \sum_{\lambda=s+1}^{\infty} h_\lambda z_1^{\lambda-s-1} \Big\|_{\mathbf{t}_2} < t_1^s \cdot \frac{\varepsilon}{2},$$

for all $\mathbf{t} = (t_1, \mathbf{t}') \le \mathbf{t}_2$. And since $h_\lambda(\mathbf{0}') = 0$ for $\lambda = 0, \ldots, s-1$, for every small t_1 there exists a suitable $\mathbf{t}'$ such that

$$\Big\| \sum_{\lambda=0}^{s-1} h_\lambda z_1^\lambda \Big\|_{\mathbf{t}} = \sum_{\lambda=0}^{s-1} \|h_\lambda\|_{\mathbf{t}'} t_1^\lambda < t_1^s \cdot \frac{\varepsilon}{2}.$$

Consequently, $\|h\|_{\mathbf{t}} \le \varepsilon \cdot t_1^s$. ∎

Remark. In a similar manner one can show that if $g_1, \ldots, g_N \in \mathbb{C}[\![\mathbf{z}]\!]$ are convergent power series and each g_i is z_1-regular of order s_i, then for every $\varepsilon > 0$ there is an arbitrary small $\mathbf{t} \in \mathbb{R}_+^n$ for which

$$g_i \in B_{\mathbf{t}},\ (g_i)_{s_i} \text{ is a unit in } B_{\mathbf{t}'} \text{ and } \|z_1^{s_i} - g_i (g_i)_{s_i}^{-1}\| \le \varepsilon \cdot t_1^{s_i}.$$

Weierstrass Preparation Theorem

2.6 Theorem (Weierstrass division formula). *Let $g \in H_n$ be z_1-regular of order s. Then for every $f \in H_n$ there are uniquely determined elements $q \in H_n$ and $r \in H_{n-1}[z_1]$ with* $\deg(r) < s$ *such that*

$$f = q \cdot g + r.$$

If f and g are polynomials in z_1 with $\deg(g) = s$, *then q is also a polynomial.*

PROOF: There exists a $\mathbf{t} \in \mathbb{R}_+^n$ and an ε with $0 < \varepsilon < 1$ such that f and g lie in $B_{\mathbf{t}}$, g_s is a unit in $B_{\mathbf{t}'}$, and $\|z_1^s - g g_s^{-1}\|_{\mathbf{t}} \le \varepsilon \cdot t_1^s$. It then follows from the division formula in $B_{\mathbf{t}}$ that there exist q and r with $f = q \cdot g + r$.

Let two decompositions of f be given:

$$f = q_1 \cdot g + r_1 = q_2 \cdot g + r_2.$$

We can find a $\mathbf{t} \in \mathbb{R}_+^n$ such that f, q_1, q_2, r_1, r_2 lie in $B_{\mathbf{t}}$ and g satisfies the W-condition in $B_{\mathbf{t}}$. From the Weierstrass formula in $B_{\mathbf{t}}$ it follows that $q_1 = q_2$ and $r_1 = r_2$. ∎

2.7 Theorem (Weierstrass preparation theorem). *Let $g \in H_n$ be z_1-regular of order s. Then there exists a uniquely determined unit $e \in H_n$ and a Weierstrass polynomial $\omega \in H_{n-1}[z_1]$ of degree s such that*

$$g = e \cdot \omega.$$

If g is a polynomial in z_1, then e is also a polynomial in z_1.

PROOF: There exists a $\mathbf{t} \in \mathbb{R}_+^n$ such that g satisfies the W-condition in $B_{\mathbf{t}}$. The existence of the decomposition $g = e \cdot \omega$ with a unit e and a normalized polynomial ω of degree s therefore follows directly from the preparation theorem in $B_{\mathbf{t}}$. Since g is z_1-regular of order s, the same is true for ω. So ω is a Weierstrass polynomial.

Now, ω has the form $\omega = z_1^s - r$, where $r \in H_{n-1}[z_1]$ and $\deg(r) < s$. Thus, if there exist two representations $g = e_1(z_1^s - r_1) = e_2(z_1^s - r_2)$, it follows that $z_1^s = e_1^{-1} \cdot g + r_1 = e_2^{-1} \cdot g + r_2$. The Weierstrass formula implies that $e_1 = e_2$, $r_1 = r_2$ and therefore $\omega_1 = \omega_2$. ∎

Exercises

1. Write a computer program to do the following: Given two polynomials $f(w, x, y)$ (of degree n in w and degree m in x and y) and $g(w, x, y)$ with $g(w, 0, 0) = w^s$, the program uses the Weierstrass algorithm to determine q and r (up to order m in x and y) such that $f = q \cdot g + r$.
2. Let $f : \mathsf{P}^{n-1} \times \mathsf{D} \to \mathbb{C}$ be a holomorphic function and $0 < r < 1$ be a real number such that $\zeta \mapsto f(\mathbf{z}', \zeta)$ has no zero for $\mathbf{z}' \in \mathsf{P}^{n-1}$ and $r \le |\zeta| < 1$. Then prove that there is a number k such that for every $\mathbf{z}' \in \mathsf{P}^{n-1}$ the function $\zeta \mapsto f(\mathbf{z}', \zeta)$ has exactly k zeros (with multiplicity) in D. Use this statement to give an alternative proof for the uniqueness in the Weierstrass preparation theorem.
3. Show that the implicit function theorem for a holomorphic function $f : \mathbb{C}^n \times \mathbb{C} \to \mathbb{C}$ with $f(\mathbf{0}) = 0$ and $f_{z_n}(\mathbf{0}) \ne 0$ follows from the Weierstrass preparation theorem.

3. Prime Factorization

Unique Factorization. Let I be an arbitrary integral domain with 1. Then $I^* := I - \{0\}$ is a commutative monoid with respect to the ring multiplication, and the set $I^\times$ of units of I is an abelian group.

Let a, b be elements of I^*. We say that a *divides* b (symbolically $a\,|\,b$) if there exists $c \in I^*$ with $b = a \cdot c$. We can also allow the case $b = 0$. Then every element of I^* divides 0, and a unit divides every element of I.

Definition. Consider an element $a \in I^* - I^\times$.

1. a is called *irreducible* (or *indecomposable*) if from $a = a_1 \cdot a_2$ (with $a_1, a_2 \in I^*$) it follows that $a_1 \in I^\times$ or $a_2 \in I^\times$.
2. a is called *prime* if $a\,|\,a_1 a_2$ implies that $a\,|\,a_1$ or $a\,|\,a_2$.

Irreducible and prime elements can be defined in an arbitrary commutative monoid. In I^* every prime element is irreducible, and in some rings (for example, in $\mathbb{Z}$ or in $\mathbb{R}[X]$) it is also the case that every irreducible element is prime. In $\mathbb{Z}[\sqrt{-5}]$ one can find irreducible elements that are not prime.

Definition. I is called a *unique factorization domain* (UFD) if every element $a \in I^\times$ can be written as a product of finitely many primes.

One can show that the decomposition into primes is uniquely determined up to order and multiplication by units. In a UFD every irreducible element is prime and any two elements have a *greatest common divisor* (gcd).

Every *principal ideal domain*[2] is a UFD, and in this case the greatest common divisor of two elements a and b can be written as a linear combination of a and b. For example, $\mathbb{Z}$ and $K[X]$ (with an arbitrary field K) are principal ideal domains. So in particular, $\mathbb{C}[X]$ is a UFD.

Gauss's Lemma. Let I be an integral domain. Two pairs $(a, b), (c, d) \in I \times I^*$ are called equivalent if $ad = bc$. The equivalence class of a pair (a, b) is called a *fraction* and is denoted by a/b. The set of all fractions has the structure of a field and is denoted by $Q(I)$. We call it the *quotient field* of I.

The set of polynomials $f(u) = a_0 + a_1 u + \cdots + a_n u^n$ in u with coefficients $a_i \in I$ constitutes the polynomial ring $I[u]$. The set $I^0[u]$ of monic polynomials in $I[u]$ is a commutative monoid. Therefore, we can speak of factorization and irreducibility in $I^0[u]$.

3.1 Gauss's lemma). *Let I be a unique factorization domain and $Q = Q(I)$. If ω_1, ω_2 are elements of $Q^0[u]$ with $\omega_1\omega_2 \in I^0[u]$, then $\omega_\lambda \in I^0[u]$ for $\lambda = 1, 2$.*

PROOF: For $\lambda = 1, 2$, $\omega_\lambda = a_{\lambda,0} + a_{\lambda,1}u + \cdots + a_{\lambda,s_\lambda - 1}u^{s_\lambda - 1} + u^{s_\lambda}$ with $a_{\lambda,\nu} \in Q$. Therefore, there exist elements $d_\lambda \in I$ such that $d_\lambda \cdot \omega_\lambda \in I[u]$. We can choose d_λ in such a way that the coefficients of $d_\lambda \cdot \omega_\lambda$ have no common divisor (such polynomials are called *primitive*).

[2] A principal ideal domain is an integral domain in which every ideal is generated by a single element.

We define $d := d_1 d_2$ and assume that there is a prime element p with $p \mid d$. Then p doesn't divide all coefficients $d_\lambda a_{\lambda,\nu}$ of $d_\lambda \cdot \omega_\lambda$. Let μ_λ be minimal such that $p \nmid d_\lambda a_{\lambda,\mu_\lambda}$. Then

$$(d_1\omega_1)(d_2\omega_2) = \cdots + u^{\mu_1+\mu_2}(da_{1,\mu_1}a_{2,\mu_2} + \text{ something divisible by } p) + \cdots .$$

Since I is a UFD, p doesn't divide $(d_1 a_{1,\mu_1})(d_2 a_{2,\mu_2})$. So the coefficient of $u^{\mu_1+\mu_2}$ is not divisible by p. But since $\omega_1\omega_2$ has coefficients in I, every divisor of d must divide every coefficient of $d \cdot \omega_1\omega_2 = (d_1\omega_1)(d_2\omega_2)$. This is a contradiction![3]

When d has no prime divisor, it must be a unit. But then d_1 and d_2 are also units, and $\omega_\lambda = d_\lambda^{-1}(d_\lambda\omega_\lambda)$ belongs to $I^0[u]$. ∎

3.2 Corollary. *Let I be a unique factorization domain.*

1. *If $a \in I^0[u]$ is prime in $Q[u]$, then it is also prime in $I^0[u]$.*
2. *If $a \in I^0[u]$ is reducible in $Q[u]$, then it is reducible in $I^0[u]$.*
3. *Every element of $I^0[u]$ is a product of finitely many prime elements.*
4. *If $a \in I^0[u]$ is irreducible, it is also prime.*

PROOF: 1. Let $a \in I^0[u]$ be a prime element in $Q[u]$. If a divides a product $a'a''$ in $I^0[u]$, then it does so in $Q[u]$. Therefore, it divides one of the factors in $Q[u]$. Assume that there is an element $b \in Q[u]$ with $a' = ab$. By Gauss's lemma $b \in I^0[u]$. This shows that a is prime in $I^0[u]$.

2. Let $a \in I^0[u]$ be a product of nonunits $a_1, a_2 \in Q[u]$. If $c_i \in Q$ is the highest coefficient of a_i, then $c_1c_2 = 1$, $c_i^{-1}a_i \in Q^0[u]$ and $a = (c_1^{-1}a_1)(c_2^{-1}a_2)$. By Gauss $c_i^{-1}a_i \in I^0[u]$, and these elements cannot be units there. So a is reducible in $I^0[u]$.

3. Every element $a \in I^0[u]$ is a finite product $a = a_1 \cdots a_l$ of prime elements of $Q[u]$. One can choose the a_i monic, as in (2). Using Gauss's lemma several times one shows that the a_i belong to $I^0[u]$. By (1) they are also prime in $I^0[u]$.

4. Let $a \in I^0[u]$ be irreducible. Since it is a product of prime elements, it must be prime itself. ∎

Remark. In the proof we didn't use that I is a unique factorization domain. We needed only the fact that $Q[u]$ is a UFD (since Q is a field) and the statement of Gauss's lemma: If $a_1, a_2 \in Q^0[u]$ and $a_1a_2 \in I^0[u]$, then $a_i \in I^0[u]$ for $i = 1, 2$.

---

[3] The original version of Gauss's lemma states that the product of primitive polynomials is again primitive. The reader may convince himself that this fact can be derived from our proof.

Factorization in H_n. Now the above results will be applied to the case $I = H_n$.

> **Definition.** Let $f \in H_n$, $f = \sum_{\lambda=0}^{\infty} p_\lambda$ be the expansion of f as a series of homogeneous polynomials. One defines the *order* of f by the number
>
> $$\operatorname{ord}(f) := \min\{\lambda \in \mathbb{N}_0 \,:\, p_\lambda \neq 0\} \quad \text{and} \quad \operatorname{ord}(0) := \infty.$$
>
> (See also Exercise 1.4 in this chapter)

Then the following hold:

1. $\operatorname{ord}(f) \geq 0$ for every $f \in H_n$.
2. $\operatorname{ord}(f) = 0 \iff f$ is a unit.
3. $\operatorname{ord}(f_1 \cdot f_2) = \operatorname{ord}(f_1) + \operatorname{ord}(f_2)$.

3.3 Theorem. *H_n is a unique factorization domain.*

PROOF: We proceed by induction on n.

For $n = 0$, $H_n = \mathbb{C}$ is a field, and every nonzero element is a unit. In this case there is nothing to show.

Now suppose that the theorem has been proved for $n-1$. Let $f \in H_n$ be a nonunit, $f \neq 0$. If f is decomposable and $f = f_1 \cdot f_2$ is a proper decomposition, then $\operatorname{ord}(f) = \operatorname{ord}(f_1) + \operatorname{ord}(f_2)$, and the orders of the factors are strictly smaller than the order of f. Therefore, f can be decomposed into a finite number of irreducible factors.

It remains to show that an irreducible f is prime. Assume that $f \,|\, f_1 f_2$, with $f_\lambda \in (H_n)^*$ for $\lambda = 1, 2$. There exists a shear σ such that $f_1 \circ \sigma$, $f_2 \circ \sigma$ and $f \circ \sigma$ are z_1-regular. If we can show that $f \circ \sigma$ divides one of the $f_\lambda \circ \sigma$, then the same is true for f and f_λ. Therefore, we may assume that f_1, f_2, and f are z_1-regular.

By the preparation theorem there are units e_1, e_2, e and Weierstrass polynomials $\omega_1, \omega_2, \omega$ such that $f_1 = e_1 \cdot \omega_1$, $f_1 = e_2 \cdot \omega_2$, and $f = e \cdot \omega$. Then ω divides $\omega_1 \omega_2$. If $\omega_1 \omega_2 = q \cdot \omega$ with $q \in H_n$, then the division formula says that q is uniquely determined and a polynomial in z_1. So ω divides $\omega_1 \omega_2$ in $H_{n-1}^0[z_1]$. Since ω is irreducible in H_n, it must also be irreducible in $H_{n-1}^0[z_1]$. By the induction hypothesis H_{n-1} is a UFD, and therefore ω is prime in $H_{n-1}^0[z_1]$. It follows that $\omega \,|\, \omega_1$ or $\omega \,|\, \omega_2$ in $H_{n-1}^0[z_1]$ and consequently in H_n. This means that $f \,|\, f_1$ or $f \,|\, f_2$ in H_n. ∎

Hensel's Lemma. Let $\omega \in H_n[u]$ be a monic polynomial of degree s. There is a polydisk P around $\mathbf{0} \in \mathbb{C}^n$ where all the coefficients of ω converge to holomorphic functions. Therefore, we can look at ω as a parametrized family of polynomials in one variable u. By the fundamental theorem of algebra $\omega(\mathbf{0}, u)$ splits into linear factors, and the same is true for every $\omega(\mathbf{z}, u)$ with $\mathbf{z} \in P$. We now show that such splittings are coherently induced by some splitting of ω in $H_n^0[u]$, at least in a neighborhood of $\mathbf{0}$.

3.4 Hensel's lemma. *Let $\omega(\mathbf{0}, u) = \prod_{\lambda=1}^{l}(u - c_\lambda)^{s_\lambda}$ be the decomposition into linear factors (with $c_\nu \neq c_\mu$ for $\nu \neq \mu$ and $s_1 + \cdots + s_l = s$). Then there are uniquely determined polynomials $\omega_1, \ldots, \omega_l \in H_n^0[u]$ with the following properties:*

1. *$\deg(\omega_\lambda) = s_\lambda$, for $\lambda = 1, \ldots, l$.*
2. *$\omega_\lambda(\mathbf{0}, u) = (u - c_\lambda)^{s_\lambda}$.*
3. *$\omega = \omega_1 \cdot \ldots \cdot \omega_l$.*

PROOF: We proceed by induction on the number l. The case $l = 1$ is trivial. We assume that the theorem has been proved for $l - 1$.

First consider the case $\omega(\mathbf{0}, 0) = 0$. Without loss of generality we can assume that $c_1 = 0$. Then $\omega(\mathbf{0}, u) = u^{s_1} \cdot h(u)$, where h is a polynomial over $\mathbb{C}$ with $\deg(h) = s - s_1$ and $h(0) \neq 0$. So ω is u-regular of order s_1, and there exists a unit $e \in H_n^0[u]$ and a Weierstrass polynomial ω_1 with $\omega = e \cdot \omega_1$. Since $\omega_1(\mathbf{0}, u) = u^{s_1}$, it follows that

$$e(\mathbf{0}, u) = h(u) = \prod_{\lambda=2}^{l}(u - c_\lambda)^{s_\lambda}.$$

By induction there are elements $\omega_2, \ldots, \omega_l \in H_n^0[u]$ with $\deg(\omega_\lambda) = s_\lambda$, $\omega_\lambda(\mathbf{0}, u) = (u - c_\lambda)^{s_\lambda}$ and $e = \omega_2 \cdots \omega_l$. Then $\omega = \omega_1 \omega_2 \cdots \omega_l$ is the desired decomposition.

If $\omega(\mathbf{0}, 0) \neq 0$, then we replace ω by $\omega'(\mathbf{z}, u) := \omega(\mathbf{z}, u + c_1)$ and obtain a decomposition $\omega' = \omega_1' \cdots \omega_l'$ as above. Define

$$\omega_\lambda(\mathbf{z}, u) := \omega_\lambda'(\mathbf{z}, u - c_1).$$

This gives a decomposition $\omega = \omega_1 \cdots \omega_l$ in the sense of the theorem. The uniqueness statement also follows by induction. ∎

The Noetherian Property. Let R be a commutative ring with 1. An *R-module* is an abelian group M (additively written) together with a composition $R \times M \to M$ that satisfies the following rules:

1. $r(x_1 + x_2) = rx_1 + rx_2$ for $r \in R$ and $x_1, x_2 \in M$.
2. $(r_1 + r_2)x = r_1 x + r_2 x$ for $r_1, r_2 \in R$ and $x \in M$.
3. $r_1(r_2 x) = (r_1 r_2)x$ for $r_1, r_2 \in R$ and $x \in M$.

4. $1 \cdot x = x$ for $x \in M$.

These are the same rules as those for vector spaces (and the elements of a module are sometimes also called vectors). However, it may happen that $rx = 0$ even if $r \neq 0$ and $x \neq 0$. Therefore, in general, an R-module has no basis. So-called *free* modules have bases by definition. An example is the free module $R^q := R \times \cdots \times R$ (q times), with a basis of unit vectors. An example of a nonfree module is the $\mathbb{Z}$-module $M := \mathbb{Z}/6\mathbb{Z}$, where $2 \cdot \overline{3} = \overline{2 \cdot 3} = \overline{0}$.

If M is an R-module, then a submodule of M is a subset $N \subset M$ with the following properties:

1. $x, y \in N \implies x + y \in N$.
2. $r \in R$ and $x \in N \implies rx \in N$.

A submodule of an R-module is itself an R-module.

Example

The ring R is also an R-module. The composition is the ordinary ring multiplication. In this case the submodules of R are exactly the ideals in R. An R-module M is called *finite* if there is a finite set $\{x_1, \ldots, x_n\} \subset M$ such that every $x \in M$ is a linear combination of the x_i with coefficients in R. The free module R^q is obviously finite. But $\mathbb{Z}/6\mathbb{Z}$ is also finite, being generated by the class $\overline{1}$.

Definition. An R-module M is called *noetherian* if every submodule $N \subset M$ is finite.

A ring R is called *noetherian* if it is a noetherian R-module. This means that every ideal in R is finitely generated (in the sense of a module).

3.5 Proposition. *Let R be a noetherian ring. Then any ascending chain of ideals*

$$I_0 \subset I_1 \subset I_2 \subset \cdots \subset R$$

becomes stationary, i.e., there is a k_0 such that $I_k = I_{k_0}$ for $k \geq k_0$.

PROOF: The set $J := \bigcup_{k=0}^{\infty} I_k$ is obviously an ideal. Since R is noetherian, J is generated by finitely many elements $f_1, \ldots, f_N$. Each f_ν lies in an ideal I_{k_ν}. If $k_0 = \max(k_1, \ldots, k_N)$, then all f_ν are elements of I_{k_0}. So $I_k = I_{k_0}$ for $k \geq k_0$. ∎

3.6 Theorem. *If R is a noetherian ring, then R^q is a noetherian R-module.*

PROOF: We proceed by induction on q.

The case $q = 1$ is trivial. Assume that $q \geq 2$ and the theorem has been proved for $q - 1$. Let $M \subset R^q$ be an R-submodule. Then

$$I := \{r \in R : \exists \mathbf{r}' \in R^{q-1} \text{ with } (r, \mathbf{r}') \in M\}$$

is an ideal in R and as such is finitely generated by elements $r_1, \ldots, r_l$. For every r_λ there is an element $\mathbf{r}'_\lambda \in R^{q-1}$ such that $\mathbf{r}_\lambda := (r_\lambda, \mathbf{r}'_\lambda)$ lies in M.

The set $M' := M \cap (\{0\} \times R^{q-1})$ can be identified with an R-submodule of R^{q-1}, and by the induction assumption it is finite. Let $\mathbf{r}_\lambda = (0, \mathbf{r}'_\lambda)$, $\lambda = l+1, \ldots, p$, be generators of M'.

An arbitrary element $\mathbf{x} \in M$ can be written in the form $\mathbf{x} = (x_1, \mathbf{x}')$ with $x_1 \in I$. Then $x_1 = \sum_{\lambda=1}^{l} a_\lambda r_\lambda$, $a_\lambda \in R$, and

$$\mathbf{x} - \sum_{\lambda=1}^{l} a_\lambda \mathbf{r}_\lambda = \big(0, \mathbf{x}' - \sum_{\lambda=1}^{l} a_\lambda \mathbf{r}'_\lambda\big) \in M'.$$

That is, there are elements $a_{l+1}, \ldots, a_p \in R$ such that

$$\mathbf{x} - \sum_{\lambda=1}^{l} a_\lambda \mathbf{r}_\lambda = \sum_{\lambda=l+1}^{p} a_\lambda \mathbf{r}_\lambda.$$

Hence $\{\mathbf{r}_1, \ldots, \mathbf{r}_p\}$ is a system of generators for M. ∎

3.7 Rückert basis theorem. *The ring H_n of convergent power series is noetherian.*

PROOF: We proceed by induction on n. For $n = 0$, $H_n = \mathbb{C}$, and the statement is trivial. We now assume that $n \geq 1$ and that the theorem has been proved for $n - 1$. Let $I \subset H_n$ be a nonzero ideal and $g \neq 0$ an element of I. Without loss of generality we can further assume that g is z_1-regular of order s.

Let $\Phi = \Phi_g : H_n \to (H_{n-1})^s$ be the *Weierstrass homomorphism*, which is defined in the following manner: For every $f \in H_n$ there are uniquely defined elements $q \in H_n$ and $r = r_0 + r_1 z_1 + \cdots + r_{s-1} z_1^{s-1} \in H_{n-1}[z_1]$ such that $f = q \cdot g + r$. Let $\Phi(f) := (r_0, \ldots, r_{s-1})$.

Now, Φ is an H_{n-1}-module homomorphism. By the induction hypothesis H_{n-1} is noetherian, and so $(H_{n-1})^s$ is a noetherian H_{n-1}-module. Since $M := \Phi(I)$ is an H_{n-1}-submodule, it is finitely generated. Let $\mathbf{r}_\lambda = (r_0^{(\lambda)}, \ldots, r_{s-1}^{(\lambda)})$, $\lambda = 1, \ldots, l$, be generators of M.

If $f \in I$ is arbitrary, then $f = q \cdot g + r$ with $r = r_0 + r_1 z_1 + \cdots + r_{s-1} z_1^{s-1}$, and there are elements $a_1, \ldots, a_l \in H_{n-1}$ such that $(r_0, r_1, \ldots, r_{s-1}) = \Phi_g(f) = \sum_{\lambda=1}^{l} a_\lambda \mathbf{r}_\lambda$. Hence we obtain the representation

$$f = q \cdot g + \sum_{\lambda=1}^{l} a_\lambda \cdot \left(r_0^{(\lambda)} + r_1^{(\lambda)} z_1 + \cdots + r_{s-1}^{(\lambda)} z_1^{s-1}\right).$$

The set $\{g, r^{(1)}, \ldots, r^{(l)}\}$ with $r^{(\lambda)} = r_0^{(\lambda)} + r_1^{(\lambda)} z_1 + \cdots + r_{s-1}^{(\lambda)} z_1^{s-1}$ is a system of generators of I. ∎

Exercises

1. Prove that $\mathscr{O}(\mathbb{C})$ is not a UFD.
2. Let M be a finite H_n-module, and $\mathfrak{m} \subset H_n$ the maximal ideal. If $M = \mathfrak{m} \cdot M$, then $M = 0$.
3. Let $f : \mathsf{P}^{n-1} \times \mathsf{D} \to \mathbb{C}$ be a holomorphic function such that for every $\mathbf{z}' \in \mathsf{P}^{n-1}$ there is a unique solution $z_n = \varphi(\mathbf{z}') \in \mathsf{D}$ of the equation $f(\mathbf{z}', z_n) = 0$. Use function theory of one variable to show that φ is continuous, and use Hensel's lemma to show that φ is holomorphic.
4. Let $f \in H_n$ be z_1-regular, $f = e \cdot \omega$ with a unit e and a Weierstrass polynomial $\omega \in H_{n-1}[z_1]$. Prove that f is irreducible in H_n if and only if ω is irreducible in $H_{n-1}[z_1]$.
5. Show that $f(z, w) := z^2 - w^2(1 - w)$ is irreducible in the polynomial ring $\mathbb{C}[z, w]$ and reducible in H_n.
6. Let $f \in H_n$ be given with $f_{z_i}(\mathbf{0}) \neq 0$ for some i. Prove that f is irreducible in H_n.

4. Branched Coverings

Germs. Let $B \subset \mathbb{C}^n$ be an open set and $\mathbf{z}_0 \in B$ a fixed point. A *local holomorphic function* at $\mathbf{z}_0$ is a pair (U, f) consisting of an arbitrary neighborhood $U = U(\mathbf{z}_0) \subset B$ and a holomorphic function f on U. Two such functions $f : U \to \mathbb{C}$ and $g : V \to \mathbb{C}$ are called equivalent if there is a neighborhood $W = W(\mathbf{z}_0) \subset U \cap V$ such that $f|W = g|W$. The equivalence class of a local holomorphic function (U, f) at $\mathbf{z}_0$ is called a *germ* (of holomorphic functions) and is denoted by $f_{\mathbf{z}_0}$. The value $f(\mathbf{z}_0)$ as well as all derivatives of f at $\mathbf{z}_0$ (and therefore the Taylor series of f at $\mathbf{z}_0$) are uniquely determined by the germ. On the other hand, if a convergent power series at $\mathbf{z}_0$ is given, then this series converges in an open neighborhood of $\mathbf{z}_0$ to a holomorphic function f, and the germ of f determines the given power series. So the set $\mathscr{O}_{\mathbf{z}_0}$ of all germs of holomorphic functions at $\mathbf{z}_0$ can be identified with the $\mathbb{C}$-algebra of all convergent power series of the form $\sum_{\nu \geq 0} (\mathbf{z} - \mathbf{z}_0)^\nu$. This algebra is isomorphic to the algebra H_n and has the same algebraic properties.

Let $f_{\mathbf{z}_0} \neq 0$ be any element of $\mathscr{O}_{\mathbf{z}_0}$ with $f(\mathbf{z}_0) = 0$. Then there are a neighborhood $U(\mathbf{z}_0) \subset B$, a neighborhood $W(\mathbf{0}) \subset \mathbb{C}^{n-1}$, a holomorphic function e on U, and holomorphic functions $a_1, \ldots, a_s$ on W such that after a suitable change of coordinates the following hold:

1. $e(\mathbf{z}) \neq 0$ for every $\mathbf{z} \in U$.
2. $f(\mathbf{z}) = e(\mathbf{z}) \cdot \omega(\mathbf{z} - \mathbf{z}_0)$ for $\omega(w_1, \mathbf{w}') = w_1^s + a_1(\mathbf{w}')z_1^{s-1} + \cdots + a_s(\mathbf{w}')$.

Pseudopolynomials. A *pseudopolynomial* of degree s over a domain $G \subset \mathbb{C}^n$ is a holomorphic function ω in $G \times \mathbb{C}$ that is given by an expression

$$\omega(u, \mathbf{z}) = u^s + h_1(\mathbf{z})u^{s-1} + \cdots + h_s(\mathbf{z}),$$

with $h_1, \ldots, h_s \in \mathcal{O}$, where $\mathcal{O} = \mathcal{O}(G)$ denotes the ring of holomorphic functions on G. The set of pseudopolynomials of any degree over G will be written as $\mathcal{O}^0[u]$.

We begin with several remarks on the algebraic structure.

4.1 Proposition. *If G is a domain, i.e., a connected open set, then the ring $\mathcal{O} = \mathcal{O}(G)$ is an integral domain.*

PROOF: We need to show only that $\mathcal{O}$ has no zero divisors. Assume that f_1, f_2 are two holomorphic functions on G with both $f_i \not\equiv 0$. Since G is a domain, their zero sets are both nowhere dense in G, and there is a point $\mathbf{z} \in G$ with $f_1(\mathbf{z}) \cdot f_2(\mathbf{z}) \neq 0$. So $f_1 \cdot f_2 \not\equiv 0$. ∎

It also follows that $\mathcal{O}^0[u]$ is free of zero divisors. We denote by Q the quotient field of $\mathcal{O}$. Then the group $Q[u]^\times$ of units in the integral domain $Q[u]$ consists of the nonzero polynomials of degree 0. If $\mathcal{O}^* \subset \mathcal{O}$ is the multiplicative subgroup of not identically vanishing holomorphic functions on G, then $Q[u]^\times \cap \mathcal{O} = \mathcal{O}^*$.

4.2 Proposition. *If $\omega_1, \omega_2 \in Q^0[u]$ are pseudopolynomials with $\omega_1 \cdot \omega_2 \in \mathcal{O}^0[u]$, then $\omega_1, \omega_2 \in \mathcal{O}^0[u]$.*

PROOF: If $\omega = u^s + (f_1/g_1)u^{s-1} + \cdots + (f_s/g_s)$ is an arbitrary element of $Q^0[u]$, then for all $\mathbf{z} \in G$ the germs $g_{i,\mathbf{z}}$ are not 0.

For a moment we omit the i. If the quotient of $f_\mathbf{z}$ and $g_\mathbf{z}$ is holomorphic, i.e., $f_\mathbf{z} = h_\mathbf{z} \cdot g_\mathbf{z}$ with $h_\mathbf{z} \in \mathcal{O}_\mathbf{z}$, then $h_\mathbf{z}$ is uniquely determined and there is a ball B around $\mathbf{z}$ in G such that $h_\mathbf{z}$ comes from a holomorphic function h on B and the equation $f = h \cdot g$ is valid in B. If we take another point $\mathbf{z}' \in B$, the germ of h at this point is the quotient of the germs of f and g at this point. So $\mathbf{z} \mapsto h_\mathbf{z}(\mathbf{z})$ defines a global holomorphic function h on G. We write $h = f/g$.

Thus, if $\omega_\mathbf{z} := u^s + ((f_1)_\mathbf{z}/(g_1)_\mathbf{z})u^{s-1} + \cdots + ((f_s)_\mathbf{z}/(g_s)_\mathbf{z})$ lies in $\mathcal{O}_\mathbf{z}^0[u]$ for every $\mathbf{z} \in G$, then $\omega \in \mathcal{O}^0[u]$.

Now we apply Gauss's lemma in the unique factorization domain $\mathcal{O}_\mathbf{z} \cong H_n$. Let $\omega := \omega_1 \cdot \omega_2$. Then $(\omega_1)_\mathbf{z} \cdot (\omega_2)_\mathbf{z} = \omega_\mathbf{z} \in \mathcal{O}_\mathbf{z}^0[u]$ for every $\mathbf{z} \in G$. Consequently, the coefficients of $(\omega_i)_\mathbf{z}$ are holomorphic, and by the remarks above this means that $\omega_i \in \mathcal{O}^0[u]$. ∎

The following is an immediate consequence of the above two propositions:

4.3 Theorem. *Let $G \in \mathbb{C}^n$ be a domain. Then $\mathcal{O}^0[u]$ is a factorial monoid; i.e., every element is a product of finitely many primes.*

Euclidean Domains

Definition. An integral domain I is called a *Euclidean domain* if there is a function $N : I^* \to \mathbb{N}_0$ with the following property (division with remainder): For all $a, b \in I, b \neq 0$, there exist $q, r \in I$ with

1. $a = q \cdot b + r$,
2. $r = 0$ or $N(r) < N(b)$.

The function N is called the *norm* of the Euclidean domain.

Examples

1. $\mathbb{Z}$ is a Euclidean domain with $N(a) := |a|$.
2. If k is a field, then $k[x]$ is a Euclidean domain, by $N(f) := \deg(f)$.

Every Euclidean domain I is a principal ideal domain and thus factorial. If a, b are elements of I, then the set of all linear combinations

$$r \cdot a + s \cdot b \neq 0, \quad r, s \in I,$$

has an element d with $N(d)$ minimal. The element d generates the ideal $\mathfrak{a} = \{ra + sb \, : \, r, s \in I\}$ and is a greatest common divisor of a and b. It is determined up to multiplication by a unit.

Now assume again that G is a domain in $\mathbb{C}^n$, $\mathcal{O} = \mathcal{O}(G)$, and $Q = Q(\mathcal{O})$. Then $Q[u]$ is a Euclidean domain. If ω_1, ω_2 are pseudopolynomials in $\mathcal{O}^0[u]$, there is a linear combination in $\omega = r_1\omega_1 + r_2\omega_2 \not\equiv 0$ in $Q[u]$ with minimal degree. It can be multiplied by the product of the denominators of the coefficients in r_1 and r_2. Then r_1, r_2, and ω are in $\mathcal{O}[u]$, and ω is a greatest common divisor of ω_1, ω_2.

The Algebraic Derivative. Let $\mathcal{O}$ and Q be as above. If $\omega \in \mathcal{O}^0[u]$ has positive degree, then it has a unique prime decomposition $\omega = \omega_1 \cdots \omega_l$. The degree of each ω_i is positive. We say that ω *is (a pseudopolynomial) without multiple factors* if all the ω_i are distinct.

The *(algebraic) derivative* of a pseudopolynomial is defined as follows. If $\omega = \sum_{\nu=0}^{s} a_\nu u^\nu$, then $D(\omega) := \sum_{\nu=1}^{s} \nu \cdot a_\nu \cdot u^{\nu-1}$. Thus

$$\begin{aligned} D(\omega_1 + \omega_2) &= D(\omega_1) + D(\omega_2), \\ D(\omega_1 \cdot \omega_2) &= D(\omega_1) \cdot \omega_2 + \omega_1 \cdot D(\omega_2). \end{aligned}$$

4.4 Theorem. *An element $\omega \in \mathcal{O}^0[u]$ is without multiple factors if and only if a greatest common divisor of ω and $D(\omega)$ is a function $h \in \mathcal{O}^*$.*

PROOF: If ω has the irreducible ω_i as a multiple factor, then $D(\omega)$ is also divisible by ω_i. This is also true in $Q[u]$. So a greatest common divisor is certainly not a function $h \in \mathcal{O}^*$.

Assume now that $\omega = \prod_i \omega_i$ has no multiple factor. Then

$$D(\omega) = \sum_i \omega_1 \cdots D(\omega_i) \cdots \omega_l.$$

If the degree of the greatest common divisor γ of ω and $D(\omega)$ is positive, then γ is a product of certain ω_i. So at least one ω_i divides both ω and $D(\omega)$. Then ω_i divides $\omega_1 \cdots D(\omega_i) \cdots \omega_l$ and hence $D(\omega_i)$. This is not possible, because $D(\omega_i)$ has lower degree. So the degree of the greatest common divisor is 0, and therefore it is a function $h \in \mathcal{O}^*$. ■

Symmetric Polynomials.

Definition. A polynomial $p \in \mathbb{Z}[u_1, \dots, u_s]$ is called *symmetric* if for all i, j we have $p(u_1, \dots, u_i, \dots u_j, \dots, u_s) = p(u_1, \dots, u_j, \dots, u_i, \dots, u_s)$.

There are the *elementary symmetric polynomials* $\sigma_1, \dots, \sigma_s$ defined as follows:

$$\begin{aligned} \sigma_1(u_1, \dots, u_s) &= u_1 + \cdots + u_s, \\ \sigma_2(u_1, \dots, u_s) &= u_1(u_2 + \cdots + u_s) + u_2(u_3 + \cdots + u_s) + \cdots + u_{s-1}u_s, \\ &\vdots \\ \sigma_s(u_1, \dots, u_s) &= u_1 \cdots u_s. \end{aligned}$$

The following result is proved, e.g., in the book [vdW66].

4.5 Theorem. *If $p \in \mathbb{Z}[u_1, \dots, u_s]$ is symmetric, then there is exactly one polynomial $Q(y_1, \dots, y_s) \in \mathbb{Z}[y_1, \dots, y_s]$ such that $p = Q(\sigma_1, \dots, \sigma_s)$.*

The Discriminant. Consider the special symmetric polynomial

$$p_V(u_1, \dots, u_s) = \prod_{i<j} (u_i - u_j)^2$$

(square of the Vandermonde determinant). Since it is symmetric, there is a uniquely determined polynomial $Q_V(y_1, \dots, y_s)$ with integral coefficients such that

$$p_V(u_1, \dots, u_s) = Q_V(\sigma_1(u_1, \dots, u_s), \dots, \sigma_s(u_1, \dots, u_s)).$$

Definition. If $\omega = u^s + h_1(\mathbf{z})u^{s-1} + \cdots + h_s(\mathbf{z})$ is a pseudopolynomial in $\mathcal{O}^0[u]$, then $\Delta_\omega = Q_V(-h_1, h_2, \ldots, (-1)^s h_s)$ is called the *discriminant* of ω. It is a holomorphic function in G, and we denote its zero set by D_ω.

It is well known from the theory of polynomials that

$$(-1)^i h_i(\mathbf{z}) = \sigma_i(w_1, \ldots, w_s),$$

where $w_1, \ldots, w_s$ are the zeros of the polynomial $u \mapsto \omega(u, \mathbf{z})$. So $\Delta_\omega(\mathbf{z}) = 0$ if and only if there is a pair $i \neq j$ with $w_i = w_j$.

Assume now that ω is without multiple factors. Then there is a linear combination of ω and $D(\omega)$ that is a function $h \in \mathcal{O}^*$. We restrict to a point $\mathbf{z} \in G$ with $h(\mathbf{z}) \neq 0$. Then the greatest common divisor of $\omega(u, \mathbf{z})$ and $D(\omega)(u, \mathbf{z})$ is 1. This means that $\omega(u, \mathbf{z})$ has no multiple factors; i.e., the zeros of $\omega(u, \mathbf{z})$ are all distinct. So $\Delta_\omega(\mathbf{z}) \neq 0$, and D_ω is nowhere dense.

Example

Let $G \subset \mathbb{C}^n$ be a domain, a, b holomorphic functions in G, and $\omega(u, \mathbf{z}) := u^2 - a(\mathbf{z}) \cdot u + b(\mathbf{z})$. In this case

$$p_V(u_1, u_2) = \prod_{i<j}(u_i - u_j)^2 = (u_1 - u_2)^2 = (u_1 + u_2)^2 - 4 \cdot u_1 \cdot u_2.$$

So $Q_V(y_1, y_2) = y_1^2 - 4 \cdot y_2$, and

$$\Delta_\omega(\mathbf{z}) = Q_V(a(\mathbf{z}), b(\mathbf{z})) = a(\mathbf{z})^2 - 4b(\mathbf{z}).$$

If $\mathbf{z} \in G$ and $\Delta_\omega(\mathbf{z}) \neq 0$, there are two different solutions of $\omega(u, \mathbf{z}) = 0$.

Hypersurfaces. We use the theory of pseudopolynomials to study analytic hypersurfaces. Such analytic sets are locally the zero set of one holomorphic function. Assume that f is a holomorphic function in a connected neighborhood of the origin in $\mathbb{C}^{n+1}$ that is not identically 0. Without loss of generality we may assume that $A = N(f)$ contains the origin. Then a generic complex line ℓ through $\mathbf{0}$ meets A in a neighborhood of $\mathbf{0}$ only at the origin. After a linear coordinate transformation, $\ell = \{(u, \mathbf{z}) : \mathbf{z} = \mathbf{0}\}$ is the first coordinate axis. By the Weierstrass preparation theorem $f_{(0,\mathbf{0})} = e_{(0,\mathbf{0})} \cdot \omega_{(0,\mathbf{0})}$ in the ring H_{n+1}, where $e_{(0,\mathbf{0})}$ is a unit in H_{n+1}, and $\omega_{(0,\mathbf{0})} \in H_n[u]$ a Weierstrass polynomial. We can represent the germs locally by holomorphic functions. Thus there is a domain $G \subset \mathbb{C}^n$ containing $\mathbf{0}$, and a disk $D = \{u \in \mathbb{C} : |u| < r\}$ such that in $U := D \times G$ there are a holomorphic function e that does not vanish in U and a pseudopolynomial ω over G with $f = e \cdot \omega$ in U. We may assume that $A \cap (\partial D \times G) = \varnothing$. Therefore, the zero set of f in U is that of ω.

We can decompose ω into prime factors. Using the fact that any power of a prime factor vanishes at the same points as the prime factor does, we may assume that ω is without multiple factors. Then the discriminant Δ_ω is not identically zero in G. We set $D_\omega = \{\mathbf{z} \in G : \Delta_\omega(\mathbf{z}) = 0\}$.

4.6 Theorem (on branched coverings). *If $\mathbf{z}_0 \in G - D_\omega$, there are a neigborhood $W = W(\mathbf{z}_0) \subset G - D_\omega$ and holomorphic functions $f_1, \ldots, f_s$ in W with $f_i(\mathbf{z}) \neq f_j(\mathbf{z})$ for $i \neq j$ and $\mathbf{z} \in W$ such that*

$$\omega(u, \mathbf{z}) = (u - f_1(\mathbf{z})) \cdots (u - f_s(\mathbf{z})) \text{ in } \mathbb{C} \times W.$$

There are fewer than s points over any point $\mathbf{z}_0 \in D_\omega$ (see Figure III.1).

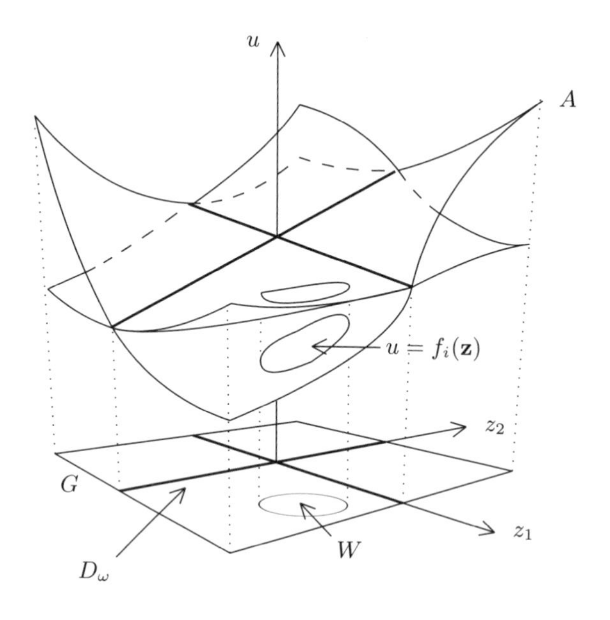

Figure III.1. A branched covering over G

A point $\mathbf{z} \in G$ above which there are fewer than s points is called a *branch point.* All points of the discriminant set D_ω are branch points. Over all other points our set A is locally the union of disjoint graphs of holomorphic functions, and is therefore regular.

PROOF: For $\mathbf{z}_0 \in G - D_\omega$ the polynomial $\omega(u, \mathbf{z}_0)$ has s distinct roots. We write $\omega(u, \mathbf{z}_0) = (u - c_1) \cdots (u - c_s)$, where the c_i all are distinct. If $\omega(u, \mathbf{z}) = u^s + h_1(\mathbf{z})u^{s-1} + \cdots + h_s(\mathbf{z})$, then the germ

$$\omega_{\mathbf{z}_0} := u^2 + (h_1)_{\mathbf{z}_0} u^{s-1} + \cdots + (h_s)_{\mathbf{z}_0}$$

is a polynomial over $\mathcal{O}_{\mathbf{z}_0} \cong H_n$. By Hensel's lemma it has a decomposition $\omega_{\mathbf{z}_0} = \omega_{1,\mathbf{z}_0} \cdots \omega_{s,\mathbf{z}_0}$ with the following properties:

1. $\omega_{i,\mathbf{z}_0}(u, \mathbf{z}_0) = u - c_i$ for $i = 1, \dots, s$.
2. $\deg(\omega_{i,\mathbf{z}_0}) = 1$.

We have $\omega_{i,\mathbf{z}_0} = u - r_i$, with $r_i \in H_n$. There are a connected neighborhood $W(\mathbf{z}_0) \subset G - D_\omega$ and holomorphic functions $f_1, \dots, f_s$ in W such that the power series r_i converge to f_i. Since the germs of ω and $(u - f_1) \cdots (u - f_s)$ coincide at $\mathbf{z}_0$, it follows from the identity theorem that

$$\omega|_{\mathbb{C}\times W} = (u - f_1) \cdots (u - f_s),$$

and since $W \subset G - D_\omega$, it also follows that $f_i(\mathbf{z}) \neq f_j(\mathbf{z})$ for $i \neq j$ and $\mathbf{z} \in W$.
■

Examples

1. Let $G = \mathbb{C}$ and $\omega = z_1^2 - z_2$. Then the discriminant is given by $\Delta_\omega(z_2) = 4z_2$, and $D_\omega = \{0\} \subset \mathbb{C}$. If $z_2 \in \mathbb{C}^*$, there is a neighborhood $W \subset \mathbb{C} - D_\omega$ where $\sqrt{z_2}$ is well defined. There we have $\omega = (z_1 - \sqrt{z_2}) \cdot (z_1 + \sqrt{z_2})$. This gives a surface above $\mathbb{C}$ that is a connected unbranched 2-sheeted covering over $\mathbb{C} - \{0\}$. The point 0 is a branch point. This is the *(branched) Riemann surface* of $\sqrt{z}$. The unbranched part was already discussed in Section II.8.
2. A completely different situation is obtained if we take $\omega = z_1^2 - z_2^2 = (z_1 - z_2) \cdot (z_1 + z_2)$. The discriminant is $4z_2^2$ in this case, and the discriminant set D_ω is again the origin in $\mathbb{C}$. The set A consists of two distinct sheets, which intersect above 0, and both are projected biholomorphically onto $\mathbb{C}$. The set $A - \{\mathbf{0}\}$, i.e., A without the branch point, is no longer connected.
3. In higher dimensions the situation is even more complicated. Let us consider the analytic set $A = N(f)$, where $f(z_1, \dots, z_n) = z_1^{s_1} + \cdots + z_n^{s_n}$ with $s_i \geq 2$ for $i = 1, \dots, n$. This is a very simple holomorphic function. The derivatives are $f_{z_i} = s_i \cdot z_i^{s_i - 1}$, and their joint zero set consists only of $\mathbf{0} \in \mathbb{C}^n$. So all other points of A are regular.

   Every line ℓ through the origin lies completely in A, or f has a zero of order s with $s \geq \min(s_1, \dots, s_n)$ on ℓ at the origin. Therefore, there is no line that intersects A in $\mathbf{0}$ transversally. From this one can conclude that $\mathbf{0}$ is in fact a singular point of A (see, for example, Exercise 8.2 in Chapter I).

   Now we look on $f : \mathbb{C}^n \to \mathbb{C}$ as a fibration with general fiber

   $$A_t = \{\mathbf{z} \in \mathbb{C}^n \,:\, z_1^{s_1} + \cdots + z_n^{s_n} = t\}.$$

   Then $A = A_0$ has an isolated singularity, while all other sets A_t are regular everywhere. We call the family $(A_t)_{t \in \mathbb{C}}$ a *deformation* of A.

The Unbranched Part. We assume that $G \subset \mathbb{C}^n$ is a domain and $\omega(u, \mathbf{z})$ a pseudopolynomial over G of degree s. We set $G' = G - D_\omega$,

$$A = \{(u, \mathbf{z}) \in \mathbb{C} \times G : \omega(u, \mathbf{z}) = 0\},$$

and $A' = A|G'$ the part of A over G'. Then A' is an unbranched covering of G'. It is an n-dimensional submanifold of $\mathbb{C} \times G'$, and we have the canonical projection $\pi : A \to G$. If $(u_0, \mathbf{z}_0) \in A'$ is a point, there is a small neighborhood $B = B(u_0, \mathbf{z}_0) \subset A'$ that is mapped by π holomorphically and topologically onto a ball around $\mathbf{z}_0$ in G'. We also call B a *ball.* The holomorphic map $(\pi|_B)^{-1} : \pi(B) \to \mathbb{C}^{n+1}$ is a local parametrization of A'. A complex function f in B is called *holomorphic* if $f \circ (\pi|_B)^{-1}$ is holomorphic. In particular, the components of π itself are holomorphic functions on B.

For holomorphic functions in B we have the same properties as for holomorphic functions in a domain of $\mathbb{C}^n$. For example, the identity theorem remains valid, and we obtain the following results:

4.7 Proposition. *Assume that A_1 is a connected component of A' and that M is an analytic subset of A_1. Then $M = A_1$, or M is nowhere dense in A_1.*

4.8 Proposition. *If f is a holomorphic function on A', and A_1 a connected component of A', then either f vanishes identically on A_1 or its zero set is nowhere dense in A_1.*

4.9 Proposition. *Let A_1 again be a connected component of A'. Assume that M is a nowhere dense analytic set in A_1 and that f is a holomorphic function in $A_1 - M$ that is bounded along M. Then f has a unique holomorphic extension to A_1.*

Decompositions. We consider the interaction between the decomposition of a pseudopolynomial into irreducible factors and the decomposition of its zero set into "irreducible" components.

4.10 Proposition. *Let $G \subset \mathbb{C}^n$ be a domain and $\omega(u, \mathbf{z})$ a pseudopolynomial over G without multiple factors. Then ω is irreducible if and only if the intersection of its zero set A with $\mathbb{C} \times (G - E)$ is connected for every nowhere dense analytic subset $E \subset G$ which contains the discriminant set D_ω.*

PROOF: Since locally over $G' = G - D_\omega$ the set $A' = A|G'$ looks like a domain in $\mathbb{C}^n$, a nowhere dense analytic set does not disconnect A', locally and globally. Therefore, we may assume that $E = D_\omega$.

If ω is not irreducible, every factor ω_i defines an analytic set A_i over $G - D_\omega$. The intersection of different A_i is empty. So $A|(G - D_\omega)$ is not connected.

If, on the other hand, $A|G'$ has the connected components A_i, $i = 1, \dots, s$, with $s > 0$, then for any point $\mathbf{z} \in G - D_\omega$ there is a ball $B \subset G - D_\omega$ around $\mathbf{z}$ such that $A_i|B$ splits into graphs of holomorphic functions $f_j : j = 1, \dots, s_i$. In each case we form the pseudopolynomial $\omega_i = (u - f_1) \cdots (u - f_{s_i})$. The zero set of this ω_i is exactly $A_i|B$, and it determines ω_i and vice versa. So over the intersection of two different balls the pseudopolynomials must be the same, and thus we obtain global holomorphic functions ω_i in $G - D_\omega$. If $\mathbf{z} \in D_\omega$, then there is a neighborhood W of $\mathbf{z}$ such that $A_i|(W - D_\omega) \subset A|W$ is a bounded set. So the coefficients of ω_i are bounded over this neighborhood and extend holomorphically to G. We also denote this extension by ω_i, and for reasons of continuity it follows that $\omega = \omega_1 \cdots \omega_s$. ∎

If the ω_i are the irreducible factors of ω, we call their zero sets A_i the *irreducible components* of A. The sets $A_i' = A_i|G'$ are the connected components of $A|G'$.

4.11 Proposition. *Assume that ω^*, ω are pseudopolynomials without multiple factors over a domain G and that $A^* = \{\omega^* = 0\} \subset A = \{\omega = 0\}$. Then ω^* is a factor of ω.*

Proof: Let D denote the union of the discriminants of ω^* and ω. It is a nowhere dense analytic set in G. Over $G - D$ we decompose the two unbranched coverings into connected components. There we have $A^* = A_1 \cup \cdots \cup A_{s^*}$ and $A = A_1 \cup \cdots \cup A_s$ with $s^* \le s$. This yields pseudopolynomials over $G - D$ that extend to pseudopolynomials $\omega_1, \dots, \omega_s$ over G, with $\omega^* = \omega_1 \cdots \omega_{s^*}$ and $\omega = \omega_1 \cdots \omega_s$. This implies the result. ∎

The following result is proved analogously.

4.12 Proposition. *Assume that ω is free of multiple factors and that $A = \{\omega = 0\}$ is the disjoint union of two nonempty sets M', M'' that are closed in $\mathbb{C} \times G$. Then there are pseudopolynomials ω', ω'' over G with $M' = \{\omega' = 0\}$, $M'' = \{\omega'' = 0\}$, and $\omega' \cdot \omega'' = \omega$.*

Proof: The construction is first carried out outside D_ω. We set $G' = G - D_\omega$ and use the fact that every nonempty open subset of $A' = A|G'$ must be a union of connected components of A'. If $\omega = \omega_1 \cdots \omega_s$ is the decomposition into irreducible factors, then we may assume that there is an s^* with $0 \le s^* \le s$ such that for $\omega' = \omega_1 \cdots \omega_{s^*}$ and $\omega'' = \omega_{s^*+1} \cdots \omega_s$ we have $M'|G' = \{(u, \mathbf{z}) \in \mathbb{C} \times G' : \omega'(u, \mathbf{z}) = 0\}$ and $M''|G' = \{(u, \mathbf{z}) \in \mathbb{C} \times G' : \omega''(u, \mathbf{z}) = 0\}$.

It is now essential that in a continuous family $f(u, \mathbf{z})$ of holomorphic functions of one variable u the zeros depend continuously on the family parameter $\mathbf{z}$ ("continuity of roots"). We omit the proof here. If we apply this fact (and the equations $A = M' \cup M''$, $\omega = \omega' \cdot \omega''$), we get that the sets $M'|G'$ and

$M''|G'$ are not empty (i.e., $0 < s^* < s$) and that their closures in $\mathbb{C} \times G$ are M', respectively M''. ∎

Projections. In the next section we will investigate zero sets of several holomorphic functions. Here we begin with the simplest case, the common zero set N of a pseudopolynomial ω over $G \subset \mathbb{C}^n$ and an additional holomorphic function f in a neighborhood of $A = \{(u, \mathbf{z}) \in \mathbb{C} \times G : \omega(u, \mathbf{z}) = 0\}$. Our method involves the projection of N to G.

4.13 Proposition. *Assume that $f = f(u, \mathbf{z})$ is a continuous function on A that is holomorphic outside of $\mathbb{C} \times D_\omega$ and does not vanish identically in a neighborhood of any point of A. Then the projection of $N = \{f = \omega = 0\}$ to G is an analytic set $N' = \{\underline{f} = 0\}$, where $\underline{f}$ is a holomorphic function in G that does not vanish identically.*

PROOF: If $\mathbf{z} \in G - D_\omega$, we have a ball $B \subset G - D_\omega$ around it such that over B our ω has the form $\omega(u, \mathbf{z}) = (u - f_1(\mathbf{z})) \cdots (u - f_s(\mathbf{z}))$. The function f does not vanish identically on any graph $u = f_i$. Consequently,

$$\underline{f}(\mathbf{z}) := f(f_1(\mathbf{z}), \mathbf{z}) \cdots f(f_s(\mathbf{z}), \mathbf{z})$$

does not vanish identically. In the usual way we obtain the holomorphic function $\underline{f}$ in the entire set $G - D_\omega$. It is bounded along D_ω. So it extends to a holomorphic function in G. ∎

Now consider the following situation: Assume that G is a domain in $\mathbb{C}^n$ and that ω is a pseudopolynomial over G without multiple factors. Let f be a holomorphic function in a neighborhood of $A = \{\omega(u, \mathbf{z}) = 0\} \subset \mathbb{C} \times G$ that does not vanish identically on any open subset of A and define

$$N := \{(u, \mathbf{z}) \in \mathbb{C} \times G : \omega(u, \mathbf{z}) = f(u, \mathbf{z}) = 0\}.$$

Denote by N' the projection of N to G.

We want to give a definition for "unbranched points" of N. The difficulty is that there may exist such unbranched points of N lying in the set of branch points of ω.

4.14 Proposition. *For any point $\mathbf{z}_0 \in N'$ there is an arbitrarily small linear coordinate change in $z_1, \ldots, z_n$ such that thereafter the line parallel to the z_1-axis through $\mathbf{z}_0$ intersects N' in $\mathbf{z}_0$ as an isolated point.*

In such coordinates there is a neighborhood $U(\mathbf{z}_0) \subset G$, a domain G' in the space $\mathbb{C}^{n-1}$ of the variables $\mathbf{z}' = (z_2, \ldots, z_n)$, and a pseudopolynomial $\omega'(z_1, \mathbf{z}')$ over G' such that $\{(z_1, \mathbf{z}') \in \mathbb{C} \times G' : \omega'(z_1, \mathbf{z}') = 0\} = N' \cap U$.

PROOF: A "small" linear change of the coordinates $z_1, \ldots, z_n$ means here that the transformation is very near to the identity. Since $\underline{f}$ does not vanish identically, after a small generic coordinate transformation the line parallel to the z_1-axis through $\mathbf{z}_0$ intersects N' in $\mathbf{z}_0$ as an isolated point. And then it is also clear that U, G', and ω' with the desired properties exist. ∎

Let us now assume that we have chosen a point $\mathbf{z}_0 \in N'$ and suitable coordinates as above, and that U, G', and ω' have also been chosen.

Definition. In the given situation, a point $(u, \mathbf{z}) \in N \cap (\mathbb{C} \times U)$ is called an *unbranched point* of N if $\mathbf{z} \in N' - \mathbb{C} \times D_{\omega'}$ and there is a neighborhood $V = V(\mathbf{z}) \subset N' - \mathbb{C} \times D_{\omega'}$ with a holomorphic function g on V such that $N \cap (\mathbb{C} \times V)$ is the graph $\{u = g(\mathbf{w}) : \mathbf{w} \in V\}$. (Figure III.2 shows the situation.)

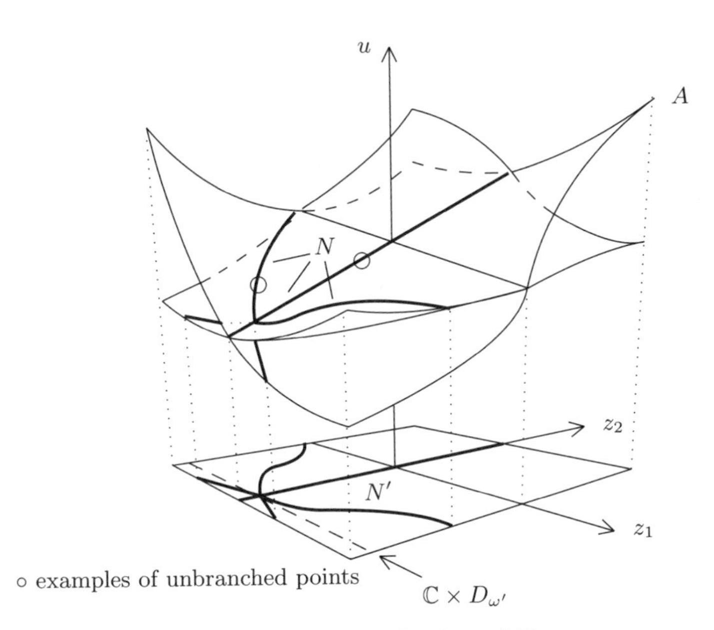

Figure III.2. Branched and unbranched points of N

4.15 Theorem. *In the given situation, in every neighborhood of an arbitrary point $(u_1, \mathbf{z}_1)$ of $N \cap (\mathbb{C} \times U)$ there are unbranched points of N.*

PROOF: We may assume that $\mathbf{z}_1 \in N' - (\mathbb{C} \times D_{\omega'})$. Then we take a small neighborhood $W = E \times U_1$ of $(u_1, \mathbf{z}_1)$ such that the following hold:

1. $U_1 \subset U - (\mathbb{C} \times D_{\omega'})$.
2. $(u_1, \mathbf{z}_1)$ is the only point of A above $\mathbf{z}_1$ in W.

3. The set $A \cap W$ is defined by a pseudopolynomial ω^* over U_1.

Setting $N_1' \subset N' \cap U_1'$ to be the image set of $W \cap N$ under the canonical projection $\pi : W \to U_1$, take U_1 so small that $N_1' = N' \cap U_1$.

We restrict ω^* to N_1' and replace possible multiple factors by one factor at a time. So we obtain a new pseudopolynomial ω_1 without multiple factors over N_1' such that

$$N \cap W = \{(u, \mathbf{z}) \in E \times N_1' \, : \, \omega_1(u, \mathbf{z}) = 0\}, \; E \text{ a suitable disk.}$$

Arbitrarily near to $(u_1, \mathbf{z}_1)$ we can find points $(u_2, \mathbf{z}_2) \in N \cap W$ lying over $N_1' - D_{\omega_1}$. All of these points are unbranched points of N, since we can find neighborhoods $W_2(u_2, \mathbf{z}_2) \subset W$ and $U_2(\mathbf{z}_2) \subset U_1$ with the same properties as W and U_1. Now choose U_2 so small that it contains no point of D_{ω_1} and that $\omega_1|N' \cap U_2$ decomposes into linear factors. Then every sheet of $A|(N' \cap U_2)$ with the property that it contains points of N is a graph over $N' \cap U_2$. ■

Exercises

1. Prove that every symmetric polynomial in $u_1, \ldots, u_s$ can be written as a polynomial in the *power sums* $S_k := u_1^k + \cdots + u_s^k$.
2. Calculate the discriminant of a cubic polynomial.
3. Let $D := \mathsf{D}_r(0) \subset \mathbb{C}$ and f be a holomorphic function in an open neighborhood of $\overline{D}$ without zeros in ∂D. If f has in D the zeros $c_1, \ldots, c_s$ (some of them may be equal), then
$$\frac{1}{2\pi \mathrm{i}} \int_{\partial D} \frac{\zeta^m f'(\zeta)}{f(\zeta)} \, d\zeta = \sum_{j=1}^{s} c_j^m \, .$$
4. Let $f : \mathsf{D} \times \mathsf{P}^n \to \mathbb{C}$ be a holomorphic function in the variables $u, z_1, \ldots, z_n$. Assume that f is u-regular of order s and that for every $\mathbf{z} \in \mathsf{P}^n$ the function $u \mapsto f(u, \mathbf{z})$ has exactly s zeros $u_1(\mathbf{z}), \ldots, u_s(\mathbf{z})$ (with multiplicity) in D. Show that the coefficients of the "pseudopolynomial" $\omega(u, \mathbf{z}) := \prod_{j=1}^{s}(u - u_j(\mathbf{z}))$ are holomorphic.
5. Prove the "continuity of roots": Let $f(u, \mathbf{z})$ be u-regular at the origin. Show that there is an $r > 0$ such that if $g(\mathbf{z})$ is a function defined in a neighborhood of $\mathbf{0}$ with $|g(\mathbf{z})| < r$ and $f(g(\mathbf{z}), \mathbf{z}) = 0$, then g is continuous at $\mathbf{0}$.
6. A complex function f on an analytic set A in a domain $G \subset \mathbb{C}^n$ is called *holomorphic* if it is locally the restriction of a holomorphic function in the ambient space.
   (a) Let $A := \{(w, z) \in \mathbb{C}^2 \, : \, w^2 = z^3\}$ be the *Neil parabola*. There is a bijective holomorphic parametrization of A given by $w = t^3$ and $z = t^2$. Describe the holomorphic functions on A as functions of the parameter t. Is there a meromorphic function on A that has a pole at ∞ with main part t^m?

(b) Show that the analytic set $A = \{(w, z_1, z_2) \in \mathbb{C}^3 : w^2 = z_1 z_2\}$ is not regular at the origin. Consider the holomorphic map $(t_1, t_2) \mapsto (t_1 t_2, t_1^2, t_2^2)$. It is a "two-to-one" map. Describe the local holomorphic functions on A in t_1, t_2.

7. Prove that there is a topological holomorphic map from $\mathbb{C}^*$ onto $A := \{(w, z) \in \mathbb{C}^2 : w^2 = z_1 z_2\}$.
8. Prove that there is no topological holomorphic map from $\mathbb{C}^*$ onto the "elliptic surface" $A := \{(w, z) \in \mathbb{C}^2 : w^2 = (z^2 - 1)(z^2 - 4)\}$.
9. Define the pseudopolynomial $\omega \in \mathcal{O}(\mathbb{C}^2)[u]$ by $\omega(u, z_1, z_2) := u^2 - u \cdot z_1$ and determine the discriminant set and the irreducible components of $A := \{\omega = 0\}$. Let f on A be defined by $f(u, z_1, z_2) := z_2 \cdot (u - 1)$. Consider $N := \{f = \omega = 0\} \subset \mathbb{C}^3$ and determine the projection $N' \subset \mathbb{C}^2$ and the set of unbranched points of N.

5. Irreducible Components

Embedded–Analytic Sets. We wish to study general analytic sets. Since it is easier to work with pseudopolynomials than with arbitrary holomorphic functions, we introduce the notion of "embedded-analytic sets." These are subsets of the common zero set of finitely many pseudopolynomials, and they are not a priori analytic by definition. But later on, it will turn out that they are indeed analytic.

Assume that $G \subset \mathbb{C}^{n-d} = \{\mathbf{z}' = (z_{d+1}, \dots, z_n)\}$ is a domain and that $\omega_i(z_i; \mathbf{z}')$, $i = 1, \dots, d$, are pseudopolynomials over G without multiple factors. The zero sets of the single ω_i intersect transversally[4] in $\mathbb{C}^d \times G$. We denote by $D \subset G$ the union of the d discriminant sets belonging to the ω_i. We call it the *union discriminant set.* We put

$$\widehat{A} := \{(z_1, \dots, z_d, \mathbf{z}') : \omega_i(z_i; \mathbf{z}') = 0, \text{ for } i = 1, \dots, d \text{ and } \mathbf{z}' \in G\}.$$

Over any ball $B \subset G - D$ the set $\widehat{A}|B$ consists of finitely many disjoint holomorphic graphs. Every graph is contained in a connected component Z of $\widehat{A}|(G - D)$. We call the closure of Z in $\widehat{A}$ an *irreducible embedded-analytic component* of $\widehat{A}$.

Definition. If $\widehat{A}$ is defined as above, any union of finitely many irreducible embedded-analytic components of $\widehat{A}$ is called an *embedded-analytic set* of dimension $n - d$.

[4] Two submanifolds $M, N \subset \mathbb{C}^n$ intersect transversally at a point $\mathbf{z} \in M \cap N$ if the entire space is spanned by vectors that are tangent to M or N at $\mathbf{z}$. In our case the common zero set of the pseudopolynomials ω_i contains enough vectors to span $\mathbb{C}^n$. Therefore, we say that these zero sets intersect transversally.

By definition any embedded-analytic set A can be decomposed into finitely many irreducible components.

The surrounding set $\widehat{A} \supset A$ is not uniquely determined. Sometimes we can make $\widehat{A}$ smaller by throwing away those irreducible factors of ω_i that do not vanish identically on A. Then the ω_i are uniquely determined by A.

5.1 Proposition. *Assume that A is an embedded-analytic set in $\widehat{A} \subset \mathbb{C}^d \times G$ and that f is a holomorphic function in a neighborhood of A that does not vanish identically on any open subset of A. If $N = \{\mathbf{z} \in A : f(\mathbf{z}) = 0\}$, then for any point $\mathbf{z}_0 \in N$ there is an arbitrarily small linear change of the coordinates $\mathbf{z}'$ such that the affine space parallel to the $(z_1, \ldots, z_{d+1})$-"axis" through $\mathbf{z}_0$ intersects N in an isolated point. If $\mathbf{z}_1 \in N$ is any point near $\mathbf{z}_0$, then there are unbranched points of N arbitrarily near $\mathbf{z}_1$. At all of these points N is a submanifold of dimension $n - d - 1$.*

PROOF: We proceed as in the proof for the last theorem of the previous section. The procedure to find unbranched points will be denoted by $(*)$.

First we construct the projection $\underline{f}$ of f, which is holomorphic in G. For this observe that if D is the union discriminant set of the polynomials $\omega_1, \ldots, \omega_d$ and $\mathbf{z}' \in G - D$, we always have the same number of points $\mathbf{z}_1, \ldots, \mathbf{z}_s$ in A over $\mathbf{z}'$. We set $\underline{f}(\mathbf{z}') = f(\mathbf{z}_1) \cdots f(\mathbf{z}_s)$ and obtain $\underline{f}$, which is holomorphic on $G - D$. Since it is bounded along D, we can extend it holomorphically to G. Therefore, the projection set is $N' = \{\underline{f}(\mathbf{z}') = 0\}$.

Assume now that $\mathbf{z}'_0 \in N'$. Then, after an arbitrarily small linear coordinate change in the variables $\mathbf{z}'$, the line L parallel to the z_{d+1}-axis through $\mathbf{z}'_0$ intersects N' in an isolated point. Then by Weierstrass's theorem we can find a neighborhood $U = U(\mathbf{z}'_0) \subset G$, a domain G', and a pseudopolynomial $\omega'(z_{d+1}, \mathbf{z}'')$ over G' without multiple factors such that $U \cap N'$ is equal to the set $\{\mathbf{z}' = (z_{d+1}, \mathbf{z}'') \in \mathbb{C} \times G' : \omega'(z_{d+1}, \mathbf{z}'') = 0\}$, with $\mathbf{z}'' = (z_{d+2}, \ldots, z_n)$.

Since the space $\mathbb{C}^d \times L$ intersects N at $\mathbf{z}_0$ in an isolated point, it remains to prove the existence of unbranched points. We apply $(*)$ to $N \cap (\mathbb{C}^d \times U)$ and prove in the same way as before that for points $\mathbf{z}_1 \in N \cap (\mathbb{C}^d \times U)$ there are unbranched points of N arbitrarily near to $\mathbf{z}_1$. Of course, at these points N is a submanifold of dimension $n - d - 1$. ∎

For the following we use the same notation and hypotheses as above.

5.2 Theorem. *Let $N' \subset G$ be the projection of N. For every point $\mathbf{z}_0 = (\widetilde{\mathbf{z}}_0, \mathbf{z}'_0) \in N \subset \mathbb{C}^d \times G$, after a suitable linear change of the coordinates $\mathbf{z}'$ there is a neigborhood $U = U(\mathbf{z}'_0) \subset G$, a domain G' in the space of the variables $z_{d+2}, \ldots, z_n$, and a pseudopolynomial ω' over G' without multiple factors such that:*

1. *$N' \cap U = \{\omega' = 0\}$.*

2. $N \cap (\mathbb{C}^d \times U)$ *is an embedded-analytic set of dimension* $n-d-1$.

PROOF: We use the notation and results from the proof above. Thus the first statement is clear. We set $\underline{\omega}_{d+1} := \omega'$. Restricting the ω_i to $\mathbb{C}^d \times (N' \cap U)$ and projecting them down to $\mathbb{C}^d \times G'$, we get pseudopolynomials $\underline{\omega}_i(z_i; \mathbf{z}'')$ over G'. Then $N \cap (\mathbb{C}^d \times U)$ is in the joint zero set of $\underline{\omega}_1, \ldots, \underline{\omega}_{d+1}$. Let $\underline{A}$ be the union of those irreducible components of this set that contain the unbranched points of N. Since N is the closure of unbranched points, it follows that $N \cap (\mathbb{C}^{d+1} \times G') \subset \underline{A}$. By the mapping theorem that we prove in the next paragraph, every irreducible component of $\underline{A}$ is in N. So we have the desired equality. ∎

Images of Embedded–Analytic Sets. Assume that $G = \mathbb{C}^d \times G' \subset \mathbb{C}^n$ is a domain and $A \subset G$ an **irreducible** embedded-analytic set over G'.

5.3 Mapping theorem. *Let* $G_1 = \mathbb{C}^{d_1} \times G'_1 \subset \mathbb{C}^{n_1}$ *be a domain,* $A_1 \subset G_1$ *an embedded-analytic set over* G'_1, $\mathbf{F}$ *a holomorphic mapping from a neighborhood of* $A \subset G$ *into* G_1 *such that* $\mathbf{F}(U) \subset A_1$ *for some nonempty open subset* $U \subset A$. *Then* $\mathbf{F}(A) \subset A_1$.

PROOF: We denote by $D \subset G$ the union of the discriminant sets of the $\omega_i(z_i, \mathbf{z}')$ that define the surrounding set $\widehat{A}$ for A. It is sufficient to prove that $\mathbf{F}(A \cap (\mathbb{C}^d \times (G' - D))) \subset A_1$. Since we can connect two points of $A \cap (\mathbb{C}^d \times (G' - D))$ by a chain of arbitrarily small balls, it is enough to give the proof for such a ball. So we may replace A by a ball in $\mathbb{C}^{n-d}$ and may assume that $\mathbf{F}$ is defined in a neighborhood of $\overline{B}$.

Let A_1 be an embedded-analytic set in

$$\widehat{A}_1 = \{\widetilde{\omega}_1(w_1, \mathbf{w}') = \cdots = \widetilde{\omega}_{d_1}(w_{d_1}, \mathbf{w}') = 0\}.$$

Then $\widetilde{\omega}_i \circ \mathbf{F}|_U \equiv 0$, and by the identity theorem $\widetilde{\omega}_i \circ \mathbf{F}|_B \equiv 0$. So $\mathbf{F}(B) \subset \widehat{A}_1$.

For an arbitrary point $\mathbf{v} \in G'_1$ we choose a small transformation of the coordinates in $\mathbb{C}^{d_1}$ such that A_1 is also embedded in a set $\widehat{A}_1^{\mathbf{v}} = \{\omega_1^{\mathbf{v}} = \cdots = \omega_{d_1}^{\mathbf{v}} = 0\}$. The transformation can be made arbitrarily small, and we can do it so that

$$\widehat{A}_1 \cap \widehat{A}_1^{\mathbf{v}} \cap (\mathbb{C}^{d_1} \times \{\mathbf{v}\}) = A_1 \cap (\mathbb{C}^{d_1} \times \{\mathbf{v}\}).$$

Then A_1 is given by the infinite set of holomorphic equations $\widetilde{\omega}_i = 0$, $\omega_i^{\mathbf{v}} = 0$, $\mathbf{v} \in G'_1$. If $\mathbf{F}$ maps a nonempty open part of B into A_1, then by the identity theorem $\widetilde{\omega}_i \circ \mathbf{F} = \omega_i^{\mathbf{v}} \circ \mathbf{F} \equiv 0$, and consequently $\mathbf{F}(B) \subset A_1$. This completes the proof. ∎

Remark. Assume that A is an analytic set in a domain $G \subset \mathbb{C}^n$ and that $\mathbf{z}_0$ is a point of A. If $\mathbf{z}_0$ is a regular point of dimension $n-d$, then there is a neighborhood $U = U(\mathbf{z}_0) \subset G$ such that $A \cap U$ is an embedded-analytic set.

In fact, there is a neighborhood U with holomorphic functions $f_1, \dots, f_d$ such that $N(f_1, \dots, f_d) = A \cap U$ and the rank of the Jacobian is d everywhere. We may assume that

$$\det \left((f_i)_{z_j}(\mathbf{z}_0) \;\middle|\; \begin{array}{l} i = 1, \dots, d \\ j = 1, \dots, d \end{array} \right) \neq 0.$$

Then the transformation $\mathbf{F}(z_1, \dots, z_n) = (f_1(\mathbf{z}), \dots, f_d(\mathbf{z}), z_{d+1}, \dots, z_n)$ maps a neighborhood of $\mathbf{z}_0$ biholomorphically onto a neighborhood of the image point. If the inverse is given by

$$\mathbf{z} = \mathbf{F}^{-1}(\mathbf{w}) = (g_1(\mathbf{w}), \dots, g_d(\mathbf{w}), w_{d+1}, \dots, w_n),$$

then A is given by the equations

$$\begin{aligned} z_1 &= g_1(0, \dots, 0, z_{d+1}, \dots, z_n), \\ &\vdots \\ z_d &= g_d(0, \dots, 0, z_{d+1}, \dots, z_n). \end{aligned}$$

So $A \cap U$ is an embedded-analytic set.

Local Decomposition. We use embedded-analytic sets to show that an arbitrary intersection of analytic sets is again an analytic set.

First we consider the following situation. Let $G \subset \mathbb{C}^n$ be a domain and $\mathbf{z}_0 \in G$ a point. Assume that at $\mathbf{z}_0$ a set $\mathscr{S}$ of local analytic functions f is given such that

1. For every $f \in \mathscr{S}$ there is a connected open neighborhood $U(\mathbf{z}_0) \subset G$ with $f \in \mathcal{O}(U)$ and $f \not\equiv 0$.
2. $f(\mathbf{z}_0) = 0$.

We want to construct a "maximal" analytic set S^* in a neighborhood $U^*(\mathbf{z}_0) \subset G$ such that for each zero set N of finitely many elements $f \in \mathscr{S}$ there is a neighborhood $V = V(\mathbf{z}_0) \subset G$ with $S^* \cap V \subset N \cap V$. Then S^* is uniquely determined near $\mathbf{z}_0$ and can be considered as the common zero set of the functions $f \in \mathscr{S}$. It may be nontrivial even if the domains of definition of the functions f tend to the point $\mathbf{z}_0$. For example, if $\mathscr{S}$ is the set of the functions $f_n(\mathbf{z}) := z_1^n/(1 - nz_1)$, defined on $U_n := \{\mathbf{z} \in \mathbb{C}^n : \operatorname{Re}(z_1) < 1/n\}$, then S^* is the analytic set $\{z_1 = 0\}$ in an arbitrary neighborhood of $\mathbf{0}$, whereas the intersection of the U_n does not contain any neighborhood of the origin.

We employ the results from the beginning of this section several times and carry out an induction on the codimension of the embedded-analytic sets obtained from the functions $f \in \mathscr{S}$.

(A) We begin with one arbitrarily chosen function $f \in \mathscr{S}$. The equation $f = 0$ gives an analytic set[5] S of codimension 1. We decompose S into irreducible components S_i in a neighborhood $U(\mathbf{z}_0)$ (given by pseudopolynomials in $\mathbb{C} \times G'$), and we choose the neighborhood U so small that the S_i stay irreducible in the whole neighborhood.

(B) Next we try to obtain codimension 2. If every function $f \in \mathscr{S}$ vanishes identically near $\mathbf{z}_0$ on S_i, we leave S_i unchanged (and have it as a codimension 1 component for our S^*). Otherwise, there is an $f' \in \mathscr{S}$ that does not vanish identically in any small neighborhood of $\mathbf{z}_0$ on S_i. We apply Theorem 5.2: After an arbitarily small linear change of the coordinates $\mathbf{z}'$ the set $S_i \cap \{f' = 0\}$ is a finite union of irreducible embedded-analytic sets S_{ij} of codimension two, which stay irreducible if we pass to some smaller neighborhood of $\mathbf{z}_0$. The S_{ij} are embedded in the zero set of two pseudopolynomials $\omega_1^{ij}(z_1; z_3, \ldots, z_n)$ and $\omega_2^{ij}(z_2; z_3, \ldots, z_n)$.

(C) Now codimension 3 follows. For this we need consider only the S_{ij}. Leave S_{ij} unchanged if every f vanishes on S_{ij} (and get codimension 2 components for S^*). Otherwise, find an $f'' \in \mathscr{S}$ not vanishing identically on S_{ij}, and (after an arbitrarily small coordinate change of the variables $\mathbf{z}'' = (z_3, \ldots, z_n)$) the set $S_{ij} \cap \{f'' = 0\}$ is the union of a finite set of irreducible embedded-analytic sets S_{ijk}, given in the zero set of three pseudopolynomials $\omega_\lambda^{ijk}(z_\lambda; z_4, \ldots, z_n)$, $\lambda = 1, 2, 3$.

(D) Continuing, it is possible to obtain components of codimension 1, 2, $\ldots$, $n-1$, and finally one reaches dimension 0. If there is a 1-dimensional component $S = S_{i_1 \ldots i_{n-1}}$ such that not every $f \in \mathscr{S}$ vanishes on S, we have to replace S by the one-point set $\{\mathbf{z}_0\}$. Then the procedure stops. Only finitely many steps were necessary.

We obtained a finite system $\mathscr{S}_0$ of local holomorphic functions f, f', f'', ... and a finite system $\mathcal{A}$ of irreducible embedded-analytic sets S_i, S_{ij}, S_{ijk}, ... and may assume that they all are defined in one neighborhood $U(\mathbf{z}_0) \subset G$, that every $S \in \mathcal{A}$ of dimension d is embedded in a set $\mathbb{C}^{n-d} \times G'_d$, and that the union discriminant sets $D_S \subset G'_d \subset \mathbb{C}^d$ belong to the embedding of S. The necessary linear coordinate change in $\mathbf{z}'$ can be made at the beginning of the procedure, i.e., once for all steps of the procedure.

If $S \in \mathcal{A}$ is an irreducible embedded-analytic set that has an open part in the union of the other sets of $\mathcal{A}$, then it also has an open part in an irreducible $S' \in \mathcal{A}$, $S' \neq S$. It follows by the mapping theorem that it is completely contained in S'. Then we simply throw it away and denote the new system again by $\mathcal{A}$. After finitely many steps we have that the intersection of every

[5] An exact definition of dimension and codimension of analytic sets will be given later. Here we use embedded-analytic sets, for which the dimension has already been defined. An analytic hypersurface is obviously embedded-analytic.

S with the union of the rest of $\mathcal{A}$ is nowhere dense in S. Moreover, the points of S over D_S are nowhere dense in S.

We denote by $S^* = \bigcup_{S \in \mathcal{A}} S$ the union of all remaining components. Then S^* is given by the finitely many holomorphic functions $f \in \mathscr{S}_0$. Therefore, it is an analytic set. If $\hat{\mathscr{S}} \subset \mathscr{S}$ is an arbitrary finite subset, then in a small neighborhood of $\mathbf{z}_0$ every $f \in \hat{\mathscr{S}}$ vanishes at every $\mathbf{z} \in S^*$. So $S^* \subset N(\hat{\mathscr{S}})$. Obviously, S^* is uniquely determined by this property.

Finally, we want to show that the decomposition into irreducible embedded-analytic sets is unique. For that we use the notion of regularity for points of embedded-analytic sets just as in the analytic case. Clearly, the intersection of two different $S_i \in \mathcal{A}$ contains no regular point. So the points of every $S \in \mathcal{A}$ are regular if they are not in such an intersection and not over D_A. We denote the *set of regular points* of S^* by $\dot{S}^*$ and set $\dot{S} = S \cap \dot{S}^*$ for $S \in \mathcal{A}$. Then for $S \in \mathcal{A}$ the sets $\dot{S}$ are the connected components of $\dot{S}^*$. Since the set S^* is uniquely determined in a neighborhood of $\mathbf{z}_0$, this is also true for its connected components. And since the closure of $\dot{S}$ is S, the irreducible embedded-analytic components S are also uniquely determined near $\mathbf{z}_0$.

5.4 Theorem. *The intersection of (even infinitely many) analytic sets is an analytic set and is locally a finite union of components that are irreducible embedded-analytic sets. This decomposition is locally uniquely determined.*

PROOF: Let $\{A_\iota : \iota \in I\}$ be a family of analytic sets in a domain $G \subset \mathbb{C}^n$, and $\mathbf{z}_0 \in A := \bigcap_{\iota \in I} A_\iota$ an arbitrary point. We consider the system $\mathscr{S}$ of all local holomorphic functions f such that:

1. f is defined in an open neighborhood U of $\mathbf{z}_0$ (depending on f).
2. $f \not\equiv 0$ near $\mathbf{z}_0$.
3. There is an $\iota \in I$ such that f vanishes near $\mathbf{z}_0$ on A_ι.

As above, $\mathbf{z}_0$ is contained in an analytic set S^* that is the union of irreducible embedded-analytic sets S and that is given by a finite subsystem $\mathscr{S}_0 \subset \mathscr{S}$.

If $\mathbf{z}$ is a point of A that is sufficiently near $\mathbf{z}_0$, then every $f \in \mathscr{S}_0$ is defined at $\mathbf{z}$ and vanishes on some A_ι and consequently at $\mathbf{z}$. This shows that $A \subset S^*$ in a neighborhood of $\mathbf{z}_0$. On the other hand, let $\mathbf{z}$ be a point in the intersection of S^* with a small neighborhood of $\mathbf{z}_0$. Any analytic set A_ι is given by finitely many holomorphic functions $f_1^\iota, \dots, f_N^\iota$ that belong to the system $\mathscr{S}$. Then by construction every f_ν^ι vanishes on every embedded-analytic component S of S^*, in particular at $\mathbf{z}$. Therefore, $S^* \subset A_\iota$ for all ι. Thus S^* is contained in the intersection A of the A_ι, and we have the equality $A = S^*$ near $\mathbf{z}_0$. Since S^* is an analytic set that has a unique decomposition into irreducible embedded-analytic sets, this completes the proof. ∎

Analyticity. Now we are able to prove the following result, which we announced at the beginning of the section:

5.5 Proposition. *Every embedded-analytic set A in a domain $\mathbb{C}^d \times G \subset \mathbb{C}^n$ is an analytic set.*

PROOF: As in the last part of the proof of the mapping theorem, it follows that the embedded-analytic set A is given as the joint zero set of infinitely many holomorphic functions. Theorem 5.4 shows that A is an analytic set. ∎

Consequently, every analytic set has locally a unique decomposition into irreducible analytic components.

The Zariski Topology. We prove that the system of all analytic sets has the properties of the system of closed sets of a topology.

5.6 Theorem. *The system $\mathcal{A}$ of all analytic sets in a domain $G \subset \mathbb{C}^n$ has the following properties:*

1. *G and the empty set belong to $\mathcal{A}$.*
2. *If $A_1, \dots, A_l \in \mathcal{A}$, then also $A = \bigcup_{i=1}^{l} A_i \in \mathcal{A}$.*
3. *If I is an index set and $\{A_\iota : \iota \in I\}$ a collection of analytic sets in G, then $A = \bigcap_{\iota \in I} A_\iota$ is also an analytic set in G.*

PROOF:

(1) G is defined by the zero function, and $\varnothing$ by the constant function 1.

(2) Let $\mathbf{z} \in A = A_1 \cup \cdots \cup A_l$. Then in a neighborhood $U(\mathbf{z})$ there are holomorphic functions $f_{i,j} : i = 1, \dots, l \; j = 1, \dots, d_i$, such that for all i we have

$$U \cap A_i = N(f_{i,1}, \dots, f_{i,d_i}).$$

It follows that $U \cap A = N(f_{1,j_1} \cdots f_{l,j_l} : j_i = 1, \dots, d_i)$.

3) This is Theorem 5.4. ∎

So the analytic sets are the closed sets of a topology in G. We call this topology the *(analytic) Zariski topology* of G. It plays an important rule in complex algebraic geometry.

Global Decompositions. Assume that $G \subset \mathbb{C}^n$ is a domain and $A \subset G$ an analytic subset. We call A *irreducible* if the set of regular points $\dot{A} \subset A$ is connected. It follows that A has the same dimension d at all regular points. This number d is called the *dimension* of A and is denoted by $\dim(A)$. Every irreducible embedded-analytic set is also an irreducible analytic set.

5.7 Theorem. *Every analytic set A has a unique decomposition into countably many irreducible analytic subsets A_i. The covering $\mathscr{A} = \{A_i : i = 1, 2, 3, \dots\}$ is locally finite.*

PROOF: We decompose $\dot{A}$ into connected components. Let A' be such a component. It has dimension d in all of its points.

We consider a point $\mathbf{z}_0 \in A$ that lies in $\overline{A'}$. In a neighborhood $U = U(\mathbf{z}_0) \subset G$ we have a decomposition of A into finitely many irreducible embedded-analytic components $A_1, \ldots, A_m$. By A_i^* we denote the set of points of A_i that are not over the union discriminant set. Some d-dimensional A_i^* meet A'. Their union A^* is contained in A' and dense in $A' \cap U$. Hence, the closure of A^* in U is equal to $\overline{A'} \cap U$. But A^* is an analytic set.

From this it follows that $\overline{A'}$ is an analytic set, that only finitely many $\overline{A'}$ intersect U, and that the union of all $\overline{A'}$ is A (as it is locally). Since the topology of G is countable, it follows that the set of the $\overline{A'}$ is countable. ■

5.8 Corollary. *If A is irreducible and $A = A_1 \cup A_2$, where A_1, A_2 are arbitrary analytic sets, then $A = A_1$ or $A = A_2$.*

Sometimes this condition is used as the definition of irreducibility.

5.9 Proposition. *Let $A, B \subset G$ be irreducible analytic sets. If there is an open set $U \subset G$ such that $A \cap U \neq \varnothing$ and $A \cap U \subset B \cap U$, then $A \subset B$.*

PROOF: This is an immediate consequence of the mapping theorem. ■

Another corollary is the following:

5.10 Identity theorem (for analytic sets). *Let $A, B \subset G$ be irreducible analytic sets. If there is a point $\mathbf{z}_0 \in A \cap B$ and an open neighborhood $U = U(\mathbf{z}_0) \subset G$ with $A \cap U = B \cap U$, then $A = B$.*

5.11 Proposition. *Let $A, B \subset G$ be analytic subsets with $A \subset B$. If A is irreducible, then A is contained in some irreducible component of B.*

PROOF: Let $B = \bigcup_{\lambda \in \Lambda} B_\lambda$ be the unique decomposition into irreducible components. We can choose an open set $U \subset G$ and a finite set $\{\lambda_1, \ldots, \lambda_l\} \subset \Lambda$ such that $U \cap A \neq \varnothing$ is irreducible and $U \cap B = (U \cap B_{\lambda_1}) \cup \cdots \cup (U \cap B_{\lambda_l})$. Then $U \cap A = U \cap A \cap B = (U \cap A \cap B_{\lambda_1}) \cup \cdots \cup (U \cap A \cap B_{\lambda_l})$. Thus there is an index j such that $U \cap A = U \cap A \cap B_{\lambda_j}$, so $U \cap A \subset U \cap B_{\lambda_j}$. It follows that A is contained in B_{λ_j}. ■

Now we can generalize the notion of the dimension to arbitrary analytic sets.

Definition. If $A \subset G$ is an analytic set with irreducible components A_i, then $\dim(A) := \sup_i \dim(A_i)$ is called the *(complex) dimension* of A.

In general, the dimension of an analytic set can be ∞. But if $G_1 \subset\subset G$ is a relatively compact subdomain, then only finitely many irreducible components intersect G_1. So the dimension of $A \cap G_1$ is finite.

An analytic set is called *pure-dimensional* of dimension d if all its irreducible components have the same dimension d.

Exercises

1. Let A be an analytic set near the origin in $\mathbb{C}^n$. Assume that every irreducible component of A has dimension ≥ 1. Show that there exists a neighborhood $U = U(\mathbf{0})$ such that $A \cap U$ is the union of irreducible one-dimensional analytic sets containing $\mathbf{0}$.
2. Consider $A := \{(z_1, z_2) \in \mathbb{C}^2 : z_2^2 = z_1^3 + z_1^2\}$. Show that A is irreducible, but A has a nontrivial decomposition into irreducible components in a small neighborhood of the origin.
3. Let $A_1, A_2 \subset \mathbb{C}^n$ be analytic sets. Prove that $\overline{A_1 - A_2}$ is analytic.
4. Let $\{A_i : i \in \mathbb{N}\}$ be a locally finite family of irreducible analytic sets in a domain $G \subset \mathbb{C}^n$. Suppose that $A_i \not\subset A_j$ for $i \neq j$ and prove that the A_i are the irreducible components of their union.

6. Regular and Singular Points

Compact Analytic Sets. Our goal is to prove the following simple proposition.

6.1 Proposition. *If $G \subset \mathbb{C}^n$ is a domain and $A \subset G$ an irreducible compact analytic set, then A consists of a single point.*

We first prove a lemma.

6.2 Lemma. *If f is a holomorphic function in a neighborhood of A, then $f|_A$ is constant.*

PROOF: We assume that the dimension of A is $n - d$. Since A is compact, there is a point $\mathbf{z}_0 \in A$ where $|f|$ takes its maximum. After a linear coordinate change there is a neighborhood $U = U(\mathbf{z}_0) \subset G$ and a domain $G' \subset \mathbb{C}^{n-d}$ such that $A \cap U$ is an embedded-analytic set over G'. Denote by $D \subset G'$ the union discriminant set. Over every $\mathbf{z}'' \in G' - D$ our $A \cap U$ has s points. The point $\mathbf{z}_0$ lies over some $\mathbf{z}_0'' \in G'$ and we may assume that it is the only point of $A \cap U$ over $\mathbf{z}_0''$.

It remains to construct the elementary symmetric functions associated with $f|_{A \cap U}$ over $G' - D$. For this, if $\mathbf{z}_1, \dots, \mathbf{z}_s$ are mapped onto $\mathbf{z}''$, we define $f_i(\mathbf{z}'') := \sigma_i(f(\mathbf{z}_1), \dots, f(\mathbf{z}_s))$. These functions are holomorphic on $G' - D$ and bounded along D. So they extend to holomorphic functions in G'. The absolute value of every extension takes on its maximum at $\mathbf{z}_0''$. By the maximum principle each such is constant in G'. Since the values of f can be reconstructed from the values of the f_i, it follows that f is constant over G', in particular in some open subset of $\dot{A}$. Since $\dot{A}$ is connected, it follows that f is constant on $\dot{A}$ and then by continuity also on A. ∎

PROOF of the proposition: All coordinate functions z_i must be constant on A. So A is a single point. ∎

A consequence is that every compact analytic subset $A \subset G$ consists of only finitely many points.

Embedding of Analytic Sets. Assume that $A \subset G$ is an analytic set in a domain $G \subset \mathbb{C}^n$, that $\mathbf{0} \in A$, and that the plane $P = \{z_{d+1} = \cdots = z_n = 0\}$ intersects A in an isolated point.

6.3 Theorem. *In a neighborhood $U(\mathbf{0})$ the set A is an analytic subset of an embedded-analytic set of dimension $n-d$ that is defined over a domain G' in the space of variables $\mathbf{z}' = (z_{d+1}, \ldots, z_n)$. If the set of $(n-d)$-dimensional regular points is dense in A, then A is itself an embedded-analytic set.*

Remark. No coordinate transformation is necessary for this statement!

PROOF: By definition, A is the zero set $N(f_1, \ldots, f_N)$ of finitely many holomorphic functions in a neighborhood of $\mathbf{0}$. Since $A \cap P = \{\mathbf{0}\}$ is isolated, there is an i such that f_i does not vanish identically in any neighborhood of $\mathbf{0}$ on the z_1-axis. Consequently, f_i is z_1-regular, and we can apply the Weierstrass preparation theorem, which implies that A is locally contained in the zero set of a pseudopolynomial $\omega(z_1; z_2, \ldots, z_n)$.

Now we proceed by induction on d. In the case $d = 1$ there is nothing to prove. If $d > 1$ we consider the projection $\pi : \mathbb{C}^n \to \mathbb{C}^{n-1}$ with $\mathbf{z} \mapsto \mathbf{z}' = (z_2, \ldots, z_n)$. To $f_2, \ldots, f_N$ there are associated as usual holomorphic functions $\underline{f}_2, \ldots, \underline{f}_N$ of $\mathbf{z}'$ such that

$$\pi(\{\omega = f_2 = \cdots = f_N = 0\}) = \underline{A} := \{\mathbf{z}' : \underline{f}_2(\mathbf{z}') = \cdots = \underline{f}_N(\mathbf{z}') = 0\}$$

in some neighborhood of $\mathbf{0}$.

The intersection of $P' = \{\mathbf{z}' : z_{d+1} = \cdots = z_n = 0\}$ with $\underline{A}$ contains $\mathbf{0}$ as an isolated point. So we have for $\underline{A}$ the same situation, but with one dimension fewer. By the induction hypothesis it follows that there are pseudopolynomials $\omega_2, \ldots, \omega_d$ such that $\underline{A}$ locally is contained in the set $\{\omega_2(z_2; \mathbf{z}'') = \cdots = \omega_d(z_d; \mathbf{z}'') = 0\}$. By exchanging z_1 with z_2, we obtain a pseudopolynomial $\omega_1(z_1; \mathbf{z}'')$ such that A is contained in $\{\omega_1 = \omega_2 = \cdots = \omega_d = 0\}$.

If the regular points of dimension $n-d$ are dense in A, we can take the union A^* of those irreducible components of the embedded analytic set that contain such a regular point. Then A^* and A are identical. This completes the proof. ∎

Remark. This result is also known as the "embedding theorem of Remmert and Stein."

Again we consider a domain $G \subset \mathbb{C}^n$, an analytic set $A \subset G$, and a domain $G' \subset \mathbb{C}^{n-d}$ such that $\pi(G) \subset G'$ (where $\pi : \mathbf{z} \mapsto \mathbf{z}' = (z_{d+1}, \dots, z_n)$). Suppose that there exists a domain $G^* \subset G$ such that for every $\mathbf{z}' \in G'$ there exists a neighborhood $U = U(\mathbf{z}') \subset\subset G'$ with $(\mathbb{C}^d \times \overline{U}) \cap G^* \subset\subset (\mathbb{C}^d \times \overline{U}) \cap G$ and $(\mathbb{C}^d \times \overline{U}) \cap A \subset G^*$. Then the following holds.

6.4 Proposition. *If the set of regular $(n-d)$-dimensional points of A is dense in A, then A is an embedded-analytic set over G'.*

PROOF: We take an arbitrary point $\mathbf{z}'_0 \in G'$. It follows from the hypotheses that the set $(\mathbb{C}^d \times \{\mathbf{z}'_0\}) \cap A$ is compact and analytic. Therefore, it consists of finitely many points. Each of these points has a neighborhood such that the restriction of the set A to this neighborhood is an embedded-analytic set over a neighborhood of $\mathbf{z}'_0$. By multiplying the pseudopolynomials by the same distinguished variable belonging to our various points we obtain d uniquely determined pseudopolynomials over a neighborhood $U'(\mathbf{z}'_0) \subset G'$ such that $A \cap (\mathbb{C}^d \times U')$ is just their joint zero set. But these pseudopolynomials glue together to form global pseudopolynomials over G'. ∎

Regular Points of an Analytic Set

Assume again that $G \subset \mathbb{C}^n$ is a domain, and $A \subset G$ an analytic set.

6.5 Theorem. *For any $\mathbf{z}_0 \in A$ there is a fixed neighborhood $U(\mathbf{z}_0) \subset G$ with finitely many holomorphic functions $f_1, \dots, f_N$ whose joint zero set is $A \cap U$ such that for all d at every regular point $\mathbf{z} \in A \cap U$ of dimension $n-d$ the rank of their Jacobian at $\mathbf{z}$ is equal to d.*

It is remarkable that this statement is also true for a singular point $\mathbf{z}_0$ of A.

PROOF: After applying a linear coordinate transformation in $\mathbb{C}^n$ we can find a neighborhood $U = U(\mathbf{z}_0) \subset G$ such that $A \cap U$ is a finite union of irreducible embedded-analytic components. To give these in the canonical form a further coordinate transformation is not necessary. We denote by A' the union of all $(n-d)$-dimensional irreducible components of $A \cap U$ and choose pseudopolynomials $\omega_1, \dots, \omega_d$ of minimal degree such that A' is contained in the common zero set of the ω_i.

Let $\widehat{D} \subset A'$ be the set of points that lie over the union discriminant set $D_{A'}$. Its dimension is equal to $n-d-1$. Now we carry out the proof in several steps and construct sets $\widehat{D}_1, \widehat{D}_2, \ldots, \widehat{D}_{n-d+1} = \varnothing$ with $\widehat{D}_{i+1} \subset \widehat{D}_i$ and $\dim(\widehat{D}_i) = n-d-i$. We begin with $\widehat{D}_1 := \widehat{D}$. The Jacobian of $\omega_1, \ldots, \omega_d$ has rank d in $A' - \widehat{D}_1$ (implicit function theorem).

Next we decompose $\widehat{D}_1$ into irreducible components C_λ and choose, if possible, in each C_λ a point $\mathbf{z}_\lambda$ where A' is regular. We can make U so small that only finitely many C_λ occur in U. Then we apply another linear transformation in $\mathbb{C}^n$ that is near the identity such that in a small neighborhood of any $\mathbf{z}_\lambda$ the set A' can be written as an $(n-d)$-dimensional holomorphic graph

$$\{\mathbf{z} : z_i = f_{\lambda,i}(z_{d+1}, \ldots, z_n) \text{ for } i = 1, \ldots, d\}.$$

Finally, we apply a linear transformation that is very near the identity to the variables $z_1, \ldots, z_d$ such that for every λ and every point $\mathbf{z}$ in A' above $(z^\lambda_{d+1}, \ldots, z^\lambda_n)$ the first d coordinates of $\mathbf{z}$ are distinct.

Now we use Proposition 6.4. In the new coordinates (and in a slightly smaller neighborhood, which we again denote by U) the set A' is again an embedded-analytic set contained in the common zero set of pseudopolynomials $\widetilde{\omega}_1, \ldots, \widetilde{\omega}_d$. We choose $\widetilde{\omega}_i$ with minimal degree. We can assume that the components of $\widehat{D}_1$ are still irreducible in U, and that the points $\mathbf{z}_\lambda$ are still in U. In a neighborhood of $\mathbf{z}_\lambda$ we have a decomposition

$$\widetilde{\omega}_i(z_i; z_{d+1}, \ldots, z_n) = (z_i - f_{\lambda,i}(z_{d+1}, \ldots, z_n)) \cdot \omega^*_{\lambda,i}(z_1, \ldots, z_n),$$

with $\omega^*_{\lambda,i}(\mathbf{z}_\lambda) \neq 0$. So the Jacobian determinant of $\widetilde{\omega}_1, \ldots, \widetilde{\omega}_d$ with respect to the variables $z_1, \ldots, z_d$ does not vanish at any $\mathbf{z}_\lambda$. We denote the zero set of this Jacobian in $\widehat{D}_1 \cap U$ by $\widehat{D}_2$. It has dimension $n - d - 2$, and $\omega_1, \ldots, \omega_d, \widetilde{\omega}_1, \ldots, \widetilde{\omega}_d$ have rank d on $A' - \widehat{D}_2$.

Now apply the same procedure to $\widehat{D}_2$ and obtain an $(n-d-3)$-dimensional $\widehat{D}_3$ and continue in this way until reaching $\widehat{D}_{n-d+1} = \varnothing$.

By putting all of the pseudopolynomials together, in a small neighborhood U of $\mathbf{z}_0$ we get holomorphic functions $f_1, \ldots, f_d, f_{d+1}, \ldots, f_N$ (with $N = (n-d+1) \cdot d$) whose rank is d in every regular point of $A' \cap U$. Since the pseudopolynomials always were chosen with minimal degree, it follows that $A' = N(f_1, \ldots, f_N)$ near $\mathbf{z}_0$.

Now set $A^+ = \overline{A \cap U - A'}$. It is the union of the remaining irreducible components of $A \cap U$. We may assume that U is so small that A^+ is the common zero set of finitely many holomorphic functions $g_1, \ldots, g_s$ in U. Multiplying the f_i by the g_j yields finitely many holomorphic functions in U that describe the set $A \cap U$. No point of $A' \cap A^+$ is a regular point of A. For every $\mathbf{z} \in A' - A^+$ there is a g_j that does not vanish there. So the rank of the Jacobian of the $f_i \cdot g_j$ is equal to d at every nonsingular point of $A' - A^+$.

The same procedure can be used for every d and the corresponding A', and consequently, in finitely many steps we obtain a neighborhood of $\mathbf{z}_0$ and holomorphic functions in this neighborhood with the desired properties. ∎

The Singular Locus. From the preceding theorem we conclude the following:

6.6 Theorem. *The set* $\text{Sing}(A)$ *of singular points of an analytic set* A *is again an analytic set.*

Proof: The intersection of two irreducible components of A belongs to $\text{Sing}(A)$. The union S of all these intersections is an analytic set.

Assume that $\mathbf{z}_0$ is a point of an irreducible component A' of A and $\dim(A') = n - d$. Then there is a neighborhood $U = U(\mathbf{z}_0) \subset G$ with holomorphic functions $f_1, \ldots, f_N$ vanishing exactly on $A' \cap U$ such that their Jacobian has rank d in each of the regular points. Let S^* be the analytic set of all points of $A' \cap U$ where all $d \times d$ minors of the Jacobian vanish. Clearly, S^* is contained in $\text{Sing}(A') \cap U$. On the other hand, at any point of $\text{Sing}(A') \cap U$ the Jacobian of $f_1, \ldots, f_N$ cannot have rank d. So $\text{Sing}(A') \cap U = S^*$, and $\text{Sing}(A')$ is analytic in G.

The union of S and the sets $\text{Sing}(A')$ for all irreducible components A' is the set $\text{Sing}(A)$. It is analytic, since the union is locally finite. ∎

The set $\text{Sing}(A)$ is called the *singular locus* of A.

Extending Analytic Sets. Let $G \subset \mathbb{C}^n$ be a domain.

6.7 Lemma. *Let* $\mathbf{z}_0 = (z_1^{(0)}, \ldots, z_n^{(0)}) \subset \mathbb{C}^n$ *be an arbitrary point and* $E = \{\mathbf{z} : z_i = z_i^{(0)} \text{ for } i = 1, \ldots, d\}$ *an affine plane of codimension* d *containing* $\mathbf{z}_0$. *If* $A \subset G$ *is an irreducible analytic set of positive dimension that is not a subset of* E, *then there is an open dense subset* $C \subset \mathbb{C}^d$ *such that*

$$f_{\mathbf{c}}(z_1, \ldots, z_n) := c_1\big(z_1 - z_1^{(0)}\big) + \cdots + c_d\big(z_d - z_d^{(0)}\big)$$

does not vanish identically on A *for every* $\mathbf{c} = (c_1, \ldots, c_d) \in C$. *In particular, for any hyperplane* $H_0 \subset \mathbb{C}^n$ *containing* E *there is a hyperplane* H *arbitrarily close to* H_0 *and also containing* E *such that* $\dim(A_i) \le \dim(A) - 1$ *for every irreducible component* A_i *of* $A \cap H$.

Proof: We define $\varphi : \mathbb{C}^d \to \mathcal{O}(\mathbb{C}^n)$ by $\varphi(\mathbf{c}) := f_{\mathbf{c}}$. This is a $\mathbb{C}$-linear map, and $V := \{\mathbf{c} \in \mathbb{C}^d : f_{\mathbf{c}}|_A \equiv 0\}$ is a linear subspace. Suppose that $V = \mathbb{C}^d$. Then $(z_i - z_i^{(0)})|_A \equiv 0$ for $i = 1, \ldots, d$, and therefore $A \subset E$. This is a contradiction, and consequently, V must be a proper subspace of $\mathbb{C}^d$. For any $\mathbf{c}$ in the open dense subset $C := \mathbb{C}^d - V$, $f_{\mathbf{c}}$ does not vanish identically on A. Thus $H_{\mathbf{c}} := \{\mathbf{z} : f_{\mathbf{c}}(\mathbf{z}) = 0\}$ is a hyperplane containing E. ∎

Our main tool for extending analytic sets is the following.

6.8 Proposition. *If $E = \{\mathbf{z} \in \mathbb{C}^n : z_1 = \cdots = z_d = 0\}$ is an $(n-d)$-dimensional plane and A an analytic set in $G - E$, whose irreducible components all have dimension $n-d+l$ with $0 < l < d$, then the closure $\overline{A}$ of A in G is an analytic set in G.*

PROOF: The proposition is of a local nature. We may assume that A is irreducible and $\mathbf{0} \in E \cap G$. It is enough to construct a continuation of A into a neighborhood of $\mathbf{0}$.

Let $\mathbf{c} = (0, \ldots, 0, c_{d+1}, \ldots, c_n)$ be an arbitrary point of $E \cap G$. We consider the following family of $(d-l)$-dimensional planes through $\mathbf{c}$: For

$$\mathbf{A} = \left(a_{ij} \;\middle|\; \begin{array}{l} i = 1, \ldots, n-d+l \\ j = 1, \ldots, d-l \end{array} \right) \in M_{n-d+l,d-l}(\mathbb{C})$$

we have the linear map $\mathbf{L_A} : \mathbb{C}^{d-l} \to \mathbb{C}^{n-d+l}$ and define

$$P(\mathbf{c}, \mathbf{A}) := \mathbf{c} + \{(\mathbf{w}', \mathbf{L_A}(\mathbf{w}')) \,:\, \mathbf{w}' \in \mathbb{C}^{d-l}\}.$$

So $P(\mathbf{c}, \mathbf{A})$ consists of vectors $\mathbf{w} = (w_1, \ldots, w_{d-l}, w_{d-l+1}, \ldots, w_n)$ with

$$\begin{aligned} w_{d-l+i} &= \sum_{j=1}^{d-l} a_{ij} \cdot w_j & &\text{for } i = 1, \ldots, l, \\ w_{d-l+i} &= c_{d-l+i} + \sum_{j=1}^{d-l} a_{ij} \cdot w_j & &\text{for } i = l+1, \ldots, n-d+l. \end{aligned}$$

Then every $P(\mathbf{c}, \mathbf{A})$ meets E exactly in $\mathbf{c} = (0, \ldots, 0, c_{d+1}, \ldots, c_n)$. If $\mathbf{O}$ is the zero matrix, then $P(\mathbf{c}) := P(\mathbf{c}, \mathbf{O})$ is the plane $\mathbb{C}^{d-l} \times \{(0, \ldots, 0, c_{d+1}, \ldots, c_n)\}$.

In the next chapter we will introduce Grassmannian manifolds and a topology on the set of linear subspaces (with fixed dimension) of a given vector space. In our case it follows that a neighborhood of $\mathbf{c} + P_0$ is given by the set of all planes $P = \mathbf{c} + \widetilde{P}$ with $\widetilde{P} \oplus (\mathbf{0} \times \mathbb{C}^{n-d+l}) = \mathbb{C}^n$. This shows that every $(d-l)$-dimensional plane through $\mathbf{c}$ that is near $P(\mathbf{c}, \mathbf{O})$ is of the form $P(\mathbf{c}, \mathbf{A})$.

We choose real numbers $0 < r_1 < r_2$ and $r > 0$ so small that the "shell"

$$S = \{\mathbf{z} = (\mathbf{z}', \mathbf{z}'') \in \mathbb{C}^{d-l} \times \mathbb{C}^{n-d+l} \,:\, r_1 < \|\mathbf{z}'\| < r_2 \text{ and } |\mathbf{z}''| < r\}$$

is a relatively compact open subset of G. Only finitely many irreducible components A_i of A enter S. We can find a hyperplane H_1 containing E that intersects the A_i not at all or in codimension 1. Let A_{ij} be the finitely many irreducible components of $H_1 \cap A_i$ that enter S. We can find another hyperplane H_2 containing E that intersects all A_{ij} at most in codimension 2. We continue this procedure running through the irreducible components of the $A_{ij} \cap H_2$. After finitely many steps we have $n-d+l$ hyperplanes such that their intersection P is a plane $P(\mathbf{c}, \mathbf{A})$ that meets $A \cap S$ in at most finitely many points. By the above lemma we can choose this plane arbitrarily near $P_0 := P(\mathbf{c})$.

Now apply a linear coordinate transformation very near to the identity that leaves E invariant and maps our plane $P(\mathbf{c}, \mathbf{A})$ onto P_0 and replace the transformed shell S by a new shell S' (in the new coordinates) that is a little bit smaller such that $\overline{S'}$ is contained in the old (transformed) S. This can be done so that $\partial S' \cap P_0 \cap A = \varnothing$.

Earlier we proved that A is an embedded-analytic set in a neighborhood of the points of the intersection $A \cap P_0 \cap S'$ over a domain G' in the space of variables $z_{d-l+1}, \dots, z_n$. So there is a small closed ball $\overline{B} \subset \mathbb{C}^{n-d+l}$ around the origin such that $\left(\mathbb{C}^{d-l} \times \overline{B}\right) \cap \partial S' \cap A$ remains empty, every irreducible component of $\left(\mathbb{C}^{d-l} \times \overline{B}\right) \cap S' \cap A$ enters $P_0 \cap S'$, and every plane through a point of $\overline{B}$ and parallel to P_0 intersects $A \cap S'$ in at most finitely many points.

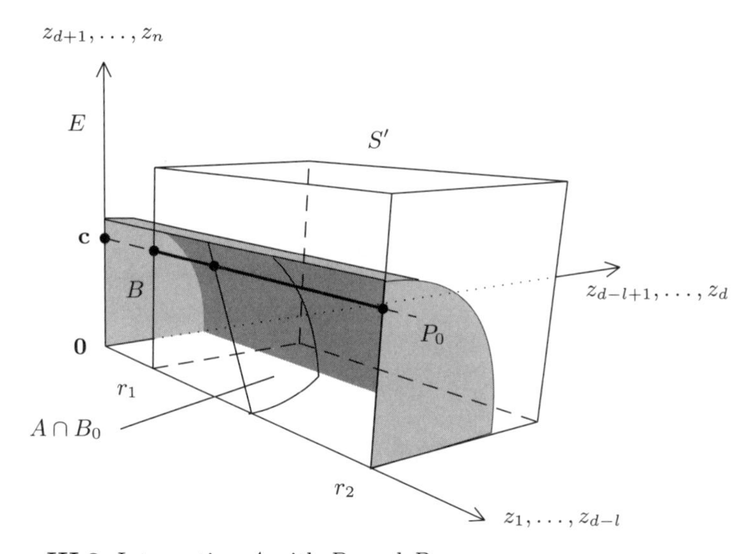

Figure III.3. Intersecting A with B_0 and P_0

Now, the set $B_0 := \{\mathbf{z}' \in \mathbb{C}^{d-l} : \|\mathbf{z}'\| < r_2\} \times \overline{B}$ is a neighborhood of the origin in $\mathbb{C}^n$, and each of our parallel planes through points of $\overline{B} - E$ meets $A \cap B_0$ in a compact analytic set and therefore in at most finitely many points (cf. Figure III.3). At all of these points the set A is locally an embedded-analytic set over $\overline{B}$. By multiplying the pseudopolynomials by the same distinguished variable that we obtained for the different intersection points over the same base point in $\overline{B} - E$, we have that $A \cap \left(\mathbb{C}^{d-l} \times \left(\overline{B} - E\right)\right)$ is an embedded-analytic set over $\overline{B} - E$. The coefficients of the corresponding pseudopolynomials over $\overline{B} - E$ are bounded along E. Hence, they can be analytically extended to $\overline{B}$. This means that $A \cap (B_0 - E)$ has a unique analytic continuation to B_0. ∎

The proposition just proved is also true if $A \subset G - E$ is an analytic set, whose irreducible components all have dimension greater than $n - d$, since we can write A as a finite union of pure-dimensional analytic sets. As a consequence we have the following theorem.

6.9 Theorem of Remmert–Stein. *Assume that $G \subset \mathbb{C}^n$ is a domain, $K \subset G$ an $(n-d)$-dimensional analytic subset, and A an analytic subset of $G - K$ all components of which have dimension $> n - d$. Then the closure $\overline{A}$ of A in G is an analytic set in G.*

PROOF: If $\mathbf{z}_0 \in K$ is a regular point, then K can be transformed in a neighborhood of $\mathbf{z}_0$ to a plane E. So $\overline{A}$ is analytic in a neighborhood of $\mathbf{z}_0$ and therefore in all regular points of K. We can replace K by the set K_1 of singular points of K, which is analytic again and has lower dimension. By the same argument we show that $\overline{A}$ is analytic at all regular points of K_1. Continuing in this way we prove that $\overline{A}$ is analytic in G. ∎

This theorem first was proved by R. Remmert and K. Stein; see [ReSt53].

The Local Dimension. We show how our results are linked with the classical dimension theory of analytic sets.

Let $G \subset \mathbb{C}^n$ be a domain, $A \subset G$ an analytic set, and $\mathbf{z}_0 \in A$ a point. There is an open neighborhood $U = U(\mathbf{z}_0) \subset G$ such that $U \cap A$ is a finite union of irreducible analytic components $A_1, \dots, A_l$. If we choose U small enough, then the A_i are uniquely determined.

Definition. In the given situation the uniquely determined number

$$\dim_{\mathbf{z}_0}(A) := \max_{\lambda=1,\dots,l} \dim(A_\lambda)$$

is called the *(local) dimension* of A at $\mathbf{z}_0$.

The set A has dimension 0 at $\mathbf{z}_0 \in A$ if and only if there is an open neighborhood $U = U(\mathbf{z}_0) \subset G$ such that $U \cap A = \{\mathbf{z}_0\}$.

6.10 Proposition. *Let $k := \dim_{\mathbf{z}_0}(A)$ be positive. Then k is the smallest number with the property that there are holomorphic functions $f_1, \dots, f_k$ in a small neighborhood U of $\mathbf{z}_0$ such that $\mathbf{z}_0$ is isolated in $A \cap N(f_1, \dots, f_k)$.*

PROOF: If $\dim_{\mathbf{z}_0}(A) = k$, then there must be at least one k-dimensional irreducible component A' of A at $\mathbf{z}_0$.

If f is any holomorphic function near $\mathbf{z}_0$, then either $f|_{A'} \equiv 0$ (and therefore $A \cap N(f)$ still k-dimensional) or $A' \cap N(f)$ is $(k-1)$-dimensional. So at least k functions are required.

On the other hand, by Lemma 6.7 we can find a holomorphic function f_1 near $\mathbf{z}_0$ that does not vanish identically on any irreducible component A' of dimension k at $\mathbf{z}_0$. It follows that $A' \cap N(f_1)$ has dimension $k-1$ for all such components A'. We can repeat this process, and after k steps we reach dimension zero, so that $\mathbf{z}_0$ is isolated in $A \cap N(f_1, \dots, f_k)$. ∎

Definition. If A has dimension k at $\mathbf{z}_0$, then any system $\{f_1, \dots, f_k\}$ of holomorphic functions with $A \cap N(f_1, \dots, f_k) = \{\mathbf{z}_0\}$ is called a *parameter system* for A in $\mathbf{z}_0$.

6.11 Ritt's lemma. *Let $B \subset A$ be closed analytic sets in a domain $G \subset \mathbb{C}^n$. Then B is nowhere dense in A if and only if $\dim_{\mathbf{z}}(B) < \dim_{\mathbf{z}}(A)$ for every $\mathbf{z} \in B$.*

PROOF: Let the criterion be fulfilled, and $\mathbf{z}_0$ be an arbitrary point of B. Then there exists an open neighborhood U of $\mathbf{z}_0$ in G and a parameter system $\{f_1, \dots, f_k\}$ on U for B at $\mathbf{z}_0$. Since $\dim_{\mathbf{z}_0}(A) > k$, it is not possible that $\mathbf{z}_0$ is isolated in $A \cap N(f_1, \dots, f_k)$. This means that $(A - B) \cap N(f_1, \dots, f_k) \cap W \neq \varnothing$ for every neighborhood $W = W(\mathbf{z}_0)$. So B is nowhere dense in A.

On the other hand, let B be nowhere dense in A, and $\mathbf{z}_0$ a point of B. In a small neighborhood U of $\mathbf{z}_0$ we have unique decompositions into irreducible components:

$$B \cap U = B_1 \cup \dots \cup B_m \quad \text{and} \quad A = A_1 \cup \dots \cup A_l.$$

Every component B_i is contained in a component $A_{j(i)}$, and for any open neighborhood $W = W(\mathbf{z}_0)$ we have $(A_{j(i)} - B_i) \cap W \neq \varnothing$, because otherwise, there would exist points $\mathbf{z} \in W \cap B_i$ where B is dense in A. So $\dim(B_i) < \dim(A_{j(i)})$ for all i. It follows that $\dim_{\mathbf{z}_0}(B) < \dim_{\mathbf{z}_0}(A)$. ∎

Let $G \subset \mathbb{C}^n$ be a domain, $A \subset G$ an analytic set, and $\mathbf{z}_0 \in A$ a point. If $\dim_{\mathbf{z}_0}(A) = k$, then the number $n-k$ is called the *codimension* of A at $\mathbf{z}_0$.

6.12 Second Riemann extension theorem. *Suppose that $n \geq 2$ and that the analytic set $A \subset G$ has everywhere at least codimension 2. Then any holomorphic function f on $G - A$ has a holomorphic extension to G.*

PROOF: We may assume that A is irreducible of codimension $d \geq 2$. If $\mathbf{z}_0$ is a regular point of A, then there is a neighborhood U of $\mathbf{z}_0$ such that $U \cap A$

is biholomorphically equivalent to an open subset of a linear subspace E of codimension d. By the theorem on removable singularities (see Section II.1) f can be holomorphically extended to $\mathbf{z}_0$.

We repeat this procedure. Beginning with the set $\operatorname{Sing}(A)$, which has codimension $d+1$, after finitely many steps only a set of isolated points remains. Since f can also be extended to these points, we obtain the desired result. ∎

Exercises

1. Let $A_1, A_2 \subset \mathbb{C}^n$ be analytic sets. Show that $\dim(A_1 \cap A_2) \geq \dim(A_1) + \dim(A_2) - n$.
2. Assume that $G \subset \mathbb{C}^n$ is a domain and f a nonconstant holomorphic function on G. Prove that there is an at most countably infinite set $Z \subset \mathbb{C}$ such that $A_c := \{\mathbf{z} \in G : f(\mathbf{z}) = c\}$ is regular for $c \in \mathbb{C} - Z$.
3. Let $G \subset \mathbb{C}^n$ be a domain and $A \subset G$ an analytic set. Prove that for any $k \geq 0$ the closure of the set $A_k := \{\mathbf{z} \in A : \dim_{\mathbf{z}}(A) = k\}$ is either empty or is a pure k-dimensional analytic subset of G.
4. Let $G \subset \mathbb{C}^n$ be a domain and $f_1, \ldots, f_m$ holomorphic functions on G. Denote by N the common zero set $N(f_1, \ldots, f_m)$. Show that if the rank of the Jacobian of $f_1, \ldots, f_m$ at some point $\mathbf{z}_0 \in N$ is equal to r, then there is a neigborhood $U = U(\mathbf{z}_0) \subset G$ and a closed submanifold $M \subset U$ of dimension less than or equal to $n - r$ such that $U \cap A \subset M$.
5. Let A be an analytic set in a domain $G \subset \mathbb{C}^n$. For every point $\mathbf{z}_0 \in A$ the set $I_{\mathbf{z}_0}(A) := \{(f)_{\mathbf{z}_0} \in \mathcal{O}_{\mathbf{z}_0} : f|_A \equiv 0\}$ is an ideal in $\mathcal{O}_{\mathbf{z}_0}$. Show that there are a neighborhood U of $\mathbf{z}_0$ and holomorphic functions $f_1, \ldots, f_k$ on U such that:
   (a) $A \cap U = N(f_1, \ldots, f_k)$.
   (b) $I_{\mathbf{z}_0}(A)$ is generated by the germs $(f_1)_{\mathbf{z}_0}, \ldots, (f_k)_{\mathbf{z}_0}$.
   Show that the vector space
   $$T_{\mathbf{z}_0}(A) = \Big\{\mathbf{w} \in \mathbb{C}^n : \sum_{\nu=1}^{n} w_\nu f_{z_\nu}(\mathbf{z}_0) = 0 \text{ for every } f \in I_{\mathbf{z}_0}(A)\Big\}$$
   has dimension $n - \operatorname{rk}_{\mathbf{z}_0}(f_1, \ldots, f_k)$. It is called the *Zariski tangential space* of A at $\mathbf{z}_0$.
6. Let A be defined as in Exercise 5, and suppose that $\mathbf{z}_0$ is a regular point of A. Show that $T_{\mathbf{z}_0}(A)$ is the set of tangent vectors $\dot\alpha(0)$ (see Section I.7), where $\alpha : I \to \mathbb{C}^n$ is differentiable, $\alpha(I) \subset A$, and $\alpha(0) = \mathbf{z}_0$.
7. Let A be an analytic set in a domain $G \subset \mathbb{C}^n$, and let $\mathbf{z}_0 \in A$ be an arbitrary point. The *embedding dimension* of A at $\mathbf{z}_0$ is the smallest integer e such that there is an open neighborhood $U = U(\mathbf{z}_0)$ and a closed submanifold $M \subset U$ of dimension e with $A \cap U \subset M$. It is denoted by $\operatorname{embdim}_{\mathbf{z}_0}(A)$. Prove that $\mathbf{z} \mapsto \operatorname{embdim}_{\mathbf{z}}(A)$ is upper semicontinuous on A, and that $\operatorname{embdim}_{\mathbf{z}_0}(A) = \dim_{\mathbb{C}}(T_{\mathbf{z}_0}(A))$.
8. Consider the analytic set $A = \{w = \exp(1/z)\} \subset \{(w, z) \in \mathbb{C}^2 : z \neq 0\}$. Determine the closure of A in $\mathbb{C}^2$.

Chapter IV

Complex Manifolds

1. The Complex Structure

Complex Coordinates. Let X be a Hausdorff space, i.e., a topological space satisfying the Hausdorff separation axiom. Sometimes such a space is also called a *separated space* or a T_2-space. Hausdorff spaces are the most common in topology (for example, every metric space is a Hausdorff space), but non-Hausdorff spaces do arise, in particular in algebraic geometry. The space $\mathbb{C}^n$ with the Zariski topology is not Hausdorff.

We think that a space X is too big if there exists a discrete subset with the cardinality of the continuum. Therefore, we demand that the topology of X have a countable base. In this case X is said to satisfy the *second axiom of countability*. Obviously, $\mathbb{C}^n$ has a countable basis. A metric space has a countable basis if and only if it contains a countable dense subset.

A Hausdorff space X is called *locally compact* if every point $x \in X$ has a compact neighborhood. If X is compact, then it is also locally compact. If X is locally compact, but not compact, then X can be made compact by adjoining just one point (Alexandrov's one-point compactification). Every Hausdorff space that is locally homeomorphic to an open subset of $\mathbb{C}^n$ is locally compact. So, for example, every Riemann domain over $\mathbb{C}^n$ is locally compact.

Definition. An open covering $\mathscr{V} = \{V_\nu : \nu \in N\}$ of a Hausdorff space X is called a *refinement* of the covering $\mathscr{U} = \{U_\iota : \iota \in I\}$ of X if there is a map $\tau : N \to I$ (the *refinement map*) with

$$V_\nu \subset U_{\tau(\nu)} \text{ for every } \nu \in N.$$

The refinement map is not uniquely determined, but we can fix it once and for all.

A covering $\mathscr{V} = \{V_\nu : \nu \in N\}$ is called *locally finite* if each $x \in X$ has a neighborhood $U = U(x)$ such that U meets only finitely many V_ν.

Definition. A Hausdorff space X is called *paracompact* if every open covering $\mathscr{U}$ of X has a locally finite open refinement $\mathscr{V}$.

Every compact space is paracompact. Furthermore, every locally compact space with countable basis is paracompact.

For the moment we only assume that X is a Hausdorff space.

Definition. An n-dimensional *complex coordinate system* (U, φ) in X consists of an open set $U \subset X$ and a topological map φ from U onto an open set $B \subset \mathbb{C}^n$.

If $p \in X$ is a point, then every coordinate system (U, φ) in X with $p \in U$ is called a *coordinate system at* p. The entries z_i in $\mathbf{z} = \varphi(p)$ are called the *complex coordinates* of p (with respect to (U, φ)).

If f is a complex function in U, we can consider it as a function of the complex coordinates $z_1, \dots, z_n$, by

$$(z_1, \dots, z_n) \mapsto f \circ \varphi^{-1}(z_1, \dots, z_n).$$

Two (n-dimensional) complex coordinate systems (U, φ) and (V, ψ) in X are called *(holomorphically) compatible* if either $U \cap V = \varnothing$ or the map

$$\varphi \circ \psi^{-1} : \psi(U \cap V) \to \varphi(U \cap V)$$

is biholomorphic (see Figure IV.1).

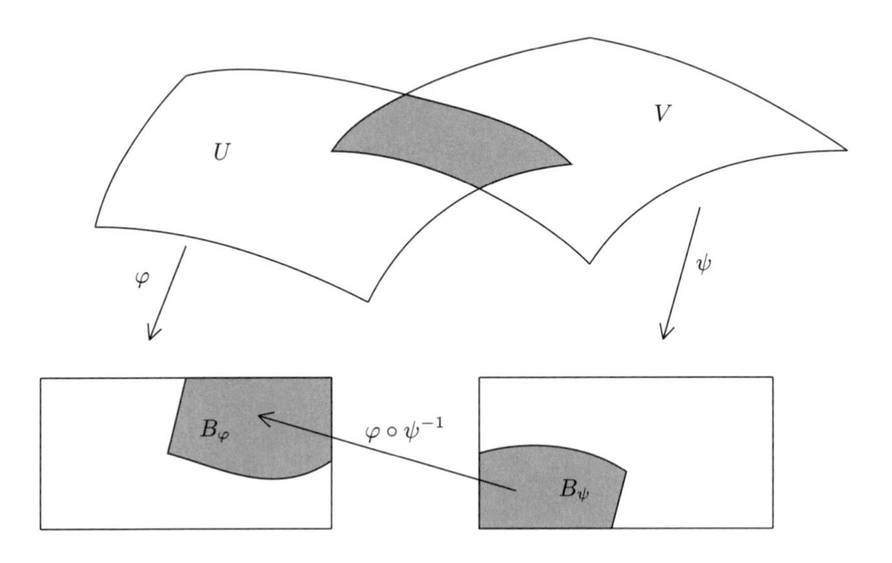

Figure IV.1. Change of coordinates

The sets $B_\psi := \psi(U \cap V)$ and $B_\varphi := \varphi(U \cap V)$ are open subsets of $\mathbb{C}^n$. If z_i (respectively w_j) are the complex coordinates with respect to φ (respectively ψ), then compatibility of the coordinate systems means that the functions $z_i = z_i(w_1, \dots, w_n)$ and $w_j = w_j(z_1, \dots, z_n)$ are holomorphic.

A covering of X with pairwise compatible n-dimensional complex coordinate systems is called an n-dimensional *complex atlas* on X. Two such atlases $\mathcal{A}_1$ and $\mathcal{A}_2$ are called *equivalent* if any two coordinate systems $(U, \varphi) \in \mathcal{A}_1$ and $(V, \psi) \in \mathcal{A}_2$ are compatible. An equivalence class of (n-dimensional) complex atlases on X is called an n-dimensional *complex structure* on X. It contains a maximal atlas that is the union of all atlases in the equivalence class.

Definition. An n-dimensional *complex manifold* is a Hausdorff space X with countable basis, equipped with an n-dimensional complex structure.

Every complex manifold is locally compact and paracompact.

Examples

1. The complex n-space $\mathbb{C}^n$ is an n-dimensional complex manifold. The complex structure is given by the coordinate system $(\mathbb{C}^n, \mathrm{id})$.
2. If X is an arbitrary n-dimensional complex manifold, then any nonempty open subset $B \subset X$ is again an n-dimensional complex manifold. For $p \in B$ there is a coordinate system (U, φ) in X at p. Then $(U \cap B, \varphi|_{U\cap B})$ is a coordinate system in B at p. All of these coordinate systems are compatible.
3. Let $G \subset \mathbb{C}^n$ be a domain and $X \subset G$ a k-dimensional complex submanifold. Of course, X is a Hausdorff space (in the relative topology) with countable basis. For $\mathbf{z}_0 \in X$ there are open neighborhoods $W = W(\mathbf{z}_0) \subset G$ and $B = B(\mathbf{0}) \subset \mathbb{C}^n$ and a biholomorphic map $\mathbf{F} : W \to B$ such that

   $$\mathbf{F}(W \cap X) = \{(w_1, \ldots, w_n) \in B \,:\, w_{k+1} = \cdots = w_n = 0\}.$$

   Let $\mathrm{pr}' : \mathbb{C}^n \to \mathbb{C}^k$ be the projection $(w_1, \ldots, w_n) \mapsto (w_1, \ldots, w_k)$. We define $U := W \cap X$ and $\varphi := \mathrm{pr}' \circ \mathbf{F} : U \to \mathbb{C}^k$. Then (U, φ) is a k-dimensional complex coordinate system in X at $\mathbf{z}_0$.

   If (V, ψ) is another coordinate system, with $\psi = \mathrm{pr}' \circ \widetilde{\mathbf{F}}$, then

   $$\varphi \circ \psi^{-1}(w_1, \ldots, w_k) = \mathrm{pr}' \circ \mathbf{F} \circ \widetilde{\mathbf{F}}^{-1}(w_1, \ldots, w_k, 0, \ldots, 0)$$

   is holomorphic. So we get a complex structure on X.
4. Finally, let (G, π) be a Riemann domain (over $\mathbb{C}^n$). Then G is a connected Hausdorff space, and for every $p \in G$ there is an open neighborhood $U = U(p)$ such that $B := \pi(U)$ is open and $\varphi := \pi|_U : U \to B$ is topological. Then (U, φ) is a complex coordinate system. If $\psi = \pi|_V$ is another coordinate system, then for $x \in U \cap V$ we have $\varphi(x) = \psi(x) = \pi(x) =: \mathbf{z}$ and

   $$\varphi \circ \psi^{-1}(\mathbf{z}) = \varphi(x) = \mathbf{z}.$$

Therefore, the coordinate systems are compatible. We get a complex structure on G. One can prove that G has a countable basis (cf. [Gr55], §2]). So every Riemann domain over $\mathbb{C}^n$ is an n-dimensional complex manifold.

Holomorphic Functions

Holomorphic Functions. Let X be an n-dimensional complex manifold.

Definition. A complex function f on an open subset $B \subset X$ is called *holomorphic* if for each $p \in B$ there is a coordinate system (U, φ) at p such that $f \circ \varphi^{-1} : \varphi(U \cap B) \to \mathbb{C}$ is holomorphic.

If $z_1, \dots, z_n$ are the complex coordinates with respect to (U, φ), then

$$(z_1, \dots, z_n) \mapsto f \circ \varphi^{-1}(z_1, \dots, z_n)$$

is a holomorphic function in the ordinary sense. If $z_\nu = z_\nu(w_1, \dots, w_n)$, where $w_1, \dots, w_n$ are the complex coordinates with respect to a coordinate system (V, ψ), then

$$f \circ \psi^{-1}(w_1, \dots, w_n) = f \circ \varphi^{-1}(z_1(w_1, \dots, w_n), \dots, z_n(w_1, \dots, w_n))$$

is also holomorphic. So the definition of holomorphy is independent of the coordinate system. We denote the set of holomorphic functions on B by $\mathcal{O}(B)$. It is a $\mathbb{C}$-algebra with unit element.

Example

Let $G \subset \mathbb{C}^n$ be a domain and $X \subset G$ a k-dimensional complex submanifold. We consider a complex coordinate system (U, φ) in X, where U is the intersection of X with an open set $W \subset G$ and $\varphi = \mathrm{pr}' \circ \mathbf{F}$, with a biholomorphic map $\mathbf{F} : W \to B \subset \mathbb{C}^n$ such that $\mathbf{F}(U) = \{\mathbf{w} \in B : w_{k+1} = \cdots = w_n = 0\}$. If f is a holomorphic function on G, then

$$f|_X \circ \varphi^{-1}(w_1, \dots, w_k) = f \circ \mathbf{F}^{-1}(w_1, \dots, w_k, 0, \dots, 0)$$

is holomorphic. Therefore, $f|_X$ is a holomorphic function on the complex manifold X.

1.1 Identity theorem. *Let X be connected. If f, g are two holomorphic functions on X that coincide in a nonempty open subset $U \subset X$, then $f = g$.*

PROOF: Let $W = \{x \in X : f(x) = g(x)\}$. Then $W^\circ \neq \varnothing$, since $U \subset W$. Assume that there exists a boundary point x_0 of W° in X and let (U, φ) be a coordinate system at x_0 with $\varphi(x_0) = \mathbf{0}$. Then all derivatives of $f \circ \varphi^{-1}$ and $g \circ \varphi^{-1}$ must coincide at $\mathbf{0}$. It follows that the power series of these functions around the origin are equal. But then $f = g$ in a whole neighborhood of x_0, and this is a contradiction.

If there were a point $x \in M := X - W^\circ$ that was not an interior point of M, then x would be a boundary point of W°. This shows that M must be open. Since X is connected, M has to be empty. ∎

1.2 Maximum principle. *Let X be connected, $f \in \mathcal{O}(X)$, and $x_0 \in X$ a point such that $|f|$ has a local maximum at x_0. Then f is constant.*

PROOF: The functions f and $g := f(x_0)$ are both holomorphic on X. If (U, φ) is a coordinate system at x_0 and $B := \varphi(U)$, then $f_0 := f \circ \varphi^{-1}$ is holomorphic on B, and $|f_0|$ has a local maximum at $\mathbf{z}_0 := \varphi(x_0)$. Thus there is an open neighborhood $B' = B'(\mathbf{z}_0) \subset B$ such that f_0 is constant on B' and f is constant on $U' := \varphi^{-1}(B')$. So $f|_U = g|_U$, and by the identity theorem $f = g$; i.e., f is constant. ∎

1.3 Corollary. *If X is compact and connected, then every holomorphic function on X is constant.*

PROOF: The continuous function $|f|$ takes its maximum at some point of X. Now the corollary follows from the maximum principle. ∎

1.4 Corollary. *There is no compact complex submanifold of positive dimension in $\mathbb{C}^n$.*

PROOF: Let $X \subset \mathbb{C}^n$ be a compact connected submanifold. Then the standard coordinate functions $z_\nu|_X$ must be constant, for $\nu = 1, \ldots, n$. This means that X is a single point. If X is not connected, it is a finite set of points. ∎

Remark. Another proof is given in Section III.6.

Riemann Surfaces

Riemann Surfaces. An *(abstract) Riemann surface* is by definition a 1-dimensional connected complex manifold.

Example

The complex plane $\mathbb{C}$ and every domain in $\mathbb{C}$ are Riemann surfaces. Recall the Riemann surface of $\sqrt{z}$,

$$X = \{(w, z) \in \mathbb{C}^2 : w^2 = z,\ z \neq 0\}.$$

Since X is a Riemann domain over $\mathbb{C}$, it is a 1-dimensional connected complex manifold. From the projection $\pi := \mathrm{pr}_2|_X : X \to \mathbb{C}$ we get complex coordinate systems (U, φ) with $\varphi := \pi|_U$ and sufficiently small U.

The function $f : X \to \mathbb{C}$ with $f(w, z) := w$ is a global holomorphic function with $f^2 = z$.

Example

The Riemann sphere $\overline{\mathbb{C}} = \mathbb{C} \cup \{\infty\}$ is a compact connected Hausdorff space. We have two coordinate systems $(\mathbb{C}, \varphi)$ and $(\overline{\mathbb{C}} - \{0\}, \psi)$ with $\varphi(z) = z$ and $\psi(z) = 1/z$. On $\mathbb{C}^* = \mathbb{C} \cap (\overline{\mathbb{C}} - \{0\})$ we have $\varphi \circ \psi^{-1}(z) = 1/z$, and this is holomorphic. So $\overline{\mathbb{C}}$ is a compact Riemann surface. Every global holomorphic function on $\overline{\mathbb{C}}$ is constant, but there are nontrivial meromorphic functions, for example $f(z) = z$ (with one pole at ∞). Here a *meromorphic function* on X is a function f that is holomorphic outside a discrete subset $P \subset X$ and satisfies

$$\lim_{x \to p} |f(x)| = \infty \qquad \text{for every } p \in P.$$

The points of P are called the *poles* of f.

Holomorphic Mappings. Let $F : X \to Y$ be a continuous map between complex manifolds.

> **Definition.** The map F defined above is called *holomorphic* if for any $p \in X$ there is a coordinate system (U, φ) in X at p and a coordinate system (V, ψ) in Y at $F(p)$ with $F(U) \subset V$ such that
>
> $$\psi \circ F \circ \varphi^{-1} : \varphi(U) \to \psi(V)$$
>
> is a holomorphic map.

1.5 Proposition. *The map $F : X \to Y$ is holomorphic if and only if for any open subset $V \subset Y$ and any $f \in \mathcal{O}(V)$ it follows that $f \circ F \in \mathcal{O}(F^{-1}(U))$.*

The proof is an easy exercise.

The *category of complex manifolds* consists of a class of *objects*, the complex manifolds, and a class of sets such that to any pair (X, Y) of objects there is assigned a set $\mathcal{O}(X, Y)$ (which may be empty), the set of holomorphic maps between X and Y. In a general category this set would be called the set of *morphisms* from X to Y.

For $(G, F) \in \mathcal{O}(Y, Z) \times \mathcal{O}(X, Y)$ we always have the composition $G \circ F \in \mathcal{O}(X, Z)$ such that the following axioms hold:

1. If $H \circ G$ and $G \circ F$ are defined, then $(H \circ G) \circ F = H \circ (G \circ F)$.
2. For any manifold X we have the identity map $\mathrm{id}_X \in \mathcal{O}(X, X)$ such that $\mathrm{id}_Y \circ F = F$ and $F \circ \mathrm{id}_X = F$, if the compositions are defined.

Another example for a category is given by the topological spaces and continuous mappings. If we replace in our definitions above the field $\mathbb{C}$ by $\mathbb{R}$ and the word "holomorphic" by "differentiable", we get the category of differentiable manifolds and differentiable mappings. From every n-dimensional complex manifold we obtain a $2n$-dimensional differentiable manifold by "forgetting the complex structure."

A holomorphic function $f : X \to \mathbb{C}$ is obviously a holomorphic mapping. More generally, in the case of a Riemann surface a meromorphic function f on X may be viewed as a holomorphic mapping $f : X \to \overline{\mathbb{C}}$.

Definition. A *biholomorphic map* $F : X \to Y$ is a topological map such that F and F^{-1} are holomorphic. If there exists a biholomorphic map between X and Y, then the manifolds are called *isomorphic* or *biholomorphically equivalent*, and we write $X \cong Y$.

Remark. If X is a complex manifold and (U, φ) a complex coordinate system with $\varphi(U) = B \subset \mathbb{C}^n$, then $\varphi : U \to B$ is a biholomorphic map.

Cartesian Products. Assume that $X_1, \dots, X_m$ are complex manifolds of dimension $n_1, \dots, n_m$. Then the set $X = X_1 \times \cdots \times X_m$ carries a natural topology generated by the sets $U = U_1 \times \cdots \times U_m$, $U_i \subset X_i$ open for $i = 1, \dots, m$. One sees immediately that X is a Hausdorff space with countable basis.

Given complex coordinate systems (U_i, φ_i) in X_i, for $i = 1, \dots, m$, one defines a coordinate system (U, φ) for X by

$$\varphi(x_1, \dots, x_m) := (\varphi_1(x_1), \dots, \varphi_m(x_m)) \in \mathbb{C}^n = \mathbb{C}^{n_1 + \cdots + n_m}.$$

It is clear that two such coordinate systems are compatible. So we obtain an n-dimensional complex atlas and a complex structure on X. The projections $p_i : X \to X_i$ are holomorphic maps for $i = 1, \dots, m$.

A simple example is $\mathbb{C}^n = \underbrace{\mathbb{C} \times \cdots \times \mathbb{C}}_{n \text{ times}}$.

The Cartesian product of two complex manifolds X_1, X_2 satisfies the following *universal property*:

Given any complex manifold Y and any two holomorphic maps $F : Y \to X_1$ and $G : Y \to X_2$, there exists exactly one holomorphic map $H : Y \to X_1 \times X_2$ with $F = p_1 \circ H$ and $G = p_2 \circ H$.

Although trivial in our case (we set $H := (F, G)$), this property becomes important in more general categories.

Analytic Subsets. Let X be an n-dimensional complex manifold.

> **Definition.** A subset $A \subset X$ is called *analytic* if for each point $p \in X$ there are a (connected) open neighborhood $U = U(p)$ and finitely many holomorphic functions $f_1, \dots, f_m$ on U such that
>
> $$U \cap A = \{q \in U : f_i(q) = 0 \text{ for } i = 1, \dots, m\}.$$
>
> We call A an *analytic hypersurface* if we can always take $m = 1$.

From the definition it follows that A is a closed subset of X. Locally, an analytic set in X is the same as an analytic set in an open set $B \subset \mathbb{C}^n$. So most properties of analytic sets in $\mathbb{C}^n$ can be transferred.

1.6 Proposition. *If X is connected and $A \subset X$ analytic, then either $A = X$ or A is nowhere dense and $X - A$ is connected.*

PROOF: Assume that $A \neq X$. If A is somewhere dense in X, then A contains interior points (because it is closed in X). Since X is connected, the interior of A has a boundary point $p \in X - A$ (same argument as in the proof of the identity theorem). We take a connected neighborhood $U = U(p)$ such that $A \cap U = \{q \in U : f_1(q) = \cdots = f_m(q) = 0\}$. Then U contains an open subset V (consisting of interior points of A) where $f_1, \dots, f_m$ vanish identically. By the identity theorem they vanish on the whole set U, and p cannot be a boundary point of the interior of A. This is a contradiction, and it follows that A is nowhere dense.

If $X - A$ is not connected, it can be decomposed into two nonempty open subsets U_1, U_2. The function $f : X - A \to \mathbb{C}$ with $f(x) \equiv 0$ on U_1 and $f(x) \equiv 1$ on U_2 is holomorphic and bounded. By Riemann's extension theorem (which can be applied locally) there exists a holomorphic function $\widehat{f}$ on X that coincides with f outside A. Since $\widehat{f}$ can take only the values 0 and 1, it is locally constant. But on the connected manifold X every locally constant function is constant. This is a contradiction. ∎

Let $f_1, \dots, f_m$ be holomorphic functions that are defined on an open subset $U \subset X$. Let $p \in U$ be a point and (V, ψ) a complex coordinate system in X at p. The mapping $\mathbf{f} = (f_1, \dots, f_m) : U \to \mathbb{C}^m$ is holomorphic, and we define

$$J_{\mathbf{f}}(p; \psi) := \left(\frac{\partial (f_i \circ \psi^{-1})}{\partial z_j}(\psi(p)) \;\middle|\; \begin{matrix} i = 1, \dots, m \\ j = 1, \dots, n \end{matrix} \right).$$

This is something like a Jacobian matrix of $\mathbf{f}$ at p, but it depends on the coordinate system ψ. Since

$$
\begin{aligned}
\frac{\partial(f_i \circ \psi^{-1})}{\partial z_j}(\psi(p)) &= \frac{\partial((f_i \circ \varphi^{-1}) \circ (\varphi \circ \psi^{-1}))}{\partial z_j}(\psi(p)) \\
&= \sum_{k=1}^{n} \frac{\partial(f_i \circ \varphi^{-1})}{\partial w_k}(\varphi(p)) \frac{\partial(w_k \circ \varphi \circ \psi^{-1})}{\partial z_j}(\psi(p)),
\end{aligned}
$$

we have

$$J_{\mathbf{f}}(p;\psi) = J_{\mathbf{f}}(p;\varphi) \cdot J_{\varphi\circ\psi^{-1}}(\psi(p)).$$

This shows that

$$\operatorname{rk}_p(f_1,\ldots,f_m) := \operatorname{rk} J_{(f_1,\ldots,f_m)}(p;\psi)$$

is independent of the chosen coordinate system.

Definition. An analytic set $A \subset X$ is called *regular (of codimension d)* at a point $p \in A$ if there are an open neighborhood $U = U(p) \subset X$ and holomorphic functions $f_1, \ldots, f_d$ on U such that:

1. $A \cap U = \{q \in U : f_1(q) = \cdots = f_d(q) = 0\}$.
2. $\operatorname{rk}_p(f_1, \ldots, f_d) = d$.

The number $n - d$ is called the *dimension* of A at p.

If A is regular at every point, A is called a *complex submanifold* of X.

1.7 Proposition. *An analytic set A is regular of codimension d at $p \in A$ if and only if there is a complex coordinate system (U, φ) in X at p with $\varphi(U) = B \subset \mathbb{C}^n$ and $\varphi(U \cap A) = \{\mathbf{w} \in B : w_{n-d+1} = \cdots = w_n = 0\}$.*

If A is a complex submanifold of X, then A itself is a complex manifold.

PROOF: Let (U, ψ) be an arbitrary coordinate system at p and $W := \psi(U)$. Then $\widetilde{A} := \psi(A \cap U)$ is an analytic subset of W that is regular of codimension d at $\mathbf{z}_0 := \psi(p)$, and there exists a biholomorphic map $\mathbf{f}$ from W onto an open neighborhood $B = B(\mathbf{0}) \subset \mathbb{C}^n$ with $\mathbf{f}(\mathbf{z}_0) = \mathbf{0}$ and $\mathbf{f}(\widetilde{A}) = \{\mathbf{w} : w_{n-d+1} = \cdots = w_n = 0\}$. We take $\varphi := \mathbf{f} \circ \psi$.

If $A \subset X$ is a submanifold, then A inherits a natural complex structure from X. This can be demonstrated in the same way as in the case $X = \mathbb{C}^n$. ∎

Example

Let $F : X \to Y$ be a holomorphic map from an n-dimensional manifold into an m-dimensional manifold. Then

$$G_F := \{(x, y) \in X \times Y : y = F(x)\}$$

is called the *graph* of F.

Let $p_0 \in X$ be a point and $q_0 := F(p_0) \in Y$. We choose coordinate systems (U, φ) in X at p_0 and (V, ψ) in Y at q_0, with $F(U) \subset V$. Then $(U \times V, \varphi \times \psi)$ is a coordinate system in $X \times Y$ at $(p_0, q_0) \in G_F$. Writing $\psi \circ F = (f_1, \dots, f_m)$ we get

$$G_F \cap (U \times V) = \{(\varphi \times \psi)^{-1}(\mathbf{z}, \mathbf{w}) \,:\, f_i \circ \varphi^{-1}(\mathbf{z}) - w_i = 0 \text{ for } i = 1, \dots, m\}.$$

So G_F is locally defined by the functions $g_i(p, q) := f_i(p) - w_i \circ \psi(q)$, for $i = 1, \dots, m$. Since $\operatorname{rk}_{(p_0,q_0)}(g_1, \dots, g_m) = m$, we see that G_F is an n-dimensional submanifold.

The *diagonal* $\Delta_X \subset X \times X$ is a special case, which is given as the graph of the identity:

$$\Delta_X = \{(x, x') \in X \times X \,:\, x = x'\}.$$

Example

Let $A = \{(w, z_1, z_2) \in \mathbb{C}^3 \,:\, w^2 = z_1 z_2\}$. The projection $p : (w, z_1, z_2) \mapsto (z_1, z_2)$ realizes A as a branched covering over $\mathbb{C}^2$ that is the zero set of the pseudopolynomial $\omega(w; z_1, z_2) = w^2 - z_1 z_2$. Outside the discriminant set $D_\omega = \{(z_1, z_2) \,:\, z_1 z_2 = 0\}$ it always has two regular leaves over $\mathbb{C}^2$. So A is everywhere 2-dimensional and regular outside D_ω. It is even regular outside the origin, since $\nabla\omega(w, z_1, z_2)$ vanishes only at $(0, 0, 0)$. One can show that $\mathbf{0}$ is, in fact, a singular point; e.g., by using Exercise 8.2 in Chapter I.

The map $\varphi : \mathbb{C}^2 \to A$ with $\varphi(t_1, t_2) := (t_1 t_2, t_1^2, t_2^2)$ is surjective. We call it a *uniformization.* The Jacobian

$$J_\varphi(t_1, t_2) = \begin{pmatrix} t_2 & t_1 \\ 2t_1 & 0 \\ 0 & 2t_2 \end{pmatrix}$$

vanishes exactly at $(0, 0)$. The image point $(0, 0, 0) = \varphi(0, 0)$ is then called a nonuniformizable point.

Analogously to the situation in $\mathbb{C}^n$ one proves that the set $\operatorname{Sing}(A)$ of singular points of an analytic set A is a nowhere dense analytic subset. The set A is called *irreducible* if $A - \operatorname{Sing}(A)$ is connected. To every analytic set $A \subset X$ there is a uniquely determined locally finite system of irreducible analytic sets $(A_\lambda)_{\lambda \in \Lambda}$ such that A is the union of all these *irreducible components* A_λ.

Differentiable Functions. Let X be an n-dimensional complex manifold, $B \subset X$ an open set. A function $f : B \to \mathbb{C}$ is called *differentiable* (respectively *smooth*), if for every complex coordinate system (U, φ) with $U \cap B \neq \varnothing$ the function $f \circ \varphi^{-1}$ is differentiable (respectively infinitely differentiable) on $\varphi(U \cap B)$. We denote the $\mathbb{R}$-algebra of real-valued smooth functions on B by $\mathscr{E}(B)$ and the $\mathbb{C}$-algebra of complex-valued smooth fuctions by $\mathscr{E}(B, \mathbb{C})$.

1.8 Proposition. *In every complex manifold X there is a sequence of compact subsets (K_n) with $K_n \subset (K_{n+1})^\circ$ and $\bigcup_{n=1}^\infty K_n = X$.*

PROOF: Since X is locally compact with countable basis, we can find a countable basis $(B_\nu)_{\nu \in \mathbb{N}}$ of the topology of X such that each $\overline{B}_\nu$ is compact. We take $K_1 := \overline{B}_1$. If n_1 is the minimal number such that $K_1 \subset \overline{B}_1 \cup \cdots \cup \overline{B}_{n_1}$, then $k_1 \geq 2$, and we take $K_2 := \overline{B}_1 \cup \cdots \cup \overline{B}_{n_1}$, and so on. ∎

We call $(K_n)_{n \in \mathbb{N}}$ a *compact exhaustion* of X.

1.9 Proposition. *Let an open covering $\mathscr{U} = \{U_\iota : \iota \in I\}$ of X be given, and two real numbers r, r' with $0 < r' < r$. Then there is a locally finite open refinement $\mathscr{V} = \{V_\lambda : \lambda \in L\}$ of $\mathscr{U}$ such that the following hold:*

1. *For each $\lambda \in L$ there exists a complex coordinate system $(V_\lambda, \varphi_\lambda)$ in X with $\varphi_\lambda(V_\lambda) = B_r(\mathbf{0})$.*
2. *The open sets $\varphi_\lambda^{-1}(B_{r'}(\mathbf{0}))$ also cover X.*

PROOF: We use a compact exhaustion (K_n) and define $M_1 := K_1$ and $M_n := K_n - (K_{n-1})^\circ$ for $n \geq 2$. Then (M_n) is a covering of X by compact sets.

We consider a fixed $M = M_n$. For each $x \in M$ there is an index $\iota = \iota(x) \in I$ and an open neighborhood $V = V(x) \subset U_\iota \cap ((K_{n+1})^\circ - K_{n-2})$. We can make V so small that there is a complex coordinate system $\varphi : V \to B_r(\mathbf{0})$ with $\varphi(x) = \mathbf{0}$, and we define $V' := \varphi^{-1}(B_{r'}(\mathbf{0}))$. The set M is covered by finitely many neighborhoods $V'_{n,1}, \ldots, V'_{n,m_n}$ like our V'. Then

$$\mathscr{V} := \{V_{n,i} : n \in \mathbb{N},\, i = 1, \ldots, m_n\}$$

is the desired covering. ∎

Definition. A *(smooth) partition of unity* on X is a family $(\varphi_\iota)_{\iota \in I}$ of smooth real-valued functions such that:

1. $\varphi_\iota \geq 0$ everywhere.
2. The system of the sets $\mathrm{supp}(\varphi_\iota)$ is locally finite.
3. $\sum_{\iota \in I} \varphi_\iota = 1$.

1.10 Theorem. *For any open covering $\mathscr{U} = \{U_\iota : \iota \in I\}$ of X there is a partition of unity (φ_ι) with* $\mathrm{supp}(\varphi_\iota) \subset U_\iota$.

PROOF: We have a locally finite refinement $\mathscr{V} = \{V_\lambda : \lambda \in L\}$ of $\mathscr{U}$ and complex coordinates $\varphi_\lambda : V_\lambda \to B_r(\mathbf{0})$ as in Proposition 1.9. If $\psi : \mathbb{C}^n \to \mathbb{R}$ is a smooth function with $0 \le \psi(\mathbf{z}) \le 1$, $\psi(\mathbf{z}) \equiv 1$ on $B_{r'}(\mathbf{0})$ and $\psi(\mathbf{z}) \equiv 0$ on $\mathbb{C}^n - B_r(\mathbf{0})$, we define a smooth function ψ_λ on X by $\psi_\lambda = \psi \circ \varphi_\lambda$ on V_λ and $\psi_\lambda(x) \equiv 0$ otherwise.

Let $\tau : L \to I$ be a refinement map (with $V_\lambda \subset U_{\tau(\lambda)}$). Then $\mathscr{W} = \{W_\iota : \iota \in I\}$ with $W_\iota := \bigcup_{\lambda \in \tau^{-1}(\iota)} V_\lambda$ is an open refinement of $\mathscr{U}$ with $W_\iota \subset U_\iota$. In addition it is locally finite, because for $x \in X$ there is a neighborhood $P = P(x)$ such that $P \cap V_\lambda \ne \varnothing$ only for $\lambda \in L_0$, $L_0 \subset L$ finite. But then $P \cap W_\iota \ne \varnothing$ only for $\iota = \tau(\lambda)$, $\lambda \in L_0$.

We define $\widetilde{\varphi}_\iota := \sum_{\lambda \in \tau^{-1}(\iota)} \psi_\lambda$. The sum is finite at every point. So $\widetilde{\varphi}_\iota$ is smooth and has its support in W_ι. Every $x \in X$ lies in a set $\varphi_\lambda^{-1}(B_{r'}(\mathbf{0}))$, where ψ_λ is positive. Therefore, $\varphi := \sum_\iota \widetilde{\varphi}_\iota$ is well defined and everywhere positive. Now we can define the partition of unity by $\varphi_\iota := \widetilde{\varphi}_\iota / \varphi$. ■

1.11 Corollary. *Let $U \subset X$ be an open set and $V \subset\subset U$ an open subset. Then there exists a function $f \in \mathscr{E}(X)$ with $f|_V = 0$ and $f|_{(X-U)} = 1$.*

PROOF: The system $\{U, X - \overline{V}\}$ is an open covering of X. Let $\{\varphi_1, \varphi_2\}$ be a partition of unity for this covering. Then $\mathrm{supp}(\varphi_1) \subset U$, $\mathrm{supp}(\varphi_2) \subset X - \overline{V}$, and $\varphi_1 + \varphi_2 = 1$. We take $f := \varphi_2$. ■

Tangent Vectors. Let X be an n-dimensional manifold and $a \in X$ an arbitrary point.

Definition. A *derivation* on X at a is an $\mathbb{R}$-linear map $v : \mathscr{E}(X) \to \mathbb{R}$ such that

$$v[f \cdot g] = v[f] \cdot g(a) + f(a) \cdot v[g] \quad \text{for } f, g \in \mathscr{E}(X).$$

If c is constant, then $v[c] = 0$ for every derivation v.

1.12 Proposition. *If $f \in \mathscr{E}(X)$ and $f|_U = 0$ for some open neighborhood $U = U(a) \subset X$, then $v[f] = 0$ for every derivation v at a.*

PROOF: We choose a function $g \in \mathscr{E}(X)$ such that $g|_V = 0$ for some open neighborhood $V = V(a) \subset\subset U$ and $g|_{(X-U)} = 1$. Then $g \cdot f = f$, and from the derivation rule it follows that

$$v[f] = v[g \cdot f] = v[g] \cdot f(a) + g(a) \cdot v[f] = 0.$$

■

1.13 Corollary. *If f, g are two functions of $\mathscr{E}(X)$ with $f|_U = g|_U$ for some open neighborhood $U = U(a) \subset X$, then $v[f] = v[g]$ for every derivation v at the point a.*

It follows from the corollary that we can restrict derivations to locally defined functions and work with coordinates. If $\varphi = (z_1, \dots, z_n) : U \to \mathbb{C}^n$ is a coordinate system at a and $z_\nu = x_\nu + \mathrm{i}\, y_\nu$, then we define partial derivatives at a by

$$\left(\frac{\partial}{\partial x_i}\right)_a [f] := (f \circ \varphi^{-1})_{x_i}(\varphi(a)) \quad \text{and} \quad \left(\frac{\partial}{\partial y_i}\right)_a [f] := (f \circ \varphi^{-1})_{y_i}(\varphi(a)),$$

for $i = 1, \dots, n$. The partial derivatives depend on the chosen coordinate system, but once we have made our choice, every derivation v at a has a unique representation[1]

$$v = \sum_{i=1}^{n} a_i \left(\frac{\partial}{\partial x_i}\right)_a + \sum_{i=1}^{n} b_i \left(\frac{\partial}{\partial y_i}\right)_a,$$

with $a_i = v[x_i]$ and $b_i = v[y_i]$ for $i = 1, \dots, n$.

In $\mathbb{C}^n$ the space of derivations is isomorphic in a natural way to the space of tangent vectors. But what is a tangent vector on a complex manifold X? We start with a differentiable path $\alpha : I \to X$, where $I \subset \mathbb{R}$ is an interval with $0 \in I$, and $\alpha(0) = a$. Let (U, φ) be a coordinate system in X at a. Then we can write $\varphi \circ \alpha = (\alpha_1, \dots, \alpha_n)$ and get the tangent vector

$$(\varphi \circ \alpha)'(0) = (\alpha_1'(0), \dots, \alpha_n'(0)).$$

Unfortunately, this vector depends on the coordinate system. But a tangent vector at a should somehow be completely determined by a pair $(\varphi, \mathbf{c})$, where φ is a coordinate system at a and $\mathbf{c} = (c_1, \dots, c_n) \in \mathbb{C}^n$ an arbitrary vector.

In this sense the tangent vector to α is given by the pair $(\varphi, (\varphi \circ \alpha)'(0))$. If we take another coordinate system ψ, then

$$(\psi \circ \alpha)'(0) = (\varphi \circ \alpha)'(0) \cdot J_{\psi \circ \varphi^{-1}}(\varphi(a))^{\,t}.$$

Therefore, we call two pairs $(\varphi, \mathbf{c})$ and $(\psi, \mathbf{c}')$ equivalent if the Jacobian of $\psi \circ \varphi^{-1}$ at $\varphi(a)$ transforms $\mathbf{c}$ into $\mathbf{c}'$, i.e., if

$$\mathbf{c} = \mathbf{c}' \cdot J_{\varphi \circ \psi^{-1}}(\psi(a))^{\,t}.$$

An equivalence class is called a *tangent vector* at a. The set $T_a(X)$ of all tangent vectors at a is called the *tangent space*. It carries the structure of a complex vector space, which can be defined on representatives:

---

[1] For the proof use the fact that every smooth function f on a domain $G \subset \mathbb{C}^n$ has near $\mathbf{z}_0 \in G$ a unique representation $f(\mathbf{z}) = \sum_{\nu=1}^n g_\nu(\mathbf{z})(x_\nu - x_\nu^0) + \sum_{\nu=1}^n h_\nu(\mathbf{z})(y_\nu - y_\nu^0)$ with $g_\nu(\mathbf{z}_0) = f_{x_\nu}(\mathbf{z}_0)$ and $h_\nu(\mathbf{z}_0) = f_{y_\nu}(\mathbf{z}_0)$.

$$\begin{aligned}(\varphi, \mathbf{c}_1) + (\varphi, \mathbf{c}_2) &:= (\varphi, \mathbf{c}_1 + \mathbf{c}_2),\\ \lambda \cdot (\varphi, \mathbf{c}) &:= (\varphi, \lambda \cdot \mathbf{c}).\end{aligned}$$

The Complex Structure on the Space of Derivations. If $f = g + \mathrm{i}h$ is a complex-valued smooth function on the open set $B \subset X$ and v a derivation at $a \in B$, then we define

$$v[f] := v[g] + \mathrm{i}\, v[h]\,.$$

1.14 Proposition. *For every $\mathbf{c} \in \mathbb{C}^n$ and every coordinate system φ at a there is a unique derivation v at a such that*

$$v[f] := \mathbf{c} \cdot \nabla(f \circ \varphi^{-1})(\varphi(a))^{\,t} \text{ for every } \textbf{holomorphic} \text{ function } f\,.$$

The derivation v depends only on the equivalence class of $(\varphi, \mathbf{c})$.

PROOF: If a coordinate system $\varphi = (z_1, \dots, z_n)$ with $z_\nu = x_\nu + \mathrm{i}y_\nu$ and a vector $\mathbf{c} = \mathbf{a} + \mathrm{i}\mathbf{b}$ are given, then v can be defined by

$$v := \sum_{\nu=1}^{n} a_\nu \left(\frac{\partial}{\partial x_\nu}\right)_a + \sum_{\nu=1}^{n} b_\nu \left(\frac{\partial}{\partial y_\nu}\right)_a .$$

If f is a holomorphic function, then $f_{y_\nu} = \mathrm{i} f_{x_\nu}$ and $f_{x_\nu} = f_{z_\nu}$. Consequently,

$$v[f] = \sum_{\nu=1}^{n} (a_\nu + \mathrm{i}b_\nu)(f \circ \varphi^{-1})_{x_\nu}(\varphi(a)) = \mathbf{c} \cdot \nabla(f \circ \varphi^{-1})(\varphi(a))^{\,t}\,.$$

The uniqueness follows from the equations $v[x_\nu] + \mathrm{i}\, v[y_\nu] = v[z_\nu] = c_\nu$.

If the pair $(\varphi, \mathbf{c})$ is equivalent to $(\psi, \mathbf{c}')$, then

$$\begin{aligned}\mathbf{c}' \cdot \nabla(f \circ \psi^{-1})(\psi(a))^{\,t} &= \mathbf{c}' \cdot J_{\varphi\circ\psi^{-1}}(\psi(a))^{\,t} \cdot \nabla(f \circ \varphi^{-1})(\varphi(a))^{\,t}\\ &= \mathbf{c} \cdot \nabla(f \circ \varphi^{-1})(\varphi(a))^{\,t}.\end{aligned}$$

Therefore, v is determined by the equivalence class of $(\varphi, \mathbf{c})$. ∎

The assignment $(\varphi, \mathbf{c}) \mapsto v$ induces a real vector space isomorphism between the tangent space $T_a(X)$ and the space of derivations at a. It follows that the tangent space has complex dimension n. The pair $(\varphi, \mathbf{e}_\nu)$ is mapped onto the derivation $(\partial/\partial x_\nu)_a$, and $(\varphi, \mathrm{i}\mathbf{e}_\nu)$ onto $(\partial/\partial y_\nu)_a$.

1.15 Proposition. *A complex structure on the space of derivations at a is given by*

$$J(v)[f] = \mathrm{i} \cdot v[f], \text{ for every holomorphic function } f\,.$$

PROOF: Obviously, J is $\mathbb{R}$-linear, and $J \circ J(v) = -v$. ∎

If v corresponds to the tangent vector given by $(\varphi, \mathbf{c})$, then $J(v)$ corresponds to $(\varphi, \mathrm{i}\mathbf{c})$. One must distinguish carefully between the real number $J(v)[f]$ and the complex number $\mathrm{i} \cdot v[f]$ if f is a real-valued smooth function.

The differential operators $(\partial/\partial z_\nu)_a$ and $(\partial/\partial \overline{z}_\nu)_a$ are not real-valued derivations, and therefore they do not correspond to tangent vectors. But they are nevertheless useful. If v is a derivation, then

$$\begin{aligned} v[f] &= \sum_{\nu=1}^{n} a_\nu (f \circ \varphi^{-1})_{x_\nu} + \sum_{\nu=1}^{n} b_\nu (f \circ \varphi^{-1})_{y_\nu} \\ &= \sum_{\nu=1}^{n} a_\nu \cdot ((f \circ \varphi^{-1})_{z_\nu} + (f \circ \varphi^{-1})_{\overline{z}_\nu}) \\ &\quad + \sum_{\nu=1}^{n} b_\nu \cdot \mathrm{i} \cdot ((f \circ \varphi^{-1})_{z_\nu} - (f \circ \varphi^{-1})_{\overline{z}_\nu}) \\ &= \sum_{\nu=1}^{n} c_\nu (f \circ \varphi^{-1})_{z_\nu} + \sum_{\nu=1}^{n} \overline{c}_\nu (f \circ \varphi^{-1})_{\overline{z}_\nu}, \end{aligned}$$

if $c_\nu := a_\nu + \mathrm{i} b_\nu$ for $\nu = 1, \ldots, n$.

Therefore, every (real-valued) derivation can be written in the form

$$v = \sum_{\nu=1}^{n} c_\nu \left(\frac{\partial}{\partial z_\nu}\right)_a + \sum_{\nu=1}^{n} \overline{c}_\nu \left(\frac{\partial}{\partial \overline{z}_\nu}\right)_a .$$

The Induced Mapping. Let $F : X \to Y$ be a holomorphic map between complex manifolds. Let $x \in X$ be an arbitrary point, $y := F(x) \in Y$.

Definition. The *tangential map* $F_* = (F_*)_x : T_x(X) \to T_y(Y)$ is defined by

$$(F_* v)[g] := v[g \circ F], \text{ for derivations } v \text{ and functions } g \in \mathscr{E}(Y).$$

The map F_* is linear, acting on tangent vectors as follows:

$$F_* : (\varphi, \mathbf{c}) \mapsto (\psi, \mathbf{c} \cdot J_{\psi \circ F \circ \varphi^{-1}}(\varphi(x))^{\,t}),$$

if φ is a coordinate system at x, and ψ is a coordinate system at y.

Now we have an assignment between the category of complex manifolds (with a distinguished point) and the category of vector spaces. To any manifold

X and any point $a \in X$ there is associated the tangent space $T_a(X)$. To any holomorphic map $F : X \to Y$ with $F(a) = b$ there is associated the homomorphism $F_* : T_a(X) \to T_b(Y)$. This assignment has the following properties:

$$\begin{aligned} (\mathrm{id}_X)_* &= \mathrm{id}_{T_x(X)}, \\ (G \circ F)_* &= G_* \circ F_* \text{ (if } G : Y \to Z \text{ is another holomorphic map).} \end{aligned}$$

Such an assignment is called a *covariant functor*. If it interchanged the order of the maps, it would be called a contravariant functor.

Remark. For historical reasons the elements of the tangent space are called *contravariant vectors* and the elements of its dual space *covariant vectors*. But the tangent functor behaves covariantly on the tangent vectors and contravariantly on the covariant tangent vectors in $T_a(X)'$. One should keep this in mind.

Immersions and Submersions. We are particularly interested in the case where the (local) Jacobian of a holomorphic map $F : X \to Y$ has maximal rank. If $n = \dim(X)$ and $m = \dim(Y)$, then the rank is bounded by $\min(n, m)$. Only two cases are possible:

> **Definition.** The holomorphic map F is called an *immersion* at x if $\mathrm{rk}(F_*) = n \le m$, and F is called a *submersion* at x if $\mathrm{rk}(F_*) = m \ge n$. In the first case $(F_*)_x$ is injective; in the second case it is surjective.

We call F an immersion (respectively submersion) if it is an immersion (respectively submersion) at every point $x \in X$.

Remark. If $F : X \to Y$ is an **injective** immersion, then for every $x \in X$ there are neighborhoods $U(x) \subset X$ and $V(F(x)) \subset Y$ such that $F(U)$ is a submanifold of V. In addition, if X is compact, then $F(X)$ is a submanifold of Y. We omit the proof here.

1.16 Theorem. *Let $x_0 \in X$ be a point, $y_0 := F(x_0)$. The following conditions are equivalent:*

1. *F is a submersion at x_0.*
2. *There are neighborhoods $U = U(x_0) \subset X$ and $V = V(y_0) \subset Y$ with $F(U) \subset V$, a manifold Z, and a holomorphic map $G : U \to Z$ such that $x \mapsto (F(x), G(x))$ defines a biholomorphic map from U to an open subset of $V \times Z$.*
3. *There is an open neigborhood $V = V(y_0) \subset Y$ and a holomorphic map $s : V \to X$ with $s(y_0) = x_0$ and $F \circ s = \mathrm{id}_V$. (Then s is called a* local section *for F.)*

PROOF: (1) $\implies$ (2) : We can restrict ourselves to a local situation and assume that $U = U(\mathbf{0}) \subset \mathbb{C}^n$ and $V = V(\mathbf{0}) \subset \mathbb{C}^m$ are open neighborhoods, and $F : U \to V$ a holomorphic map with $F(\mathbf{0}) = \mathbf{0}$ and $\operatorname{rk}(J_F(\mathbf{0})) = m$.

We write $J_F(\mathbf{0}) = (J'_F(\mathbf{0}), J''_F(\mathbf{0}))$, with $J'_F(\mathbf{0}) \in M_{m,m}(\mathbb{C})$ and $J''_F(\mathbf{0}) \in M_{m,n-m}(\mathbb{C})$. Choosing suitable coordinates we may assume that $\det J'_F(\mathbf{0}) \neq 0$. We define a new holomorphic map $\widetilde{F} : U \to V \times \mathbb{C}^{n-m} \subset \mathbb{C}^n$ by

$$\widetilde{F}(\mathbf{z}', \mathbf{z}'') := (F(\mathbf{z}', \mathbf{z}''), \mathbf{z}''), \qquad \text{for } \mathbf{z}' \in \mathbb{C}^m,\ \mathbf{z}'' \in \mathbb{C}^{n-m}.$$

Then

$$J_{\widetilde{F}}(\mathbf{0}) = \begin{pmatrix} J'_F(\mathbf{0}) & J''_F(\mathbf{0}) \\ \mathbf{0} & \mathbf{E}_{n-m} \end{pmatrix}, \qquad \text{and therefore } \det J_{\widetilde{F}}(\mathbf{0}) \neq 0.$$

By the inverse function theorem there are neighborhoods $\widetilde{U}(\mathbf{0}) \subset U$ and $W(\mathbf{0}) \subset \mathbb{C}^n$ such that $\widetilde{F} : \widetilde{U} \to W$ is biholomorphic.

We observe that $Z := \mathbb{C}^{n-m}$ is a complex manifold, and $G := \operatorname{pr}_2 : \widetilde{U} \to Z$ with $(\mathbf{z}', \mathbf{z}'') \mapsto \mathbf{z}''$ is a holomorphic map such that $(F, G) = \widetilde{F}$ is biholomorphic near $\mathbf{0}$.

(2) $\implies$ (3) : If U, V, Z, and G are given such that $F(U) \subset V$ and $(F, G) : U \to W \subset V \times Z$ is biholomorphic, then $s : V \to X$ can be defined by

$$s(y) := (F, G)^{-1}(y, G(x_0)).$$

Then $(F, G)(s(y_0)) = (y_0, G(x_0)) = (F, G)(x_0)$, and therefore $s(y_0) = x_0$. Furthermore, $(F, G) \circ s(y) = (F, g) \circ (F, G)^{-1}(y, G(x_0)) = (y, G(x_0))$. Thus $F \circ s(y) = y$.

(3) $\implies$ (1) : If s is a local section for F, with $s(y_0) = x_0$, then $F_* \circ s_*(v) = v$ for every $v \in T_{y_0}(Y)$. Thus it follows immediately that F_* is surjective. ∎

1.17 Corollary. *If $F : X \to Y$ is a submersion, then for each $y \in Y$ the fiber $F^{-1}(y)$ is empty or an $(n - m)$-dimensional submanifold of X. In the latter case $T_x(F^{-1}(y)) = \operatorname{Ker}((F_*)_x)$ for all $x \in F^{-1}(y)$.*

PROOF: We consider a point $x_0 \in X$. Let $M := F^{-1}(y_0)$ be the fiber over $y_0 := F(x_0)$. Then we can find neighborhoods $U = U(x_0) \subset X$, $V = V(y_0) \subset Y$, a manifold Z, and a holomorphic map $G : U \to Z$ such that $(F, G) : U \to W \subset V \times Z$ is biholomorphic. It follows that $M \cap U = (F, G)^{-1}(\{y_0\} \times Z) \cap U$ is a manifold of dimension $n - m$.

Since $F|M$ is constant, we have $F_*|T_{x_0}(M) \equiv 0$. This means that $T_{x_0}(M) \subset \operatorname{Ker}(F_*)$. Since these spaces have the same dimension, they must be equal. ∎

Gluing. Assume that X is a set that is the union of a countable collection $(X_\nu)_{\nu\in\mathbb{N}}$ of subsets such that the following hold:

1. For every $\nu \in \mathbb{N}$ there is a bijection $\varphi_\nu : X_\nu \to M_\nu$, where M_ν is an n-dimensional complex manifold.
2. For every pair $(\nu, \mu) \in \mathbb{N} \times \mathbb{N}$ the subset $\varphi_\nu(X_\nu \cap X_\mu)$ is open in M_ν, and the map
$$\varphi_\nu \circ \varphi_\mu^{-1} : \varphi_\mu(X_\nu \cap X_\mu) \to \varphi_\nu(X_\nu \cap X_\mu)$$
is biholomorphic.
3. For every pair of points $a \in X_\nu$ and $b \in X_\mu$ with $a \neq b$ there are open neighborhoods $U(\varphi_\nu(a)) \subset M_\nu$ and $V(\varphi_\mu(b)) \subset M_\mu$ with $\varphi_\nu^{-1}(U) \cap \varphi_\mu^{-1}(V) = \varnothing$.

1.18 Proposition. *Under the above conditions there is a unique complex structure on X such that the X_ν are open in X and the $\varphi_\nu : X_\nu \to M_\nu$ are biholomorphic.*

PROOF: We give only a sketch of the proof and leave the details as an exercise for the reader.

A subset $U \subset X$ is called *open* if $\varphi_\nu(U \cap X_\nu)$ is open in M_ν for every ν. Then the collection of open sets has the properties of a topology on X. In addition, for every open set $W \subset M_\nu$ the set $\varphi_\nu^{-1}(W)$ is open in X. Consequently, the maps $\varphi_\nu : X_\nu \to M_\nu$ are homeomorphisms. From the last hypothesis it follows that the topology on X is Hausdorff, and since the collection of the X_ν is countable, it has a countable basis.

If $U \subset M_\nu$ is open and $\psi : U \to \mathbb{C}^n$ a coordinate system, then $\widehat{\psi} := \psi \circ \varphi_\nu : \varphi_\nu^{-1}(U) \to \mathbb{C}^n$ is a coordinate system for X. One checks easily that two such coordinate systems are biholomorphically compatible. So we obtain a complex structure on X. ∎

One says that X is obtained by *gluing* the manifolds M_ν. Another way to describe this process is the following. Let there be given a collection of complex manifolds M_ν, open subsets $M_{\nu\mu} \subset M_\nu$, and biholomorphic maps $\varphi_{\nu\mu} : M_{\mu\nu} \to M_{\nu\mu}$ (including $\varphi_{\nu\nu} = \mathrm{id}_{M_\nu}$). Consider pairs (x, ν) with $x \in M_\nu$. Then (x, ν) is called equivalent to (y, μ) if

$$x \in M_{\nu\mu}, \quad y \in M_{\mu\nu} \text{ and } x = \varphi_{\nu\mu}(y).$$

The set X of equivalence classes is the result of the gluing process. Of course, one has to add a condition that ensures the Hausdorff property.

Exercises

1. Let M be a compact connected complex manifold with $\dim(M) \geq 2$ and $N \subset M$ a closed submanifold of codimension greater or equal to 2. Show that every holomorphic function $f : M - N \to \mathbb{C}$ is constant.

2. Let X be a Riemann surface.
   (a) If f is a meromorphic function on X, and P the set of poles, show that
$$\widehat{f}(x) := \begin{cases} f(x) & \text{for } x \in X - P, \\ \infty & \text{for } x \in P \end{cases}$$
   defines a holomorphic map $\widehat{f} : X \to \overline{\mathbb{C}}$.
   (b) Prove that every holomorphic map $f : X \to \overline{\mathbb{C}}$ that is not identically ∞ defines a meromorphic function on X.
3. Let $f : X \to Y$ be a nonconstant holomorphic map between Riemann surfaces, $x_0 \in X$, and $y_0 := f(x_0) \in Y$. Prove that there is a $k \geq 1$ such that there are complex coordinates $\varphi : U(x_0) \to \mathbb{C}$ and $\psi : V(y_0) \to \mathbb{C}$ with:
   (a) $\varphi(x_0) = 0$, $\psi(y_0) = 0$.
   (b) $f(U) \subset V$.
   (c) $\psi \circ f \circ \varphi^{-1}(z) = z^k$.
4. The general linear group $\mathrm{GL}_n(\mathbb{C})$ is an open subset of the vector space $M_n(\mathbb{C})$. Prove that the *special linear group* $\mathrm{SL}_n(\mathbb{C}) := \{\mathbf{A} \in M_n(\mathbb{C}) : \det(\mathbf{A}) = 1\}$ is a submanifold of $\mathrm{GL}_n(\mathbb{C})$. Calculate the tangent space $T_{\mathbf{E}}(\mathrm{SL}_n(\mathbb{C})) \subset T_{\mathbf{E}}(\mathrm{GL}_n(\mathbb{C})) = M_n(\mathbb{C})$, where $\mathbf{E} = \mathbf{E}_n$ is the identity matrix.
5. Let $f : X \to Y$ be a holomorphic map and $Z \subset Y$ a closed submanifold. Show that if
$$\mathrm{Im}((f_*)_x) + T_{f(x)}(Z) = T_{f(x)}(Y)$$
   for every $x \in f^{-1}(Z)$, then $f^{-1}(Z)$ is a submanifold of X.
6. The holomorphic maps $f : X \to Z$ and $g : Y \to Z$ are called *transversal* if for every $(x, y) \in X \times Y$ with $f(x) = g(y) =: z$ the following holds:
$$\mathrm{Im}((f_*)_x) + \mathrm{Im}((g_*)_x) = T_z(Z)\,.$$
   Prove that the *fiber product*
$$X \times_Z Y := \{(x, y) \in X \times Y \,:\, f(x) = f(y)\}$$
   is a complex submanifold of $X \times Y$.

2. Complex Fiber Bundles

Lie Groups and Transformation Groups. Assume that G is a set that has the structure of a group and at the same time that of an n-dimensional complex manifold. The inverse of $g \in G$ will be denoted by g^{-1}, the identity element by e, and the composition of two elements $g_1, g_2 \in G$ by $g_1 g_2$.

Definition. We call G a *complex Lie group* if the following two properties hold:

1. The mapping $g \mapsto g^{-1}$ (from G to G) is holomorphic.
2. The mapping $(g_1, g_2) \mapsto g_1 g_2$ (from $G \times G$ to G) is holomorphic.

There are many examples of complex Lie groups. The simplest one is the space $\mathbb{C}^n$, where the composition is vector addition. Another example is the group $\mathbb{C}^*$ with respect to ordinary multiplication of complex numbers.

The most important example is the *general linear group*

$$\mathrm{GL}_n(\mathbb{C}) := \{\mathbf{A} \in M_n(\mathbb{C}) \,:\, \det \mathbf{A} \neq 0\}.$$

Its complex structure is obtained by considering it as an open subset of $\mathbb{C}^{n^2}$. The multiplication of matrices is bilinear, and the determinants appearing in the calculation of the inverse of a matrix $\mathbf{A}$ are polynomials in the coefficients of $\mathbf{A}$.

Every matrix $\mathbf{A} \in \mathrm{GL}_n(\mathbb{C})$ defines a linear and therefore holomorphic map $\Phi_{\mathbf{A}} : \mathbb{C}^n \to \mathbb{C}^n$ by

$$\Phi_{\mathbf{A}}(\mathbf{z}) := \mathbf{z} \cdot \mathbf{A}^{\,t}.$$

Then $\Phi_{\mathbf{AB}}(\mathbf{z}) = \mathbf{z} \cdot (\mathbf{AB})^{\,t} = \mathbf{z} \cdot (\mathbf{B}^{\,t}\mathbf{A}^{\,t}) = (\mathbf{z} \cdot \mathbf{B}^{\,t}) \cdot \mathbf{A}^{\,t} = \Phi_{\mathbf{A}}(\Phi_{\mathbf{B}}(\mathbf{z}))$. If $\mathbf{E}_n$ is the identity matrix, then $\Phi_{\mathbf{E}_n} = \mathrm{id}$. Furthermore, if $\mathbf{A}$ is any matrix with $\Phi_{\mathbf{A}} = \mathrm{id}$, then $\mathbf{A}$ must be the identity matrix, since $\Phi_{\mathbf{A}}(\mathbf{e}_i) = \mathbf{e}_i \cdot \mathbf{A}^{\,t}$ is the transpose of the ith column of $\mathbf{A}$.

We want to generalize this situation. Let X be a complex manifold and G a complex Lie group.

> **Definition.** We say that G *acts analytically on* X (or is a *complex Lie transformation group* on X) if there is a holomorphic mapping $\Phi : G \times X \to X$ with
>
> $$\Phi(g_1 g_2, x) = \Phi(g_1, \Phi(g_2, x)) \text{ for } g_1, g_2 \in G, x \in X.$$

The holomorphic map $x \mapsto \Phi(g, x)$ is denoted by Φ_g. We say that G acts *effectively* or *faithfully* on X if $\Phi_g = \mathrm{id}_X$ implies that $g = e$.

Often we write gx instead of $\Phi(g, x)$ or $\Phi_g(x)$. A point $x \in X$ with $gx = x$ is called a *fixed point* of g. We say that G acts *freely* if only the identity element $e \in G$ has fixed points in X. The general linear group $\mathrm{GL}_n(\mathbb{C})$ acts analytically and faithfully on $\mathbb{C}^n$, but not freely.

Let $\{\mathbf{w}_1, \ldots, \mathbf{w}_{2n}\}$ be any basis of $\mathbb{C}^n$ over $\mathbb{R}$. Then

$$\Gamma := \mathbb{Z}\mathbf{w}_1 + \cdots + \mathbb{Z}\mathbf{w}_{2n}$$

is a subgroup of the (additive) group $\mathbb{C}^n$ generated by $\mathbf{w}_1, \ldots, \mathbf{w}_{2n}$. The group Γ acts on $\mathbb{C}^n$ by $\Phi(\mathbf{w}, \mathbf{z}) := \mathbf{z} + \mathbf{w}$. This is an example of a free action.

Fiber Bundles. Let X and F be complex manifolds and G a complex Lie group acting analytically and faithfully on F.

Definition. A *topological* (respectively *holomorphic*) *fiber bundle* over X with *structure group* G and *typical fiber* F is given by a topological space (respectively a complex manifold) P and a continuous (respectively holomorphic) map $\pi : P \to X$, together with

1. an open covering $\mathscr{U} = \{U_\iota \,:\, \iota \in I\}$ of X,
2. for any $\iota \in I$ a topological (respectively biholomorphic) map

$$\varphi_\iota : \pi^{-1}(U_\iota) \to U_\iota \times F$$

with $\mathrm{pr}_1 \circ \varphi_\iota = \pi$,

3. for any pair of indices $(\iota, \kappa) \in I \times I$ a continuous (respectively holomorphic) map $g_{\iota\kappa} : U_\iota \cap U_\kappa \to G$ with

$$\varphi_\iota \circ \varphi_\kappa^{-1}(x, p) = (x, g_{\iota\kappa}(x)p)$$

for $x \in U_{\iota\kappa} := U_\iota \cap U_\kappa$ and $p \in F$.

The maps φ_ι are called *local trivializations* and the maps $g_{\iota\kappa}$ a *system of transition functions.*

Since G is acting faithfully, we get the following *compatibility condition*:

$$g_{\iota\kappa} g_{\kappa\lambda} = g_{\iota\lambda} \quad \text{on} \quad U_{\iota\kappa\lambda} := U_\iota \cap U_\kappa \cap U_\lambda.$$

Then $g_{\iota\iota} = e$ and $g_{\kappa\iota} = g_{\iota\kappa}^{-1}$.

Now let a system of transition functions $(g_{\iota\kappa})$ be given such that the compatibility condition is satisfied. Using the gluing techniques mentioned at the end of Section 1, a suitable bundle space P, a projection $\pi : P \to X$, and local trivializations can be constructed as follows:

Identifying (x, p) and $(x, g_{\iota\kappa}(x)p)$, we can glue together the Cartesian products $U_\kappa \times F$ and $U_\iota \times F$ over $U_{\iota\kappa}$. Due to the compatibility condition this works in a unique way over $U_{\iota\kappa\lambda}$. The obvious projection from P to X is continuous. Therefore, P is a fiber bundle over X with structure group G and typical fiber F. If the transition functions $g_{\iota\kappa}$ are holomorphic, then P carries the structure of a complex manifold, and the projection and the local trivializations are holomorphic. So P becomes an analytic fiber bundle in that case.

Example

Let X be an n-dimensional complex manifold. There is an open covering $\mathscr{U} = \{U_\iota \,:\, \iota \in I\}$, together with complex coordinate systems

$$\varphi_\iota : U_\iota \to B_\iota \subset \mathbb{C}^n.$$

We take $\mathbb{C}$ as typical fiber and $\mathbb{C}^*$ as structure group, acting on $\mathbb{C}$ by multiplication. Then we can define transition functions $g_{\iota\kappa} : U_{\iota\kappa} \to \mathbb{C}^*$ by

$$g_{\iota\kappa}(x) := \det J_{\varphi_\iota \circ \varphi_\kappa^{-1}}(\varphi_\kappa(x))^{-1}.$$

The compatibility conditions are satisfied because of the chain rule and the determinant product theorem. So by the gluing procedure described above we get a holomorphic fiber bundle over X that is called the *canonical bundle* and is denoted by K_X.

Equivalence. Let $\pi_P : P \to X$ and $\pi_Q : Q \to X$ be two topological or holomorphic fiber bundles over the same manifold X, with the same fiber F and the same structure group G. We assume that there is an open covering $\mathscr{U} = \{U_\iota : \iota \in I\}$ of X such that there are trivializations $\varphi_\iota : \pi_P^{-1}(U_\iota) \to U_\iota \times F$ and $\psi_\iota : \pi_Q^{-1}(U_\iota) \to U_\iota \times F$.

> **Definition.** A *fiber bundle isomorphism* between P and Q is a topological (respectively biholomorphic) map $h : P \to Q$ with $\pi_Q \circ h = \pi_P$ such that for any $\iota \in I$ there is a continuous (respectively holomorphic) map $h_\iota : U_\iota \to G$ with
>
> $$\psi_\iota \circ h \circ \varphi_\iota^{-1}(x, p) = (x, h_\iota(x)p).$$
>
> The bundles are called *equivalent* in this case.

We give a description of bundle equivalence in the context of transition functions. Let $(g'_{\iota\kappa})$ be the system of transition functions for P with respect to $\mathscr{U}$, and $(g''_{\iota\kappa})$ the corresponding system for Q. Then we have

$$\begin{aligned}
(x, h_\iota(x)g'_{\iota\kappa}(x)p) &= \psi_\iota \circ h \circ \varphi_\iota^{-1}(x, g'_{\iota\kappa}(x)p) \\
&= \psi_\iota \circ h \circ \varphi_\kappa^{-1}(x, p) \\
&= \psi_\iota \circ \psi_\kappa^{-1}(x, h_\kappa(x)p) \\
&= (x, g''_{\iota\kappa}(x)h_\kappa(x)p).
\end{aligned}$$

Since G is acting faithfully, it follows that

$$\text{(C)} \qquad h_\iota g'_{\iota\kappa} = g''_{\iota\kappa} h_\kappa \quad \text{over } U_{\iota\kappa}.$$

Two systems of transition functions $(g'_{\iota\kappa})$ and $(g''_{\iota\kappa})$ with respect to the same covering are called *topologically (respectively analytically) equivalent* or *cohomologous* if there are continuous (respectively holomorphic) maps $h_\iota : U_\iota \to G$ satisfying condition (C).

Equivalent fiber bundles have equivalent systems of transition functions. On the other hand, it is easy to see that fiber bundles constructed from equivalent systems of transition functions are themselves equivalent. Now we will demonstrate that the latter remains valid in passing to finer coverings.

Let us assume that there are given systems of transition functions $g' = (g'_{\iota\kappa})$ and $g'' = (g''_{\iota\kappa})$ with respect to a covering $\mathscr{U} = \{U_\iota : \iota \in I\}$. We call them *equivalent* as well if there is a refinement $\mathscr{V} = \{V_\nu : \nu \in N\}$ of $\mathscr{U}$ (with refinement map $\tau : N \to I$) and a collection of maps $h_\nu : V_\nu \to G$ such that

$$h_\nu g'_{\tau(\nu)\tau(\mu)} = g''_{\tau(\nu)\tau(\mu)} h_\mu \quad \text{on } V_{\nu\mu} \subset U_{\tau(\nu)\tau(\mu)}, \text{ for all } \nu, \mu \in N.$$

To show that the systems are equivalent in the old sense, we define $\widetilde{h}_\iota : U_\iota \to G$ by

$$\widetilde{h}_\iota := g''_{\iota\tau(\mu)} h_\mu \big(g'_{\iota\tau(\mu)}\big)^{-1} \quad \text{on } U_\iota \cap V_\mu.$$

In fact, $\widetilde{h}_\iota$ is well defined, since on $U_\iota \cap V_{\nu\mu}$ we have

$$h_\nu = g''_{\tau(\nu)\tau(\mu)} h_\mu \big(g'_{\tau(\nu)\tau(\mu)}\big)^{-1}$$

and therefore $g''_{\iota\tau(\nu)} h_\nu \big(g'_{\iota\tau(\nu)}\big)^{-1} = g''_{\iota\tau(\mu)} h_\mu \big(g'_{\iota\tau(\mu)}\big)^{-1}$. Then on $U_{\iota\kappa} \cap V_\mu$ we get

$$\begin{aligned} \widetilde{h}_\iota g'_{\iota\kappa} &= g''_{\iota\tau(\mu)} h_\mu \big(g'_{\iota\tau(\mu)}\big)^{-1} g'_{\iota\kappa} \\ &= g''_{\iota\tau(\mu)} h_\mu g'_{\tau(\mu)\kappa} \\ &= g''_{\iota\kappa} g''_{\kappa\tau(\mu)} h_\mu \big(g'_{\kappa\tau(\mu)}\big)^{-1} \\ &= g''_{\iota\kappa} \widetilde{h}_\kappa. \end{aligned}$$

The bundles are equivalent!

If two bundles are given with respect to two different coverings, then they are called equivalent if they are equivalent with respect to a common refinement, for example, the intersection of the coverings. Then everything works as above.

Complex Vector Bundles. Let X be an n-dimensional complex manifold.

> **Definition.** A *complex topological* (respectively *holomorphic*) *vector bundle* of *rank* r over X is a topological (respectively holomorphic) fiber bundle V over X with $\mathbb{C}^r$ as typical fiber and $\mathrm{GL}_r(\mathbb{C})$ as structure group. In the case $r = 1$ we are speaking of a *complex line bundle.*

If $\pi : V \to X$ is the bundle projection, then we denote by V_x the fiber $\pi^{-1}(x)$. It has the structure of an r-dimensional complex vector space. A trivialization $\Phi : \pi^{-1}(U) \to U \times \mathbb{C}^r$ is also called a *vector bundle chart.* For any $x \in U$ the induced map $\Phi_x : V_x \to \mathbb{C}^r$ is a vector space isomorphism.

> **Definition.** Let V be a holomorphic vector bundle over X. If $U \subset X$ is an open subset, then a continuous (differentiable, holomorphic) *cross*

section (or simply *section*) in V over U is a continuous (differentiable, holomorphic) map $s : U \to V$ with $\pi \circ s = \mathrm{id}_U$.

We denote by $\Gamma(U, V)$ (or $\mathcal{O}(U, V)$) the vector space of holomorphic cross sections in V over U, by $\mathcal{E}(U, V)$ the space of differentiable sections, and by $\mathcal{C}^0(U, V)$ the space of continuous sections.

The vector bundle V is called *globally generated* if the canonical map $\Gamma(X, V) \to V_x$ with $s \mapsto s(x)$ is surjective for every $x \in X$.

Let $(U_\iota, \Phi_\iota)_{\iota \in I}$ be a collection of vector bundle charts $\Phi_\iota : \pi^{-1}(U_\iota) \to U_\iota \times \mathbb{C}^r$ for V, and $g_{\iota\kappa} : U_{\iota\kappa} \to \mathrm{GL}_r(\mathbb{C})$ the system of transition functions, given by

$$\Phi_\iota \circ \Phi_\kappa^{-1}(x, \mathbf{z}) = \big(x, \mathbf{z} \cdot g_{\iota\kappa}(x)^t\big) \quad \text{for} \quad (x, \mathbf{z}) \in U_{\iota\kappa} \times \mathbb{C}^r.$$

If s is a holomorphic section in V, then

$$\Phi_\iota \circ s|_{U_\iota}(x) = (x, s_\iota(x))$$

defines a system of holomorphic maps $s_\iota : U_\iota \to \mathbb{C}^r$, and we obtain the compatibility condition

$$s_\iota(x) = s_\kappa(x) \cdot g_{\iota\kappa}(x)^t \quad \text{on } U_{\iota\kappa}.$$

On the other hand, any such system (s_ι) defines a global section s.

Example

We can define vector bundles by giving a system of transition functions. The construction of the bundle space is carried out with the same gluing technique as for general fiber bundles.

If X is an arbitrary n-dimensional complex manifold, and $(U_\iota, \varphi_\iota)_{\iota \in I}$ a complex atlas for X, then

$$g_{\iota\kappa}(x) := J_{\varphi_\iota \circ \varphi_\kappa^{-1}}(\varphi_\kappa(x)) \in \mathrm{GL}_n(\mathbb{C})$$

defines a system of transition functions with respect to $\mathscr{U} = \{U_\iota,\ \iota \in I\}$. The corresponding vector bundle $T(X)$ is called the *tangent bundle* of X. It results from gluing $(x, \mathbf{c}) \in U_\kappa \times \mathbb{C}^n$ with $(x, \mathbf{c} \cdot g_{\iota\kappa}(x)^t) \in U_\iota \times \mathbb{C}^n$, for $x \in U_{\iota\kappa}$. Therefore, we can identify the fiber $(T(X))_x$ with the tangent space $T_x(X)$. The local trivializations $\Phi_\iota : T(X)|_{U_\iota} \to U_\iota \times \mathbb{C}^n$ are given as follows:

For $a \in U_\iota$ the trivialization Φ_ι maps a tangent vector $v \in T_a(X)$, represented by $(\varphi_\iota, \mathbf{c})$, onto the pair $(a, \mathbf{c}) \in U_\iota \times \mathbb{C}^n$. If we denote the equivalence class of $(\varphi_\iota, \mathbf{c})$ at a by $[\varphi_\iota, \mathbf{c}]$, we obtain

$$\Phi_\iota \circ \Phi_\kappa^{-1}(a, \mathbf{c}) = \Phi_\iota([\varphi_\kappa, \mathbf{c}]) = \Phi_\iota([\varphi_\iota, \mathbf{c} \cdot g_{\iota\kappa}(a)^t]) = (a, \mathbf{c} \cdot g_{\iota\kappa}(a)^t).$$

A holomorphic section in $T(X)$ is also called a *holomorphic vector field*.

Definition. Let V, W be two holomorphic vector bundles over X. A *vector bundle homomorphism* between V and W is a fiber preserving holomorphic map $\eta : V \to W$ such that for any $x \in X$ a linear map $\eta_x : V_x \to W_x$ is induced.

The map η is called a *vector bundle isomorphism* if η is bijective and η, η^{-1} both are vector bundle homomorphisms.

A holomorphic vector bundle V of rank r over X is called *trivial* if it is isomorphic to the bundle $X \times \mathbb{C}^r$. This is equivalent to the existence of a *frame* $\{\xi_1, \dots, \xi_r\}$ of holomorphic sections $\xi_i \in \Gamma(X, V)$ such that for every $x \in X$ the elements $\xi_1(x), \dots, \xi_r(x) \in V_x$ are linearly independent. In this case V is globally generated. But there are also nontrivial bundles that are globally generated.

2.1 Proposition. *A holomorphic map $\eta : V \to W$ is a vector bundle homomorphism if and only if for each pair of vector bundle charts $\Phi : V|_U \to U \times \mathbb{C}^r$ and $\Psi : W|_U \to U \times \mathbb{C}^s$ there is a holomorphic map $h : U \to M_{s,r}(\mathbb{C})$ with*

$$\Psi^{-1} \circ \eta \circ \Phi(x, \mathbf{z}) = (x, \mathbf{z} \cdot h(x)^t).$$

We omit the elementary proof.

Standard Constructions. Let X be an n-dimensional complex manifold. We can think of a vector bundle over X as a parametrized family of vector spaces. Therefore, numerous constructions from linear algebra carry over to the theory of vector bundles.

1. The direct sum: If V, W are two vector bundles over X, then the *direct sum*, or *Whitney sum*, $V \oplus W := V \times_X W$ carries a vector bundle structure that is defined as follows:

Let $\mathscr{U} = \{U_\iota : \iota \in I\}$ be an open covering of X such that there are vector bundle charts $\Phi_\iota : V|_{U_\iota} \to U_\iota \times \mathbb{C}^r$ and $\Psi_\iota : W|_{U_\iota} \to U_\iota \times \mathbb{C}^k$. Then a vector bundle chart

$$(V \oplus W)|_{U_\iota} = \{(v, w) \in V \times W : \pi_V(v) = \pi_W(w) \in U_\iota\} \to U_\iota \times \mathbb{C}^{r+k}$$

can be defined by

$$(v, w) \mapsto (\pi_V(v); \mathrm{pr}_2 \circ \Phi_\iota(v), \mathrm{pr}_2 \circ \Psi_\iota(w)).$$

If $g_{\iota\kappa}$, respectively $h_{\iota\kappa}$, are transition functions for V, respectively W, then the matrices

$$G_{\iota\kappa} := \begin{pmatrix} g_{\iota\kappa} & 0 \\ 0 & h_{\iota\kappa} \end{pmatrix}$$

are transition functions for $V \oplus W$. The fiber of the Whitney sum is given by $(V \oplus W)_x = V_x \times W_x$.

2. The dual bundle: Let $\pi : V \to X$ be a holomorphic vector bundle of rank r with trivializations $\Phi_\iota : \pi^{-1}(U_\iota) \to U_\iota \times \mathbb{C}^r$ and transition functions $g_{\iota\kappa}$. Gluing the sets $U_\iota \times \mathbb{C}^r$ together by means of the transition functions $g'_{\iota\kappa} := g^{\,t}_{\kappa\iota}$ leads to the *dual bundle* $\pi' : V' \to X$. Denote the associated trivializations over U_ι by Ψ_ι.

We will show that for every $x \in X$ there is a natural isomorphism

$$(V')_x \to (V_x)' = \mathrm{Hom}_{\mathbb{C}}(V_x, \mathbb{C}).$$

Given elements $x \in U_\iota$, $v \in V_x$, and $\lambda \in (V')_x$, we define

$$\lambda(v) := (\Phi_\iota)_x(v) \cdot (\Psi_\iota)_x(\lambda)^{\,t} \in \mathbb{C}\,,$$

using the vector space isomorphisms $(\Phi_\iota)_x : V_x \to \mathbb{C}^r$ and $(\Psi_\iota)_x : (V')_x \to \mathbb{C}^r$.

For $x \in U_\iota \cap U_\kappa$ we obtain

$$\begin{aligned}
&(\Phi_\kappa)_x(v) \cdot (\Psi_\kappa)_x(\lambda)^{\,t} \\
&\quad= (\Phi_\kappa)_x \circ (\Phi_\iota)_x^{-1} \circ (\Phi_\iota)_x(v) \cdot [(\Psi_\kappa)_x \circ (\Psi_\iota)_x^{-1} \circ (\Psi_\iota)_x(\lambda)]^{\,t} \\
&\quad= (\Phi_\iota)_x(v) \cdot g_{\kappa\iota}(x)^{\,t} \cdot [(\Psi_\iota)_x(\lambda) \cdot g_{\iota\kappa}(x)]^{\,t} \\
&\quad= (\Phi_\iota)_x(v) \cdot g_{\kappa\iota}(x)^{\,t} \cdot g_{\iota\kappa}(x)^{\,t} \cdot (\Psi_\iota)_x(\lambda) \\
&\quad= (\Phi_\iota)_x(v) \cdot (\Psi_\iota)_x(\lambda).
\end{aligned}$$

This shows that the definition of $\lambda(v)$ is independent of the trivializations.

3. Tensor powers of a line bundle: Let $\pi : F \to X$ be a line bundle with transition functions $g_{\iota\kappa} : U_{\iota\kappa} \to \mathbb{C}^*$.

> **Definition.** For $k \in \mathbb{N}$, the *tensor power* F^k is the line bundle defined by the transition functions $g^k_{\iota\kappa}$.

We give an interpretation of F^k using the dual bundle $\pi' : F' \to X$. Assume that there is an open subset $U \subset X$ and a holomorphic function $f : (\pi')^{-1}(U) \to \mathbb{C}$. If $\psi_\iota : (F')|_{U_\iota} \to U_\iota \times \mathbb{C}$ are trivializations (with $\psi_\iota \circ \psi_\kappa^{-1}(x, z) = (x, z \cdot g_{\iota\kappa}(x)^{-1})$), then we have a power series expansion

$$f \circ \psi_\iota^{-1}(x, z) = \sum_{\nu=0}^{\infty} a_{\nu,\iota}(x) z^\nu$$

on $\psi_\iota(F'|_{U_\iota \cap U})$ with holomorphic functions $a_{\nu,\iota}$ on $U \cap U_\iota$. Over $U_\iota \cap U_\kappa \cap U$ the following holds:

$$f \circ \psi_\kappa^{-1}(x,z) = f \circ \psi_\iota^{-1}\big(x, z \cdot g_{\iota\kappa}(x)^{-1}\big) = \sum_{\nu=0}^{\infty} \big(a_{\nu,\iota}(x) g_{\iota\kappa}(x)^{-\nu}\big) z^\nu,$$

and therefore $a_{\nu,\iota} = a_{\nu,\kappa} \cdot g_{\iota\kappa}^\nu$. This means that $a_\nu = (a_{\nu,\iota})$ is a cross section in F^k over U. So every holomorphic function f on F' that is homogeneous of degree k on the fibers is a section in F^k. In particular, F_x^k can be identified with the space $L_k(F_x', \mathbb{C})$ of k-linear functions $f : F_x' \times \cdots \times F_x' \to \mathbb{C}$.

For $e_1, \dots, e_k \in F_x$ the *tensor product* $e_1 \otimes \cdots \otimes e_k \in (F^k)_x$ is defined by

$$(e_1 \otimes \cdots \otimes e_k)(\lambda_1, \dots, \lambda_k) := \lambda_1(e_1) \cdots \lambda_k(e_k).$$

The tensor power $e \otimes \cdots \otimes e$ of an element $e \in F_x$ is denoted by e^k for short.

Finally, we define $F^{-k} := (F')^k \cong (F^k)'$.

4. The tensor product: Let $p : V \to X$ be a vector bundle of rank r with transition functions $G_{\iota\kappa} : U_{\iota\kappa} \to \mathrm{GL}_r(\mathbb{C})$, and $\pi : F \to X$ a line bundle with transition functions $g_{\iota\kappa}$ (with respect to the covering $\mathscr{U} = \{U_\iota : \iota \in I\}$).

> **Definition.** The *tensor product* $V \otimes F$ is the vector bundle of rank r given by the transition functions
>
> $$g_{\iota\kappa} \cdot G_{\iota\kappa} : U_{\iota\kappa} \to \mathrm{GL}_r(\mathbb{C}).$$

Let $\Phi_\iota : V|_{U_\iota} \to U_\iota \times \mathbb{C}^r$ and $\psi_\iota : F'|_{U_\iota} \to U_\iota \times \mathbb{C}$ be local trivializations. If $f : F'|_U \to V|_U$ is a holomorphic map which is linear on the fibers, then over $U_\iota \cap U$ we have

$$\Phi_\iota \circ f \circ \psi_\iota^{-1}(x,z) = (x, z \cdot \eta_\iota(x)),$$

where $\eta_\iota : U_\iota \cap U \to \mathbb{C}^r$ is a holomorphic map. Over $U_\iota \cap U_\kappa \cap U$ we calculate

$$\Phi_\iota \circ f \circ \psi_\kappa^{-1}(x,z) = \Phi_\iota \circ f \circ \psi_\iota^{-1}\big(x, z \cdot g_{\iota\kappa}(x)^{-1}\big) = \big(x, \eta_\iota(x) \cdot z \cdot g_{\iota\kappa}(x)^{-1}\big),$$

and on the other hand,

$$\Phi_\iota \circ f \circ \psi_\kappa^{-1}(x,z) = \Phi_\iota \circ \Phi_\kappa^{-1}(x, \eta_\kappa(x) \cdot z) = \big(x, \eta_\kappa(x) \cdot G_{\iota\kappa}(x)^t \cdot z\big).$$

It follows that

$$\eta_\iota(x) = \eta_\kappa(x) \cdot \big(g_{\iota\kappa}(x) \cdot G_{\iota\kappa}(x)\big)^t.$$

Consequently, $\eta = (\eta_\iota)$ is a cross section in $V \otimes F$, and we obtain for every x an isomorphism

$$(V \otimes F)_x \cong \mathrm{Hom}_{\mathbb{C}}(F_x', V_x).$$

For $v \in V_x$ and $e \in F_x$ the *tensor product* $v \otimes e \in (V \otimes F)_x$ is defined by

$$(v \otimes e)(\lambda) := \lambda(e) \cdot v.$$

Lifting of Bundles. Let $f : X \to Y$ be a holomorphic map between complex manifolds and $p : V \to Y$ a vector bundle of rank r.

Definition. The *lifted bundle*, or *pullback*, f^*V over X is defined by

$$f^*V := X \times_Y V := \{(x, v) \in X \times V \,:\, f(x) = p(v)\}.$$

The bundle projection $\widehat{p} : f^*V \to X$ is given by $\widehat{p}(x, v) := x$.

The fiber of f^*V over $x \in X$ is given by $(f^*V)_x = V_{f(x)}$. Therefore, the lifted bundle is trivial over the preimage sets $f^{-1}(y)$.

One has the following commutative diagram:

$$\begin{array}{ccc} f^*V & \xrightarrow{\mathrm{pr}_2} & V \\ \widehat{p} \downarrow & & \downarrow p \\ X & \xrightarrow{f} & Y \end{array}$$

If $\mathscr{U} = \{U_\iota \,:\, \iota \in I\}$ is an open covering of Y such that V is trivial over U_ι, then $\widehat{\mathscr{U}} := \bigl\{\widehat{U}_\iota = f^{-1}(U_\iota) \,:\, \iota \in I\bigr\}$ is an open covering of X such that f^*V is trivial over $\widehat{U}_\iota$.

If $\Phi_\iota : V|_{U_\iota} \to U_\iota \times \mathbb{C}^r$ is a trivialization for V, then we can define a trivialization $\widehat{\Phi}_\iota : f^*V|_{\widehat{U}_\iota} \to \widehat{U}_\iota \times \mathbb{C}^r$ by

$$\widehat{\Phi}_\iota(x, v) := \bigl(x, (\Phi_\iota)_{f(x)}(v)\bigr).$$

If $G_{\iota\kappa}$ are transition functions for V with $\Phi_\iota \circ \Phi_\kappa^{-1}(x, \mathbf{w}) = \bigl(x, \mathbf{w} \cdot G_{\iota\kappa}(x)^{\,t}\bigr)$, then

$$\widehat{\Phi}_\iota \circ \widehat{\Phi}_\kappa^{-1}(x, \mathbf{w}) = \bigl(x, (\Phi_\iota \circ \Phi_\kappa^{-1})_{f(x)}(\mathbf{w})\bigr) = \bigl(x, \mathbf{w} \cdot G_{\iota\kappa}(f(x))^{\,t}\bigr);$$

i.e., f^*V is given by the transition functions $G_{\iota\kappa} \circ f$.

If $\xi \in \Gamma(U, V)$ is a holomorphic section over some open subset $U \subset Y$, then ξ can be lifted to a section $\widehat{\xi} \in \Gamma(\widehat{U}, f^*V)$ given by

$$\widehat{\xi}(x) := (x, \xi(f(x))).$$

Subbundles and Quotients.

Definition. Let $\pi : V \to X$ be a vector bundle of rank r. A subset $W \subset V$ is called a *subbundle* (of rank p) of V if there is a p-dimensional linear subspace $E \subset \mathbb{C}^r$, and for any $x \in X$ an open neighborhood $U = U(x)$ and a trivialization $\Phi : V|_U \to U \times \mathbb{C}^r$ such that $\Phi^{-1}(U \times E) = W|_U$.

One sees immediately that:

1. $W_x = W \cap V_x$ is always a p-dimensional linear subspace of V_x.
2. W is a submanifold of V.
3. W is itself a vector bundle.

Example

Let V be a vector bundle of rank r on X and $Y \subset X$ a submanifold. If we denote the natural injection $Y \hookrightarrow X$ by j, then $V|_Y := j^*V$ is a vector bundle of rank r on Y. We apply this to the tangent bundle $T(X)$.

Choose an open covering $\mathscr{U} = (U_\iota)_{\iota \in I}$ such that there are complex coordinate systems $\varphi_\iota = (z_1^\iota, \ldots, z_n^\iota)$ for X in U_ι with the following properties:

1. $U_\iota \cap Y = \{z_{d+1}^\iota = \cdots = z_n^\iota = 0\}$.
2. $z_1^\iota, \ldots, z_d^\iota$ are complex coordinates for Y.

A trivialization $\Phi_\iota : j^*T(X)|_{U_\iota \cap Y} \to (U_\iota \cap Y) \times \mathbb{C}^n$ is given by

$$(y, [\varphi_\iota, \mathbf{c}]) \mapsto (y, \mathbf{c}).$$

A tangent vector v belongs to $T_y(Y) \subset T_y(X)$ if and only if there is a differentiable path $\alpha : I \to Y$ with $\alpha(0) = x$ and $(\varphi_\iota \circ \alpha)'(0) = \mathbf{c}$. This is equivalent to the statement that $\mathbf{c} = (c_1, \ldots, c_d, 0, \ldots, 0)$. Therefore

$$\Phi_\iota^{-1}(U_\iota \times \{\mathbf{c} \in \mathbb{C}^n \,:\, c_{d+1} = \cdots = c_n = 0\}) = T(Y)|_{U_\iota},$$

and $T(Y)$ is a subbundle of $j^*T(X)$.

If $G_{\iota\kappa} := \big(\partial z_\nu^\iota / \partial z_\mu^\kappa \;\big|\; \nu, \mu = 1, \ldots, n\big)$ are the transition functions for $T(X)$, then

$$G_{\iota\kappa} \circ j(\mathbf{z}') = \begin{pmatrix} g_{\iota\kappa}(\mathbf{z}', \mathbf{0}) & \# \\ 0 & g_{\iota\kappa}^*(\mathbf{z}', \mathbf{0}) \end{pmatrix},$$

where $g_{\iota\kappa} = \big((\partial z_\nu^\iota / \partial z_\mu^\kappa)|_Y \;\big|\; \nu, \mu = 1, \ldots, d\big)$, are the transition functions for $T(Y)$, and $g_{\iota\kappa}^*(\mathbf{z}', \mathbf{0}) = \big((\partial z_\nu^\iota / \partial z_\mu^\kappa)(\mathbf{z}', \mathbf{0}) \;\big|\; \nu, \mu = d+1, \ldots, n\big)$.

Let V be a vector bundle of rank r. If $W \subset V$ is a subbundle, then we define the *quotient bundle* V/W by $(V/W)_x := V_x / W_x$. We have to show that there are trivializations for V/W. If $E \subset \mathbb{C}^r$ is a subspace such that there are trivializations $\Phi : V|_U \to U \times \mathbb{C}^r$ with $W|_U = \Phi^{-1}(U \times E)$, then we can choose a subspace $F \subset \mathbb{C}^r$ with $E \oplus F = \mathbb{C}^r$ and define $\overline{\Phi} : (V/W)|_U \to U \times F$ by

$$\overline{\Phi}(v \mod W) := \mathrm{pr}_F(\Phi(v)),$$

where $\mathrm{pr}_F : E \oplus F \to F$ is the canonical projection.

Using trivializations as above one obtains transition functions

$$G_{\iota\kappa} = \begin{pmatrix} g_{\iota\kappa} & \# \\ 0 & h_{\iota\kappa} \end{pmatrix}$$

for V such that $g_{\iota\kappa}$ are transition functions for W, and $h_{\iota\kappa}$ are transition functions for V/W.

Example

The quotient bundle $N_X(Y) := j^*T(X)/T(Y)$ is called the *normal bundle* of Y in X.

Exercises

1. Let G be a complex Lie group that acts analytically on a complex manifold X. Then for every $x \in X$ the *stabilizer* $G_x := \{g \in G : gx = x\}$ is a closed Lie subgroup of G, i.e., a subgroup and a (closed) complex submanifold.

   Prove that there is a unique complex structure on G/G_x such that the canonical projection $\pi : G \to G/G_x$ is a holomorphic submersion.
2. Let $\pi : V \to X$ be a complex vector bundle of rank r. For $x \in X$ let $\mathscr{B}_x$ be the set of bases of V_x. Prove that the disjoint union of the $\mathscr{B}_x$, $x \in X$, carries the structure of a fiber bundle over X with structure group and typical fiber equal to $\mathrm{GL}_r(\mathbb{C})$.
3. Let $\varphi : V \to W$ be a vector bundle homomorphism over X. Suppose that $\mathrm{rk}(\varphi_x)$ is independent of $x \in X$. Prove that

$$\mathrm{Ker}(\varphi) := \dot{\bigcup_{x \in X}} \mathrm{Ker}(\varphi_x) \quad \text{and} \quad \mathrm{Im}(\varphi) := \dot{\bigcup_{x \in X}} \mathrm{Im}(\varphi_x)$$

   are subbundles of V, respectively W. Show that $\mathrm{Im}(\varphi) \cong V/\mathrm{Ker}(\varphi)$.
4. Let $X = \overline{\mathbb{C}}$ be the Riemann sphere. Determine the transition functions for $T(X)$ for the canonical bundle K_X and for the normal bundle of $\{\infty\}$ in X.
5. Let $f : X \to Y$ be a holomorphic map. Prove that there is a uniquely determined vector bundle homomorphism $f' : T(X) \to f^*T(Y)$ with $(f')_x = (f_*)_x$ for $x \in X$.
6. Let $p : V \to X$ be a holomorphic vector bundle. Show that there are vector bundle homomorphisms $h : p^*V \to T(V)$ and $k : T(V) \to p^*T(X)$ over V with $\mathrm{Im}(h) = \mathrm{Ker}(k)$.

3. Cohomology

Cohomology Groups. Let X be an n-dimensional complex manifold and $\pi : V \to X$ a complex vector bundle of rank r. Assume that there is an open covering $\mathscr{U} = \{U_\iota : \iota \in I\}$ of X.

We consider sections in V on the sets U_ι and on intersections $U_{\iota\kappa} := U_\iota \cap U_\kappa$, respectively $U_{\iota\kappa\lambda} := U_\iota \cap U_\kappa \cap U_\lambda$.

Definition. A 0-dimensional *cochain* with values in V (with respect to $\mathscr{U}$) is a function s that assigns to every $\iota \in I$ a section $s_\iota \in \Gamma(U_\iota, V)$. The set of 0-dimensional cochains is denoted by $C^0(\mathscr{U}, V)$.

A 1-dimensional cochain (with values in V) is a function ξ that assigns to every pair $(\iota, \kappa) \in I \times I$ a section $\xi_{\iota\kappa} \in \Gamma(U_{\iota\kappa}, V)$. The set of 1-dimensional cochains is denoted by $C^1(\mathscr{U}, V)$.

Finally, a 2-dimensional cochain (with values in V) is a function Λ that assigns to every triple $(\iota, \kappa, \nu) \in I \times I \times I$ a section $\Lambda_{\iota\kappa\nu} \in \Gamma(U_{\iota\kappa\nu}, V)$, and the set of all these 2-dimensional cochains is denoted by $C^2(\mathscr{U}, V)$.

Assume that a 0-dimensional cochain s is given. One may ask whether the sections $s_\iota \in \Gamma(U_\iota, V)$ can be glued together to a global section $s \in \Gamma(X, V)$. For that it is necessary and sufficient that $s_\iota = s_\kappa$ on $U_{\iota\kappa}$. This can be expressed in another way: If we assign to every 0-cochain s a 1-cochain δs by $(\delta s)_{\iota\kappa} := s_\kappa - s_\iota$, then s defines a global section if and only if $\delta s = 0$.

Definition. The *coboundary operators*

$$\delta : C^0(\mathscr{U}, V) \to C^1(\mathscr{U}, V) \quad \text{and} \quad \delta : C^1(\mathscr{U}, V) \to C^2(\mathscr{U}, V)$$

are defined by

$$\begin{aligned} (\delta s)_{\iota\kappa} &:= s_\kappa - s_\iota \qquad (\text{on } U_{\iota\kappa}), \\ (\delta \xi)_{\iota\kappa\nu} &:= \xi_{\kappa\nu} - \xi_{\iota\nu} + \xi_{\iota\kappa} \ (\text{on } U_{\iota\kappa\nu}). \end{aligned}$$

A cochain $s \in C^0(\mathscr{U}, V)$ (respectively $\xi \in C^1(\mathscr{U}, V)$) is called a *cocycle* if $\delta s = 0$ (respectively $\delta \xi = 0$). The sets of cocycles are denoted by $Z^0(\mathscr{U}, V)$ (respectively $Z^1(\mathscr{U}, V)$).

Remarks

1. The sets of cochains and the sets of cocycles are all complex vector spaces.
2. We can identify $Z^0(\mathscr{U}, V)$ with $\Gamma(X, V)$.
3. A 1-cochain ξ is a cocycle if and only if the following compatibility condition holds:
$$\xi_{\iota\nu} = \xi_{\iota\kappa} + \xi_{\kappa\nu} \quad \text{on } U_{\iota\kappa\nu}.$$
4. Sometimes we need cocycles of degree 2. We call an element $\Lambda \in C^2(\mathscr{U}, V)$ a cocycle if the following compatibility condition holds:
$$\Lambda_{\kappa\nu\mu} = \Lambda_{\iota\nu\mu} - \Lambda_{\iota\kappa\mu} + \Lambda_{\iota\kappa\nu} \text{ on } U_{\iota\kappa\nu\mu}.$$
The set of all these cocycles is denoted by $Z^2(\mathscr{U}, V)$.

A *coboundary* is an element of the image of the coboundary operator. The sets of coboundaries are again vector spaces, denoted by $B^1(\mathscr{U},V) = \delta C^0(\mathscr{U},V)$ and $B^2(\mathscr{U},V) := \delta C^1(\mathscr{U},V)$. For completeness we define $B^0(\mathscr{U},V) := 0$.

3.1 Proposition. *$B^i(\mathscr{U},V) \subset Z^i(\mathscr{U},V)$ for $i = 0,1,2$.*

PROOF: The case $i=0$ is trivial. For $\xi = \delta s \in B^1(\mathscr{U},V)$ we have

$$\xi_{\iota\kappa} + \xi_{\kappa\nu} = (s_\kappa - s_\iota) + (s_\nu - s_\kappa) = s_\nu - s_\iota = \xi_{\iota\nu}.$$

For $\Lambda = \delta\eta \in B^2(\mathscr{U},V)$ we have

$$\begin{aligned} \Lambda_{\iota\nu\mu} - \Lambda_{\iota\kappa\mu} + \Lambda_{\iota\kappa\nu} &= (\eta_{\nu\mu} - \eta_{\iota\mu} + \eta_{\iota\nu}) \\ &\quad - (\eta_{\kappa\mu} - \eta_{\iota\mu} + \eta_{\iota\kappa}) \\ &\quad + (\eta_{\kappa\nu} - \eta_{\iota\nu} + \eta_{\iota\kappa}) \\ &= \eta_{\nu\mu} - \eta_{\kappa\mu} + \eta_{\kappa\nu} = \Lambda_{\kappa\nu\mu}. \end{aligned}$$

■

Definition. $H^i(\mathscr{U},V) := Z^i(\mathscr{U},V)/B^i(\mathscr{U},V)$ is called the ith *cohomology group* of V with respect to $\mathscr{U}$.

We have $H^0(\mathscr{U},V) = \Gamma(X,V)$, independently of the covering, and we have

$$H^1(\mathscr{U},V) = \frac{\{\xi \in C^1(\mathscr{U},V) : \xi_{\iota\nu} = \xi_{\iota\kappa} + \xi_{\kappa\nu}\}}{\{\xi : \exists s \in C^0(\mathscr{U},V) \text{ with } \xi_{\iota\kappa} = s_\kappa - s_\iota\}}.$$

The canonical map from $Z^1(\mathscr{U},V)$ to $H^1(\mathscr{U},V)$ will be denoted by q.

We do not want to elaborate on H^2, because we need it only in very special cases.

Refinements. Let $\mathscr{V} = \{V_n : n \in N\}$ be a refinement of $\mathscr{U}$. Then there is a refinement map $\tau : N \to I$ with $V_n \subset U_{\tau(n)}$. It induces maps $\tau_0 : C^0(\mathscr{U},V) \to C^0(\mathscr{V},V)$ and $\tau_1 : C^1(\mathscr{U},V) \to C^1(\mathscr{V},V)$ by

$$(\tau_0 s)_n := (s_{\tau(n)})|V_n \quad \text{and} \quad (\tau_1\xi)_{nm} := (\xi_{\tau(n)\tau(m)})|V_{nm}\,.$$

Then $\delta(\tau_0 s) = \tau_1(\delta s)$, and if $\delta\xi = 0$, then also $\delta(\tau_1\xi) = 0$. Therefore, τ induces a map $\tau^* : H^1(\mathscr{U},V) \to H^1(\mathscr{V},V)$ by

$$\tau^*(q(\xi)) := q(\tau_1\xi).$$

By the remarks above it is clear that τ^* is well defined, and it is a vector space homomorphism.

3.2 Proposition. *The map τ^* is independent of the refinement map τ and is injective.*

PROOF: Let $\sigma : N \to I$ be another refinement map and ξ a cocycle with respect to $\mathscr{U}$. We have to show that $\tau_1\xi - \sigma_1\xi$ is a coboundary:

$$\begin{aligned}(\tau_1\xi - \sigma_1\xi)_{nm} &= \xi_{\tau(n)\tau(m)}|_{V_{nm}} - \xi_{\sigma(n)\sigma(m)}|_{V_{nm}} \\ &= (\xi_{\tau(n)\sigma(n)} + \xi_{\sigma(n)\tau(m)}) - (\xi_{\sigma(n)\tau(m)} + \xi_{\tau(m)\sigma(m)}) \\ &= \xi_{\tau(n)\sigma(n)}|_{V_{nm}} - \xi_{\tau(m)\sigma(m)}|_{V_{nm}}\,.\end{aligned}$$

We define $\eta \in C^0(\mathscr{V}, V)$ by

$$\eta_n := \xi_{\tau(n)\sigma(n)}|_{V_n}.$$

Then $\tau_1\xi - \sigma_1\xi = \delta\eta$.

Now we consider a cocycle ξ with respect to $\mathscr{U}$ such that $\tau_1\xi = \delta s$ for some $s \in C^0(\mathscr{V}, V)$. We want to see that ξ itself is a coboundary.

On $V_{nm} \cap U_\iota$ we have

$$\xi_{\tau(n)\tau(m)} = \xi_{\tau(n)\iota} + \xi_{\iota\tau(m)} = \xi_{\iota\tau(m)} - \xi_{\iota\tau(n)}.$$

Since $\xi_{\tau(n)\tau(m)}|_{V_{nm}} = (s_m - s_n)|_{V_{nm}}$, it follows that

$$\xi_{\iota\tau(m)} - s_m = \xi_{\iota\tau(n)} - s_n \quad \text{on } V_{nm} \cap U_\iota.$$

Therefore, we can define h_ι on U_ι by $h_\iota|_{U_\iota \cap V_n} := \xi_{\iota\tau(n)} - s_n$. This gives an element $h \in C^0(\mathscr{U}, V)$, and on $U_{\iota\kappa} \cap V_n$ we have

$$h_\iota - h_\kappa = (\xi_{\iota\tau(n)} - s_n) - (\xi_{\kappa\tau(n)} - s_n) = \xi_{\iota\tau(n)} + \xi_{\tau(n)\kappa} = \xi_{\iota\kappa}.$$

This means that $\xi = \delta(-h)$. So τ^* is injective. ∎

We have seen that if $\mathscr{V}$ is a refinement of $\mathscr{U}$, then $H^1(\mathscr{U}, V)$ can be identified with a subspace of $H^1(\mathscr{V}, V)$. Therefore, we form the union of the spaces $H^1(\mathscr{U}, V)$ over all coverings $\mathscr{U}$ and denote this union by $H^1(X, V)$. We call it the *absolute*, or *Čech, cohomology group* of X with values in V.

Acyclic Coverings. Let X be a complex manifold and $p : V \to X$ a vector bundle over X.

The covering $\mathscr{U}$ is called *acyclic*, or a *Leray covering* for V, if $H^1(U_\iota, V) = 0$ for every $\iota \in I$.

If $\mathscr{V} = \{V_n : n \in N\}$ is another covering, then $U_\iota \cap \mathscr{V} := \{U_\iota \cap V_n : n \in N\}$ is a covering of U_ι, and since $H^1(U_\iota \cap \mathscr{V}, V)$ is a subspace of $H^1(U_\iota, V)$, we have also $H^1(U_\iota \cap \mathscr{V}, V) = 0$.

3.3 Theorem. *If $\mathscr{U} = \{U_\iota : \iota \in I\}$ is acyclic and $\mathscr{V} = \{V_n : n \in N\}$ is a refinement, with refinement map $\tau : N \to I$, then $\tau^* : H^1(\mathscr{U}, V) \to H^1(\mathscr{V}, V)$ is bijective.*

PROOF: We start with an $\eta \in Z^1(\mathscr{V}, V)$ and define $\eta^{(\iota)} \in Z^1(U_\iota \cap \mathscr{V}, V)$ by

$$\eta^{(\iota)}_{nm} := \eta_{nm}|_{U_\iota \cap V_{nm}}, \text{ for all } n, m \text{ with } U_\iota \cap V_{nm} \neq \varnothing.$$

Since $\mathscr{U}$ is acyclic, there is an element $g^{(\iota)} \in C^0(U_\iota \cap \mathscr{V}, V)$ with $\eta^{(\iota)} = \delta g^{(\iota)}$. Then $\eta_{nm} + g^{(\iota)}_n = g^{(\iota)}_m$ on $U_\iota \cap V_{nm}$, and therefore $g^{(\iota)}_m - g^{(\kappa)}_m = g^{(\iota)}_n - g^{(\kappa)}_n$ on $U_{\iota\kappa} \cap V_{nm}$.

Now we define $\xi \in Z^1(\mathscr{U}, V)$ by $\xi_{\iota\kappa}|_{U_{\iota\kappa} \cap V_n} := g^{(\iota)}_n - g^{(\kappa)}_n$, and $h \in C^0(\mathscr{V}, V)$ by $h_n := g^{(\tau(n))}_n$ (on $V_n = V_n \cap U_{\tau(n)}$).

Then on V_{nm} we have

$$\begin{aligned}
(\tau_1 \xi - \eta)_{nm} &= \xi_{\tau(n)\tau(m)} - \eta_{nm} \\
&= g^{(\tau(n))}_m - g^{(\tau(m))}_m - (g^{(\tau(n))}_m - g^{(\tau(n))}_n) \\
&= g^{(\tau(n))}_n - g^{(\tau(m))}_m \\
&= h_n - h_m.
\end{aligned}$$

So $\eta = \tau_1 \xi + \delta h$, and τ^* is surjective. ∎

3.4 Corollary. *If $\mathscr{U}$ is an acyclic covering of X, then*

$$H^1(X, V) = H^1(\mathscr{U}, V).$$

PROOF: We have $H^1(\mathscr{U}, V) \subset H^1(X, V)$. If a is an element of $H^1(X, V)$, then there is a covering $\mathscr{V}$ with $a \in H^1(\mathscr{V}, V)$. Now we can find a common refinement $\mathscr{W}$ of $\mathscr{U}$ and $\mathscr{V}$. Then $H^1(\mathscr{V}, V) \subset H^1(\mathscr{W}, V) = H^1(\mathscr{U}, V)$, and therefore $a \in H^1(\mathscr{U}, V)$. ∎

Generalizations. The simplest case of a vector bundle over X is the trivial line bundle $\mathcal{O}_X := X \times \mathbb{C}$. Therefore, the associated Čech cohomology group $H^1(X, \mathcal{O}_X)$ plays an important role for the function theory on X.

The trivial fiber bundle $X \times \mathbb{C}^*$ is not a vector bundle, but it is not so far from that. If $\pi : P \to X$ is a general analytic fiber bundle, then a *section* in P over an open set $U \subset X$ is a holomorphic map $s : U \to P$ with $\pi \circ s = \mathrm{id}_U$. If the typical fiber of P is an abelian complex Lie group, then the set $\Gamma(U, P)$ of all sections in P over U carries in a natural way the structure of an abelian group. In the case of the bundle $\mathcal{O}^*_X := X \times \mathbb{C}^*$ we have a canonical isomorphism

$$\Gamma(U, \mathcal{O}^*_X) \cong \mathcal{O}^*(U) := \{f \in \mathcal{O}(U) : f(x) \neq 0 \text{ for every } x \in U\}.$$

This is a multiplicative abelian group.

If $\mathscr{U} = \{U_\iota : \iota \in I\}$ is an open covering of X, then we can form cochains $\xi : (\iota, \kappa) \mapsto \xi_{\iota\kappa} \in \mathcal{O}^*(U_{\iota\kappa})$. The set $C^1(\mathscr{U}, \mathcal{O}_X^*)$ of all these cochains forms a (multiplicative) abelian group, and we can define the subgroups $Z^1(\mathscr{U}, \mathcal{O}_X^*)$ and $B^1(\mathscr{U}, \mathcal{O}_X^*)$ of cocycles and coboundaries: A cochain ξ is called a *cocycle* if $\xi_{\iota\nu} = \xi_{\iota\kappa}\xi_{\kappa\nu}$ on $U_{\iota\kappa\nu}$, and ξ is called a *coboundary* if there are functions $s_\iota \in \mathcal{O}^*(U_\iota)$ such that $\xi_{\iota\kappa} = s_\kappa s_\iota^{-1}$ on $U_{\iota\kappa}$.

Since all the groups are abelian, we have the quotient group

$$H^1(\mathscr{U}, \mathcal{O}_X^*) := Z^1(\mathscr{U}, \mathcal{O}_X^*)/B^1(\mathscr{U}, \mathcal{O}_X^*),$$

which we call the *first cohomology group with values in* $\mathcal{O}_X^*$ (with respect to the covering $\mathscr{U}$). Just as in the case of vector bundles we can pass to finer coverings and finally form the Čech cohomology group $H^1(X, \mathcal{O}_X^*)$.

There is a nice interpretation for the elements of $H^1(X, \mathcal{O}_X^*)$. Every cocycle with values in $\mathcal{O}_X^*$ defines a line bundle over X, and this bundle is independent of the covering. Two cocycles ξ' and ξ'' define equivalent line bundles if and only if there are functions h_ι with $\xi'_{\iota\kappa} = \xi''_{\iota\kappa} h_\kappa h_\iota^{-1}$, i.e., if and only if the cohomology classes of ξ' and ξ'' are equal. Therefore, $H^1(X, \mathcal{O}_X^*)$ is the set of isomorphy classes of line bundles over X. This group is also called the *Picard group* of X and is denoted by $\text{Pic}(X)$. The group structure is induced by the tensor product. The identity element corresponds to the trivial bundle $\mathcal{O}_X$ and the inverse to the dual bundle.

In the same way as above we can form cohomology groups of any fiber bundle P whose typical fiber is an abelian group. If the fiber of P is a nonabelian group, things become a little bit more complicated. We can define cocycles with values in P, but they do not form a group. In the set $Z^1(\mathscr{U}, P)$ of cocycles we can introduce an equivalence relation by

$$\xi' \sim \xi'' :\iff \exists\, h_\iota \text{ with } \xi'_{\iota\kappa} = h_\iota^{-1} g''_{\iota\kappa} h_\kappa .$$

The set $H^1(\mathscr{U}, P)$ of all equivalence classes is called the *cohomology set with values in* P (with respect to $\mathscr{U}$). Usually it is not a group, but there is a distinguished element, represented by the cocycle ξ with $\xi_{\iota\kappa} = 1$ for all ι, κ. Passing to finer coverings and forming the cohomology set $H^1(X, P)$ causes no problems. The most important nonabelian case is the cohomology set

$$H^1(X, X \times \text{GL}_r(\mathbb{C})),$$

whose elements correspond to the isomorphy classes of vector bundles of rank r over X. The distinguished element is represented by the trivial bundle $X \times \mathbb{C}^r$.

In the definition of the cohomology groups of fiber bundles P over X with an abelian group as typical fiber we used only the fact that $\Gamma(U, P)$ is an abelian group for every open set $U \subset X$, and that we can restrict sections over U to open subsets $V \subset U$. Having this in mind we can define another sort of cohomology group.

Let $K \subset \mathbb{C}$ be a subgroup with respect to the addition, for example $K = \mathbb{Z}$, $\mathbb{R}$, or $\mathbb{C}$. For an open set $U \subset X$ we define the (abelian) group $K(U)$ by

$$K(U) := \{f : U \to K \ : \ f \text{ is locally constant }\}.$$

Then $K(U)$ is an abelian group. If $V \subset U$ is an open subset and f an element of $K(U)$, then $f|_V \in K(V)$. If U is connected, then $K(U) \cong K$.

We can define groups of cochains $C^0(\mathscr{U}, K)$, $C^1(\mathscr{U}, K)$, and $C^2(\mathscr{U}, K)$ just as we did it for vector bundles. If ξ is an element of $C^1(\mathscr{U}, K)$, then $\xi_{\iota\kappa} \in K(U_{\iota\kappa})$. Cocycles, coboundaries, and cohomology groups are defined in the usual way. For example, we have

$$H^1(\mathcal{U}, K) = \frac{\{c : c_{\iota\kappa} \in K(U_{\iota\kappa}) \text{ and } c_{\iota\nu} = c_{\iota\kappa} + c_{\kappa\nu}\}}{\{c : \exists e_\iota \in K(U_\iota) \text{ with } c_{\iota\kappa} = e_\kappa - e_\iota\}}.$$

The Singular Cohomology. We want to give a short overview of cohomology groups of topological spaces and their relation to Čech cohomology as defined above. For proofs see [Gre67].

For $q \in \mathbb{N}_0$ the set

$$\Delta_q := \Big\{\mathbf{x} = (x_0, \ldots, x_n) \in \mathbb{R}^{q+1} \ : \ \sum_{i=0}^{n} x_i = 1,\ x_i \geq 0\Big\}$$

is called the *q-dimensional standard simplex*. The 0-dimensional simplex is a point, Δ_1 a line segment, Δ_2 a triangle, and so on.

Let X be a topological space. We assume that all spaces here are connected and locally connected. A *singular q-simplex* in X is a continuous map $\sigma : \Delta_q \to X$. If X is a complex manifold and if there is an open neighborhood $U = U(\Delta_q)$ and a smooth map $\widehat{\sigma} : U \to X$ with $\widehat{\sigma}|_{\Delta_q} = \sigma$, then σ is called a *differentiable q-simplex*. A *singular q-chain* in X is a (formal) linear combination $n_1\sigma_1 + \cdots + n_k\sigma_k$ of singular q-simplices with $n_i \in \mathbb{Z}$. The set $S_q(X)$ of all singular q-chains in X is the free abelian group generated by the singular q-simplices.

For a q-simplex σ and $i = 0, \ldots, q$ the $(q-1)$-simplex $\sigma_i : \Delta_{q-1} \to X$ is defined by

$$\sigma_i(x_0, \ldots, x_{q-1}) := \sigma(x_0, \ldots, x_{i-1}, 0, x_i, \ldots, x_{q-1}).$$

It is called the ith face of σ.

The $(q-1)$-chain $\partial\sigma := \sum_{i=0}^{q}(-1)^i\sigma_i$ is called the *boundary* of σ. The boundary operator ∂ induces a homomorphism

$$\partial : S_q(X) \to S_{q-1}(X),$$

and it follows easily that $\partial \circ \partial = 0$.

Definition. The group $H_q(X) := \{c \in S_q(X) : \partial c = 0\}/\partial S_{q+1}(X)$ is called the qth *singular homology group* of X.

In general, $H_0(X)$ is isomorphic to the free abelian group generated by the connected components of X. Since we assume all spaces to be connected, we have $H_0(X) \cong \mathbb{Z}$. If X is an n-dimensional complex manifold, then $H_q(X) = 0$ for $q > 2n$.

Now for $q \geq 0$ we define the group of *singular q-cochains* to be

$$S^q(X) := \mathrm{Hom}_{\mathbb{Z}}(S_q(X), \mathbb{Z}).$$

Then the coboundary operator $\delta : S^q(X) \to S^{q+1}(X)$ is defined by

$$\delta f[c] := f[\partial c], \quad \text{for } f \in S^q(X) \text{ and } c \in S_{q+1}(X).$$

Obviously, we have $\delta \circ \delta = 0$. We can define cocycles (elements f of $S^q(X)$ with $\delta f = 0$) and coboundaries (elements of the form δg with $g \in S^{q-1}(X)$).

Definition. The group $H^q(X) := \{f \in S^q(X) : \delta f = 0\}/\delta S^{q-1}(X)$ is called the qth *singular cohomology group* of X.

From above it is clear that $H^0(X) \cong \mathbb{Z}$.

A topological space X is called *contractible* if there is a point $x_0 \in X$ and a continuous map $F : [0,1] \times X \to X$ with $F(0,x) = x$ and $F(1,x) = x_0$. In that case $H^1(X) = \{0\}$. The space X is called *locally contractible* if every point of X has arbitrarily small contractible neighborhoods. Among the connected topological spaces there is a big class of spaces (including the so-called CW-complexes) that are locally contractible and have the following properties:

1. $H^q(X) \cong H^q(X, \mathbb{Z})$ for $q = 0, 1, 2, \ldots$.
2. Every open covering of X has a refinement $\mathscr{U} = \{U_\iota : \iota \in I\}$ that is acyclic in the sense that $H^1(U_\iota, \mathbb{Z}) = 0$ for every $\iota \in I$.

We call such spaces *good topological spaces.* For example, every (connected) complex manifold is a good topological space, and also every irreducible analytic set.

Finitely generated abelian groups are classified as follows:

If G is a finitely generated abelian group, then there are uniquely determined numbers $r \in \mathbb{N}_0$ and $n_1, \ldots, n_s \in \mathbb{N}$ with $n_i \geq 2$ and $n_i | n_{i+1}$ for $i = 1, \ldots, s-1$ such that

$$G \cong \mathbb{Z}^r \oplus (\mathbb{Z}/n_1\mathbb{Z}) \oplus \cdots \oplus (\mathbb{Z}/n_s\mathbb{Z}).$$

The number r is called the rank of G, and $\mathbb{Z}^r$ the free part of G. The sum of the $\mathbb{Z}/n_i\mathbb{Z}$ is called the torsion part of G.

In many cases cohomology can be computed from homology:

3.5 Theorem. *If X is a compact good topological space (for example, a compact connected complex manifold), then $H_q(X)$ is finitely generated for every q, and for $q \geq 1$ there is an isomorphism*

$$H^q(X) \cong (\text{free part of } H_q(X)) \oplus (\text{torsion part of } H_{q-1}(X)).$$

The rank of $H^1(X)$ is called the *first Betti number* of X, and is denoted by $b_1(X)$.

Now we give an application of Čech cohomology methods to singular cohomology. Let X and Y be good topological spaces.

3.6 Theorem (Künneth formula).

$$H^1(X \times Y, \mathbb{Z}) \cong H^1(X, \mathbb{Z}) \oplus H^1(Y, \mathbb{Z}).$$

PROOF: We choose open coverings $\mathscr{U} = \{U_\iota : \iota \in I\}$ of X and $\mathscr{V} = \{V_n : n \in N\}$ of Y such that all the U_ι, V_ν and all their pairwise intersections are connected. We write $W_{\iota n} := U_\iota \times V_n$ and $W_{\iota n, \kappa m} := W_{\iota n} \cap W_{\kappa m} = U_{\iota\kappa} \times V_{nm}$. A cocycle $\psi \in Z^1(\mathscr{U} \times \mathscr{V}, \mathbb{Z})$ is given by constant maps $\psi_{\iota n, \kappa m} : W_{\iota n, \kappa m} \to \mathbb{Z}$.

We identify any cocycle $\xi \in Z^1(\mathscr{U}, \mathbb{Z})$ with a cocycle $\widehat{\xi} \in Z^1(\mathscr{U} \times \mathscr{V}, \mathbb{Z})$ by $\widehat{\xi}_{\iota n, \kappa m} = \xi_{\iota\kappa}$, and also any $\eta \in Z^1(\mathscr{V}, \mathbb{Z})$ with an $\widehat{\eta} \in Z^1(\mathscr{U} \times \mathscr{V}, \mathbb{Z})$. This induces natural injections

$$j_1 : H^1(\mathscr{U}, \mathbb{Z}) \to H^1(\mathscr{U} \times \mathscr{V}, \mathbb{Z}) \quad \text{and} \quad j_2 : H^1(\mathscr{V}, \mathbb{Z}) \to H^1(\mathscr{U} \times \mathscr{V}, \mathbb{Z})$$

with $\operatorname{Im}(j_1) \cap \operatorname{Im}(j_2) = \{0\}$, and therefore an injective map

$$j : H^1(\mathscr{U}, \mathbb{Z}) \times H^1(\mathscr{V}, \mathbb{Z}) \to H^1(\mathscr{U} \times \mathscr{V}, \mathbb{Z})$$

by $j(a, b) := j_1(a) + j_2(b)$.

Given a cocycle $\psi \in Z^1(\mathscr{U} \times \mathscr{V}, \mathbb{Z})$, for any $\iota \in I$ we define $\psi_\iota \in Z^1(\mathscr{V}, \mathbb{Z})$ by $(\psi_\iota)_{nm} := \psi_{\iota n, \iota m}$. The cohomology class of ψ_ι in $H^1(\mathscr{V}, \mathbb{Z})$ is independent of the index ι. We see this as follows.

(a) If $U_{\iota\kappa} \neq \varnothing$, then by the cocycle property we have

$$\begin{aligned}\psi_{\iota n,\iota m} - \psi_{\kappa n,\kappa m} &= \psi_{\iota n,\kappa n} + \psi_{\kappa n,\kappa m} + \psi_{\kappa m,\iota m} - \psi_{\kappa n,\kappa m}\\ &= \psi_{\iota n,\kappa n} - \psi_{\iota m,\kappa m}.\end{aligned}$$

Setting $(\varphi_{\iota\kappa})_n := \psi_{\iota n,\kappa n}$ we get a 0-cochain $\varphi_{\iota\kappa} \in C^0(\mathscr{V}, \mathbb{Z})$ with $\delta\varphi_{\iota\kappa} = \psi_\kappa - \psi_\iota$. So the cohomology classes of ψ_ι and ψ_κ are equal in this case.

(b) If $U_{\iota\kappa} = \varnothing$, one can find a chain of sets $U_{\lambda_1}, \dots, U_{\lambda_N}$ with $U_{\iota\lambda_1} \neq \varnothing$, $U_{\lambda_i\lambda_{i+1}} \neq \varnothing$, and $U_{\lambda_N\kappa} \neq \varnothing$ (since X is connected), and from (a) it again follows that the cohomology classes of ψ_ι and ψ_κ are equal.

Let ψ_0 be a representative of the common cohomology class of the ψ_ι. The assignment $\psi \mapsto \psi_0$ induces a map $p : H^1(\mathscr{U} \times \mathscr{V}, \mathbb{Z}) \to H^1(\mathscr{V}, \mathbb{Z})$. We will prove that for every class $c \in H^1(\mathscr{U} \times \mathscr{V}, \mathbb{Z})$ there is a class $a \in H^1(\mathscr{U}, \mathbb{Z})$ with $j_1(a) = c - j_2(p(c))$, and consequently $j(a, p(c)) = c$.

For each ι there is an $\eta_\iota \in C^0(\mathscr{V}, \mathbb{Z})$ with $\psi_0 = \psi_\iota - \delta_{\mathscr{V}}(\eta_\iota)$, and we define $\eta \in C^0(\mathscr{U} \times \mathscr{V}, \mathbb{Z})$ by $\eta_{\iota n} := (\eta_\iota)_n$.

Then $\gamma := \psi - \widehat{\psi_0} - \delta\eta \in Z^1(\mathscr{U} \times \mathscr{V}, \mathbb{Z})$ and

$$\gamma_\iota = \psi_\iota - (\psi_\iota - \delta_{\mathscr{V}}(\eta_\iota)) - \delta_{\mathscr{V}}(\eta_\iota) = 0 \text{ for every } \iota.$$

Now we construct a $\varrho \in Z^1(\mathscr{U}, \mathbb{Z})$ with $\widehat{\varrho} = \gamma$:

For $n \in N$ define $\varrho_n \in Z^1(\mathscr{U}, \mathbb{Z})$ by

$$(\varrho_n)_{\iota\kappa} := \gamma_{\iota n,\kappa n}.$$

Since $\gamma_{\iota n,\iota m} = 0$ for all n, m, and ι, in the case $V_{nm} \neq \varnothing$ we have

$$\gamma_{\iota m,\kappa m} = \gamma_{\iota n,\iota m} + \gamma_{\iota m,\kappa m} + \gamma_{\kappa m,\kappa n} = \gamma_{\iota n,\kappa n},$$

and therefore $\varrho_m = \varrho_n$. If $V_{nm} = \varnothing$, we can argue in the same way as above in (b), because Y is connected.

So we have a $\varrho \in Z^1(\mathscr{U}, \mathbb{Z})$ with $\varrho_n = \varrho$ for every n and

$$\varrho_{\iota\kappa} = \gamma_{\iota n,\kappa n} = \gamma_{\iota n,\kappa n} + \gamma_{\kappa n,\kappa m} = \gamma_{\iota n,\kappa m} \quad \text{(for arbitrary } n, m).$$

Therefore, $\widehat{\varrho} = \gamma$ and $\psi = \widehat{\varrho} + \widehat{\psi_0} + \delta\eta$. This shows that j is surjective. ∎

Exercises

1. Let $\mathscr{U} = \{U_1, U_2, U_3\}$ be the covering of $X := \{z \in \mathbb{C} : 1 < |z| < 2\}$ given by $U_1 := \{x + \mathrm{i}y : y < x\}$, $U_2 := \{x + \mathrm{i}y : y < -x\}$, and $U_3 := \{x + \mathrm{i}y : y > 0\}$. Calculate $H^1(\mathscr{U}, \mathbb{Z})$.
2. Prove that $\dim_{\mathbb{C}} H^1(\mathbb{C}^2 - \{\mathbf{0}\}, \mathcal{O}) = \infty$, where $\mathcal{O}$ denotes the trivial line bundle over $\mathbb{C}^2$. Hint: Use Laurent series.

3. Show that $H^1(\overline{\mathbb{C}}, \mathcal{O}) = 0$.
4. Show that $\mathscr{U} = \{U_i : i = 1, \dots, n\}$ with $U_i := \{\mathbf{z} : z_i \neq 0\}$ is an acyclic covering of $\mathbb{C}^n - \{\mathbf{0}\}$ for $\mathcal{O}$.
5. Let X be an n-dimensional complex manifold such that $H_q(X) = \mathbb{Z}$ for q even, $0 \le q \le 2n$, and $H_q(X) = 0$ otherwise. Calculate the singular cohomology of X.

4. Meromorphic Functions and Divisors

The Ring of Germs. Let X be an n-dimensional complex manifold and $x \in X$ a point. Two holomorphic functions f, g defined near x are called *equivalent* at x if there exists a neighborhood $U = U(x)$ with $f|_U = g|_U$. The equivalence class of f at x is called the *germ* of f at x. We denote the germ by f_x and the set of all germs by $\mathcal{O}_x$.

Having fixed a complex coordinate system $\varphi : U \to B \subset \mathbb{C}^n$ at x, we may identify the set $\mathcal{O}_x$ with the ring H_n of convergent power series by

$$f_x \mapsto \text{Taylor series of } f \circ \varphi^{-1} \text{ at } \varphi(x).$$

So $\mathcal{O}_x$ has the structure of a local $\mathbb{C}$-algebra.[2] An element $f_x \in \mathcal{O}_x$ is a unit if and only if $f(x) \neq 0$. Of course, $\mathcal{O}_x$ is also noetherian and a unique factorization domain.

4.1 Proposition. *Let f, g be holomorphic functions near $x_0 \in X$. If the germs f_{x_0}, g_{x_0} are relatively prime in $\mathcal{O}_{x_0}$, then there is an open neigborhood $U = U(x_0) \subset X$ such that f_x, g_x are relatively prime in $\mathcal{O}_x$ for $x \in U$.*

PROOF: We can work in $\mathbb{C}^n$ and assume that $x_0 = \mathbf{0}$ and that $f_\mathbf{0}$ and $g_\mathbf{0}$ are Weierstrass polynomials in z_1. Since $f_\mathbf{0}$ and $g_\mathbf{0}$ are relatively prime in H_n, they are also relatively prime in $H^0_{n-1}[z_1]$. If Q is the quotient field of H_{n-1}, it follows from Gauss's lemma that $f_\mathbf{0}$ and $g_\mathbf{0}$ are relatively prime in $Q[z_1]$.

We can find a linear combination

$$h = a \cdot f_\mathbf{0} + b \cdot g_\mathbf{0},$$

where $a, b \in H_{n-1}[z_1]$, and $h \in (H_{n-1})^*$ is the greatest common divisor of $f_\mathbf{0}$ and $g_\mathbf{0}$. If $U = U' \times U'' \subset \mathbb{C} \times \mathbb{C}^{n-1}$ is a sufficiently small neighborhood of the origin, the power series a, b converge to pseudopolynomials over U'' and h to a holomorphic function on U'' that does not vanish identically.

[2] A commutative $\mathbb{C}$-algebra A with unity is called a *local* $\mathbb{C}$-*algebra* if the set $\mathfrak{m}$ of nonunits forms an ideal in A and the composition of canonical homomorphisms $\mathbb{C} \hookrightarrow A \twoheadrightarrow A/\mathfrak{m}$ is surjective.

Let $\mathbf{w} = (w_1, \mathbf{w}')$ be a point in U. If $\varphi_{\mathbf{w}}$ is a common factor of $f_{\mathbf{w}}$ and $g_{\mathbf{w}}$, then it divides $h_{\mathbf{w}}$, which has degree 0 as a polynomial in z_1. So $\varphi_{\mathbf{w}}$ does not depend on z_1 and does not vanish identically in $z_2, \ldots, z_n$. Therefore, $\varphi_{\mathbf{w}}$ is z_1-regular of order 0, and by the preparation theorem it is, up to a unit, a Weierstrass polynomial of degree 0; i.e., $\varphi_{\mathbf{w}}$ is already a unit.

This shows that $f_{\mathbf{w}}$ and $g_{\mathbf{w}}$ are relatively prime. ∎

Analytic Hypersurfaces. Let X be an n-dimensional complex manifold. We consider analytic hypersurfaces $A \subset X$. Then locally A is given as the zero set of one holomorphic function f. We always assume that f does not vanish identically and therefore A is nowhere dense in X. If locally $A = \{(z_1, \ldots, z_n) : z_n = 0\}$, then every holomorphic function g vanishing on A is of the form $g(z_1, \ldots, z_n) = z_n \cdot \widetilde{g}(z_1, \ldots, z_n)$. We will generalize this result to the arbitrary case.

4.2 Proposition. *Every hypersurface $A \subset X$ is a pure-dimensional analytic set of dimension $n-1$.*

PROOF: After choosing appropriate coordinates, we may assume that A is contained in an open set $U \subset \mathbb{C}^n$ and f is a Weierstrass polynomial ω in z_1 without multiple factors such that

$$A = N(\omega) = \{\mathbf{z} \in U : \omega(\mathbf{z}) = 0\}.$$

Since $N(\omega)$ is a branched covering over some domain $G \subset \mathbb{C}^{n-1}$, every irreducible component of $N(\omega)$ has dimension $n-1$. ∎

4.3 Theorem (Nullstellensatz for hypersurfaces). *Let $A \subset X$ be an analytic hypersurface and x_0 an arbitrary point of A.*

1. *There exists an open neighborhood $U = U(x_0) \subset X$ and a holomorphic function f on U such that:*
   (a) *$U \cap A = \{x \in U : f(x) = 0\}$.*
   (b) *If h is a holomorphic function on a neighborhood $V = V(x_0) \subset X$ with $h|_A = 0$, then there is a neighborhood $W = W(x_0) \subset U \cap V$ and a holomorphic function q on W such that $h|_W = q \cdot (f|_W)$.*
2. *If $\widetilde{f}$ is any holomorphic function defining A in U and h is again a holomorphic function on a neighborhood $V = V(x_0)$ vanishing on A, then there exists a $k \in \mathbb{N}$ and a holomorphic function $\widetilde{q}$ on a neighborhood $W = W(x_0) \subset U \cap V$ such that $h^k|_W = \widetilde{q} \cdot (\widetilde{f}|_W)$.*

PROOF: Again we work in $\mathbb{C}^n$ and assume that $x_0 = \mathbf{0}$. Let $\widetilde{f}$ be an arbitrary defining function for A near the origin. After choosing appropriate coordinates we find a unit e and a Weierstrass polynomial $\widetilde{\omega}$ in z_1 such that $\widetilde{f} = e \cdot \widetilde{\omega}$. We choose a neighborhood $U = U' \times U''$ of the origin such that:

1. $A \cap U = \{(z_1, \mathbf{z}') \in U \, : \, \widetilde{\omega}(z_1, \mathbf{z}') = 0\}$.
2. There is a prime factorization $\widetilde{\omega} = \omega_1^{k_1} \cdots \omega_l^{k_l}$ on U.

We define $\omega := \omega_1 \cdots \omega_l$. This is a pseudopolynomial without multiple factors that also defines A in U.

If h is a function vanishing on A, which we may assume to be defined on U, then by the division formula there is a holomorphic function q near the origin and a pseudopolynomial r with $\deg(r) < \deg(\omega)$ such that near $\mathbf{0}$,

$$h = q \cdot \omega + r\,.$$

Since ω has no multiple factors, the greatest common divisor of ω and $\partial\omega/\partial z_1$ is a not identically vanishing holomorphic function g of $z_2, \ldots, z_n$, and we can find pseudopolynomials q_1, q_2 with

$$g = q_1 \cdot \omega + q_2 \cdot \frac{\partial \omega}{\partial z_1}\,.$$

We may assume that everything is defined on U. Suppose that there is a $\mathbf{z}'_0 \in U''$ such that $\omega(\zeta, \mathbf{z}'_0) \in \mathbb{C}[\zeta]$ has a multiple zero ζ_0. Then $\omega(\zeta_0, \mathbf{z}'_0) = \partial\omega/\partial z_1(\zeta_0, \mathbf{z}'_0) = 0$, and therefore $g(\mathbf{z}'_0) = 0$. Hence, if $g(\mathbf{z}') \neq 0$, then $\omega(\zeta, \mathbf{z}')$ has exactly $s := \deg(\omega)$ distinct zeros. Since $h|_{N(\omega)} = 0$, $h(\zeta, \mathbf{z}')$ has at least these s distinct zeros. Using this fact and the division formula, it follows that $r(\mathbf{z}') = 0$ for $\mathbf{z}' \in U'' - N(g)$. Therefore, by the identity theorem $r = 0$ and $h = q \cdot \omega$. Taking $f = \omega$ yields the first part of the theorem.

Let $k := \max(k_1, \ldots, k_l)$. Then

$$h^k = \omega^k \cdot q^k = \widetilde{\omega} \cdot \widetilde{q}.$$

This proves the second part of the theorem. ∎

Every local holomorphic function f that satisfies the conditions of the first part of the Nullstellensatz will be called a *minimal defining function* for A.

Now let A be an *irreducible* analytic hypersurface in X, and h a holomorphic function on some open subset $U \subset X$ with $h|_{(A \cap U)} = 0$. For $x_0 \in U \cap A$ there exists a neighborhood $V = V(x_0) \subset U$ and a minimal defining function f for A on V. Then

$$\operatorname{ord}_{A,x_0}(h) := \max\{m \in \mathbb{N} \, : \, \exists q \text{ with } h = f^m \cdot q \text{ near } x_0\}.$$

It follows from the Nullstellensatz that $\operatorname{ord}_{A,x_0}(h) \geq 1$, and from the unique factorization into primes that it is finite. Furthermore, it is independent of f, because if f_1, f_2 are two minimal defining functions, then we have equations

$f_1 = q_1 f_2$ and $f_2 = q_2 f_1$. It follows that $f_1 = q_1 q_2 f_1$ and therefore $f_1(1 - q_1 q_2) = 0$. So q_1 and q_2 must be units.

For every point $x_0 \in A$ there is a neighborhood $U = U(x_0)$ such that $\mathrm{ord}_{A,x}(h) \geq \mathrm{ord}_{A,x_0}(h)$ for $x \in U \cap A$.

4.4 Proposition. *If A is irreducible, h holomorphic in a neighborhood of A, and $h|_A = 0$, then the number $\mathrm{ord}_{A,x}(h)$ is independent of $x \in A$.*

PROOF: Let $x_0 \in A$ be an arbitrary point. In a neighborhood U of x_0 there exist a decomposition $A \cap U = A_1 \cup \cdots \cup A_l$ into irreducible components and minimal defining functions f_λ for A_λ. Since h vanishes on every A_λ, there exist $k_1, \ldots, k_l \in \mathbb{N}$ and a holomorphic function q on a neighborhood $V = V(x_0) \subset U$ such that

$$h = f_1^{k_1} \cdots f_l^{k_l} \cdot q, \text{ and } (f_\lambda)_{x_0} \nmid q_{x_0} \text{ for } \lambda = 1, \ldots, l.$$

Since $(f_1)_{x_0}, \ldots, (f_l)_{x_0}$ are irreducible, it follows that $(f_\lambda)_{x_0}$ and q_{x_0} are relatively prime for $\lambda = 1, \ldots, l$. But then $(f_\lambda)_x$ and q_x remain relatively prime for x sufficiently close to x_0, say in a neighborhood $W(x_0) \subset V$.

Let $n(x) := \mathrm{ord}_{A,x}(h)$. It is necessary to consider two cases:

(a) If A is irreducible at x_0, then $l = 1$, and it is clear that $n(x) = n(x_0)$ for $x \in W \cap A$.

Consequently, $x \mapsto n(x)$ is a locally constant integer-valued function on the set $\dot{A}$ of regular points of A. Since A is globally irreducible, $\dot{A}$ is connected and $n(x)$ globally constant on $\dot{A}$. Let $n^* \in \mathbb{N}$ be the value of this function.

(b) If $l > 1$, then $f := f_1 \cdots f_l$ is a minimal defining function for A at x_0. With $m := \min(k_1, \ldots, k_l)$ we have

$$h = f_1^{k_1} \cdots f_l^{k_l} \cdot q = f^m \cdot \varrho,$$

where ϱ is a holomorphic function near x_0. Therefore, $n(x_0) \geq m$.

We assume that $m = k_\lambda$. In every small neighborhood of x_0 there are regular points $x \in A_\lambda$ that do not belong to A_μ for $\mu \neq \lambda$. Then $n(x) = n^*$, and f_λ is a minimal defining function for A at x. Since $h = f_\lambda^{k_\lambda} \cdot \widetilde{q}$, with a holomorphic function $\widetilde{q}$, and $(f_\lambda)_x \nmid \widetilde{q}_x$, it follows that $n(x) = k_\lambda$.

So $m \leq n(x_0) \leq n(x) = n^* = k_\lambda = m$, and therefore $n(x_0) = n^*$. ∎

Now we define

$$\mathrm{ord}_A(h) := \begin{cases} \text{constant value of } \mathrm{ord}_{A,x}(h) & \text{if } h|_A = 0, \\ 0 & \text{otherwise.} \end{cases}$$

One easily sees that

$$\mathrm{ord}_A(h_1 h_2) = \mathrm{ord}_A(h_1) + \mathrm{ord}_A(h_2).$$

Meromorphic Functions. Let X be an n-dimensional complex manifold. We consider holomorphic functions that are defined outside of an analytic hypersurface. In the 1-dimensional case these are holomorphic functions with isolated singularities.

> **Definition.** Let $A \subset X$ be an analytic hypersurface. A complex-valued function m on $X - A$ is called a *meromorphic function* on X if for any point $x \in X$ there are holomorphic functions g, h on an open neighborhood $U = U(x) \subset X$ such that $N(h) \subset A \cap U$ and $m = g/h$ on $U - A$.

Obviously, m is holomorphic on $X - A$. In particular, every holomorphic function f on X is also meromorphic on X.

Different meromorphic functions may be given outside of different analytic hypersurfaces. If $m_\lambda : X - A_\lambda \to \mathbb{C}$ are meromorphic functions on X, then $m_1 \pm m_2$ and $m_1 \cdot m_2$ are meromorphic functions on X, given as holomorphic functions on $X - (A_1 \cup A_2)$.

If $m : X - A \to \mathbb{C}$ is a meromorphic function, for $p \in A$ we have two possibilities:

(a) There is a neighborhood $U = U(p) \subset X$ such that m is bounded on $U - A$. Then there is a holomorphic function $\widehat{m}$ on U with $\widehat{m}|_{U-A} = m|_{U-A}$, and p is called a *removable singularity* for m.

(b) For any neighborhood $V = V(p) \subset X$ and any $n \in \mathbb{N}$ there is a point $x \in V - A$ with $|m(x)| > n$. If $m = g/h$ near p, then h must vanish at p, because otherwise we would be in situation (a). Now there are again two possibilities:

   (i) If $g(p) \neq 0$, then $\lim_{x\to p}|m(x)| = +\infty$, and we have a *pole* at p.

   (ii) The other possibility is $g(p) = 0$. This cannot occur in the case $n = 1$, since it may be assumed that the germs g_p and h_p are relatively prime, but it is possible for $n > 1$. The behavior of m is extremely irregular in that case: We take any $c \in \mathbb{C}$. Then $g_p - c \cdot h_p$ and h_p are relatively prime, and therefore there exists a sequence (x_ν) of points in $N(g - ch) - N(h)$ with $\lim_{\nu\to\infty} x_\nu = p$. This means that $m(x_\nu) = c$ for every ν. We call p a *point of indeterminacy* in this case.

In the case $n = 1$ a meromorphic function is a function that is holomorphic except for a discrete set of poles. For $n > 1$ we have the *polar set*

$$P(m) := \{p \in X \;:\; m \text{ is unbounded in any neighborhood of } p\}.$$

The polar set consists of poles and points of indeterminacy. We show that $P(m)$ is an analytic hypersurface.

Let $p \in X$ be an arbitrary point and $U = U(p) \subset X$ a connected neighborhood, where m is the quotient of g and h and $N(h) \subset A$. We may assume that $p \in A$ and g_p, h_p are relatively prime. Then

$$\begin{aligned} m \text{ bounded near } p &\iff \exists\,\varphi \text{ (holomorphic near } p) \text{ with } g = \varphi \cdot h \\ &\iff h_p \,|\, g_p \\ &\iff h_p \text{ is a unit} \\ &\iff h(p) \neq 0. \end{aligned}$$

So $P(m) \cap U = \{x \in U \,:\, h(x) = 0\}$.

If $Z \subset X$ is an irreducible hypersurface, then we define $\text{ord}_Z(m)$ as follows: If $m = g/h$ near x, then $\text{ord}_{Z,x}(m) := \text{ord}_{Z,x}(g) - \text{ord}_{Z,x}(h)$. If we choose g_x, h_x relatively prime, this definition is independent of g and h. Now it follows exactly as above that $\text{ord}_{Z,x}(m)$ is constant on Z.

4.5 Identity theorem for meromorphic functions. *Let X be connected, $m : X - A \to \mathbb{C}$ a meromorphic function, and $U \subset X$ a nonempty open set such that $m|_{U-A} = 0$. Then $P(m) = \varnothing$ and $m = 0$.*

PROOF: The set $X - P(m)$ is connected, m is holomorphic there, and $U - (A \cup P(m))$ is a nonempty open subset of $X - P(m)$. By the identity theorem for holomorphic functions it follows that $m = 0$ on $X - P(m)$. But then m is globally bounded and $P(m) = \varnothing$. ∎

The set $\mathscr{M}(X)$ of meromorphic functions on X has the structure of a ring with the function $m = 0$ as zero element. We set

$$\begin{aligned} \mathscr{M}(X)^* &:= \mathscr{M}(X) - \{0\} \\ &= \{m \in \mathscr{M}(X) \,:\, m \text{ vanishes nowhere identically}\}. \end{aligned}$$

If $m \in \mathscr{M}(X)^*$ has a local representation $m = g/h$, the zero set $N(g)$ is independent of this representation. Therefore, we can define the global zero set $N(m)$, which is an analytic hypersurface in X. Outside of $P(m) \cup N(m)$, m is holomorphic and without zeros. Therefore, $1/m$ is also holomorphic there and has local representations $1/m = h/g$. So $1/m$ is also meromorphic, and consequently $\mathscr{M}(X)$ is a field. For this it is essential that X is connected!

4.6 Levi's extension theorem. *Let $A \subset X$ be an analytic set that has at least codimension 2, and let m be a meromorphic function on $X - A$. Then there exists a meromorphic function $\widehat{m}$ on X with $\widehat{m}|_{X-A} = m$.*

PROOF: Since the statement is true for holomorphic functions, we may assume that $P(m) \neq \varnothing$. So it is an analytic hypersurface in $X - A$. By the theorem of Remmert–Stein, $Q := \overline{P(m)}$ is an analytic set in X. By Riemann's second extension theorem the holomorphic function m on $(X - Q) - A$ has a holomorphic extension $\widehat{m}$ to $X - Q$.

Let $p \in A \cap Q$ be a point. We have to show that $\widehat{m}$ is locally meromorphic at p. We choose an open neighborhood $U = U(p) \subset X$ and a function $g \in \mathcal{O}(U)$ such that:

1. $Q \cap U \subset N(g)$.
2. $N(g) = N_1 \cup \cdots \cup N_k$ is a decomposition into irreducible components.

Since $\dim(A) < \dim(N_i)$, there are points $a_i \in N_i - A$ and neighborhoods $V_i = V_i(a_i) \subset U - A$ such that $\widehat{m} = p_i/q_i$ on $V_i - Q$ and $N(q_i) \subset Q \cap V_i \subset N(g)$; i.e., $g|_{N(q_i)} = 0$.

From the Nullstellensatz it follows that there is a number $s_i \in \mathbb{N}$ and a holomorphic function r_i such that $g^{s_i} = r_i \cdot q_i$. Then $\widehat{m} = p_i r_i g^{-s_i}$ near a_i. This means that there exists an $s \in \mathbb{N}$ such that $g^s \cdot \widehat{m}$ is holomorphic near $a_1, \dots, a_k$.

Thus $N := (U - A) \cap P(g^s \widehat{m})$ is empty or an analytic hypersurface that is contained in $N(g) - A$. In the latter case it is a union of irreducible components of $N(g) - A$, and this is impossible, since every such component contains a point a_i. So N must be empty, and $g^s \widehat{m}$ is holomorphic in $U - A$. By Riemann's second extension theorem there is a holomorphic extension h of $g^s \widehat{m}$ on U. Then $g^{-s} h$ is meromorphic on U with $(g^{-s}h)|_{U-A} = \widehat{m}$. ∎

Divisors. Let X be a connected complex manifold, $m \in \mathscr{M}(X)^*$, and $Z \subset X$ an irreducible analytic hypersurface. If $Z \subset P(m)$, then $\operatorname{ord}_Z(m)$ is a negative integer, and if $Z \subset N(m)$, then $\operatorname{ord}_Z(m) \in \mathbb{N}$. In all other cases we have $\operatorname{ord}_Z(m) = 0$.

If $P(m) = \bigcup_{\iota \in I} P_\iota$ and $N(m) = \bigcup_{\lambda \in L} N_\lambda$ are the decompositions of the polar set and the zero set into irreducible components, then the formal sums

$$(m)_\infty := \sum_{\iota \in I} (-\operatorname{ord}_{P_\iota}(m)) \cdot P_\iota \quad \text{and} \quad (m)_0 := \sum_{\lambda \in L} \operatorname{ord}_{N_\lambda}(m) \cdot N_\lambda$$

are called respectively the *divisor of poles* and the *divisor of zeros* of m. Finally, $\operatorname{div}(m) := (m)_0 - (m)_\infty$ is called the *divisor* of m. From the remarks above it is clear that

$$\operatorname{div}(m) = \sum_{Z \subset X} \operatorname{ord}_Z(m) \cdot Z,$$

where the sum is over all irreducible hypersurfaces Z in X.

Definition. Let $(Z_\iota)_{\iota \in I}$ be a locally finite system of irreducible analytic hypersurfaces $Z_\iota \in X$. If for every $\iota \in I$ a number $n_\iota \in \mathbb{Z}$ is given, then the formal linear combination

$$D = \sum_{\iota \in I} n_\iota \cdot Z_\iota$$

is called a *divisor* on X.

The divisor is called *positive* or *effective* if $n_\iota \geq 0$ for every $\iota \in I$.

Divisors can be added or multiplied with integer constants in an obvious way. Therefore, the set $\mathscr{D}(X)$ of all divisors on X has the structure of an abelian group.

As we just have seen, there is a map $\operatorname{div} : \mathscr{M}(X)^* \to \mathscr{D}(X)$. Since we have $\operatorname{div}(m_1 m_2) = \operatorname{div}(m_1) + \operatorname{div}(m_2)$, div is a group homomorphism.

Sometimes it is useful to generalize the notion of a divisor a little bit. Let $(A_\iota)_{\iota \in I}$ be a locally finite system of (arbitrary) analytic hypersurfaces in X, and $(n_\iota)_{\iota \in I}$ a system of integers. Then for every $\iota \in I$ we have a decomposition

$$A_\iota = \bigcup_{\lambda_\iota \in L_\iota} A^\iota_{\lambda_\iota}$$

into irreducible components. The system $\{A^\iota_{\lambda_\iota} : \iota \in I, \lambda_\iota \in L_\iota\}$ is again locally finite, and we define $\sum_{\iota \in I} n_\iota \cdot A_\iota := \sum_{\iota \in I} \sum_{\lambda_\iota \in L_\iota} n_\iota \cdot A^\iota_{\lambda_\iota}$.

With this notation it is possible to restrict divisors to open subsets: If $U \subset X$ is open and $D = \sum_{\iota \in I} n_\iota \cdot Z_\iota$ a divisor on X, then

$$D_U := \sum_{\iota \text{ with } Z_\iota \cap U \neq \varnothing} n_\iota \cdot Z_\iota \cap U.$$

4.7 Proposition. *If $A \subset X$ is an analytic hypersurface and f a minimal defining function for A in an open set U with $U \cap A \neq \varnothing$, then*

$$\operatorname{div}(f^k) = k \cdot A \cap U.$$

The proof is more or less straightforward.

Now let an arbitrary divisor D on X be given. Then for any point $p \in X$ there is an open neighborhood $U = U(p) \subset X$, a finite system $\{Z_i : i = 1, \ldots, N\}$ of irreducible hypersurfaces $Z_i \subset U$, and a system of numbers $n_i \in \mathbb{Z}$ such that

$$D|_U = \sum_{i=1}^N n_i \cdot Z_i.$$

In addition, there is a neighborhood $V = V(p) \subset U$ such that there exist minimal defining holomorphic functions f_i for Z_i in V. Then

$$\operatorname{div}\Bigl(\prod_{i=1}^N f_i^{n_i}\Bigr) = \sum_{i=1}^N n_i \cdot Z_i \cap V = D|_V.$$

In this way every divisor is locally the divisor of a meromorphic function.

Associated Line Bundles. Let X be a connected n-dimensional complex manifold. If $Z \subset X$ is an analytic hypersurface, then there is an open covering $\mathscr{U} = \{U_\iota : \iota \in I\}$ of X with the following property:

If $U_\iota \cap Z \neq \varnothing$, then there is a minimal defining function $f_\iota \in \mathcal{O}(U_\iota)$ for Z. Setting $f_\iota := 1$ if $U_\iota \cap Z = \varnothing$, in $U_{\iota\kappa}$ we get the two relations

$$f_\iota = g_{\iota\kappa} \cdot f_\kappa \quad \text{and} \quad f_\kappa = g_{\kappa\iota} \cdot f_\iota$$

with suitable holomorphic functions $g_{\iota\kappa}$ and $g_{\kappa\iota}$. Then

$$f_\iota \cdot (1 - g_{\iota\kappa} g_{\kappa\iota}) = 0 \text{ on } U_{\iota\kappa}.$$

Since f_ι does not vanish identically, it follows that $g_{\iota\kappa} g_{\kappa\iota} = 1$ on $U_{\iota\kappa}$. This shows that $g_{\iota\kappa} \in \mathcal{O}^*(U_{\iota\kappa})$ and $g_{\kappa\iota} = g_{\iota\kappa}^{-1}$. Furthermore, on $U_{\iota\kappa\lambda}$ we have the compatibility condition

$$g_{\iota\kappa} g_{\kappa\lambda} = g_{\iota\lambda}.$$

The system of the nowhere vanishing functions $g_{\iota\kappa} = f_\iota / f_\kappa$ defines a holomorphic line bundle on X, which we denote by $[Z]$. It is easy to show that this definition does not depend on the covering and the functions f_ι.

4.8 Proposition.

1. *There is a section $s_Z \in \Gamma(X, [Z])$ with $Z = \{x \in X : s_Z(x) = 0\}$.*
2. *$[Z]$ is trivial over $X - Z$.*

PROOF: The system of holomorphic functions f_ι defines a global section s_Z with $\{x \in U_\iota : s_Z(x) = 0\} = \{x \in U_\iota : f_\iota(x) = 0\} = U_\iota \cap Z$. Then it is clear that $[Z]|_{X-Z}$ is trivial. ∎

We can generalize the concept of associated line bundles to the case of divisors. If D is a divisor on X, then there is an open covering $\mathscr{U} = \{U_\iota : \iota \in I\}$ of X, and meromorphic functions m_ι on U_ι with $D|_{U_\iota} = \operatorname{div}(m_\iota)$. It follows that the functions

$$g_{\iota\kappa} := \frac{m_\iota}{m_\kappa}$$

are nowhere vanishing holomorphic functions on $U_{\iota\kappa}$. They define a line bundle, which we denote by $[D]$. If $D = k \cdot Z$, then $[D] = [Z]^k$. If $D = D_1 + D_2$, then $[D] = [D_1] \otimes [D_2]$. Thus the map

$$\delta : \mathscr{D}(X) \to \operatorname{Pic}(X), \quad \delta(D) := \text{isomorphy class of } [D],$$

is a homomorphism of groups.

4.9 Theorem. *The sequence of group homomorphisms*

$$\mathscr{M}(X)^* \xrightarrow{\operatorname{div}} \mathscr{D}(X) \xrightarrow{\delta} \operatorname{Pic}(X)$$

is exact.

PROOF: (1) Let $m \neq 0$ be a meromorphic function. Then $[\operatorname{div}(m)]$ is given by only one transition function $m/m = 1$. Therefore, $\delta \circ \operatorname{div}(m) = 1$.

(2) Let D be a divisor on X with $\delta(D) = 1$. We assume that $D|_{U_\iota} = \operatorname{div}(m_\iota)$ and $[D]$ is represented by $g_{\iota\kappa} = m_\iota/m_\kappa$. Since $[D]$ is trivial, there are nowhere vanishing holomorphic functions h_ι with $h_\iota \cdot g_{\iota\kappa} = 1 \cdot h_\kappa$ on $U_{\iota\kappa}$. Then

$$h_\iota \cdot m_\iota = h_\kappa \cdot m_\kappa \quad \text{on } U_{\iota\kappa}.$$

Therefore, a meromorphic function m on X can be defined by $m|_{U_\iota} := h_\iota \cdot m_\iota$. Obviously, $\operatorname{div}(m)|_{U_\iota} = \operatorname{div}(m_\iota) = D|_{U_\iota}$, and therefore $\operatorname{div}(m) = D$. ■

Meromorphic Sections. Let X be a connected n-dimensional complex manifold. Any analytic hypersurface $Z \subset X$ leads to a line bundle $[Z]$, together with a global holomorphic section s_Z that vanishes exactly on Z. The construction of s_Z fails in the case of an arbitrary divisor D and its associated line bundle $[D]$. Therefore, we introduce the notion of a meromorphic section.

> **Definition.** Let $\pi : L \to X$ be an analytic line bundle and $A \subset X$ an analytic hypersurface. A holomorphic section $s \in \Gamma(X - A, L)$ is called a *meromorphic section* over X in L if for every point $x \in X$ there is an open neighborhood $U = U(x) \subset X$, a function $h \in \mathcal{O}(U)$, and a holomorphic section $t \in \Gamma(U, L)$ such that:
> 1. $h \cdot s = t$ over $U - A$.
> 2. $N(h) \subset A \cap U$.

If we have a system of trivializations $\varphi_\iota : \pi^{-1}(U_\iota) \to U_\iota \times \mathbb{C}$ and transition functions $g_{\iota\kappa}$, then we have the following description of s.

For $x \in U_\iota - A$ we define $s_\iota(x)$ by $\varphi_\iota \circ s(x) = (x, s_\iota(x))$. Then $s_\iota = g_{\iota\kappa} \cdot s_\kappa$ on $U_{\iota\kappa} - A$. If we choose the U_ι small enough, there are holomorphic functions h_ι, t_ι on U_ι such that $h_\iota \cdot s_\iota = t_\iota$ over $U_\iota - A$ and $N(h_\iota) \subset A \cap U_\iota$. This means that s_ι is a meromorphic function on U_ι. We could have as well said that a meromorphic section is a system (s_ι) of meromorphic functions with $s_\iota = g_{\iota\kappa} \cdot s_\kappa$.

If Z is an irreducible hypersurface, then $\operatorname{ord}_Z(s_\iota) = \operatorname{ord}_Z(s_\kappa)$, and we denote this number by $\operatorname{ord}_Z(s)$. The sum

$$\operatorname{div}(s) := \sum_{Z \subset X \text{ irreducible hypersurface}} \operatorname{ord}_Z(s) \cdot Z$$

is called the *divisor* of the meromorphic section s.

Now we have the solution for our problem: Let D be the divisor on X given by $D|_{U_\iota} = \text{div}(m_\iota)$. Then $[D]$ is described by the transition functions $g_{\iota\kappa} = m_\iota m_\kappa^{-1}$, and the system of the m_ι defines a global meromorphic section s_D of $[D]$ with $\text{div}(s_D) = D$.

So far our definitions seem to be purely tautological. But for example, they allow us to determine the space of holomorphic sections of $[D]$ in terms of meromorphic functions on X. For that we need the following notation: If D_1, D_2 are two divisors on X, then $D_1 \geq D_2$ if and only if $D_1 - D_2$ is a positive divisor.

4.10 Theorem. *Let $D = \sum_Z n_Z \cdot Z$ be a divisor on X. Then there is a natural isomorphism $\{m \in \mathscr{M}(X)^* : \text{div}(m) \geq -D\} \xrightarrow{\cong} \Gamma(X, [D])$.*

PROOF: Let s_D be the global meromorphic section of $[D]$ with $\text{div}(s_D) = D$. Then for any meromorphic function m on X also $t := m \cdot s_D$ is a meromorphic section of $[D]$.

If m is a meromorphic function with $\text{div}(m) \geq -D$, then

$$\text{div}(t) = \text{div}(m) + \text{div}(s_D) = \text{div}(m) + D \geq -D + D = 0.$$

This means that t is a holomorphic section. The map $m \mapsto m \cdot s_D$ is obviously injective.

Let $t \in \Gamma(X, [D])$ be given. If $D|_{U_\iota} = \text{div}(m_\iota)$, then

$$t_\iota = \frac{m_\iota}{m_\kappa} \cdot t_\kappa.$$

Hence $t_\iota m_\iota^{-1} = t_\kappa m_\kappa^{-1}$ on $U_{\iota\kappa}$, and there exists a meromorphic function m on X with $m|_{U_\iota} = t_\iota m_\iota^{-1}$. Therefore, $\text{div}(m)|_{U_\iota} = \text{div}(t_\iota) - \text{div}(m_\iota) \geq -D|_{U_\iota}$. So the map is an isomorphism. ∎

Example

Let $X = \overline{\mathbb{C}}$ be the Riemannian sphere and $D = n \cdot \infty$. Then

$$\Gamma(X, [D]) = \{m \in \mathscr{M}(X)^* : \text{ord}_\infty(m) \geq -n \text{ and } \text{ord}_p(m) \geq 0 \text{ otherwise}\}.$$

The holomorphic sections in $[D]$ are just the meromorphic functions on X that have a pole of order at most n at ∞.

Exercises

1. Let x be a point in a complex manifold X and let f_x, g_x be nonunits in $\mathcal{O}_x$. Prove that f_x and g_x are relatively prime if and only if $\dim(N(f, g)) \leq n - 2$.

2. Let $G \subset \mathbb{C}^n$ be a domain, $A \subset G$ an analytic hypersurface, $U \subset G$ an open subset with $U \cap A \neq \varnothing$, and $f : U \to \mathbb{C}$ a minimal defining function for A. Prove that $\text{Sing}(A) \cap U = \{\mathbf{z} \in U : f(\mathbf{z}) = 0 \text{ and } \nabla f(\mathbf{z}) = 0\}$.
3. Consider the meromorphic function $m(z_1, z_2) := z_2/z_1$ on $\mathbb{C}^2$. Show that the closure X of the graph $\{(z_1, z_2, w) : z_1 \neq 0 \text{ and } w = m(z_1, z_2)\}$ in $\mathbb{C}^2 \times \overline{\mathbb{C}}$ is an analytic hypersurface. Determine a minimal defining function for X at $(0, 1, \infty)$.
4. Classify the singularities of $m(z_1, z_2) := \sin(z_1)/\sin(z_1 z_2)$.
5. Let $L \to X$ be a holomorphic line bundle. Prove that $L = [D]$ for some divisor D on M if and only if L has a global meromorphic section $s \neq 0$.
6. Let X be a compact Riemann surface. Show that for every nonconstant meromorphic function f on X the numbers of zeros and poles are equal (counted with multiplicity).

5. Quotients and Submanifolds

Topological Quotients. Let X be an n-dimensional complex manifold and $\sim$ an equivalence relation on X. If $x, y \in X$ are equivalent, we write $x \sim y$ or $R(x, y)$. For $x \in X$ let

$$X(x) := \{y \in X : y \sim x\} = \{y \in X : R(y, x)\}$$

be the equivalence class of x in X. These classes give a decomposition of X into pairwise disjoint sets. The set X/R of all equivalence classes is called the *topological quotient* of X *modulo* R.

Let $\pi : X \to X/R$ be the canonical projection given by $\pi : x \mapsto X(x)$. Then X/R will be endowed with the finest topology such that π is continuous. This means that $U \subset X/R$ is open if and only if $\pi^{-1}(U) \subset X$ is open. We call this topology the *quotient topology*.

A set $A \subset X$ is called *saturated* with respect to the relation R if

$$\pi^{-1}(\pi(A)) = A.$$

5.1 Proposition.

1. *A saturated $\iff$ $A = \bigcup_{x \in A} X(x)$.*
2. *If $U \subset X/R$ is open, then $\pi^{-1}(U)$ is open and saturated.*
3. *If $W \subset X$ is open and saturated, then $\pi(W) \subset X/R$ is open.*

The proof is trivial.

5.2 Proposition. *Let Z be an arbitrary topological space. A map $f : X/R \to Z$ is continuous if and only if $f \circ \pi : X \to Z$ is continuous.*

This statement is also trivial, since $(f \circ \pi)^{-1}(U) = \pi^{-1}(f^{-1}(U))$.

Analytic Decompositions. If X is an n-dimensional complex manifold and R an equivalence relation on X, one can ask whether X/R carries the structure of a complex manifold such that π is a holomorphic map. Assume that such a structure exists. Then X/R must be a Hausdorff space. If $\varphi : U \to \mathbb{C}^k$ is a complex coordinate system for X/R, then $\widehat{U} := \pi^{-1}(U)$ is an open saturated set in X, and $\mathbf{f} := \varphi \circ \pi : \widehat{U} \to \mathbb{C}^k$ a holomorphic map with $\mathbf{f}^{-1}(\mathbf{f}(x)) = \pi^{-1}(\pi(x)) = X(x)$. So the fibers of $\mathbf{f}$ are equivalence classes, and the equivalence classes must be analytic sets. If additionally π is a submersion, then $\text{rk}_x(\mathbf{f}) = k$ for every $x \in \widehat{U}$, and the fibers are $(n-k)$-dimensional manifolds. We now show that these conditions are also sufficient for the existence of a complex structure.

Let X be an n-dimensional complex manifold and $\mathcal{Z} = \{Z_\iota : \iota \in I\}$ a decomposition of X into d-dimensional analytic sets. For $x \in X$ let $\iota(x) \in I$ be the uniquely determined index with $x \in Z_{\iota(x)}$. Then there is an equivalence relation R on X such that the equivalence class $X(x)$ is exactly the analytic set $Z_{\iota(x)}$. We consider the topological quotient X/R and the canonical projection $\pi : X \to X/R$ and assume that the following conditions are fulfilled:

1. X/R is a Hausdorff space.
2. For any $x_0 \in X$ there exists a saturated open neighborhood $\widehat{U}$ of $X(x_0)$ in X and a holomorphic map $\mathbf{f} : \widehat{U} \to \mathbb{C}^{n-d}$ such that
   (a) $\mathbf{f}^{-1}(\mathbf{f}(x)) = X(x)$ for all $x \in \widehat{U}$.
   (b) $\text{rk}_x(\mathbf{f}) = n - d$ for $x \in \widehat{U}$.

5.3 Theorem. *Under the conditions above, X/R carries a unique structure of an $(n-d)$-dimensional complex manifold such that $\pi : X \to X/R$ is a holomorphic submersion.*

PROOF: Let $x_0 \in X$ be given. Then there is an open neighborhood $\widehat{U}$ of $X(x_0)$ in X with $\pi^{-1}(\pi(\widehat{U})) = \widehat{U}$, and a submersion $\mathbf{f} : \widehat{U} \to \mathbb{C}^{n-d}$ whose fibers are equivalence classes $X(x)$. If $\mathbf{z}_0 := \mathbf{f}(x_0)$, then there is an open neighborhood $W = W(\mathbf{z}_0) \subset \mathbb{C}^{n-d}$ and a holomorphic section $s : W \to \widehat{U}$ (with $s(\mathbf{z}_0) = x_0$ and $\mathbf{f} \circ s = \text{id}_W$). For $\mathbf{z} \in W$ we have $\mathbf{f}^{-1}(\mathbf{z}) = X(s(\mathbf{z}))$, and therefore

$$\pi^{-1}(\pi(s(W))) = \bigcup_{\mathbf{z} \in W} X(s(\mathbf{z})) = \bigcup_{\mathbf{z} \in W} \mathbf{f}^{-1}(\mathbf{z}) = \mathbf{f}^{-1}(W).$$

This is an open set, so $\pi(s(W)) \subset X/R$ is open as well. We define a complex coordinate system $\varphi : \pi(s(W)) \to \mathbb{C}^{n-d}$ by

$$\varphi(\pi(s(\mathbf{z}))) := \mathbf{z}.$$

Then $\varphi(\pi(x)) = \mathbf{f}(x)$. This shows that φ is well defined and continuous. It is also bijective, with $\varphi^{-1}(\mathbf{z}) = \pi(s(\mathbf{z}))$, and therefore a homeomorphism.

Now let ψ be another coordinate system given by $\psi(\pi(t(\mathbf{z}))) := \mathbf{z}$, where t is a local section for some suitable submersion $\mathbf{g}$. Then

$$\varphi \circ \psi^{-1}(\mathbf{z}) = \varphi(\pi(t(\mathbf{z}))) = \mathbf{f}(t(\mathbf{z})).$$

The coordinate transformations are holomorphic. ∎

Properly Discontinuously Acting Groups. Let G be a complex Lie group acting analytically on an n-dimensional complex manifold X. Then

$$R(x, y) :\iff \exists\, g \in G \text{ with } y = gx$$

defines an equivalence relation on X. The equivalence class $X(x) = \{y \in X : \exists\, g \in G \text{ with } y = gx\}$ is called the *orbit* of x under the group action and is also denoted by Gx. The topological quotient X/R is called the *orbit space* and is also denoted by X/G.

We consider a very special case.

> **Definition.** The group G acts *properly discontinuously* if for all $x, y \in X$ there are open neighborhoods $U = U(x)$ and $V = V(y)$ such that
>
> $$\{g \in G : gU \cap V \neq \varnothing\}$$
>
> is empty or a finite set.

Here the orbits Gx are discrete subsets of X and are therefore 0-dimensional analytic subsets. If the action is free, we want to show that all conditions are fulfilled for X/G to be a complex manifold and $\pi : X \to X/G$ a holomorphic submersion (which means in this case that π is an unbranched covering).

5.4 Lemma. *Let G act freely and properly discontinuously on X and let $x_0, y_0 \in X$ be given.*

1. *If there is a $g_0 \in G$ with $y_0 = g_0 x_0$, then there are neighborhoods $U = U(x_0)$ and $V = V(y_0)$ such that $gU \cap V = \varnothing$ for $g \neq g_0$. In the case $y_0 = x_0$ and $g_0 = e$ one can choose $V = U$.*
2. *If $gx_0 \neq y_0$ for every $g \in G$, then there are neighborhoods $U = U(x_0)$ and $V = V(y_0)$ such that $gU \cap V = \varnothing$ for every $g \in G$.*

PROOF: At first we choose neighborhoods $U_0(x_0)$ and $V_0(y_0)$ such that

$$M := \{g \in G : gU_0 \cap V_0 \neq \varnothing\}$$

is finite or empty. There is nothing to prove if $M = \{g_0\}$ in the first case or $M = \varnothing$ in the second case. Therefore we assume that there are elements $g_1, \ldots, g_N$, $N \geq 1$, with $M = \{g_0, g_1, \ldots, g_N\}$ in the first case and $M = \{g_1, \ldots, g_N\}$ in the second case. Then we define $y_\lambda := g_\lambda x_0$, for $\lambda = 1, \ldots, N$. Since G acts freely, $y_\lambda \neq y_0$ for $\lambda = 1, \ldots, N$.

We choose neighborhoods $W_\lambda = W_\lambda(y_\lambda)$ and $V = V(y_0) \subset V_0$ such that $W_\lambda \cap V = \varnothing$, and we choose a neighborhood $U = U(x_0) \subset U_0$ such that $g_\lambda U \subset W_\lambda$, for $\lambda = 1, \ldots, N$. Then $gU \cap V = \varnothing$ for $g \neq g_0$ in the first case, and $g \in G$ in the second case. ∎

5.5 Theorem. *Let G act freely and properly dicontinuously on X. Then X/G has the unique structure of an n-dimensional complex manifold, so that $\pi : X \to X/G$ is an unbranched holomorphic covering.*

PROOF: Let $U \subset X$ be an open set. Then $\pi^{-1}(\pi(U)) = \bigcup_{g \in G} gU$ is an open set, and therefore $\pi(U)$ is also open. For $x_0 \in X$ we can choose an open neigborhood $U = U(x_0)$ such that $gU \cap U = \varnothing$ for $g \neq e$. Then $\pi : U \to \pi(U)$ is bijective.

(1) We have to show that X/G is a Hausdorff space. Let $x_1, x_2 \in X$ be given, with $\pi(x_1) \neq \pi(x_2)$. Then $gx_1 \neq x_2$ for every $g \in G$. There are open neighborhoods $U = U(x_1)$ and $V = V(x_2)$ with $gU \cap V \neq \varnothing$ for every $g \in G$. Then $\pi(U)$ and $\pi(V)$ are disjoint open neighborhoods of $\pi(x_1)$ and $\pi(x_2)$.

(2) We verify the other conditions. Let $x_0 \in X$ be given and choose a small open neigborhood $U = U(x_0) \subset X$ such that $\pi : U \to \pi(U)$ is a homeomorphism and such that there exists a complex coordinate system $\varphi : U \to \mathbb{C}^n$. Then $\mathbf{f} : \widehat{U} := \pi^{-1}(\pi(U)) \to \mathbb{C}^n$ can be defined by $\mathbf{f}(gx) := \varphi(x)$, for $x \in U$ and $g \in G$. It is clear from above that $\mathbf{f}$ is well defined. The fibers of $\mathbf{f}$ are the G-orbits, and on gU we have $\mathbf{f}(y) = \varphi(g^{-1}(y))$. This shows that $\mathbf{f}$ is holomorphic, and $\mathrm{rk}_y(\mathbf{f}) = n$ for every $y \in \widehat{U}$.

If U is small enough, then $\pi^{-1}(\pi(U)) = \bigcup_{g \in G} gU$, with pairwise disjoint sets gU that are topologically equivalent to $\pi(U)$. So π is an unbranched covering. ∎

Complex Tori. Let $\{\omega_1, \ldots, \omega_{2n}\}$ be a real basis of $\mathbb{C}^n$. Then the discrete group $\Gamma := \mathbb{Z}\omega_1 + \cdots + \mathbb{Z}\omega_{2n}$ acts freely on $\mathbb{C}^n$ by translation. The set

$$A_{\mathbf{w}} := \Gamma + \mathbf{w} = \{\omega + \mathbf{w} \,:\, \omega \in \Gamma\}$$

is the orbit of $\mathbf{w}$.

The group Γ acts properly discontinuously on $\mathbb{C}^n$: Let $\mathbf{z}_0, \mathbf{w}_0 \in \mathbb{C}^n$ be given. If $\mathbf{w}_0 = \omega_0 + \mathbf{z}_0$ for some $\omega_0 \in \Gamma$, choose

$$\varepsilon < \frac{1}{2} \cdot \inf\{\|\omega\| \, : \, \omega \in \Gamma - \{0\}\,\}.$$

Then $(\omega + B_\varepsilon(\mathbf{z}_0)) \cap B_\varepsilon(\mathbf{w}_0) = \varnothing$, unless $\omega = \omega_0$.

If $\mathbf{w}_0 - \mathbf{z}_0 \notin \Gamma$ and

$$\varepsilon < \frac{1}{2} \cdot \operatorname{dist}(\mathbf{w}_0, \Gamma + \mathbf{z}_0),$$

then $(\omega + B_\varepsilon(\mathbf{z}_0)) \cap B_\varepsilon(\mathbf{w}_0) = \varnothing$ for every ω.

The n-dimensional complex manifold $T^n = T^n_\Gamma := \mathbb{C}^n/\Gamma$ is called a *complex torus*, and Γ is called the *lattice* of the torus.

The set $P := \{\mathbf{z} = t_1\omega_1 + \cdots + t_{2n}\omega_{2n} \, : \, 0 \le t_i \le 1\}$ contains a complete system of representatives for the equivalence classes. Therefore, $T^n = \pi(P)$ is a compact space. The map

$$t_1\omega_1 + \cdots + t_{2n}\omega_{2n} \mapsto (e^{2\pi \mathrm{i} t_1}, \ldots, e^{2\pi \mathrm{i} t_{2n}})$$

induces a homeomorphism $T^n \to \underbrace{S^1 \times \cdots \times S^1}_{2n \text{ times}}$.

5.6 Proposition. $H^1(S^1, \mathbb{Z}) = \mathbb{Z}$.

PROOF: Let $U_1 := \{e^{2\pi \mathrm{i} t} \, : \, -\frac{1}{4} < t < \frac{1}{4}\}$, $U_2 := \{e^{2\pi \mathrm{i} t} \, : \, 0 < t < \frac{3}{4}\}$, and $U_3 := \{e^{2\pi \mathrm{i} t} \, : \, \frac{1}{4} < t < 1\}$. Then $\mathscr{U} := \{U_1, U_2, U_3\}$ is an acyclic open covering for S^1 with $U_{123} = \varnothing$. Therefore, every triple $\xi = (a, b, c) \in \mathbb{Z}^3$ is a cocycle in $Z^1(\mathscr{U}, \mathbb{Z})$. It is a coboundary if and only if there is a triple $(u, v, w) \in \mathbb{Z}^3$ with

$$a = v - u, \quad b = w - u, \quad \text{and} \quad c = w - v.$$

This is the case if and only if $a + c = b$. Since every cocycle has the form $(a, b, c) = (0, b - a - c, 0) + (a, a + c, c) = (b - a - c) \cdot (0, 1, 0) + \delta(0, a, a + c)$, it follows that $H^1(\mathscr{U}, \mathbb{Z})$ is generated by the class of $(0, 1, 0)$. ∎

From the Künneth formula it follows that $H^1(T^n, \mathbb{Z}) \cong \mathbb{Z}^{2n}$, and therefore the first Betti number of T^n is equal to $2n$.

Hopf Manifolds. Let $\varrho > 1$ be a fixed real number and $n > 1$. Then the (multiplicative) group $\Gamma := \{\varrho^k \, : \, k \in \mathbb{Z}\}$ acts freely on $\mathbb{C}^n - \{\mathbf{0}\}$ by $\mathbf{z} \mapsto \varrho^k \cdot \mathbf{z}$.

The action is properly discontinuous. To see this, we define the sets

$$U_r := \{\mathbf{z} \in \mathbb{C}^n \, : \, r < \|\mathbf{z}\| < \varrho r\}, \text{ for } r > 0.$$

Then the sets $\varrho^k U_r$ are pairwise disjoint. If two points $\mathbf{z}_1, \mathbf{z}_2 \in \mathbb{C}^n - \{\mathbf{0}\}$ are given, one can find an $r > 0$ and a $k \in \mathbb{Z}$ such that $\mathbf{z}_1 \in U := U_r$ and $\mathbf{z}_2 \in V := \varrho^k U_r$. The case $k = 0$ is allowed. Now $\varrho^s U \cap V = \varnothing$, unless $s = k$.

So $H = H_\Gamma := (\mathbb{C}^n - \{\mathbf{0}\})/\Gamma$ is an n-dimensional complex manifold, and the canonical projection $\pi : \mathbb{C}^n - \{\mathbf{0}\} \to H$ is an unbranched holomorphic covering. H is called a *Hopf manifold.*

The map

$$\mathbf{z} \mapsto \Big(\exp\Big(2\pi\mathrm{i}\frac{\ln\|\mathbf{z}\|}{\ln\varrho}\Big), \frac{\mathbf{z}}{\|\mathbf{z}\|}\Big)$$

induces a diffeomorphism $H \to S^1 \times S^{2n-1}$. Here S^{2n-1} is the $(2n-1)$-dimensional sphere in $\mathbb{R}^{2n} = \mathbb{C}^n$ with $2n - 1 \geq 3$.

5.7 Proposition. *For $k \geq 2$, $H^1(S^k, \mathbb{Z}) = 0$.*

PROOF: We have $S^k = \{\mathbf{x} = (x_1, \ldots, x_{k+1}) : \|\mathbf{x}\| = 1\}$. Then $\mathscr{U} = \{U_1, U_2\}$ with

$$\begin{aligned} U_1 &:= \{\mathbf{x} \in S^k : -\varepsilon < x_{k+1} \leq 1\}, \\ U_2 &:= \{\mathbf{x} \in S^k : -1 \leq x_{k+1} < \varepsilon\} \end{aligned}$$

is an open covering of S^k with contractible sets, and

$$U_{12} = \{\mathbf{x} \in S^k : -\varepsilon < x_{k+1} < \varepsilon\}$$

is connected. Therefore, $C^1(\mathscr{U}, \mathbb{Z}) = Z^1(\mathscr{U}, \mathbb{Z}) = \mathbb{Z}$ and $C^0(\mathscr{U}, \mathbb{Z}) = \mathbb{Z}^2$. The coboundary map $\delta : C^0(\mathscr{U}, \mathbb{Z}) \to C^1(\mathscr{U}, \mathbb{Z})$ is given by $\delta(a, b) := b - a$. Then obviously, $B^1(\mathscr{U}, \mathbb{Z}) = \mathbb{Z}$ and $H^1(\mathscr{U}, \mathbb{Z}) = 0$. ∎

It follows that $H^1(H, \mathbb{Z}) = H^1(S^1 \times S^{2k-1}, \mathbb{Z}) = \mathbb{Z}$, and $b_1(H) = 1$.

The Complex Projective Space. In $X := \mathbb{C}^{n+1} - \{\mathbf{0}\}$ we consider the equivalence relation

$$R(\mathbf{z}, \mathbf{w}) :\iff \exists\, \lambda \in \mathbb{C}^* \text{ with } \mathbf{w} = \lambda\mathbf{z}.$$

The equivalence class $L_\mathbf{z}$ of $\mathbf{z}$ is the set $L_\mathbf{z} = \mathbb{C}\mathbf{z} - \{\mathbf{0}\}$, the complex line through $\mathbf{z}$ and $\mathbf{0}$ without the origin. So we have a decomposition of X into 1-dimensional analytic sets. We can also look at these sets as the orbits of the canonical action of $\mathbb{C}^*$ on X by scalar multiplication.

> **Definition.** The topological quotient $\mathbb{P}^n := X/R = (\mathbb{C}^{n+1} - \{\mathbf{0}\})/\mathbb{C}^*$ is called the n-dimensional *complex projective space.*

Let $\pi : X = \mathbb{C}^{n+1} - \{\mathbf{0}\} \to \mathbb{P}^n$ be the canonical projection, with $\pi(\mathbf{z}) := L_\mathbf{z}$, and let two points $\mathbf{z} = (z_0, \ldots, z_n)$, $\mathbf{w} = (w_0, \ldots, w_n)$ be given. We have

$$\begin{aligned} \pi(\mathbf{z}) = \pi(\mathbf{w}) &\iff \exists\, \lambda \in \mathbb{C}^* \text{ with } w_i = \lambda z_i \text{ for } i = 0, \ldots, n \\ &\iff \frac{w_i}{w_j} = \frac{z_i}{z_j} \text{ for all } i, j \text{ where the fractions are defined.} \end{aligned}$$

So $\pi(\mathbf{z})$ does not determine the entries z_j, but the ratios $z_i : z_j$. Therefore, we denote the point $x = \pi(z_0, \ldots, z_n)$ by $(z_0 : \ldots : z_n)$ and call $z_0, \ldots, z_n$ the *homogeneous coordinates* of x. If $z_0, \ldots, z_n$ are homogeneous coordinates of x, then so are $\lambda z_0, \ldots, \lambda z_n$ for every $\lambda \in \mathbb{C}^*$.

If $W \subset X$ is an open set, then $\pi^{-1}(\pi(W)) = \bigcup_{\lambda \in \mathbb{C}^*} \lambda \cdot W$ is a saturated open set in X, and therefore $\pi(W)$ is open in $\mathbb{P}^n$. For example, this is true for

$$\widehat{U}_i := \{\mathbf{z} = (z_0, \ldots, z_n) \in \mathbb{C}^{n+1} - \{\mathbf{0}\} \,:\, z_i \neq 0\} \subset X, \quad i = 0, \ldots, n.$$

The sets $U_i := \pi(\widehat{U}_i)$ form an open covering of $\mathbb{P}^n$.

We show that $\mathbb{P}^n$ is a Hausdorff space: Let $\mathbf{z}, \mathbf{w} \in X$ be given, with $L_{\mathbf{z}} \neq L_{\mathbf{w}}$. Then

$$\mathbf{z}^* := \frac{\mathbf{z}}{\|\mathbf{z}\|} \quad \text{and} \quad \mathbf{w}^* := \frac{\mathbf{w}}{\|\mathbf{w}\|}$$

are distinct points of $S^{2n+1} = \{\mathbf{x} \in \mathbb{R}^{2n+2} = \mathbb{C}^{n+1} \,:\, \|\mathbf{x}\| = 1\}$. Therefore, we can find an $\varepsilon > 0$ such that $B_\varepsilon(\mathbf{z}^*) \cap B_\varepsilon(\mathbf{w}^*) = \varnothing$. Then $U := \pi(B_\varepsilon(\mathbf{z}^*))$ and $V := \pi(B_\varepsilon(\mathbf{w}^*))$ are disjoint open neighborhoods of $\pi(\mathbf{z})$, respectively $\pi(\mathbf{w})$.

Now let a point $\mathbf{z}_0 = \left(z_0^{(0)}, \ldots, z_n^{(0)}\right) \in X$ be given. Then there exists an index i with $z_i^{(0)} \neq 0$, and $\mathbf{z}_0$ lies in $\widehat{U}_i$. We define $\mathbf{f}_i : \widehat{U}_i \to \mathbb{C}^n$ by

$$\mathbf{f}_i(z_0, \ldots, z_n) := \left(\frac{z_0}{z_i}, \ldots, \frac{z_{i-1}}{z_i}, \frac{z_{i+1}}{z_i}, \ldots, \frac{z_n}{z_i}\right).$$

Then

$$\begin{aligned} \mathbf{f}_i^{-1}(\mathbf{f}_i(\mathbf{z})) &= \left\{\mathbf{w} \in \widehat{U}_i \,:\, \frac{w_j}{w_i} = \frac{z_j}{z_i} \text{ for } j \neq i\right\} \\ &= \left\{\mathbf{w} \in \widehat{U}_i \,:\, \mathbf{w} = \frac{w_i}{z_i} \cdot \mathbf{z}\right\} \\ &= \pi^{-1}(\pi(\mathbf{z})). \end{aligned}$$

If a point $\mathbf{u} = (u_0, \ldots, u_n) \in \widehat{U}_i$ is given, we define a holomorphic section $s : \mathbb{C}^n \to \widehat{U}_i$ by $s(z_1, \ldots, z_n) := (u_i z_1, \ldots, u_i z_i, u_i, u_i z_{i+1}, \ldots, u_i z_n)$. Then

$$s\left(\frac{u_0}{u_i}, \ldots, \frac{u_{i-1}}{u_i}, \frac{u_{i+1}}{u_i}, \ldots, \frac{u_n}{u_i}\right) = (u_0, \ldots, u_n),$$

and

$$\mathbf{f}_i \circ s(z_1, \ldots, z_n) = (z_1, \ldots, z_n).$$

Therefore, $\mathbf{f}_i$ is a submersion, and $\operatorname{rk}_{\mathbf{z}}(\mathbf{f}_i) = n$ for every $\mathbf{z}$.

Altogether this shows that $\mathbb{P}^n$ is an n-dimensional complex manifold and $\pi : \mathbb{C}^{n+1} - \{\mathbf{0}\} \to \mathbb{P}^n$ a holomorphic submersion. Since every equivalence class $L_{\mathbf{z}}$ has a representative in the sphere S^{2n+1}, $\mathbb{P}^n = \pi(S^{2n+1})$ is compact.

Local coordinates are given by the maps $\varphi_i : U_i \to \mathbb{C}^n$ with $\varphi_i \circ \pi = \mathbf{f}_i$. This means that

$$\varphi_i(z_0 : \ldots : z_n) = \Big(\frac{z_0}{z_i}, \ldots, \widehat{\frac{z_i}{z_i}}, \ldots, \frac{z_n}{z_i}\Big).$$ [3]

The set

$$U_0 = \{(z_0 : \ldots : z_n) \in \mathbb{P}^n \,:\, z_0 \neq 0\} = \{(1 : t_1 : \ldots : t_n) \,:\, (t_1, \ldots, t_n) \in \mathbb{C}^n\}$$

is biholomorphically equivalent to $\mathbb{C}^n$. We call it an *affine part* of $\mathbb{P}^n$. If we remove U_0 from $\mathbb{P}^n$, we get the so-called *(projective) hyperplane at infinity*

$$\begin{aligned} H_0 &= \{(z_0 : \ldots : z_n) \in \mathbb{P}^n \,:\, z_0 = 0\} \\ &= \{(0 : t_1 : \ldots : t_n) \,:\, (t_1, \ldots, t_n) \in \mathbb{C}^n - \{\mathbf{0}\}\}. \end{aligned}$$

It has the structure of an $(n-1)$-dimensional complex projective space. If we continue this process we get

$$\begin{aligned} \mathbb{P}^n &= \mathbb{C}^n \cup \mathbb{P}^{n-1}, \\ \mathbb{P}^{n-1} &= \mathbb{C}^{n-1} \cup \mathbb{P}^{n-2}, \\ &\vdots \\ \mathbb{P}^2 &= \mathbb{C}^2 \cup \mathbb{P}^1. \end{aligned}$$

It remains to study $\mathbb{P}^1 = \{(z_0 : z_1) \,:\, (z_0, z_1) \in \mathbb{C}^2 - \{\mathbf{0}\}\}$. But this is the union of $\mathbb{C} = \{(1 : t) \,:\, t \in \mathbb{C}\}$ and $\infty := (0 : 1)$, with $t = z_1/z_0$. In a neighborhood of ∞ we have the complex coordinate $z_0/z_1 = 1/t$. So we see that $\mathbb{P}^1 = \overline{\mathbb{C}} = \mathbb{C} \cup \{\infty\}$ is the well-known Riemann sphere.

The hyperplane H_0 is a regular analytic hypersurface, given by

$$H_0 \cap U_i = \Big\{(z_0 : \ldots : z_n) \in U_i \,:\, \frac{z_0}{z_i} = 0\Big\}.$$

Therefore, U_0 is dense in $\mathbb{P}^n$.

It should be remarked that there is no reason to distinguish between U_0 and the other sets U_i. Everything above could have been done as well with the affine part U_i and the hyperplane $H_i := \{\pi(\mathbf{z}) \in \mathbb{P}^n \,:\, z_i = 0\}$.

Meromorphic Functions. On a compact complex manifold every global holomorphic function is constant. But we know already from the example of the Riemann sphere that there may exist nonconstant meromorphic functions. In this regard we consider the compact manifolds defined above, beginning with the complex projective space $\mathbb{P}^n$.

A nonconstant polynomial $p(\mathbf{t}) = \sum_{|\nu|=0}^{k} a_\nu \mathbf{t}^\nu$ is a holomorphic function on the set $U_0 = \{(1 : t_1 : \ldots : t_n) \mid \mathbf{t} = (t_1, \ldots, t_n) \in \mathbb{C}^n\}$. In fact, it defines a meromorphic function on $\mathbb{P}^n$ with polar set H_0. We see this as follows:

[3] The hat signalizes that the ith term is to be left out.

The functions $t_\mu = z_\mu/z_0$, $\mu \geq 1$, are holomorphic coordinates on U_0, and $w_\lambda := z_\lambda/z_i$, $\lambda \neq i$ likewise on U_i. Therefore, on $U_i - H_0 = U_i \cap U_0$ we have

$$\begin{aligned} w_0^k \cdot p(t_1, \ldots, t_n) &= \left(\frac{z_0}{z_i}\right)^k \cdot \sum_{|\nu|=0}^{k} a_\nu \left(\frac{z_1}{z_0}\right)^{\nu_1} \cdots \left(\frac{z_n}{z_0}\right)^{\nu_n} \\ &= \sum_{|\nu|=0}^{k} a_\nu w_0^{k-|\nu|} w_1^{\nu_1} \cdots w_n^{\nu_n}; \end{aligned}$$

i.e., $p = g/h$ on $U_i - H_0$, where $g(\mathbf{w}) := \sum_{|\nu|=0}^{k} a_\nu w_0^{k-|\nu|} w_1^{\nu_1} \cdots w_n^{\nu_n}$ and $h(\mathbf{w}) := w_0^k$ are holomorphic functions on U_i with

$$N(h) = \{\mathbf{w} \in U_i \,:\, w_0 = 0\} = U_i \cap H_0.$$

So there are numerous global meromorphic functions on projective space.

Now let $T = \mathbb{C}^n/\Gamma$ be an n-dimensional complex torus, and $\pi : \mathbb{C}^n \to T$ the canonical covering. If m is a meromorphic function on T, then $m \circ \pi$ is a meromorphic function on $\mathbb{C}^n$, which is periodic with respect to the generators $\omega_1, \ldots, \omega_{2n}$ of the lattice Γ. In the case $n = 1$ such meromorphic functions always exist; they are the Γ-elliptic functions. We shall later see that for $n \geq 2$ the existence of Γ-periodic functions depends on the lattice Γ. In fact there are complex tori with no nonconstant meromorphic functions.

Finally, consider the Hopf manifold $H = (\mathbb{C}^n - \{\mathbf{0}\})/\Gamma$ with $\Gamma = \{\varrho^k \,:\, k \in \mathbb{Z}\}$ and $n > 1$. Let m be a meromorphic function on H. Since the canonical projection $\pi : \mathbb{C}^n - \{\mathbf{0}\} \to H$ is a covering, $\widehat{m} := m \circ \pi$ is meromorphic on $\mathbb{C}^n - \{\mathbf{0}\}$. Since $n > 1$, it follows from Levi's extension theorem that $\widehat{m}$ can be extended to a meromorphic function on $\mathbb{C}^n$. On any line L through the origin in $\mathbb{C}^n$, $\widehat{m}$ must have isolated poles or be identically ∞. But since $\widehat{m}$ comes from H, poles on L must have a cluster point at the origin, which is impossible unless $\widehat{m}$ is constant on L. The same argument works for any other value of $\widehat{m}$. A meromorphic function on $\mathbb{C}^n - \{\mathbf{0}\}$ that is constant on every line through the origin comes from a meromorphic function on the projective space $\mathbb{P}^{n-1}$. This means that if $h : H \to \mathbb{P}^{n-1}$ is the canonical map, then a bijection $\mathscr{M}(\mathbb{P}^{n-1}) \to \mathscr{M}(H)$ is defined by $m \mapsto m \circ h$. On the n-dimensional Hopf manifold there are not "more" meromorphic functions than on $(n-1)$-dimensional projective space.

Grassmannian Manifolds. The set of 1-dimensional complex subvector spaces of $\mathbb{C}^{n+1}$ can be identified with the n-dimensional projective space, and we have given it a complex structure. Now we do the same for the set $G_{k,n}$ of k-dimensional subspaces of $\mathbb{C}^n$. The idea is the following: If $V_0 \subset \mathbb{C}^n$ is a fixed element of $G_{k,n}$, then we choose an $(n-k)$-dimensional subspace

$W_0 \subset \mathbb{C}^n$ such that $V_0 \oplus W_0 = \mathbb{C}^n$. We are looking for a topology on $G_{k,n}$ such that the set of all k-dimensional subspaces V with $V \oplus W_0 = \mathbb{C}^n$ is a neighborhood of V_0 in $G_{k,n}$.

But how to get complex coordinates? In the case $G_{1,n+1} = \mathbb{P}^n$ we consider, for example, $V_0 = \mathbb{C}\mathbf{e}_0$ with $\mathbf{e}_0 = (1, 0, \dots, 0)$ and $W_0 = \{(z_0, \dots, z_n) : z_0 = 0\}$. Then $V_0 \oplus W_0 = \mathbb{C}^{n+1}$, and a vector $\mathbf{z} = (z_0, \dots, z_n) \neq \mathbf{0}$ generates a 1-dimensional space V with $V \oplus W_0 = \mathbb{C}^{n+1}$ if and only if $z_0 \neq 0$. Multiplication by a nonzero complex scalar does not change the space V. Therefore, V is uniquely determined by

$$z_0^{-1} \cdot \mathbf{z} = z_0^{-1} \cdot (z_0, \widetilde{\mathbf{z}}) = (1, z_0^{-1} \cdot \widetilde{\mathbf{z}}) \quad \text{with } \widetilde{\mathbf{z}} = (z_1, \dots, z_n).$$

The map $\mathbf{f} : V \mapsto z_0^{-1} \cdot \widetilde{\mathbf{z}} \in \mathbb{C}^n$ gives the familiar local coordinates.

When we try to transfer this procedure to higher k, we use another viewpoint. Every V with $V \oplus W_0 = \mathbb{C}^{n+1}$ has the form $\text{Graph}(\varphi_V)$ of a linear map $\varphi_V : \mathbb{C} \to \mathbb{C}^n$ given by $\mathbf{f}(V) = \varphi_V(1)$. If $V_0 \subset \mathbb{C}^n$ is a k-dimensional subspace and $V_0 \oplus W_0 = \mathbb{C}^n$, then every other k-dimensional subspace $V \subset \mathbb{C}^n$ with $V \oplus W_0 = \mathbb{C}^n$ has the form $\text{Graph}(\varphi_V)$ for $\varphi_V \in \text{Hom}_{\mathbb{C}}(V_0, W_0)$. Fixing bases of V_0 and W_0, the matrix of φ_V with respect to these bases gives local coordinates in $M_{k,n-k}(\mathbb{C}) \cong \mathbb{C}^{k(n-k)}$.

Now we will do this job in detail. An ordered k-tuple of linearly independent vectors $\mathbf{a}_1, \dots, \mathbf{a}_k \in \mathbb{C}^n$ can be combined in a matrix

$$\mathbf{A} = \mathbf{A}(\mathbf{a}_1, \dots, \mathbf{a}_k) := \begin{pmatrix} \mathbf{a}_1 \\ \vdots \\ \mathbf{a}_k \end{pmatrix} = \begin{pmatrix} a_{11} & \cdots & a_{1n} \\ \vdots & & \vdots \\ a_{k1} & \cdots & a_{kn} \end{pmatrix}$$

with $\text{rk}(\mathbf{A}) = k$. The set

$$\text{St}(k, n) := \{\mathbf{A} \in M_{k,n}(\mathbb{C}) : \text{rk}(\mathbf{A}) = k\}$$

is called the *complex Stiefel manifold* of type (k, n). Since its complement in $M_{k,n}(\mathbb{C}) \cong \mathbb{C}^{kn}$ is an analytic set given by the vanishing of all $(k \times k)$ minors of $\mathbf{A}$, $\text{St}(k, n)$ is an open set in $M_{k,n}(\mathbb{C})$ and therefore a complex manifold. The group $\text{GL}_k(\mathbb{C})$ acts on $\text{St}(k, n)$ by multiplication from the left, and every orbit of this group action represents exactly one k-dimensional subspace of $\mathbb{C}^n$. The topological quotient

$$G_{k,n} = \text{St}(k, n) / \text{GL}_k(\mathbb{C})$$

is called the *complex Grassmannian* of type (k, n).

If, for example, $W_0 = \{\mathbf{w} = (w_1, \dots, w_n) : w_1 = \cdots = w_k = 0\}$, then a matrix $\mathbf{A} \in \text{St}(k, n)$ represents a basis of a k-dimensional space V with $V \oplus W_0 = \mathbb{C}^n$ if and only if $\mathbf{A} = (\mathbf{A}_0 | \widetilde{\mathbf{A}})$ with $\mathbf{A}_0 \in \text{GL}_k(\mathbb{C})$ and $\widetilde{\mathbf{A}} \in M_{k,n-k}(\mathbb{C})$.

In this case V has the form $\text{Graph}(\varphi_V)$ for a linear map $\varphi_V : \mathbb{C}^k \to \mathbb{C}^{n-k}$. Of course, V is uniquely represented by the matrix $\mathbf{A}_0^{-1} \cdot \mathbf{A} = (\mathbf{E}_k | \mathbf{A}_0^{-1} \cdot \widetilde{\mathbf{A}})$, and $\mathbf{A}_0^{-1} \cdot \widetilde{\mathbf{A}}$ is the matrix of φ_V with respect to the standard bases.

Now we consider the set of multi-indices

$$\mathscr{I}_{k,n} := \{I = (i_1, \dots, i_k) \in \mathbb{N}^k \,:\, 1 \le i_1 < \cdots < i_k \le n\}.$$

For any $\mathbf{A} \in \text{St}(k,n)$ there exists an $I = (i_1, \dots, i_k) \in \mathscr{I}_{k,n}$ such that

$$\mathbf{A}_I := \begin{pmatrix} a_{1i_1} & \cdots & a_{1i_k} \\ \vdots & & \vdots \\ a_{ki_1} & \cdots & a_{ki_k} \end{pmatrix} \in \text{GL}_k(\mathbb{C}).$$

Then there is a permutation matrix $\mathbf{P}_I \in \text{GL}_n(\mathbb{C})$ such that $\mathbf{A} \cdot \mathbf{P}_I = (\mathbf{A}_I | \widetilde{\mathbf{A}}_I)$.

For fixed I we define

$$V_I := \{\mathbf{A} \in \text{St}(k,n) \,:\, \det \mathbf{A}_I \neq 0\}.$$

We remark that $(\mathbf{G} \cdot \mathbf{A})_I = \mathbf{G} \cdot \mathbf{A}_I$ and $\widetilde{(\mathbf{G} \cdot \mathbf{A})}_I = \mathbf{G} \cdot \widetilde{\mathbf{A}}_I$ for $\mathbf{G} \in \text{GL}_k(\mathbb{C})$. Therefore, V_I is invariant under the action of $\text{GL}_k(\mathbb{C})$.

5.8 Lemma. *Let $\pi_{k,n} : \text{St}(k,n) \to G_{k,n}$ be the canonical projection. Then*

$$\pi_{k,n}^{-1}(\pi_{k,n}(V_I)) = V_I \textit{ for every } I \in \mathscr{I}_{k,n}.$$

PROOF: Let $\mathbf{A} \in \pi_{k,n}^{-1}(\pi_{k,n}(V_I))$ be given. Then there is an $\mathbf{A}^* \in V_I$ with $\pi_{k,n}(\mathbf{A}) = \pi_{k,n}(\mathbf{A}^*)$. This means that there is a matrix $\mathbf{G} \in \text{GL}_k(\mathbb{C})$ with $\mathbf{A} = \mathbf{G} \cdot \mathbf{A}^*$. Since V_I is invariant under the action of $\text{GL}_k(\mathbb{C})$, $\mathbf{A}$ lies in V_I. The converse inclusion is trivial. ∎

So V_I is a saturated open subset of $\text{St}(k,n)$, and $U_I := \pi_{k,n}(V_I)$ is open in $G_{k,n}$. We leave it to the reader to show that $G_{k,n}$ is a Hausdorff space.

If $E_I \subset \mathbb{C}^n$ is generated by $\mathbf{e}_{i_1}, \dots, \mathbf{e}_{i_k}$, and $F_I \subset \mathbb{C}^n$ by the remaining $\mathbf{e}_j$, then $E_I \oplus F_I = \mathbb{C}^n$, and every k-dimensional subspace $V \subset \mathbb{C}^n$ with $V \oplus F_I = \mathbb{C}^n$ is represented by a matrix $\mathbf{A} \in V_I$. The uniquely determined matrix $\mathbf{A}_I^{-1} \cdot \widetilde{\mathbf{A}}_I \in M_{k,n-k}(\mathbb{C})$ describes the linear map $\varphi_V : E_I \to F_I$. Therefore, we define the holomorphic map $\mathbf{f}_I : V_I \to M_{k,n-k}(\mathbb{C}) \cong \mathbb{C}^{k(n-k)}$ by

$$\mathbf{f}_I(\mathbf{A}) := \mathbf{A}_I^{-1} \cdot \widetilde{\mathbf{A}}_I.$$

It is clear that $\mathbf{f}_I$ is holomorphic, and since

$$\mathbf{f}_I(\mathbf{G} \cdot \mathbf{A}) = (\mathbf{G} \cdot \mathbf{A}_I)^{-1} \cdot (\mathbf{G} \cdot \widetilde{\mathbf{A}}_I) = \mathbf{f}_I(\mathbf{A}),$$

$\mathbf{f}_I$ respects the fibers of $\pi_{k,n}$. It remains to show that $\mathbf{f}_I$ has maximal rank. For that we construct sections:

Let $\mathbf{A} \in V_I$ be given. Then we define $s : M_{k,n-k}(\mathbb{C}) \to V_I$ by

$$s(\mathbf{B}) := \mathbf{A}_I \cdot (\mathbf{E}_k|\mathbf{B}) \cdot \mathbf{P}_I^{-1}.$$

This is a holomorphic map with $s(\mathbf{f}_I(\mathbf{A})) = (\mathbf{A}_I|\widetilde{\mathbf{A}}_I) \cdot \mathbf{P}_I^{-1} = \mathbf{A}$ and

$$\mathbf{f}_I \circ s(\mathbf{B}) = \mathbf{f}_I(\mathbf{A}_I \cdot (\mathbf{E}_k|\mathbf{B}) \cdot \mathbf{P}_I^{-1}) = \mathbf{A}_I^{-1} \cdot (\mathbf{A}_I \cdot \mathbf{B}) = \mathbf{B}.$$

So $\mathbf{f}_I$ is a submersion and $G_{k,n}$ a complex manifold of dimension $k(n-k)$. Complex coordinates are given by $\varphi_I : U_I \to M_{k,n-k}(\mathbb{C})$ with

$$\varphi_I(\pi_{k,n}(\mathbf{A})) := \mathbf{A}_I^{-1} \cdot \widetilde{\mathbf{A}}_I.$$

Let $S_{k,n} := \{\mathbf{A} \in \mathrm{St}(k,n) \,:\, \mathbf{A} \cdot \overline{\mathbf{A}}^t = \mathbf{E}_k\}$ be the set of orthonormal systems of k vectors in $\mathbb{C}^n$. Then $S_{k,n}$ is a compact set, and $\pi_{k,n} : S_{k,n} \to G_{k,n}$ is surjective. So $G_{k,n}$ is compact.

Submanifolds and Normal Bundles. Let X be an n-dimensional complex manifold. A holomorphic map $f : Y \to X$ is called an *embedding* if there is a submanifold $Z \subset X$ such that f induces a biholomorphic map from Y onto Z. Every embedding is an immersion, but in general not every (injective) immersion is an embedding. The following proposition has been already mentioned in Section 1.

5.9 Proposition. *Let $f : Y \to X$ be a holomorphic map between complex manifolds. If Y is compact and f an injective immersion, then f is an embedding.*

PROOF: Every immersion defines a **local** embedding. To see this, we may consider a holomorphic map $\mathbf{f}$ from a neighborhood $V = V(\mathbf{0}) \subset \mathbb{C}^m$ into a neighborhood $U = U(\mathbf{0}) \subset \mathbb{C}^m \times \mathbb{C}^{n-m}$ with $\mathbf{f}(\mathbf{0}) = (\mathbf{0},\mathbf{0})$ and $\mathrm{rk}\, J_\mathbf{f}(\mathbf{0}) = m$. We write $\mathbf{f} = (\mathbf{f}_1, \mathbf{f}_2)$ and assume that already $\mathrm{rk}\, J_{\mathbf{f}_1}(\mathbf{0}) = m$. Then for the map $\mathbf{F} : V \times \mathbb{C}^{n-m} \to \mathbb{C}^n$ with

$$\mathbf{F}(\mathbf{z},\mathbf{w}) := (\mathbf{f}_1(\mathbf{z}), \mathbf{f}_2(\mathbf{z}) + \mathbf{w})$$

we have $\det J_\mathbf{F}(\mathbf{0},\mathbf{0}) = \det J_{\mathbf{f}_1}(\mathbf{0}) \neq 0$. So there exist neighborhoods $V^* = V^*(\mathbf{0}) \subset V$ and $W = W(\mathbf{0},\mathbf{0}) \subset U$ such that $\mathbf{F} : V^* \times V^* \to W$ is biholomorphic. Since $\mathbf{F}(\mathbf{z},\mathbf{0}) = \mathbf{f}(\mathbf{z})$, the image $\mathbf{f}(V^*) = \mathbf{F}(V^* \times \{\mathbf{0}\})$ is a submanifold of W.

We have proved that for every point $x_0 \in Y$ there are neighborhoods $V = V(x_0) \subset Y$ and $W = W(f(x_0)) \subset X$ such that $f(V)$ is a closed submanifold of W. We have to show that there is a small open neighborhood $U = U(f(x_0)) \subset W$ such that $f(Y) \cap U = f(V) \cap U$. Suppose that there is a sequence $x_n \in Y - V$ with $f(x_n) \to f(x_0)$. Since Y is compact, we can assume that (x_n) converges to an element $x^* \in Y - V$. Since f is injective, $f(x^*) \neq f(x_0)$. But $f(x_n)$ must converge to $f(x^*)$. This is impossible. ∎

If Y is a regular hypersurface in a complex manifold X, then there are two line bundles associated to Y, namely the normal bundle $N_X(Y)$, which is defined on Y, and the bundle $[Y]$ on X. We will show that these bundles coincide on Y.

Choose an open covering $\mathscr{U} = (U_\iota)_{\iota \in I}$ of Y in X such that there are complex coordinates $z_1^\iota, \dots, z_n^\iota$ for X in U_ι with the following properties:

1. $Y \cap U_\iota = \{x \in U_\iota : z_n^\iota(x) = 0\}$.
2. $z_1^\iota, \dots, z_{n-1}^\iota$ are complex coordinates for Y.

We have already seen in Section 2 that the normal bundle $N_X(Y)$ is given with respect to $U_\iota \cap Y$ by the transition functions

$$h_{\iota\kappa}(z_1^\kappa, \dots, z_{n-1}^\kappa, 0) = \frac{\partial z_n^\iota}{\partial z_n^\kappa}(z_1^\kappa, \dots, z_{n-1}^\kappa, 0).$$

On the other hand, the line bundle $[Y]$ is defined by the transition functions $f_{\iota\kappa} := z_n^\iota / z_n^\kappa$. But $z_n^\iota(z_1^\kappa, \dots, z_n^\kappa) = f_{\iota\kappa}(z_1^\kappa, \dots, z_n^\kappa) \cdot z_n^\kappa$ implies

$$\frac{\partial z_n^\iota}{\partial z_n^\kappa}(z_1^\kappa, \dots, z_{n-1}^\kappa, 0) = f_{\iota\kappa}(z_1^\kappa, \dots, z_{n-1}^\kappa, 0).$$

This yields the so-called first adjunction formula.

5.10 First adjunction formula.

$[Y]|_Y = N_X(Y)$ *for every regular hypersurface* $Y \subset X$.

The normal bundle $N_X(Y)$ is naturally related to the canonical bundles K_X and K_Y. If $g_{\iota\kappa}$, respectively $G_{\iota\kappa}$ are the transition functions for $T(Y)$, respectively $T(X)$, then

$$G_{\iota\kappa} \circ j = \begin{pmatrix} g_{\iota\kappa} & \# \\ 0 & h_{\iota\kappa} \end{pmatrix}.$$

Therefore, $\det G_{\iota\kappa} \circ j = \det g_{\iota\kappa} \cdot h_{\iota\kappa}$, and $\det g_{\iota\kappa}^{-1} = (\det G_{\iota\kappa}^{-1} \circ j) \cdot h_{\iota\kappa}$, which is equivalent to the second adjunction formula:

5.11 Second adjunction formula.

$K_Y = j^* K_X \otimes N_X(Y)$ *for every regular hypersurface* $Y \subset X$.

This formula can be generalized to higher codimensions; see, for example, [GriHa78].

Projective Algebraic Manifolds. If $\pi : \mathbb{C}^{n+1} - \{\mathbf{0}\} \to \mathbb{P}^n$ is the canonical projection and $x \in \mathbb{P}^n$, we define

$$\ell(x) := \pi^{-1}(x) \cup \{\mathbf{0}\}.$$

This is a complex line through the origin in $\mathbb{C}^{n+1}$, and we have $\ell(\pi(\mathbf{z})) = \mathbb{C}\mathbf{z}$ for $\mathbf{z} \in \mathbb{C}^{n+1} - \{\mathbf{0}\}$.

A set $\widehat{X} \subset \mathbb{C}^{n+1}$ is called a *conical set* or a *cone* if it is the union of a family of complex lines through the origin. This means that

$$\mathbf{z} \in \widehat{X} \implies \lambda\mathbf{z} \in \widehat{X} \text{ for } \lambda \in \mathbb{C}.$$

If X is an arbitrary subset of $\mathbb{P}^n$, then

$$\widehat{X} := \bigcup_{x \in X} \ell(x) = \pi^{-1}(X) \cup \{\mathbf{0}\}$$

is a conical set.

5.12 Lemma. *Let $\widehat{X} \subset \mathbb{C}^{n+1}$ be a conical set, f a holomorphic function near the origin, and $f = \sum_{\nu=0}^{\infty} p_\nu$ its expansion into homogeneous polynomials. If there is an $\varepsilon > 0$ such that $f|_{B_\varepsilon(\mathbf{0})\cap\widehat{X}} \equiv 0$, then $p_\nu|_{\widehat{X}} = 0$ for every ν.*

PROOF: Let $\mathbf{z} \neq \mathbf{0}$ be an arbitrary point of $B_\varepsilon(\mathbf{0}) \cap \widehat{X}$. Then

$$\lambda \mapsto f(\lambda\mathbf{z}) = \sum_{\nu=0}^{\infty} p_\nu(\mathbf{z})\lambda^\nu$$

vanishes identically for $|\lambda| < 1$. So $p_\nu(\mathbf{z}) = 0$ for every ν, and since $\widehat{X}$ is conical, $p_\nu|_{\widehat{X}} \equiv 0$ for every ν. ∎

Now let $F_1, \ldots, F_k$ be homogeneous polynomials in the variables $z_0, \ldots, z_n$. Then the analytic set

$$\widehat{X} := \{\mathbf{z} \in \mathbb{C}^{n+1} : F_1(\mathbf{z}) = \cdots = F_k(\mathbf{z}) = 0\}$$

is a *cone*. If we set $\widehat{X}' := \widehat{X} - \{\mathbf{0}\}$, then the image $X := \pi(\widehat{X}') \subset \mathbb{P}^n$ is the set

$$X = \{(z_0 : \ldots : z_n) : F_1(z_0, \ldots, z_n) = \cdots = F_k(z_0, \ldots, z_n) = 0\}.$$

In $U_i = \{(z_0 : \ldots : z_n) : z_i \neq 0\}$, we can define holomorphic functions $f_{i,\nu}$ by

$$f_{i,\nu}(z_0 : \ldots : z_n) := F_\nu\Big(\frac{z_0}{z_i}, \ldots, \frac{z_n}{z_i}\Big).$$

Then $X \cap U_i = \{x \in U_i : f_{i,1}(x) = \cdots = f_{i,k}(x) = 0\}$, and consequently X is an analytic set.

Definition. An analytic set $X \subset \mathbb{P}^n$ that is the zero set of finitely many homogeneous polynomials is called a *(projective) algebraic* set. The subsets $X \cap U_i$ are called *(affine) algebraic.*

A complex manifold X is called *projective algebraic* if there is an $N \in \mathbb{N}$ and a holomorphic embedding $j : X \to \mathbb{P}^N$ such that $j(X)$ is a regular algebraic set.

5.13 Theorem of Chow. *Every analytic set X in projective space is the zero set of finitely many homogeneous polynomials $F_1, \ldots, F_s$ such that if $x \in X$ is a regular point of codimension d, then* $\operatorname{rk}_{\mathbf{z}}(F_1, \ldots, F_s) = d$ *for every* $\mathbf{z} \in \pi^{-1}(x)$.

PROOF: If $X \subset \mathbb{P}^n$ is a nonempty analytic set, then $\widehat{X}' = \pi^{-1}(X)$ is also analytic. Since $\dim_{\mathbf{z}}(\widehat{X}') \geq 1$ for all $\mathbf{z} \in \mathbb{C}^{n+1} - \{\mathbf{0}\}$, by the extension theorem of Remmert–Stein its closure $\widehat{X} = \widehat{X}' \cup \{\mathbf{0}\}$ is analytic in $\mathbb{C}^{n+1}$.

By Theorem 6.5 in Chapter III we can find an open neighborhood $U = U(\mathbf{0}) \subset \mathbb{C}^{n+1}$ and finitely many holomorphic functions $f_1, \ldots, f_m$ on U with $N(f_1, \ldots, f_m) = U \cap \widehat{X}$ and $\operatorname{rk}_{\mathbf{z}}(f_1, \ldots, f_m) = d$ at any regular point $\mathbf{z}$ of dimension $n+1-d$ in $U \cap \widehat{X}$. Now we expand f_i into homogeneous polynomials $p_{i,\nu}$. Then $p_{i,\nu}|_{\widehat{X}} \equiv 0$ for all i, ν.

Let $I_k \subset \mathcal{O}_{\mathbf{0}} \cong H_{n+1}$ be the ideal that is generated by all $p_{i,\nu}$, $\nu \leq k$, $i = 1, \ldots, m$. Since

$$I_1 \subset I_2 \subset \cdots \subset \mathcal{O}_{\mathbf{0}}$$

is an ascending chain of ideals in a noetherian ring, it must become stationary. Thus there are homogeneous polynomials $F_1, \ldots, F_s$ such that every $p_{i,\nu}$ is a finite linear combination of the F_σ. But then every f_i is also a linear combination of the F_σ:

$$f_i = \sum_{\sigma=1}^{s} a_{i,\sigma} F_\sigma.$$

It is clear that $N(f_1, \ldots, f_m) = N(F_1, \ldots, F_s)$ near the origin. But since $\widehat{X}$ is a cone, even $\widehat{X} = N(F_1, \ldots, F_s)$. Setting

$$A(\mathbf{z}) = \left(a_{i,\sigma}(\mathbf{z}) \;\middle|\; \begin{array}{l} i = 1, \ldots, m \\ \sigma = 1, \ldots, s \end{array} \right),$$

we have

$$J_{(f_1,\ldots,f_m)}(\mathbf{z}) = A(\mathbf{z}) \cdot J_{(F_1,\ldots,F_s)}(\mathbf{z}) \text{ for } \mathbf{z} \in \widehat{X} \text{ near } \mathbf{0}.$$

Therefore, $d = \operatorname{rk}_{\mathbf{z}}(f_1, \ldots, f_m) \leq \operatorname{rk}_{\mathbf{z}}(F_1, \ldots, F_s)$. If X is regular of codimension d at x, then $\widehat{X}$ is also regular of codimension d at every $\mathbf{z} \in \pi^{-1}(x)$, because π is a submersion. It follows that $\text{rk}_{\mathbf{z}}(F_1, \dots, F_s)$ cannot be greater than d at these points. Since the rank is constant along $\pi^{-1}(x)$, it must be equal to d at every $\mathbf{z} \in \pi^{-1}(x)$. ∎

Examples

1. Let $L \subset \mathbb{C}^{n+1}$ be a complex linear subspace of codimension q. Then there are linear forms $\varphi_1, \dots, \varphi_q$ on $\mathbb{C}^{n+1}$ such that

$$L = \{\mathbf{z} \in \mathbb{C}^{n+1} \,:\, \varphi_1(\mathbf{z}) = \cdots = \varphi_q(\mathbf{z}) = 0\}.$$

Since the linear forms are homogeneous polynomials of degree 1,

$$\mathbb{P}(L) := \{(z_0 : \ldots : z_n) \in \mathbb{P}^n \,:\, \varphi_\mu(z_0, \dots, z_n) = 0 \text{ for } \mu = 1, \dots, q\}$$

is a regular algebraic set. We call $\mathbb{P}(L)$ a *(projective) linear subspace*. It is isomorphic to $\mathbb{P}^{n-q}$.

2. We now show that the Grassmannian manifolds are projective algebraic. Let $0 \le k \le n$ be given, and $N := \binom{n}{k} - 1$. We identify $\bigwedge^k \mathbb{C}^n$ with $\mathbb{C}^{N+1}$ and define the *Plücker embedding* $\text{pl} : G_{k,n} \to \mathbb{P}^N$ as follows: If a subspace $V \subset \mathbb{C}^n$ has the basis $\{\mathbf{a}_1, \dots, \mathbf{a}_k\}$, then

$$\text{pl}(V) := \pi(\mathbf{a}_1 \wedge \cdots \wedge \mathbf{a}_k).$$

It is clear that this is a well-defined injective map. To see that it is a holomorphic immersion, we choose another descripton. As above, when we introduced the Grassmannian, we use the set of multi-indices

$$\mathscr{I}_{k,n} = \{I = (i_1, \dots, i_k) \,:\, 1 \le i_1 < \cdots < i_k \le n\}.$$

To any $I \in \mathscr{I}_{k,n}$ there corresponds a permutation matrix $\mathbf{P}_I \in \text{GL}_n(\mathbb{C})$ such that for $\mathbf{A} \in \text{St}(k,n)$ we get $\mathbf{A} \cdot \mathbf{P}_I = (\mathbf{A}_I | \widetilde{\mathbf{A}}_I)$. We define $\widetilde{p} : \text{St}(k,n) \to \mathbb{C}^{N+1} - \{\mathbf{0}\}$ by

$$\widetilde{p}(\mathbf{A}) := (\det \mathbf{A}_I \,|\, I \in \mathscr{I}_{k,n}).$$

Then $\widetilde{p}(\mathbf{G} \cdot \mathbf{A}) = \det \mathbf{G} \cdot \widetilde{p}(\mathbf{A})$, so $\widetilde{p}$ induces a map $p : G_{k,n} \to \mathbb{P}^N$ such that the following diagram commutes:

$$\begin{array}{ccc} \text{St}(k,n) & \xrightarrow{\widetilde{p}} & \mathbb{C}^{n+1} - \{\mathbf{0}\} \\ {\scriptstyle \pi_{k,n}}\downarrow & & \downarrow{\scriptstyle \pi} \\ G_{k,n} & \xrightarrow{p} & \mathbb{P}^n \end{array}$$

For $I = (i_1, \dots, i_k)$ set $e_I := \mathbf{e}_{i_1} \wedge \cdots \wedge \mathbf{e}_{i_k}$. Then

$$\mathbf{a}_1 \wedge \cdots \wedge \mathbf{a}_k = \sum_{I \in \mathscr{I}_{k,n}} \det \mathbf{A}_I \cdot e_I,$$

and therefore $p = \text{pl}$.

Let $\varphi_I : U_I \to M_{k,n-k}(\mathbb{C})$ be a complex coordinate system for $G_{k,n}$. Then $\varphi_I^{-1}(\mathbf{B}) = \pi_{k,n}\big((\mathbf{E}_k|\mathbf{B}) \cdot \mathbf{P}_I^{-1}\big)$ and

$$p \circ \varphi_I^{-1}(\mathbf{B}) = \pi \circ \widetilde{p}\big((\mathbf{E}_k|\mathbf{B}) \cdot \mathbf{P}_I^{-1}\big) = \pi\big(\det\big((\mathbf{E}_k|\mathbf{B}) \cdot \mathbf{P}_I^{-1}\big)_J \,:\, J \in \mathscr{I}_{k,n}\big).$$

Obviously, p is a holomorphic map.

Every matrix $\mathbf{B} \in M_{k,n-k}(\mathbb{C})$ can be written in the form

$$\mathbf{B} = \big(\mathbf{b}_{k+1}^t, \ldots, \mathbf{b}_n^t\big)$$

with $\mathbf{b}_\mu \in \mathbb{C}^k$ for $\mu = k+1, \ldots, n$. We have $\det\big((\mathbf{E}_k|\mathbf{B}) \cdot \mathbf{P}_I^{-1}\big)_I = 1$, and if $J = (I - \{i_\nu\}) \cup \{\mu\}$ for some $\nu \in \{1, \ldots, k\}$ and $\mu \in \{k+1, \ldots, n\}$, then

$$\big((\mathbf{E}_k|\mathbf{B}) \cdot \mathbf{P}_I^{-1}\big) \cdot \mathbf{P}_J = \big(\mathbf{e}_1^t, \ldots, \mathbf{e}_{\nu-1}^t, \mathbf{b}_\mu^t, \mathbf{e}_{\nu+1}^t, \ldots, \mathbf{e}_k^t\big),$$

and therefore $\det\big((\mathbf{E}_n|\mathbf{B}) \cdot \mathbf{P}_I^{-1}\big)_J = b_{\nu\mu}$. So $p \circ \varphi_I^{-1}(\mathbf{B})$ contains the components 1, $b_{\nu\mu}$ for all ν, μ and some other components. It follows that pl is an immersion. Since $G_{k,n}$ is compact, pl is an embedding.

Projective Hypersurfaces. The simplest example of an analytic hypersurface in $\mathbb{P}^n$ is the hyperplane $H_0 = \{z_0 = 0\}$.

If $\mathbf{S} \in M_{n+1}(\mathbb{C})$ is a symmetric matrix, then $q_\mathbf{S}(\mathbf{z}) := \mathbf{z} \cdot \mathbf{S} \cdot \mathbf{z}^t$ is a homogeneous polynomial of degree 2. The hypersurface

$$Q_\mathbf{S} := \{(z_0 : \ldots : z_n) \,:\, q_\mathbf{S}(z_0, \ldots, z_n) = 0\}$$

is also called a *hyperquadric*. It follows easily from the classification of symmetric matrices that $Q_\mathbf{S} \cong Q_\mathbf{T}$ (biholomorphic) if and only if $\text{rk}(\mathbf{S}) = \text{rk}(\mathbf{T})$. In particular, every quadric of rank $n+1$ is biholomorphically equivalent to the standard hyperquadric $Q_{n-1} = \{(z_0 : \ldots : z_n) \,:\, z_0^2 + \cdots + z_n^2 = 0\}$. Since

$$Q_{n-1} \cap U_0 = \{(1 : t_1 : \ldots : t_n) \,:\, t_1^2 + \cdots + t_n^2 = -1\},$$

Q_{n-1} has no singularity in U_0. The same works in every U_i, so Q_{n-1} is a projective algebraic manifold.

Now consider the Grassmannian $G_{2,4} \hookrightarrow \mathbb{P}^5$. Since $\dim(G_{2,4}) = 4$, it is a hypersurface. From multilinear algebra one knows that a 2-vector $\omega \in \bigwedge^2 \mathbb{C}^4$ is decomposable (i.e., of the form $\omega = \mathbf{a} \wedge \mathbf{b}$) if and only if $\omega \wedge \omega = 0$. So

$$G_{2,4} = \{\pi(\omega) \,:\, \omega_{12}\omega_{34} - \omega_{13}\omega_{24} + \omega_{14}\omega_{23} = 0\}$$

is a hyperquadric that is isomorphic to Q_4. This means that there is a 1 to 1 correspondence between the set of projective lines in $\mathbb{P}^3$ (which is the same as the set $G_{2,4}$ of planes in $\mathbb{C}^4$) and the set of points of a 4-dimensional quadric.

5.14 Theorem. *Every analytic hypersurface $Z \subset \mathbb{P}^n$ is the zero set of a single homogeneous polynomial.*

PROOF: If $Z \subset \mathbb{P}^n$ is an analytic hypersurface, then $\widehat{Z} = \pi^{-1}(Z) \cup \{\mathbf{0}\} \subset \mathbb{C}^{n+1}$ is also an analytic hypersurface. Therefore, there exists an open neighborhood $U = U(\mathbf{0}) \subset \mathbb{C}^{n+1}$, a point $\mathbf{z} \in U \cap \widehat{Z}$, and a holomorphic function $f : U \to \mathbb{C}$ with $\nabla f(\mathbf{z}) \neq \mathbf{0}$ and $\widehat{Z} \cap U \subset N(f)$. Making U smaller if necessary, we can find a holomorphic function g dividing f such that $U \cap \widehat{Z} = N(g)$. Without loss of generality we may assume that $U = U' \times U'' \subset \mathbb{C} \times \mathbb{C}^n$ and

$$g(z_1, \mathbf{z}') = z_1^k + a_{k-1}(\mathbf{z}')z_1^{k-1} + \cdots + a_0(\mathbf{z}')$$

is a Weierstrass polynomial.

If $a_{\kappa,\nu}$ is the homogeneous part of a_κ of degree ν, then

$$p_k(z_1, \mathbf{z}') := z_1^k + q_{k-1,1}(\mathbf{z}')z_1^{k-1} + \cdots + q_{0,k}(\mathbf{z}')$$

is the homogeneous part of g of degree k. Since $\widehat{Z}$ is conical, it follows that $p_k|_{\widehat{Z}} = 0$.

Now there is a dense open subset $V \subset U''$ such that $\{t \in U' : g(t, \mathbf{z}') = 0\}$ consists of exactly k points, for every $\mathbf{z}' \in V$. Since $N(g) \cap U \subset N(p_k) \cap U$ and $\deg(p_k) = k$, we have $g = p_k$ over V and then everywhere in U, by the identity theorem. So $\widehat{Z} = N(p_k)$. ∎

We can choose a polynomial p with minimal degree such that

$$Z = \{(z_0 : \ldots : z_n) \in \mathbb{P}^n : p(z_0, \ldots, z_n) = 0\}.$$

Then by the *degree* of Z we understand the number $\deg(p)$. For example, $\deg(H) = 1$ for any hyperplane, and $\deg(Q) = 2$ for any hyperquadric.

Now, let $Z \subset \mathbb{P}^n$ be an arbitrary hypersurface of degree k, defined by some homogeneous polynomial p of degree k. Then

$$Z \cap U_i = \{(z_0 : \ldots : z_n) \in U_i : z_i^{-k} \cdot p(z_0, \ldots, z_n) = 0\},$$

and the line bundle $[Z]$ is given by the transition functions $g_{ij} = (z_j/z_i)^k$. In particular, for every hyperplane H we have the same line bundle $[H]$ with transition functions z_j/z_i.

Definition. If $H \subset \mathbb{P}^n$ is a hyperplane, then the line bundle $\mathcal{O}(1) := [H]$ is called the *hyperplane bundle.*

The kth tensor power of the hyperplane bundle is denoted by $\mathcal{O}(k)$.

If $Z \subset \mathbb{P}^n$ is a hypersurface of degree k, then $[Z] = \mathcal{O}(k)$.

A homogeneous polynomial F of degree k induces a global section $s_F \in \Gamma(\mathbb{P}^n, \mathcal{O}(k))$ by

$$(s_F)_i(z_0 : \ldots : z_n) := z_i^{-k} F(z_0, \ldots, z_n) \quad \text{for } z_i \neq 0.$$

In fact, $(s_F)_i$ is a holomorphic function on U_i, with

$$(s_F)_i = (s_F)_j \cdot \left(\frac{z_j}{z_i}\right)^k \text{ on } U_{ij}.$$

Obviously,

$$\{x \in \mathbb{P}^n \,:\, s_F(x) = 0\} = \{(z_0 : \ldots : z_n) \,:\, F(z_0, \ldots, z_n) = 0\}.$$

On the other hand, let s be an arbitrary global holomorphic section of $\mathcal{O}(k)$. Then s can be represented by holomorphic functions $s_i : U_i \to \mathbb{C}$ with

$$z_j^k \cdot s_j(z_0 : \ldots : z_n) = z_i^k \cdot s_i(z_0 : \ldots : z_n) \quad \text{on } U_i \cap U_j,$$

and we obtain a holomorphic function $f : \mathbb{C}^{n+1} - \{\mathbf{0}\} \to \mathbb{C}$ with

$$f(\mathbf{z}) = z_i^k \cdot s_i(\pi(\mathbf{z})) \quad \text{on } \pi^{-1}(U_i).$$

There is a holomorphic continuation F of f on $\mathbb{C}^{n+1}$ with $F(\lambda \mathbf{z}) = \lambda^k \cdot F(\mathbf{z})$. This means that F is a homogeneous polynomial of degree k, with $s_F = s$. Consequently, we have the following result.

5.15 Proposition. *For $k \in \mathbb{N}$ the vector space $\Gamma(\mathbb{P}^n, \mathcal{O}(k))$ is isomorphic to the space of homogeneous polynomials of degree k in the variables $z_0, \ldots, z_n$. Analytic hypersurfaces in $\mathbb{P}^n$ are exactly the zero sets of global holomorphic sections of $\mathcal{O}(k)$.*

For any $\mathbf{z}_0 \in \mathbb{C}^{n+1} - \{\mathbf{0}\}$ and any $k \in \mathbb{N}$ there is a homogeneous polynomial F of degree k with $F(\mathbf{z}_0) \neq 0$. Therefore $\mathcal{O}(k)$ is generated by global sections.

The bundle $\mathcal{O}(1)$ can be described in purely geometric terms. For this we note that if

$$p_0 := (0 : \ldots : 0 : 1) \in \mathbb{P}^{n+1},$$

then a projection $\pi_0 : \mathbb{P}^{n+1} - \{p_0\} \to \mathbb{P}^n$ is given by

$$\pi_0(z_0 : \ldots : z_n : z_{n+1}) := (z_0 : \ldots : z_n).$$

We obtain local trivializations $\varphi_i : \pi_0^{-1}(U_i) \to U_i \times \mathbb{C}$ by

$$\varphi_i(z_0 : \ldots : z_n : z_{n+1}) := \left((z_0 : \ldots : z_n), \frac{z_{n+1}}{z_i}\right),$$

with $\varphi_i^{-1}((z_0 : \ldots : z_n), c) = (z_0 : \ldots : z_n : cz_i)$. Obviously this gives the transition functions $z_j z_i^{-1}$. If F is a linear form on $\mathbb{C}^{n+1}$, then the section s_F is given at $x = \pi(\mathbf{z}) = (z_0 : \ldots : z_n)$ by

$$s_F(x) = \varphi_i^{-1}(x, z_i^{-1} F(\mathbf{z})) = (z_0 : \ldots : z_n : F(\mathbf{z})).$$

The Euler Sequence. Here we discuss the tangent bundle of the projective space in more detail. Let $\pi : \mathbb{C}^{n+1} - \{\mathbf{0}\} \to \mathbb{P}^n$ be the canonical projection. Then for any point $\mathbf{z} \in \mathbb{C}^{n+1} - \{\mathbf{0}\}$ we have a linear map $\varphi_{\mathbf{z}} : \mathbb{C}^{n+1} \to T_{\pi(\mathbf{z})}(\mathbb{P}^n)$ defined by

$$\varphi_{\mathbf{z}}(\mathbf{w})[f] := \frac{d}{dt}\Big|_0 (f \circ \pi(\mathbf{z} + t\mathbf{w})).$$

The vector $\mathbf{w}$ can be interpreted as tangent vector $\dot{\alpha}(0) = \alpha_{*,0}((d/dt)_0)$ with $\alpha(t) := \mathbf{z} + t\mathbf{w}$. It follows that

$$\pi_{*,\mathbf{z}}(\mathbf{w})[f] = (\pi \circ \alpha)_{*,0}(d/dt)_0[f] = (d/dt)_0[f \circ \pi \circ \alpha] = \varphi_{\mathbf{z}}(\mathbf{w})[f].$$

Since π is a submersion and $\varphi_{\mathbf{z}}(\mathbf{z}) = \mathbf{0}$, the linear map $\varphi_{\mathbf{z}}$ is surjective with $\operatorname{Ker}(\varphi_{\mathbf{z}}) = \mathbb{C}\mathbf{z}$.

If we use the local coordinates $t_\nu := z_\nu / z_i$ in $U_i = \{(z_0 : z_1 : \ldots : z_n) \mid z_i \neq 0\}$, then for any $\mathbf{z}$ with $z_i \neq 0$ it follows that

$$\varphi_{\mathbf{z}}(\mathbf{w}) = \sum_{\nu \neq i} \Big(\frac{z_i w_\nu - w_i z_\nu}{z_i^2} \Big) \frac{\partial}{\partial t_\nu}.$$

Thus a trivialization $\psi_i : T(\mathbb{P}^n)|_{U_i} \to U_i \times \mathbb{C}^n$ is given by

$$\psi_i(\varphi_{\mathbf{z}}(\mathbf{w})) := \Big(\pi(\mathbf{z}), \frac{1}{z_i}(a_0, \ldots, \widehat{a_i}, \ldots, a_n)\Big),$$

with $a_\nu = w_\nu - (w_i z_i^{-1}) z_\nu$ for $\nu \neq i$.

Let $\mathcal{O}(1)^{\oplus(n+1)}$ be the direct sum of $n+1$ copies of the hyperplane bundle. Recall that every linear form F on $\mathbb{C}^{n+1}$ defines a global section s_F of $\mathcal{O}(1)$ by $(s_F)_i = z_i^{-1} F$. Then two vector bundle homomorphisms $j : \mathcal{O}_{\mathbb{P}^n} = \mathbb{P}^n \times \mathbb{C} \to \mathcal{O}(1)^{\oplus(n+1)}$ and $q : \mathcal{O}(1)^{\oplus(n+1)} \to T(\mathbb{P}^n)$ can be defined by

$$j(x, c) := c \cdot (s_{z_0}(x), \ldots, s_{z_n}(x))$$

and

$$q(s_{F_0}(\pi(\mathbf{z})), \ldots, s_{F_n}(\pi(\mathbf{z}))) := \varphi_{\mathbf{z}}(F_0(\mathbf{z}), \ldots, F_n(\mathbf{z})),$$

where $F_0, \ldots, F_n$ are linear forms on $\mathbb{C}^{n+1}$.

Using the canonical trivializations, j and q can be given over U_i by

$$j : ((z_0 : \ldots : z_n), c) \mapsto ((z_0 : \ldots : z_n), c \cdot (z_0 z_i^{-1}, \ldots, z_n z_i^{-1}))$$

and

$$q : ((z_0 : \ldots : z_n), (w_0, \ldots, w_n)) \mapsto ((z_0 : \ldots : z_n), (a_0, \ldots, \widehat{a_i}, \ldots, a_n)),$$

with $a_\nu = w_\nu - (w_i z_i^{-1}) z_\nu$. This shows that both of these homomorphisms have constant rank. Since there is always an index i such that $s_{z_i}(x) \neq 0$, j is injective. Since $\mathcal{O}(1)$ is globally generated, q is well defined and surjective.

Definition. A sequence $0 \to V' \xrightarrow{j} V \xrightarrow{q} V'' \to 0$ of vector bundles is called *exact* if the following hold:

1. j and q have constant rank.
2. j is injective and q is surjective.
3. $\operatorname{Im}(j) = \operatorname{Ker}(q)$.

5.16 Theorem. *We have an exact sequence of vector bundles*

$$0 \longrightarrow \mathscr{O}_{\mathbb{P}^n} \xrightarrow{\ j\ } \mathscr{O}(1)^{\oplus(n+1)} \xrightarrow{\ q\ } T(\mathbb{P}^n) \longrightarrow 0\,.$$

The sequence is called the *Euler sequence.*

PROOF: It is clear that $q \circ j(\pi(\mathbf{z}), c) = \varphi_{\mathbf{z}}(c \cdot \mathbf{z}) = 0$. If, on the other hand, $x = \pi(\mathbf{z})$ and $q(s_{F_0}(x), \ldots, s_{F_n}(x)) = 0$, then $(F_0(\mathbf{z}), \ldots, F_n(\mathbf{z}))$ must be a multiple of $\mathbf{z}$. So there exists a $c \in \mathbb{C}$ such that $(s_{F_0}(x), \ldots, s_{F_n}(x)) = c \cdot (s_{z_0}(x), \ldots, s_{z_n}(x))$. ∎

Rational Functions. We consider some connections with complex algebraic geometry.

A meromorphic function m on $\mathbb{P}^n$ is called *rational* if $m = 0$, or if there are homogeneous polynomials F and G of the same degree such that $F \neq 0$ and

$$m(z_0 : \ldots : z_n) = \frac{F(z_0, \ldots, z_n)}{G(z_0, \ldots, z_n)}.$$

5.17 Theorem. *Every meromorphic function on $\mathbb{P}^n$ is rational.*

PROOF: If $m \in \mathscr{M}(\mathbb{P}^n)$, then its divisor has a finite representation

$$\operatorname{div}(m) = \sum_i n_i \cdot Z_i\,,$$

with $n_i \in \mathbb{Z}$ and irreducible hypersurfaces $Z_i \subset \mathbb{P}^n$. By the theorem of Chow every Z_i is the zero set of a homogeneous polynomial F_i of degree $d_i \geq 1$. Then $F := \prod_i F_i^{n_i}$ is a rational function on $\mathbb{C}^{n+1}$ that is homogeneous of degree $d := \sum_i n_i d_i \in \mathbb{Z}$.

Since $\operatorname{div}(m \circ \pi) = \operatorname{div}(F)$ on $\mathbb{C}^{n+1} - \{\mathbf{0}\}$, the function $f := (m \circ \pi) \cdot F^{-1}$ is holomorphic there without zeros. It has a holomorphic extension $\widehat{f}$ to $\mathbb{C}^{n+1}$ with $\widehat{f}(\mathbf{0}) \neq 0$, because $\widehat{f}$ cannot have isolated zeros.

For $c \in \mathbb{C}^*$, the equation $\widehat{f}(c \cdot \mathbf{z}) = c^{-d} \cdot \widehat{f}(\mathbf{z})$ is valid on an open subset of $\mathbb{C}^{n+1}$, and then (by the identity theorem) everywhere on $\mathbb{C}^{n+1}$. Thus $d \leq 0$

and $c^{-d} = 1$. From $\widehat{f}(c \cdot \mathbf{z}) = \widehat{f}(\mathbf{z})$ it follows that $\widehat{f}(\mathbf{z})$ is a constant w_0, and $m \circ \pi(\mathbf{z}) = w_0 \cdot F(\mathbf{z})$ a rational function. ■

A *rational function* on a submanifold $X \subset \mathbb{P}^n$ is the restriction $m|_X$ of a rational function m on $\mathbb{P}^n$. We have already seen that every polynomial p on $U_0 \cong \mathbb{C}^n$ can be extended to a rational function m on $\mathbb{P}^n$. Therefore, on every algebraic manifold there is a large number of rational functions.

If $A \subset \mathbb{P}^n$ is a projective algebraic set, then $A_i := A \cap U_i$ is called an *affine algebraic set.* A complex manifold is called an *affine algebraic manifold* if it is biholomorphically equivalent to an affine algebraic set. A *regular function* on an affine algebraic manifold $j : X \hookrightarrow U_i$ is a holomorphic function $f : X \to \mathbb{C}$ such that there exists a polynomial p on $U_i \cong \mathbb{C}^n$ with $f = p \circ j$. It can be shown that a rational function on an affine algebraic manifold is always a quotient of regular functions.

In algebraic geometry there is a more general definition for regular functions, which coincides with our notation in the case of affine algebraic manifolds. On projective algebraic manifolds all regular functions are constant, whereas there are many rational functions. In the affine case the field of rational functions is exactly the quotient field of the ring of regular functions.

If $Z \subset \mathbb{P}^n$ is a hypersurface of degree k, then $X := \mathbb{P}^n - Z$ is an affine algebraic manifold. We can see this as follows:

Let I be the set of multi-indices $\nu = (\nu_0, \ldots, \nu_n)$ with $\nu_0 + \cdots + \nu_n = k$. Then $\#(I) = \binom{n+k}{k}$ is the number of monomials $\mathbf{z}^\nu = z_0^{\nu_0} \cdots z_n^{\nu_n}$, $\nu \in I$. We set $N := \#(I) - 1$ and define the *Veronese map* $v_{k,n} : \mathbb{P}^n \to \mathbb{P}^N$ by

$$v_{k,n}(z_0 : \ldots : z_n) := (\mathbf{z}^\nu \mid \nu \in I).$$

One can show that $v_{k,n}$ is an embedding, and so its image is $V_{k,n} = v_{k,n}(\mathbb{P}^n)$ an algebraic submanifold of $\mathbb{P}^N$. If p is a homogeneous polynomial of degree k with zero set Z, then there are complex numbers a_ν, $\nu \in I$, such that $p = \sum_{\nu \in I} a_\nu \mathbf{z}^\nu$. It follows that

$$v_{k,n}(Z) = V_{k,n} \cap \Big\{ (w_\nu)_{\nu \in I} \, : \, \sum_{\nu \in I} a_\nu w_\nu = 0 \Big\}$$

is the intersection of $V_{k,n}$ with a hyperplane $H \subset \mathbb{P}^N$. Therefore, $\mathbb{P}^n - Z \cong v_{k,n}(\mathbb{P}^n - Z) = V_{k,n} \cap (\mathbb{P}^N - H)$ is affine algebraic.

Exercises

1. Show that $M := \{(z_0 : z_1 : z_2 : z_3) \in \mathbb{P}^3 \, : \, z_0^5 + \cdots + z_3^5 = 0\}$ is a projective manifold. Consider the group $G := \{g^m \, : \, 0 \le m \le 4\}$, where $g : \mathbb{P}^3 \to \mathbb{P}^3$ is defined by

$$g(z_0 : z_1 : z_2 : z_3) := (\varrho z_0 : \varrho^2 z_1 : \varrho^3 z_2 : \varrho^4 z_3), \quad \varrho = e^{2\pi i/5}.$$

Prove that $X := M/G$ is a complex manifold (called a *Godeaux surface*).

2. Let $\mathbf{F} : \mathbb{C}^3 \to \mathbb{C}^3$ be defined by

$$\mathbf{F}(t, z_1, z_2) := (t, \alpha z_1 + t z_2, \alpha z_2), \quad \alpha \in \mathbb{R},\ 0 < |\alpha| < 1.$$

Set $G = \{\mathbf{F}^m : m \in \mathbb{Z}\}$ and prove that $X := (\mathbb{C} \times (\mathbb{C}^2 - \{\mathbf{0}\}))/G$ is a compact complex manifold. Consider the holomorphic map $\pi : X \to \mathbb{C}$ that is induced by the projection onto the first component. Then $X_t := \pi^{-1}(t)$ is a 2-dimensional complex manifold. Show that X_t is biholomorphically equivalent to X_1 for every $t \neq 0$, but that there is no biholomorphic map between X_0 and X_1. Show that X_0 is a Hopf manifold.

3. For $\omega_1, \dots, \omega_{2n} \in \mathbb{C}^n$ define $\boldsymbol{\Omega} := \left(\omega_1^t, \dots, \omega_{2n}^t\right) \in M_{n,2n}(\mathbb{C})$ and prove that

$$\omega_1, \dots, \omega_{2n} \text{ linearly independent} \iff \det \begin{pmatrix} \boldsymbol{\Omega} \\ \overline{\boldsymbol{\Omega}} \end{pmatrix} \neq 0.$$

Let $T_{\boldsymbol{\Omega}}$ be the torus defined by the lattice $\Gamma := \mathbb{Z}\omega_1 + \cdots + \mathbb{Z}\omega_{2n}$. Prove that $T_{\boldsymbol{\Omega}}$ is biholomorphically equivalent to $T_{\boldsymbol{\Omega}'}$ if and only if there are matrices $\boldsymbol{G} \in \mathrm{GL}_n(\mathbb{C})$ and $\boldsymbol{M} \in \mathrm{GL}_{2n}(\mathbb{Z})$ such that $\boldsymbol{\Omega}' = \boldsymbol{G} \cdot \boldsymbol{\Omega} \cdot \boldsymbol{M}$.

4. The *Segre map* $\sigma_n : \mathbb{P}^1 \times \mathbb{P}^n \to \mathbb{P}^{2n+1}$ is defined by

$$\sigma_n((x_0 : x_1), (y_0 : \dots : y_n)) := (x_0 y_0 : \dots : x_0 y_n : x_1 y_0 : \dots : x_1 y_n).$$

Show that $\Sigma_n := \sigma_n(\mathbb{P}^1 \times \mathbb{P}^n)$ is a complex manifold and σ_n an injective holomorphic immersion.

5. Show that for $n \geq 2$, two irreducible hypersurfaces in $\mathbb{P}^n$ always have a nonempty intersection.

6. Decompose

$$A := \{(z_0 : z_1 : z_2) : z_0^4 + 3z_0^2 z_1^2 + 2z_0^4 + 4z_0^2 z_2^2 + 5z_1^2 z_2^2 + 3z_2^4 = 0\}$$

into irreducible components.

7. Let $d > 0$ and $n > 0$. Prove the following theorem of Bertini. If $H_{n+1}(d)$ is the vector space of homogeneous polynomials of degree d in $z_0, \dots, z_n$, then there is a dense open subset $U \subset H_{n+1}(d)$ such that

$$N(F) = \{(z_0 : \dots : z_n) \in \mathbb{P}^n : F(z_0, \dots, z_n) = 0\}$$

is a projective manifold for every $F \in U$.

8. Use the notation from the introduction to Grassmannian manifolds. For $I \in \mathscr{I}_{k,n}$ set $V_I := \{\mathbf{A} \in \mathrm{St}(k,n) : \det \mathbf{A}_I \neq 0\}$ and $U_I := \pi_{k,n}(V_I) \subset G_{k,n}$. Then

$$g_{IJ}(\pi_{k,n}(\mathbf{A})) := \mathbf{A}_I^{-1} \cdot \mathbf{A}_J \text{ on } V_I \cap V_J$$

give the transition functions of a vector bundle U_k on $G_{k,n}$. Show that there are n independent global holomorphic sections in U_k.

9. Show that $H^1(\mathbb{P}^n, \mathcal{O}) = 0$ for $n \geq 1$ and $H^1(\mathbb{P}^1, \mathcal{O}(k)) = 0$ for $k \geq 0$. Prove that there is an exact sequence of vector spaces

$$H^1(\mathbb{P}^n, \mathcal{O}(k-1)) \to H^1(\mathbb{P}^n, \mathcal{O}(k)) \to H^1(\mathbb{P}^{n-1}, \mathcal{O}(k))$$

for $n \geq 2$ and $k \geq 1$. Conclude that $H^1(\mathbb{P}^n, \mathcal{O}(k)) = 0$ for $n \geq 1$ and $k \geq 0$.
10. Let X be a compact complex manifold and $\pi : V \to X$ a holomorphic vector bundle of rank $r \geq 2$. Show that there is a holomorphic fiber bundle $P \to X$ with typical fiber $\mathbb{P}^{r-1}$ and a line bundle $L \to P$ such that $P_x = \mathbb{P}(V_x)$ and $L|_{P_x} \cong \mathcal{O}_{P_x}(1)$.
11. Prove that for every *automorphism* f of $\mathbb{P}^n$ (i.e., every biholomorphic map $f : \mathbb{P}^n \to \mathbb{P}^n$) there is an $\mathbf{A} \in \mathrm{GL}_{n+1}(\mathbb{C})$ with $f(\pi(\mathbf{z})) = \pi(\mathbf{z} \cdot \mathbf{A}^t)$. This is formulated as $\mathrm{Aut}(\mathbb{P}^n) = \mathrm{PGL}_{n+1}(\mathbb{C}) := \mathrm{GL}_{n+1}(\mathbb{C})/\mathbb{C}^*$.
12. Let $\Gamma = \mathbb{Z}\omega_1 + \mathbb{Z}\omega_2$ be a lattice in $\mathbb{C}$ and $\wp$ the Weierstrass function for this lattice. Show that the map $z \mapsto (1 : \wp(z) : \wp'(z))$ induces an embedding φ of the torus $T = \mathbb{C}/\Gamma$ into $\mathbb{P}^2$. Determine the equation of the image $\varphi(T)$ in $\mathbb{P}^2$. Determine transition functions for $\varphi^*\mathcal{O}(1)$.

6. Branched Riemann Domains

Branched Analytic Coverings. Let $f : X \to Y$ be a continuous map between Hausdorff spaces. Then the image of compact sets is compact, and the preimage of closed sets is closed.

> **Definition.** The map f is called *closed* if the image of closed subsets of X is closed in Y. It is called *proper* if the preimage of compact subsets of Y is compact.

6.1 Proposition.

1. *If $f : X \to Y$ is closed and $A \subset X$ a closed subset, then $f|_A : A \to Y$ is closed.*
2. *If f is closed, y a point of $f(X)$, and U an open neighborhood of the fiber $f^{-1}(y)$, then there is an open neighborhood $W = W(y) \subset Y$ such that $f^{-1}(W) \subset U$.*
3. *Let X and Y be additionally locally compact.*
   (a) *If every point $y \in Y$ has a neighborhood W such that $f^{-1}(W)$ is compact, then f is proper.*
   (b) *If f is proper, then f is closed.*
   (c) *If f is closed and every fiber $f^{-1}(y)$ compact, then f is proper.*

PROOF: (1) is trivial.

(2) Let f be closed and U be an open neighborhood of some nonempty fiber $f^{-1}(y)$. Then $W := Y - f(X - U)$ is an open subset of Y. If y were not in W, then there would be a point $x \in f^{-1}(y) - U$. This is impossible, and therefore W is an open neighborhood of y with $f^{-1}(W) \subset U$.

(3a) Let Y be locally compact and $K \subset Y$ compact. It follows from the criterion that there are finitely many sets $W_1, \ldots, W_m$ such that $f^{-1}(K)$ is contained in the compact set $f^{-1}(\overline{W_1}) \cup \cdots \cup f^{-1}(\overline{W_m})$. Since $f^{-1}(K)$ is closed, it must itself be compact.

(3b) Let f be proper and $A \subset X$ be a closed set. For every $y \in \overline{f(A)}$ there is a sequence of points $x_\nu \in A$ with $\lim_\nu f(x_\nu) = y$. The set $N := \{y\} \cup \{f(x_\nu) : \nu \in \mathbb{N}\}$ is compact (using the local compactness of Y), and consequently $f^{-1}(N)$ is compact. Thus there is a subsequence $(x_{\nu\mu})$ converging to some point $x \in f^{-1}(N) \cap A$. It follows that $y = f(x) \in f(A)$.

(3c) Let f be closed and assume that every fiber is compact. For $y \in Y$ choose a compact neighborhood U of $f^{-1}(y)$. This is possible, since X is locally compact. By (2) there exists an open neighborhood W of y in Y with $f^{-1}(W) \subset U$. Then $f^{-1}(\overline{W}) \subset U$ as well. As a closed subset of a compact set it is itself compact, and from (3a) it follows that f is proper. ∎

Definition. A continuous map $f : X \to Y$ between locally compact Hausdorff spaces is called *finite* if it is closed and if each fiber has only finitely many elements.

Obviously, every finite map is proper. Conversely, if a proper map has only discrete fibers, then it must be finite.

Definition. A holomorphic map $\pi : X \to Y$ between n-dimensional complex manifolds is called a *branched (analytic) covering* if the following hold:

1. The map π is open, finite, and surjective.
2. There is a closed subset $D \subset Y$ with the following properties:
   (a) For every $y \in Y$ there is an open neighborhood $U = U(y) \subset Y$ and a nowhere dense analytic subset $A \subset U$ with $D \cap U \subset A$.
   (b) $\pi : X - \pi^{-1}(D) \to Y - D$ is locally biholomorphic.

The set D is called the *critical locus*. A point $x \in X$ is called a *branch point* if π is not locally biholomorphic at x. The set B of branch points is called the *branch locus*. The covering is called *unbranched* if B is empty.

The branch locus B is nowhere dense in X.

Now we consider a domain $G \subset \mathbb{C}^n$ and d pseudopolynomials $\omega_i(w_i, \mathbf{z})$ over G and define

$$\widehat{A} := \{(\mathbf{w}, \mathbf{z}) \in \mathbb{C}^d \times G : \omega_i(w_i, \mathbf{z}) = 0 \text{ for } i = 1, \ldots, d\}.$$

Let $\pi : \widehat{A} \to G$ be the restriction of the projection $\mathrm{pr}_2 : \mathbb{C}^d \times G \to G$.

6.2 Proposition. *The map π is surjective, finite, and open.*

PROOF: (1) The surjectivity follows from the fact that, for fixed $\mathbf{z} \in G$, each $\omega_i(w_i, \mathbf{z})$ has at least one zero.

(2) Let $M \subset \widehat{A}$ be a closed subset and $\mathbf{z}_0$ a point of $\overline{\pi(M)}$. Then there is a sequence $(\mathbf{w}_\nu, \mathbf{z}_\nu)$ in M with $\mathbf{z}_\nu \to \mathbf{z}_0$. Since the coefficients of the ω_i are bounded near $\mathbf{z}_0$, the components of the zeros $\mathbf{w}_\nu$ are also bounded. Consequently, there is a subsequence $(\mathbf{w}_{\nu\mu})$ of $(\mathbf{w}_\nu)$ that converges in $\mathbb{C}^d$ to some $\mathbf{w}_0$. It is clear that $(\mathbf{w}_0, \mathbf{z}_0)$ lies in M. Thus $\mathbf{z}_0$ belongs to $\pi(M)$, and π is closed.

Since π has finite fibers, it is a finite (and in particular a proper) map.

(3) It remains to show that π is open. Let $(\mathbf{w}_0, \mathbf{z}_0) \in \widehat{A}$ be an arbitrary point. If $V(\mathbf{w}_0) \subset \mathbb{C}^d$ and $W(\mathbf{z}_0) \subset G$ are open neighborhoods, we have to find an open neighborhood $W'(\mathbf{z}_0) \subset \pi\big(\widehat{A} \cap (V \times W)\big)$.

By Hensel's lemma there are pseudopolynomials ω_i^* and ω_i^{**} such that

$$\omega_i = \omega_i^* \cdot \omega_i^{**} \text{ and } \omega_i^*(w_i, \mathbf{z}_0) = (w_i - w_i^{(0)})^{m_i} \text{ with } m_i \geq 1.$$

Define

$$A := \{(\mathbf{w}, \mathbf{z}) \in V \times W : \omega_1^*(w_1, \mathbf{z}) = \cdots = \omega_d^*(w_d, \mathbf{z}) = 0\}.$$

Then $A \subset \widehat{A} \cap (V \times W)$, and $(\pi|_A)^{-1}(\mathbf{z}_0) = \{(\mathbf{w}_0, \mathbf{z}_0)\}$. Since $\pi|_A$ is a closed map, there is a neighborhood $W'(\mathbf{z}_0) \subset W$ such that $A \cap (\mathbb{C}^d \times W') = (\pi|_A)^{-1}(W') \subset \widehat{A} \cap (V \times W)$. Since also $\pi|_A : A \to W$ is surjective, it follows that

$$W' = \pi(A \cap (\mathbb{C}^d \times W')) \subset \pi\big(\widehat{A} \cap (V \times W)\big).$$

Thus π is open. ∎

Remark. If $X \subset \widehat{A}$ is an irreducible component that is everywhere regular of dimension n, then the same proof shows that $\pi|_X$ is also surjective, finite, and open. If $D \subset G$ is the union discriminant set for $\widehat{A}$, then $\pi : X \to G$ is a branched analytic covering with critical locus D.

Branched Domains. A continuous map $f : X \to Y$ between Hausdorff spaces is called *discrete at* $x \in X$ if the fiber $f^{-1}(f(x))$ is a discrete subset of X. The map f is called *discrete* if it is discrete at every point $x \in X$.

A holomorphic map $f : X \to Y$ between complex manifolds is discrete at x_0 if and only if there is an open neighborhood $U = U(x_0) \subset\subset X$ with $f(x_0) \notin f(\partial U)$.

Now let M be an n-dimensional complex manifold.

Definition. A *branched domain* over M is a pair (X, π) with the following properties:

1. X is an n-dimensional connected complex manifold.
2. $\pi : X \to M$ is a discrete open holomorphic mapping.

Let $\pi : X \to M$ be a branched domain and $x_0 \in X$ an arbitrary point. We can choose a coordinate neighborhood B of $\pi(x_0)$ in M such that B is biholomorphically equivalent to a ball in $\mathbb{C}^n$ and $\pi(x_0) = \mathbf{0}$. Since π is discrete and open, there is a connected open neighborhood $U = U(x_0) \subset\subset X$ such that $\pi(U) \subset B$ is open and $\pi^{-1}(\mathbf{0}) \cap \overline{U} = \{x_0\}$. We assume that $\pi(U) = B$.

Define $j : U \to U \times B$ by $j(x) := (x, \pi(x))$. Then j is a holomorphic embedding, and

$$A := j(U) = \{(x, \mathbf{z}) \in U \times B \,:\, \mathbf{z} = \pi(x)\}$$

is a regular analytic set in $U \times B$. We have a factorization $\pi = \widetilde{\pi} \circ j$, with $\widetilde{\pi} := \mathrm{pr}_2|_A$. Let $w_1, \ldots, w_n$ be complex coordinates near x_0 in U. It follows from the results of Section III.6 that over a small neighborhood of $\mathbf{0}$ in B there are pseudopolynomials $\omega_i(w_i, \mathbf{z})$ such that A is an irreducible component of the embedded analytic set

$$\widehat{A} = \{(\mathbf{w}, \mathbf{z}) \,:\, \omega_i(w_i, \mathbf{z}) = 0 \text{ for } i = 1, \ldots, n\}.$$

If $D \subset B$ is the union of the discriminant sets of the ω_i, then $\widehat{A}$ (and therefore also A) is unbranched over the complement of D. Therefore, if $\pi : X \to M$ is a branched domain, then for every $x \in X$ there is a neighborhood $U(x) \subset X$ and a neighborhood $W(f(x)) \subset M$ with $\pi(U) = W$ such that $\pi|_U : U \to W$ is a branched analytic covering.

If $M = \mathbb{C}^n$ or $M = \mathbb{P}^n$, then a branched domain over M is also called a *branched Riemann domain*.

Definition. Two branched domains (X_1, π_1) (over M) and (X_2, π_2) (over N) are called *equivalent* if there are holomorphic maps $\varphi : M \to N$ and $\widehat{\varphi} : X_1 \to X_2$ such that the following diagram commutes:

$$\begin{array}{ccccc} & X_1 & \xrightarrow{\widehat{\varphi}} & X_2 & \\ \pi_1 & \downarrow & & \downarrow & \pi_2 \\ & M & \xrightarrow{\varphi} & N & \end{array}$$

Torsion Points. As an example we consider a polydisk $P = \mathsf{P}^n(\mathbf{0}, r) \subset \mathbb{C}^n$, a positive integer b, and the set

$$X_0 := \{(\mathbf{z}, w) \in P \times \mathbb{C} \,:\, w^b - z_1 = 0\},$$

together with the projection $\pi_0 : (\mathbf{z}, w) \mapsto \mathbf{z}$. Then X_0 is a regular analytic hypersurface in $P \times \mathbb{C}$ and consequently an n-dimensional complex manifold, which is the Cartesian product of a polydisk $P' \subset \mathbb{C}^{n-1}$ and a neighborhood of the origin in the Riemann surface of $w = \sqrt[b]{z}$. It is unbranched outside $\{\mathbf{z} : z_1 = 0\}$.

> **Definition.** Let (X, π) be a branched domain over some manifold M. A point $x_0 \in X$ is called a *torsion point* (or *winding point*) of order b if there is an open neighborhood $U = U(x_0) \subset X$ such that $\pi : U \to \pi(U)$ is equivalent to $\pi_0 : X_0 \to P$. In this case we say that $\pi|_U$ is a *winding covering.*

We consider the local case, i.e., a branched analytic covering $\pi : X \to Q$ over some polydisk $Q \subset \mathbb{C}^n$ around the origin, and we assume that it is branched of order b over $\mathbf{0}$ and unbranched outside $\{\mathbf{z} \in Q : z_1 = 0\}$. Then $X' := X - \pi^{-1}(\{z_1 = 0\})$ is still connected, and the covering $\pi : X \to Q$ is equivalent to $\pi_0 : X_0 \to P$. For a proof see [GrRe58], §2.5, Hilfssatz 2. One shows that the two coverings outside the branching locus have the same fundamental group. Therefore, they must be equivalent there. Using the theorem of Remmert–Stein it follows that X and X_0 are equivalent.

The following is now immediate.

6.3 Proposition. *Let (X, π) be a branched analytic covering over M, with branching locus D. If $x_0 \in X$ is a point such that $\mathbf{z}_0 = \pi(x_0)$ is a regular point of D, then there are neigborhoods $W = W(x_0) \subset X$ and $U = U(\mathbf{z}_0) \subset M$ such that $\pi : W \to U$ is a winding covering.*

The proof follows simply from the fact that one can find coordinates such that locally the situation above is at hand, and then X is locally equivalent to X_0.

Concrete Riemann Surfaces. A *concrete Riemann surface* is a branched Riemann domain $\pi : X \to \mathbb{C}$ or $\pi : X \to \mathbb{P}^1$. It is an abstract connected Riemann surface, and in the latter case it may be compact. The infinitely sheeted Riemann surface of $w = \log z$ is not a concrete Riemann surface in our sense.

Here is a method for constructing concrete Riemann surfaces. Let X be a Hausdorff space, $M = \mathbb{C}$ or $M = \mathbb{P}^1$, and $\pi : X \to M$ a continuous mapping such that for every $x_0 \in X$ there is an open neighborhood $U = U(x_0) \subset X$, a domain $V \subset \mathbb{C}$ (in the case $M = \mathbb{C}$), respectively a domain V in an affine part $\cong \mathbb{C}$ of M (in the case $M = \mathbb{P}^1$) and a topological map $\psi : V \to U$ with the following properties:

1. $\pi \circ \psi : V \to \mathbb{C}$ is holomorphic.
2. $(\pi \circ \psi)'$ does not vanish identically.

The map ψ is called a *local uniformization.*

Obviously, (U, ψ^{-1}) is a complex coordinate system. We have to show that two such systems (U_1, ψ_1^{-1}) and (U_2, ψ_2^{-1}) are holomorphically compatible. We consider only the case $M = \mathbb{C}$. The map

$$\psi := \psi_1^{-1} \circ \psi_2 : \psi_2^{-1}(U_1 \cap U_2) \to \psi_1^{-1}(U_1 \cap U_2)$$

is a homeomorphism. We denote the set $\psi_\lambda^{-1}(U_1 \cap U_2)$ by V_λ^*. Then $D_1 := \{t \in V_1^* : (\pi \circ \psi_1)'(t) = 0\}$ is discrete in V_1^*, and therefore $D_2 := \psi^{-1}(D_1) \subset V_2^*$ is also discrete. Let t_0 be a point of $V_2^* - D_2$. Then

$$(\pi \circ \psi_1)'(\psi(t_0)) \neq 0.$$

Therefore, there are open neighborhoods $U = U(\psi(t_0)) \subset V_1^*$ and $W = W(\mathbf{z}_0) \subset \mathbb{C}$ (for $\mathbf{z}_0 := \pi(\psi_2(t_0))$) such that $\pi \circ \psi_1 : U \to W$ is biholomorphic. Then $\pi|_{\psi_1(U)} : \psi_1(U) \to W$ is a homeomorphism, and

$$\psi = \psi_1^{-1} \circ (\pi|_{\psi_1(U)})^{-1} \circ \pi \circ \psi_2 = (\pi \circ \psi_1)^{-1} \circ (\pi \circ \psi_2), \text{ on } V := \psi^{-1}(U).$$

Since V is an open neighborhood of t_0, we see that ψ is holomorphic on $V_2^* - D_2$. As a continuous map it is bounded at the points of D_2, and therefore it must be holomorphic everywhere in V_2^*.

From the above it follows that X is a Riemann surface, and obviously π is a holomorphic map. We still want to see that it is a finite branched covering. If $x_0 \in X$, then there is an open neighborhood $U = U(x_0) \subset X$ and a local uniformization $\psi : V \to U$. We define $\Phi : U \to V \times \pi(U) \subset \mathbb{C}^2$ by

$$\Phi(x) := (\psi^{-1}(x), \pi(x)).$$

Then $\Phi(U) = \{(\mathbf{z}, \mathbf{w}) \in V \times \pi(U) : \pi \circ \psi(\mathbf{z}) = \mathbf{w}\}$. Restricting the projection $(\mathbf{z}, \mathbf{w}) \mapsto \mathbf{w}$, we obtain a branched covering $p : \Phi(U) \to \pi(U)$ that is equivalent to $\pi|_U : U \to \pi(U)$. Let be $f := \pi \circ \psi$ and $\mathbf{z}_0 := \psi^{-1}(x_0)$. If $f'(\mathbf{z}_0) \neq 0$, then p is unbranched. If f' has an isolated zero of order k at $\mathbf{z}_0$, then

$$f(\mathbf{z}) = \pi(x_0) + (\mathbf{z} - \mathbf{z}_0)^k \cdot h(\mathbf{z}),$$

where h is a holomorphic function with $h(\mathbf{z}_0) \neq 0$. It is clear that then p is branched of order k, equivalent to the Riemann surface of $\sqrt[k]{z}$.

Hyperelliptic Riemann Surfaces. For $g \geq 2$ choose $2g+2$ different points $z_1, z_2, \ldots, z_{2g+1}, z_{2g+2}$ in $X_0 := \mathbb{P}^1$. We assume that they are all real, in natural order, and not equal to ∞.

Now we take two copies X_1, X_2 of $\mathbb{P}^1$ regarded as lying above X_0 and cut them along the lines ℓ_i between z_{2i+1} and z_{2i+2}, $i = 0, \ldots, g$. We define

$$X_\lambda^- := \{t \in X_\lambda : \operatorname{Im}(t) \leq 0\} \quad \text{and} \quad X_\lambda^+ := \{t \in X_\lambda : \operatorname{Im}(t) > 0\}$$

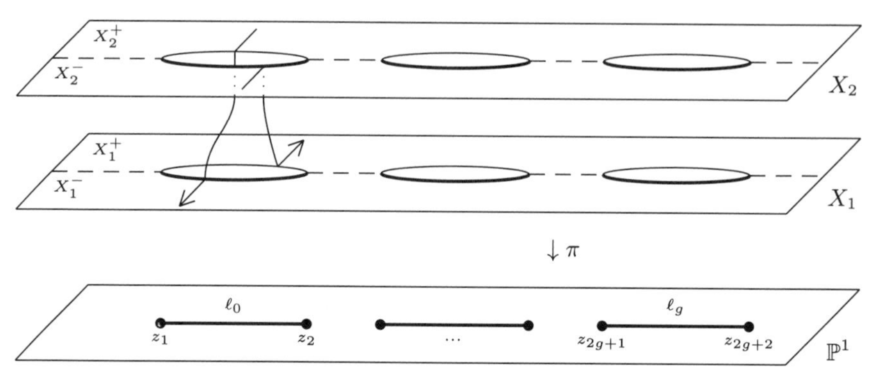

Figure IV.2. The hyperelliptic surface

and glue X_2^+ to X_1^- along the cuts ℓ_i, and similarly X_2^- to X_1^+.

We obtain a topological (Hausdorff) space X that lies as a branched covering of order 2 over $\mathbb{P}^1$ (see Figure IV.2). It is branched only at the points $z_1, \ldots, z_{2g+2}$, and at each of these points it looks exactly like the Riemann surface of $\sqrt{z}$ over the origin. Therefore, it is clear that X, together with the canonical projection $\pi : X \to \mathbb{P}^1$, is a concrete Riemann surface. It is called a *hyperelliptic surface.*[4]

A curve going in X_1^- from z_{2i+1} to z_{2i+2} and then in X_2^- back to z_{2i+1} defines a cycle C_i' in X, $i = 0, \ldots, g$. Similarly a curve in X_1 which goes from z_{2i+2} to z_{2i+3} and then in X_2 back to z_{2i+2} gives a cycle C_i'' in X, $i = 0, \ldots, g-1$. Finally we define C_g'' by going in X_1 from z_{2g+2} through ∞ to z_1 and then in X_2 back to z_{2g+2}. So $H_1(X, \mathbb{Z})$ is generated by the $2g+2$ cycles $C_0', \ldots, C_g', C_0'', \ldots, C_g''$. But $C' := C_0' + \cdots + C_g'$ is the boundary of X_2, and $C'' := C_0'' + \cdots + C_g''$ is the boundary of $X_1^+ \cup X_2^-$. Therefore, $H_1(X, \mathbb{Z}) = Z_1(X, \mathbb{Z})/B_1(X, \mathbb{Z}) \cong \mathbb{Z}^{2g}$ (and $H_0(X, \mathbb{Z}) = \mathbb{Z}$, because X is connected). It follows that $H^1(X, \mathbb{Z}) \cong \mathbb{Z}^{2g}$. The number

$$g = g(X) := \frac{1}{2} b_1(X) = \frac{1}{2} \operatorname{rk}(H^1(X, \mathbb{Z}))$$

is called the *genus* of X. Recalling that $\mathbb{P}^1$ is the Riemann sphere, we replace X_2 by the complex conjugate $\overline{X}_2$ and glue X_1 and $\overline{X}_2$ inserting tubes into the cuts ℓ_i. Thus we realize X as a sphere with g handles.

Now we give a more analytical description of the hyperelliptic surface. Define

$$X' := \Big\{(s,t) \in \mathbb{C}^2 \,:\, s^2 = \prod_{i=0}^{2g+2} (t - z_i)\Big\},$$

[4] In the case $g = 1$ we would get an *elliptic surface* that is isomorphic to a 1-dimensional complex torus.

and $\pi' : X' \to \mathbb{C}$ by $\pi(s,t) := t$. This is a branched covering with branching locus $D = \{z_1, \dots, z_{2g+2}\}$. For $F(t) := \prod_{i=0}^{2g+2}(t - z_i)$, the function $s = \sqrt{F(t)}$ is well defined on X'. Over $\{t \in \mathbb{C} : |t| > R\}$, X' consists of two disjoint punctured disks. Therefore, we can complete X' to a compact Riemann surface X by filling the points at infinity, i.e., the punctures. The map π' can be extended continuously to a map $\pi : X \to \mathbb{P}^1$. It follows by the Riemann extension theorem that π is holomorphic, and X is a branched covering of order 2 over $\mathbb{P}^1$ with branching locus D.

Remark. The Riemann surface X is **not** the completion of X' in $\mathbb{P}^2$. The projective analytic set $\overline{X'}$, which has $\{s^2 = F(t)\}$ as affine part, has a singularity at infinity.

The functions t and $s = \sqrt{F(t)}$ extend to meromorphic functions f and g on X that have poles of order 1, respectively $g + 1$, over ∞.

We now calculate $H^1(X, \mathcal{O})$. For this the following is an essential tool.

6.4 Serre's duality theorem. *If X is a compact Riemann surface and V an analytic vector bundle over X, then $H^1(X, V)$ and $H^0(X, K_X \otimes V')$ are finite-dimensional vector spaces of equal dimension.*

PROOF: See [Nar92], page 47. ∎

In the case $V = \mathcal{O}_X = X \times \mathbb{C}$ we have $H^1(X, \mathcal{O}) \cong H^0(X, K_X)$. So we need some remarks about the canonical bundle K_X.

1. Let $\mathcal{U} = (U_\iota)_{\iota \in I}$ be an open covering of X such that there are complex coordinates $\varphi_\iota : U_\iota \to \mathbb{C}$. We denote the complex coordinate on U_ι by t_ι. Then the line bundle K_X is given by the transition functions
$$g_{\iota\kappa} = \frac{dt_\kappa}{dt_\iota}.$$
A *holomorphic 1-form* on X is a global section $\omega \in \Gamma(X, T(X)') = H^0(X, K_X)$. We can write $\omega|_{U_\iota} = \omega_\iota dt_\iota$. Then $\omega_\iota dt_\iota = \omega_\kappa dt_\kappa$ on $U_{\iota\kappa}$, and therefore $\omega_\iota = g_{\iota\kappa} \cdot \omega_\kappa$.
2. On $\mathbb{P}^1$ we have two systems of complex coordinates, namely, $t_0 := z_1/z_0$ on $U_0 = \{(z_0 : z_1) : z_0 \neq 0\}$ and $t_1 := z_0/z_1$ on U_1. We denote t_0 by t and $-t_1$ by s. Then on U_{01} we have $s = -1/t$. Using s and t as complex coordinates, the canonical bundle on $\mathbb{P}^1$ is defined by
$$g_{01} = \frac{ds}{dt} = \frac{d}{dt}\left(-\frac{1}{t}\right) = \frac{1}{t^2} = \left(\frac{z_0}{z_1}\right)^2.$$
This is also the transition function of the bundle $[D]$ associated to the divisor $D = -2 \cdot \infty$. Therefore,
$$\begin{aligned} H^0(\mathbb{P}^1, K_{\mathbb{P}^1}) &\cong H^0(\mathbb{P}^1, [D]) \\ &= \{m \in \mathcal{M}(\mathbb{P}^1)^* : \operatorname{ord}_\infty(m) \geq 2 \text{ and } \operatorname{ord}_p(m) \geq 0 \text{ otherwise}\}. \end{aligned}$$

But such a function m is holomorphic and vanishes at ∞, and consequently is identically 0. Thus $H^0(\mathbb{P}^1, K_{\mathbb{P}^1}) = 0$.

3. Let X be a compact Riemann surface, and $\omega_1, \omega_2 \in H^0(X, K_X)$ two holomorphic 1-forms. In local coordinates $\omega_\lambda|_{U_\iota} = \omega_\iota^{(\lambda)} dt_\iota$, for $\lambda = 1, 2$. Then $f|_{U_\iota} := \omega_\iota^{(2)} / \omega_\iota^{(1)}$ defines a global meromorphic function f on X such that $\omega_2 = f \cdot \omega_1$.

Now we return to the hyperelliptic surface X. A holomorphic map $j : X \to X$ is defined by $(s,t) \mapsto (-s,t)$. Clearly, j permutes the two leaves, and $j^2 = \text{id}$. This map is called the *hyperelliptic involution*, and it induces a linear map $j^* : H^0(X, K_X) \to H^0(X, K_X)$ with $(j^*)^2 = \text{id}$. We get a decomposition of $H^0(X, K_X)$ into the eigenspaces associated with the eigenvalues ± 1. But if $j^*\omega = \omega$, then $\omega = \pi^*(\varphi)$, where φ is a holomorphic 1-form on $\mathbb{P}^1$. Since $H^0(\mathbb{P}^1, K_{\mathbb{P}^1}) = 0$, it follows that $\omega = 0$. Thus $j^*\omega = -\omega$ for every $\omega \in H^0(X, K_X)$.

The equation $s^2 = F(t)$ implies that $2s\, ds = F'(t)\, dt$. If $s(t) = 0$, then $t = z_k$ for some k. But $F'(z_k) = \prod_{i \neq k}(z_k - z_i) \neq 0$. Therefore,

$$\omega_0 := \frac{dt}{s} = \frac{2\, ds}{F'(t)}$$

is a holomorphic 1-form on X'. Near ∞ we have $s(t) \approx t^{g+1}$ and therefore $\omega_0 \approx (1/t)^{g-1} d(1/t)$. This shows that ω_0 is a holomorphic 1-form on X that has a zero of order $g-1$ over ∞. The same argument shows that the 1-forms $\omega_\nu := t^\nu \cdot (dt/s)$ are holomorphic on X, for $\nu = 1, \ldots, g-1$.

Now let ω be an arbitrary element of $H^0(X, K_X)$. Then there is a meromorphic function f on X with $\omega = f \cdot \omega_0$. Since ω_0 has no zero over $\mathbb{P}^1 - \{\infty\}$, f is holomorphic there. And since $j^*\omega = -\omega$ and $j^*\omega_0 = -\omega_0$, we have $j^*f = f$. This means that $f = h \circ \pi$ for some meromorphic function h on $\mathbb{P}^1$ that is holomorphic outside ∞. Consequently, h is a polynomial, say of degree d. It has d zeros in $\mathbb{C}$ and a pole of order d at ∞. It follows that f also has a pole of order d at every point above ∞. Since $\omega = f \cdot \omega_0$ is holomorphic, $d \leq g-1$. Therefore, ω is a linear combination of the 1-forms

$$\frac{dt}{s}, \quad t\,\frac{dt}{s}, \quad \ldots, \quad t^{g-1}\,\frac{dt}{s}.$$

It follows that $\dim_{\mathbb{C}} H^0(X, K_X) = g$.

6.5 Proposition. *If X is a hyperelliptic Riemann surface, then the genus $g = g(X)$ is equal to $\dim H^0(X, K_X)$.*

We will later see that the same result is true for arbitrary compact Riemann surfaces.

Exercises

1. Let $\pi : X \to M$ be a branched analytic covering with critical locus $D \subset M$. Show that if there is a point $y_0 \in D$ such that D has codimension 2 at y_0, then there is no branch point in $\pi^{-1}(y_0)$.
2. Let $\pi : X \to M$, and $\pi' : X' \to M$, be branched analytic coverings and D, respectively D', the critical locus. Prove that if $X - \pi^{-1}(D)$ is equivalent to $X' - (\pi')^{-1}(D')$, then X is equivalent to X'.
3. Consider the holomorphic map $\mathbf{F} : \mathbb{C}^2 \to \mathbb{C}^3$ defined by

$$(w, z_1, z_2) = \mathbf{F}(t_1, t_2) = (t_1 \cdot t_2, t_1^2, t_2^2).$$

It is the uniformization of a 2-dimensional analytic set $A \subset \mathbb{C}^3$. Show that $A - \{\mathbf{0}\}$ is a branched domain over $\mathbb{C}^2 - \{\mathbf{0}\}$. What are the torsion points? Describe A as the zero set of a pseudopolynomial $\omega(w, \mathbf{z})$. Is A also a branched domain in our sense?

4. Let F be a homogeneous polynomial of degree k and

$$X = \left\{(z_0 : \ldots : z_n : z_{n+1}) \in \mathbb{P}^{n+1} \,:\, z_{n+1}^k - F(z_0, \ldots, z_n) = 0\right\}.$$

Assume that $D := \{(z_0 : \ldots : z_n) \in \mathbb{P}^n \,:\, F(z_0, \ldots, z_n) = 0\}$ is regular of codimension 1 and prove that the canonical projection $\pi : X \to \mathbb{P}^n$ is a branched Riemann domain. Determine the torsion points and their orders.

5. Define $\varphi : \mathbb{P}^1 \to \mathbb{P}^n$ by $\varphi(z_0 : z_1) := (z_0^n : z_0^{n-1} z_1 : \ldots : z_1^n)$ and prove that the image $C := \varphi(\mathbb{P}^1)$ is a regular complex curve in $\mathbb{P}^n$. Define

$$X := \{(p, H) \in C \times G_{n,n+1} \,:\, p \in H\},$$

where an element $H \in G_{n,n+1}$ is identified with a projective hyperplane in $\mathbb{P}^n$. Show that the projection $\pi = \mathrm{pr}_2 : X \to G_{n,n+1}$ is an n-sheeted branched covering.

6. Let $X \subset \mathbb{P}^n$ be a d-dimensional connected complex submanifold. Prove that there is an $(n-d-1)$-dimensional complex projective plane $E \subset \mathbb{P}^n$ not intersecting X, a d-dimensional complex projective plane $F \subset \mathbb{P}^n$ with $E \cap F = \varnothing$, and a canonical projection $\pi : \mathbb{P}^n - E \to F$ such that $\pi|_X : X \to F$ is a branched domain.

7. Modifications and Toric Closures

Proper Modifications. Assume that X is an n-dimensional connected complex manifold and $A \subset X$ a compact submanifold. Is it possible to cut out A and replace it by another submanifold A' such that $X' = (X - A) \cup A'$ is again a complex manifold? In general, the answer is no, but sometimes such complex surgery is possible. Then we call the new manifold X' a modification of X.

Definition. Let $f : X \to Y$ be a proper surjective holomorphic map between two n-dimensional connected complex manifolds. The map f is called a *(proper) modification* of Y into X if there are nowhere dense analytic subsets $E \subset X$ and $S \subset Y$ such that the following hold:

1. $f(E) \subset S$.
2. f maps $X - E$ biholomorphically onto $Y - S$.
3. Every fiber $f^{-1}(y)$, $y \in S$, consists of more than one point.

The set S is called the *center* of the modification and $E = f^{-1}(S)$ the *exceptional set.*

Assume that we have a proper modification $f : X \to Y$ with center S and define

$$\widehat{X} := \{(y,x) \in Y \times X \,:\, y = f(x)\} \quad \text{and} \quad \pi := \mathrm{pr}_1|_{\widehat{X}} : \widehat{X} \to Y.$$

Then $\widehat{X}$ is an n-dimensional connected closed submanifold of $Y \times X$. If $K \subset Y$ is compact, then $\pi^{-1}(K) = (K \times X) \cap \widehat{X}$ is a closed subset of $K \times f^{-1}(K)$ and therefore compact; $\pi : \widehat{X} - \pi^{-1}(S) \to Y - S$ is biholomorphic, with inverse $y \mapsto (y, f^{-1}(y))$; and for $y_0 \in S$ we have $\pi^{-1}(y_0) = \{(y_0, x) \,:\, f(x) = y_0\} \cong f^{-1}(y_0)$. This is a set with more than one element.

Therefore, we can generalize the notion of a proper modification in the following way.

Let X and Y be two connected complex manifolds. A *generalized (proper) modification* of Y in X with center S is given by an irreducible analytic subset $\widehat{X}$ of the Cartesian product $Y \times X$ and a nowhere dense analytic subset $S \subset Y$ such that

1. $\pi := \mathrm{pr}_1|_{\widehat{X}} : \widehat{X} \to Y$ is proper.
2. $\pi^{-1}(S)$ is nowhere dense in $\widehat{X}$.
3. $\widehat{X} - \pi^{-1}(S)$ is a complex manifold.
4. π maps $\widehat{X} - \pi^{-1}(S)$ biholomorphically onto $Y - S$.
5. $\widehat{X}$ has more than one point over each point of S.

The center S is sometimes also called the *set of indeterminacy.*

If $\widehat{X}$ is a manifold, then $\pi : \widehat{X} \to X$ is an ordinary modification of X into $\widehat{X}$ with center S.

7.1 Proposition. *$\pi^{-1}(S)$ has codimension 1 in $\widehat{X}$.*

PROOF: We consider only the case where $\widehat{X}$ is regular. A proof of the general case can be found in [GrRe55].

Assume that $x_0 \in \pi^{-1}(S)$ is isolated in the fiber of π over $y_0 := \pi(x_0)$. Then there is a neighborhood U of x_0 in $Y \times X$ such that $U \cap \widehat{X}$ is a branched covering over Y. Since y_0 lies in S, there must be at least one additional point $x_1 \in \widehat{X}$ in the fiber over y_0. But then in any small neighborhood of y_0 there are points $y \in Y - S$ with several points in $\pi^{-1}(y)$. This is a contradiction. Therefore, $\dim_{x_0} \pi^{-1}(\pi(x_0)) > 0$. This is possible if and only if $\mathrm{rk}_{x_0}(\pi) < n := \dim(Y)$, and this shows that $\pi^{-1}(S)$ in local coordinates is given as the zero set of $\det J_\pi$. Therefore, it has codimension 1. ∎

Blowing Up. Let $U = U(\mathbf{0})$ be a small convex neighborhood around the origin in $\mathbb{C}^{n+1}$. We want to replace the origin in U by an n-dimensional complex projective space. If $\pi : \mathbb{C}^{n+1} - \{\mathbf{0}\} \to \mathbb{P}^n$ is the canonical projection, then every line $\mathbb{C}\mathbf{v}$ through the origin determines an element $x = \pi(\mathbf{v})$ in the projective space, and x determines the line $\ell(x) = \pi^{-1}(x) \cup \{\mathbf{0}\}$ such that $\mathbb{C}\mathbf{v} = \ell(\pi(\mathbf{v}))$. Now we insert $\mathbb{P}^n$ in such a way that we reach the point x by approaching the origin along $\ell(x)$.

We define

$$X := \{(\mathbf{w}, x) \in U \times \mathbb{P}^n \,:\, \mathbf{w} \in \ell(x)\}.$$

This is a so-called *incidence set* (another example is considered in Exercise 6.5). We first show that it is an $(n+1)$-dimensional complex manifold. In fact, we have

$$\begin{aligned}(\mathbf{w}, \pi(\mathbf{z})) \in X \quad &\iff \quad \mathbf{z} \neq \mathbf{0}, \text{ and } \exists\, \lambda \in \mathbb{C} \text{ with } \mathbf{w} = \lambda \mathbf{z} \\ &\iff \quad \exists\, i \text{ with } z_i \neq 0 \text{ and } w_j = \frac{w_i}{z_i} \cdot z_j \text{ for } j \neq i \\ &\iff \quad \mathbf{z} \neq \mathbf{0} \text{ and } z_i w_j - w_i z_j = 0 \text{ for all } i, j.\end{aligned}$$

So X is an analytic subset of $U \times \mathbb{P}^n$, with

$$X \cap (U \times U_0) \cong \{(\mathbf{w}, \mathbf{t}) \in U \times \mathbb{C}^n \,:\, w_j = w_0 t_j \text{ for } j = 1, \ldots, n\}.$$

In $U \times U_i$ (for $i > 0$) there is a similar representation. It follows that X is a submanifold of codimension n in the $(2n+1)$-dimensional manifold $U \times \mathbb{P}^n$. The map $q := \mathrm{pr}_1|_X : X \to U$ is holomorphic, mapping $X - q^{-1}(\mathbf{0})$ biholomorphically onto $U - \{\mathbf{0}\}$ by $q : (\mathbf{w}, x) \mapsto \mathbf{w}$ and $q^{-1} : \mathbf{w} \mapsto (\mathbf{w}, \pi(\mathbf{w}))$. Obviously, q is a proper map.

The preimage $q^{-1}(\mathbf{0})$ is the exceptional set $\{(\mathbf{0}, x) \,:\, \mathbf{0} \in \ell(x)\} = \{\mathbf{0}\} \times \mathbb{P}^n$. So $q : X \to U$ is a proper modification. It is called *Hopf's σ-process* or the *blowup* of U at the origin.

If $\mathbf{w} \neq \mathbf{0}$ is a point of U and (λ_n) a sequence of nonzero complex numbers converging to 0, then $q^{-1}(\lambda_n \mathbf{w}) = (\lambda_n \mathbf{w}, \pi(\mathbf{w}))$ converges to $(\mathbf{0}, \pi(\mathbf{w}))$. This is the desired property.

The Tautological Bundle. We consider the case $U = \mathbb{C}^{n+1}$, i.e., the manifold

$$F := \{(\mathbf{w}, x) \in \mathbb{C}^{n+1} \times \mathbb{P}^n \,:\, \mathbf{w} \in \ell(x)\}$$

and the **second** projection

$$p := \mathrm{pr}_2|_F : F \to \mathbb{P}^n.$$

Then $p^{-1}(x) = \ell(x)$ is always the complex line determined by x. The manifold F looks like a line bundle over the projective space. In fact, we have local trivializations $\varphi_i : p^{-1}(U_i) \to U_i \times \mathbb{C}$ defined by $\varphi_i(\mathbf{w}, x) := (x, w_i)$. Clearly, φ_i is holomorphic. In fact, it is biholomorphic with inverse map

$$\varphi_i^{-1} : (\pi(\mathbf{z}), c) \mapsto \left(\frac{c}{z_i} \cdot \mathbf{z}, \pi(\mathbf{z})\right).$$

So over U_{ij},

$$\varphi_i \circ \varphi_j^{-1}(\pi(\mathbf{z}), c) = \varphi_i\left(\frac{c}{z_j} \cdot \mathbf{z}, \pi(\mathbf{z})\right) = \left(\pi(\mathbf{z}), c \cdot \frac{z_i}{z_j}\right).$$

Hence F has the transition functions $g_{ij} = z_i/z_j$, and consequently, it is the dual bundle of the hyperplane bundle $\mathcal{O}(1)$. It is denoted by $\mathcal{O}(-1)$ and is called the *tautological bundle*, because the fiber over $x \in \mathbb{P}^n$ is the line $\ell(x)$, which is more or less the same as x. Sometimes F is also called the *Hopf bundle*, because it lies in $\mathbb{C}^{n+1} \times \mathbb{P}^n$ and the projection onto the first component is Hopf's σ-process (see Figure IV.3).

Let $j : F \to \mathbb{P}^n \times \mathbb{C}^{n+1}$ be defined by $j(\mathbf{w}, x) := (x, \mathbf{w})$, and $J_i : U_i \times \mathbb{C} \to U_i \times \mathbb{C}^{n+1}$ by

$$J_i(\pi(\mathbf{z}), c) := \left(\pi(\mathbf{z}), \frac{c}{z_i} \cdot \mathbf{z}\right).$$

Then we have the following commutative diagram:

$$\begin{array}{ccc} F|_{U_i} & \xrightarrow{\varphi_i} & U_i \times \mathbb{C} \\ j \downarrow & & \downarrow J_i \\ U_i \times \mathbb{C}^{n+1} & \cong & U_i \times \mathbb{C}^{n+1} \end{array}$$

This shows that F is a subbundle of the trivial vector bundle $\mathbb{P}^n \times \mathbb{C}^{n+1}$. In particular, it follows that $\Gamma(\mathbb{P}^n, \mathcal{O}(-1)) = 0$; i.e., there is no global holomorphic section in the tautological bundle.

There is an interesting geometrical connection between F and F'. Consider the point $x_0 := (0 : \ldots : 0 : 1) \in \mathbb{P}^{n+1}$ and the hyperplane

$$H_0 := \{\pi(\mathbf{z}) \in \mathbb{P}^{n+1} \ : \ z_{n+1} = 0\}.$$

Then the hyperplane bundle $F' = \mathcal{O}(1)$ is given by the projection $\pi_+ : \mathbb{P}^{n+1} - \{x_0\} \to H_0 \cong \mathbb{P}^n$, with

$$\pi_+(z_0 : \ldots : z_n : z_{n+1}) := (z_0 : \ldots : z_n : 0).$$

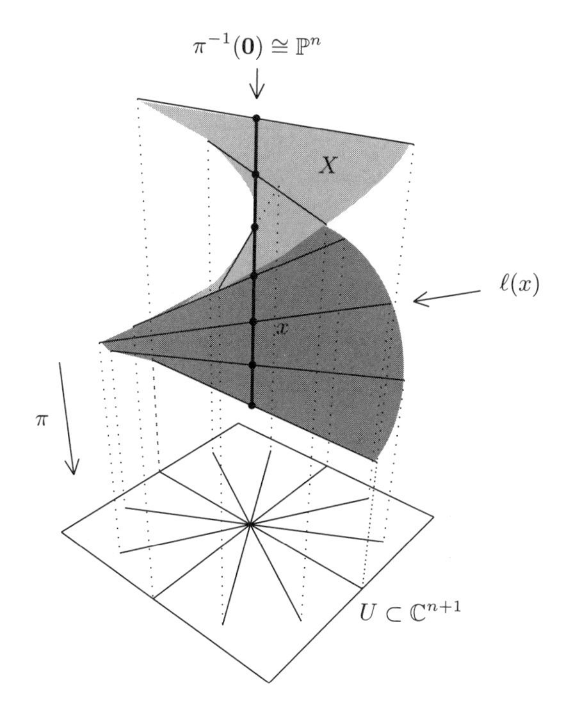

Figure IV.3. Hopf's σ-process

Here H_0 is the zero section of $\mathcal{O}(1)$. Removing the zero section gives a manifold isomorphic to $\mathbb{C}^{n+1} - \{\mathbf{0}\}$, and the fibers of the hyperplane bundle correspond to the lines through the origin. Blowing up the point x_0 gives us an additional point to each fiber and so a projective bundle $\overline{F'}$ over $\mathbb{P}^n$ with fiber $\mathbb{P}^1$. But looking in the other direction, $\overline{F'} - H_0$ is nothing but the tautological bundle $\mathcal{O}(-1)$ over the exceptional set of the blowing up.

Quadratic Transformations. We consider the case $n = 1$. Let M be a 2-dimensional connected complex manifold and $p \in M$ a point. Let $U = U(p) \subset M$ be a small neigborhood with complex coordinates z, w such that $(z(p), w(p)) = (0, 0)$. Let $X \subset U \times \mathbb{P}^1$ be the blowup of U at the origin. Then

$$Q_p(M) := (M - U) \cup X = (M - \{p\}) \cup \mathbb{P}^1$$

is again a 2-dimensional complex manifold, with a nonsingular compact analytic subset $N \cong \mathbb{P}^1$ and a proper holomorphic map $q_p : Q_p(M) \to M$ such that $q_p^{-1}(p) = N$ and $q_p : Q_p(M) - N \to M - \{p\}$ is biholomorphic. We call $Q_p(M)$ the *quadratic transformation* of M at p.

Let $F : M_1 \to M_2$ be a biholomorphic map between 2-dimensional complex manifolds and $F(p_1) = p_2$. Let $q_1 : Q_{p_1}(M_1) \to M_1$ and $q_2 : Q_{p_2}(M_2) \to M_2$

be the quadratic transformations. Then there exists a biholomorphic map $\widehat{F} : Q_{p_1}(M_1) \to Q_{p_2}(M_2)$ such that $q_2 \circ \widehat{F} = F \circ q_1$. This follows directly from the construction, and it shows that the quadratic transformation is a canonical process.

In [Ho55] Hopf proved that every proper modification with a single point as center is a finite sequence of quadratic transformations.[5]

We consider the quadratic transformation $q : Q_{\mathbf{0}}(U) \to U$, where

$$U = \{(z, w) \in \mathbb{C}^2 \, : \, |z| < r_1 \text{ and } |w| < r_2\}.$$

Then $Q_{\mathbf{0}}(U) = \{(z, w, (\zeta : \eta)) \in U \times \mathbb{P}^1 \, : \, z\eta - w\zeta = 0\} = U_0 \cup U_1$, with

$$\begin{aligned} U_0 &\cong \{(\eta, z) \in \mathbb{C}^2 \, : \, |z\eta| < r_2 \text{ and } |z| < r_1\}, \\ U_1 &\cong \{(\zeta, w) \in \mathbb{C}^2 \, : \, |w\zeta| < r_1 \text{ and } |w| < r_2\} \end{aligned}$$

and

$$w = z\eta \quad \text{and} \quad \zeta = \frac{1}{\eta} \quad \text{on } U_0 \cap U_1.$$

Thus q is given by

$$(\eta, z) \mapsto (z, z\eta) \quad \text{and} \quad (\zeta, w) \mapsto (w\zeta, w).$$

Consider, for example, the curve $C := \{(z, w) \in U \, : \, w^2 = z^3\}$, which has a singularity at the origin. Then

$$\begin{aligned} q^{-1}(C - \{\mathbf{0}\}) \cap U_0 &= \{(\eta, z) \in U_0 \, : \, z \neq 0 \text{ and } z - \eta^2 = 0\}, \\ q^{-1}(C - \{\mathbf{0}\}) \cap U_1 &= \{(\zeta, w) \in U_1 \, : \, w \neq 0 \text{ and } 1 - w\zeta^3 = 0\}. \end{aligned}$$

In $U_1 - U_0$ we have $\zeta = 0$. So $q^{-1}(C - \{\mathbf{0}\})$ lies completely in U_0, and its closure in $Q_{\mathbf{0}}(U)$ is the singularity-free curve $C' := \{(\eta, z) \, : \, z = \eta^2\}$. We call it the *strict transform* of C. The map $q : C' \to C$ is called a *resolution of the singularities* of C. One can show that for any curve in a 2-dimensional manifold the resolution of the singularities is obtained by successive quadratic transformations. A proof can be found, for example, in [Lau71].

If a 2-dimensional manifold M contains a compact submanifold $Y \cong \mathbb{P}^1$, one can ask whether Y is the exceptional set of a blowup process (then one says that Y can be "blown down" to a point). Here we give a necessary condition.

We determine the transition functions for the normal bundle of the fiber $F_0 := q^{-1}(\mathbf{0})$ in $Q_{\mathbf{0}}(U)$. The divisor F_0 is the zero set of the functions $f_0 : U_0 \to \mathbb{C}$ with $f_0(\eta, z) := z$ and $f_1 : U_1 \to \mathbb{C}$ with $f_1(\zeta, w) := w$. Therefore, the associated line bundle $[F_0]$ on $Q_{\mathbf{0}}$ is given by the transition function

[5] In algebraic geometry this was already known at the beginning of the twentieth century, for example by the italiens G. Castelnuovo and F. Enriques.

$$g_{01}(z, w, (\zeta : \eta)) = \frac{z}{w} = \frac{\zeta}{\eta} \quad \text{on } U_0 \cap U_1.$$

But ζ, η are the homogeneous coordinates on $F_0 \cong \mathbb{P}^1$. So $N_{Q_0}(F_0) = [F_0]|_{F_0}$ is described by the same transition function as the tautological bundle $\mathcal{O}(-1)$. It follows that a submanifold $Y \cong \mathbb{P}^1$ in a complex surface M can be exceptional only if $N_M(Y) \cong \mathcal{O}(-1)$.

We give a topological interpretation of this criterion. We set

$$s_0 := 1 \text{ on } U_0 \quad \text{and} \quad s_1 := \frac{z_1}{z_0} \text{ on } U_1,$$

which defines a **meromorphic** section s in $\mathcal{O}(-1)$ that has no zero and only one pole at $\infty = (0 : 1)$. We can use this section to calculate the self-intersection number $Y \cdot Y$. Since more knowledge about topology is needed for the complete calculation, we can give only a sketch.

The *self-intersection number* of a compact submanifold $Y \subset M$ can be described as follows. Take a copy Y' of Y (which need be only a topological or piecewise differentiable submanifold of M) such that it intersects Y transversally. Then the intersection $Y \cap Y'$ is a zero-dimensional submanifold, i.e., a finite set. Now count the number of these intersection points respecting orientation. The result is the self-intersection number. Furthermore, one can show that if $Z \subset N_M(Y)$ is the zero section, then $Y \cdot Y = Z \cdot Z$, and the number $Z \cdot Z$ can be calculated by intersecting Z with some not identically vanishing (continuous) section Z'. We obtain such a Z' in the following way. Let s be the meromorphic section in $N = N_M(Y)$ mentioned above. Choosing appropriate coordinates near ∞, we can there write $N = \{(z, w) \in \mathbb{C}^2 : |z| < 1\}$ and $s(z) = (z, 1/z)$ (since s has a pole of order one). For $0 < \varepsilon < 1$ and $|z| = \varepsilon$ we have $s(z) = (z, \overline{z}/\varepsilon^2)$. Then we define $\widehat{s}$ by

$$\widehat{s}(z) := \begin{cases} s(z) & \text{for } |z| > \varepsilon, \\ \left(z, \dfrac{\overline{z}}{\varepsilon^2}\right) & \text{for } |z| \le \varepsilon. \end{cases}$$

So $\widehat{s}$ is a piecewise differentiable section, homologous and transversal to the zero section with one zero of order 1. Since $\overline{z}$ gives an orientation opposite to that defined by z, we have the self-intersection $Y \cdot Y = -1$.

If $Y \cong \mathbb{P}^1$ is exceptional, then $Y \cdot Y = -1$. It is a deeper result that the converse of this statement is also true (cf. [Gr62]).

Monoidal Transformations. We consider a domain $G \subset \mathbb{C}^n$ and holomorphic functions $f_0, \dots, f_k$ on G such that $A := N(f_0, \dots, f_k) \subset G$ is a singularity-free analytic subset of codimension $k + 1$. Then we define $\mathbf{f} := (f_0, \dots, f_k) : G \to \mathbb{C}^{k+1}$ and

$$\begin{aligned} X &:= \{(\mathbf{w}, x) \in G \times \mathbb{P}^k \,:\, \mathbf{f}(\mathbf{w}) \in \ell(x)\} \\ &= \{(\mathbf{w}, \pi(\mathbf{z})) \in G \times \mathbb{P}^k \,:\, z_i f_j(\mathbf{w}) - z_j f_i(\mathbf{w}) = 0 \text{ for } i, j = 0, \ldots, k\}. \end{aligned}$$

The map $p := \mathrm{pr}_1|_X : X \to G$ is holomorphic and proper. For $\mathbf{w}_0 \in G - A$ we have $p^{-1}(\mathbf{w}_0) = \{\mathbf{w}_0\} \times \mathbb{P}^k$, whereas $p : X - p^{-1}(A) \to G - A$ is biholomorphic with $p^{-1} : \mathbf{w} \mapsto (\mathbf{w}, (f_0(\mathbf{w}) : \ldots : f_k(\mathbf{w})))$.

The set X is called the *monoidal transformation* of G with center A. The exceptional set $E(\mathbf{f}) := p^{-1}(A)$ is the Cartesian product $A \times \mathbb{P}^k$. In the case $k = n - 1$ and $f_i = z_i$ for $i = 0, \ldots, k$ this is the σ-process.

One can show that the monoidal transformation is also a canonical process. Therefore, it can be generalized to an n-dimensional connected complex manifold M and a submanifold $A \subset M$ of codimension $k + 1$. Then the monoidal transformation of M with center A is an n-dimensional manifold $\widetilde{M}$ with a hypersurface $E \subset \widetilde{M}$ such that $\widetilde{M} - E \cong M - A$. The exceptional set E is a fiber bundle over A with fiber $\mathbb{P}^k$. The fiber over $x \in A$ is the projective space $\mathbb{P}(N_x)$, where $N = N_M(A)$ is the normal bundle of A in X. The divisor E determines a line bundle $L := [E]$ on $\widetilde{M}$ with $L|_{\mathbb{P}(N_x)} = \mathcal{O}_{\mathbb{P}(N_x)}(-1)$. We leave the details to the reader (see also Exercise 5.10).

Meromorphic Maps. Let X be an n-dimensional complex manifold and f a meromorphic function on X with polar set P and $S \subset P$ its set of indeterminacy points. Then f is holomorphic on $X - P$, and we can extend it to a holomorphic map $f : X - S \to \mathbb{P}^1$ by setting $f(x) := \infty$ for $x \in P - S$. The set

$$G_f := \{(x, y) \in (X - S) \times \mathbb{P}^1 \,:\, y = f(x)\}$$

is called the *graph* of f. Let $\overline{G}_f \subset X \times \mathbb{P}^1$ be the closure of the graph in $X \times \mathbb{P}^1$.

7.2 Proposition. *The graph $\overline{G}_f$ is an irreducible analytic subset of $X \times \mathbb{P}^1$ and defines a proper modification of X in $\mathbb{P}^1$ with S as set of indeterminacy.*

PROOF: Let $x_0 \in P$ be an arbitrary point. Then there is a neighborhood $U = U(x_0) \subset X$ and holomorphic functions $g, h : U \to \mathbb{C}$ such that $f(x) = g(x)/h(x)$ for $x \in U - P$. We can assume that g_x and h_x are relatively prime for $x \in U$. Then $P \cap U = \{x \in U \,:\, h(x) = 0\}$ and $S \cap U = \{x \in U \,:\, g(x) = h(x) = 0\}$. Hence P is a hypersurface and S has codimension 2 in X. Then $\dim(S \times \mathbb{P}^1) < n$ and $G_f \subset (X \times \mathbb{P}^1) - (S \times \mathbb{P}^1)$ is an n-dimensional analytic set. By the theorem of Remmert and Stein its closure $\overline{G}_f$ in $X \times \mathbb{P}^1$ is also analytic. In fact, we have

$$\overline{G}_f \cap (U \times \mathbb{P}^1) = \{(x, (z_0 : z_1)) \in U \times \mathbb{P}^1 \,:\, z_0 g(x) - z_1 h(x) = 0\}.$$

The map $\pi := \mathrm{pr}_1 : \overline{G}_f \to X$ is obviously a proper holomorphic map, with $\pi^{-1}(x) = \{x\} \times \mathbb{P}^1$ for $x \in S$ and

$$\pi^{-1}(x) = \begin{cases} (x, (1 : f(x))) & \text{for } x \in X - P, \\ (x, (0 : 1)) & \text{for } x \in P - S. \end{cases}$$

So $\pi : \overline{G}_f - \pi^{-1}(S) \to X - S$ is biholomorphic, and $\overline{G}_f$ defines a proper modification of X in $\mathbb{P}^1$ with indeterminacy set S. As usual, X is assumed to be connected. Then $\operatorname{Reg}(\overline{G}_f)$ is also connected, and $\overline{G}_f$ irreducible. ∎

Now we consider the notion of a meromorphic map. To develop the correct point of view, we start with a very special case. Let

$$\mathbb{O}^m := \underbrace{\mathbb{P}^1 \times \cdots \times \mathbb{P}^1}_{m \text{ times}}$$

be the so-called *Osgood space*. If $f_1, \ldots, f_m$ are meromorphic functions on X without points of indeterminacy, then they define a holomorphic map $\mathbf{f} = (f_1, \ldots, f_m) : X \to \mathbb{O}^m$. If f_i has a set S_i as set of indeterminacy, we define $S := S_1 \cup \cdots \cup S_m$. Then $\mathbf{f}$ is a holomorphic map from $X - S$ into $\mathbb{O}^m$. Again we have its graph:

$$G_{\mathbf{f}} = \{(x; y_1, \ldots, y_m) \in (X - S) \times \mathbb{O}^m \,:\, y_i = f_i(x) \text{ for } i = 1, \ldots, m\}.$$

For $x_0 \in S$ we define

$$\tau_{\mathbf{f}}(x_0) := \{\mathbf{y} \in \mathbb{O}^m \,:\, \exists x_\nu \in X - S \text{ with } x_\nu \to x_0 \text{ and } \mathbf{f}(x_\nu) \to \mathbf{y}\}.$$

Then $\tau_{\mathbf{f}}(x_0)$ is a nonempty compact subset of $\mathbb{O}^m$. Setting $\tau_{\mathbf{f}}(x) := \{\mathbf{f}(x)\}$ for $x \in X - S$, we get a map $\tau_{\mathbf{f}}$ from X into the power set $P(\mathbb{O}^m)$.[6] We define

$$\overline{G}_{\mathbf{f}} := \{(x, \mathbf{y}) \in X \times \mathbb{O}^m \,:\, \mathbf{y} \in \tau_{\mathbf{f}}(x)\}.$$

One sees easily that $\overline{G}_{\mathbf{f}}$ is the topological closure of $G_{\mathbf{f}}$ in $X \times \mathbb{O}^m$. We cannot apply the Remmert–Stein theorem to show that $\overline{G}_{\mathbf{f}}$ is an analytic set, but locally it is an irreducible component of the analytic set A with

$$\begin{aligned} A \cap (U \times \mathbb{O}^m) \;=\; & \{(x; (z_0^1 : z_1^1), \ldots, (z_0^m : z_1^m)) \in U \times \mathbb{O}^m \,: \\ & z_0^1 g_1(x) - z_1^1 h_1(x) = \cdots = z_0^m g_m(x) - z_1^m h_m(x) = 0\}, \end{aligned}$$

where $f_i = g_i/h_i$ on $U - P_i$, for $i = 1, \ldots, m$. We could carry out the simple proof here, but we refer to [Re57], Satz 33.

Definition. A *meromorphic map* between complex manifolds X and Y is given by a nowhere dense analytic subset $S \subset X$, a holomorphic map $f : X - S \to Y$, and an irreducible analytic set $\widehat{X} \subset X \times Y$ such that

$$f \circ \mathrm{pr}_1 = \mathrm{pr}_2 \quad \text{on } \widehat{X} \cap ((X - S) \times Y)$$

and $\pi := \mathrm{pr}_1|_{\widehat{X}} : \widehat{X} \to X$ is proper and surjective. The set S is called the *set of indeterminacy* of f. We also write $f : X \rightharpoonup Y$.

[6] As usual, the *power set* $P(M)$ of a set M is the set of all subsets of M.

We define $\tau_f : X \to P(Y)$ by $\tau_f(x) := \text{pr}_2(\pi^{-1}(x))$. Then $\tau_f(x)$ is always a nonempty compact set, and

$$\widehat{X} = \{(x, y) \in X \times Y \,:\, y \in \tau_f(x)\}.$$

For $x \in X - S$ we have $\pi^{-1}(x) = \{(x, f(x))\}$, and for $x \in S$ we have $\pi^{-1}(x) = \widehat{X} \cap (\{x\} \times Y)$. This is a compact analytic set. If S is minimal, then $\pi : \widehat{X} \to X$ is a proper modification with center S.

So alternatively we could have defined a meromorphic map between X and Y to be a map $\tau : X \to P(Y)$ such that:

1. $\widehat{X} := \{(x, y) \in X \times Y \,:\, y \in \tau(x)\}$ is an irreducible analytic subset of $X \times Y$.
2. $\pi := \text{pr}_1|_{\widehat{X}} : \widehat{X} \to X$ is a proper modification.

This definition is due to Remmert.

A meromorphic map is called *surjective* if $\text{pr}_2(\widehat{X}) = Y$. If f is a surjective meromorphic map between X and Y and g a meromorphic map between Y and Z, then it is possible to define the composition $g \circ f$ as a meromorphic map between X and Z (cf. [Ku60]). But it is not a composition of maps in the set-theoretic sense!

Any meromorphic function f on X defines a meromorphic map between X and $\mathbb{P}^1$, any m-tuple $(f_1, \dots, f_m)$ of meromorphic functions a meromorphic map between X and $\mathbb{O}^m$.

7.3 Proposition. *Let $f : X \to Y$ be a proper modification of Y with center S. Then $f^{-1} : Y \rightharpoonup X$ is a surjective meromorphic map with S as set of indeterminacy.*

PROOF: We use the analytic set

$$\widehat{Y} := \{(y, x) \in Y \times X \,:\, y = f(x)\} = \{(y, x) \in Y \times X \,:\, x \in f^{-1}(y)\}.$$

Then $f^{-1} : Y - S \to X$ is holomorphic with $f^{-1} \circ \text{pr}_1(y, x) = f^{-1}(y) = x = \text{pr}_2(y, x)$ for $(y, x) \in \widehat{Y} \cap ((Y - S) \times X)$. Obviously, $\pi := \text{pr}_1|_{\widehat{Y}} : \widehat{Y} \to Y$ is a proper holomorphic map. It is surjective, since f is surjective. Finally, we have $\text{pr}_2(\widehat{Y}) = X$. So f^{-1} defines a surjective meromorphic map. ∎

Toric Closures. Let X be an n-dimensional complex manifold. A *closure* of X is an n-dimensional compact complex manifold M such that X is biholomorphically equivalent to an open subset of M. We are interested in closures of $X = \mathbb{C}^n$.

In the case $n = 1$ the situation is simple. Assume that we have a compact Riemann surface M that is a closure of $X = \mathbb{C}$. Then we define $f : M - \{0\} \to \mathbb{C}$ by

$$f(z) := \begin{cases} 1/z & \text{for } z \in \mathbb{C} - \{0\}, \\ 0 & \text{otherwise.} \end{cases}$$

The function f is continuous and is holomorphic at every point z with $f(z) \neq 0$. By the theorem of Radó–Behnke–Stein–Cartan (cf. [Hei56]) f is everywhere holomorphic. But then the zero set $N(f)$ is discrete. We can consider f as a meromorphic function on M with exactly one pole of order one at $z = 0$. So f can have only one zero (cf. [Fo81], Section I.4, Proposition 4.24 and corollary), and M must be the Riemann sphere $\mathbb{P}^1$, the well-known one-point compactification of $\mathbb{C}$.

In higher dimensions one can find several closures of $\mathbb{C}^n$. In [Bie33] an injective holomorphic map $\beta : \mathbb{C}^2 \to \mathbb{C}^2$ is constructed whose functional determinant equals 1 everywhere and whose image $U = \beta(\mathbb{C}^2)$ has the property that there exist interior points in $\mathbb{C}^2 - U$. We can regard U as an open subset of $\mathbb{P}^2$. Then $\mathbb{P}^2$ is a closure of $\mathbb{C}^2 \cong U$, but this closure contains interior points in its boundary. We want to avoid such "pathological" situations, even if they may be interesting for those working in dynamical systems.

Definition. A *closure* of $\mathbb{C}^n$ is a triple (X, U, Φ) where:

1. X is an n-dimensional connected compact complex manifold.
2. $U \subset X$ is an open subset with $\overline{U} = X$.
3. $\Phi : \mathbb{C}^n \to U$ is a biholomorphic map.

The closure (X, U, Φ) is called *regular* if for any polynomial p the holomorphic function $p \circ \Phi^{-1}$ extends to a meromorphic function on X.

7.4 Proposition. *If (X, U, Φ) is a regular closure of $\mathbb{C}^n$, then $X_\infty := X - U$ is an analytic hypersurface in X.*

PROOF: Let f_i be the meromorphic extension of $z_i \circ \Phi^{-1}$, for $i = 1, \dots, n$. The polar set S_i of f_i is a hypersurface in X. We set $P := P_1 \cup \cdots \cup P_n$ and claim that $P = X - U$.

One inclusion being trivial, we consider a point $x_0 \in X - U$. There is a sequence of points $x_\nu \in U$ with $\lim_{\nu \to \infty} x_\nu = x_0$. If we write $\mathbf{z}_\nu := \Phi^{-1}(x_\nu)$ as $\mathbf{z}_\nu = (z_1^{(\nu)}, \dots, z_n^{(\nu)})$, there must be an index k with $|z_k^{(\nu)}| \to \infty$. So $|f_k(x_\nu)| \to \infty$ for $\nu \to \infty$. This means that $x_0 \in P_k \subset P$. ∎

Example

The manifold $X = \mathbb{P}^n$ is a closure of $\mathbb{C}^n$. As open subset U with $\overline{U} = X$ we can take the set $U := U_0 = \{(z_0 : \ldots : z_n) \mid z_0 \neq 0\}$. Then $\Phi_0 : \mathbb{C}^n \to U$ with $\Phi_0(t_1, \dots, t_n) := (1 : t_1 : \ldots : t_n)$ is biholomorphic. We have seen in Section 5 that any polynomial on U extends to a meromorphic function f on $\mathbb{P}^n$ with polar set $H_0 = \{(z_0 : \ldots : z_n) \mid z_0 = 0\}$. So $(\mathbb{P}^n, U_0, \Phi_0)$ is a regular closure of $\mathbb{C}^n$, with $X_\infty = H_0$.

Another example is the Osgood space $\mathbb{O}^n$, which contains $\mathbb{C}^n$ via

$$\Phi : (t_1, \ldots, t_n) \mapsto ((1 : t_1), \ldots, (1 : t_n)).$$

Also, $(\mathbb{O}^n, \mathbb{C}^n, \Phi)$ is a regular closure, but here

$$X_\infty = (\{\infty\} \times \mathbb{P}^1 \times \cdots \times \mathbb{P}^1) \cup \cdots \cup (\mathbb{P}^1 \times \cdots \times \mathbb{P}^1 \times \{\infty\})$$

is an analytic hypersurface with singularities. The complex Lie group $\mathbb{G}_n := \mathbb{C}^* \times \cdots \times \mathbb{C}^*$ (n times) acts in a natural way on $\mathbb{C}^n$ and is contained in $\mathbb{C}^n$ as an open orbit of this action.

> **Definition.** A complex manifold X is called a *toric variety* or *torus embedding* if $\mathbb{G}_n$ acts holomorphically on X and there is an open dense subset $U \subset X$ biholomorphically equivalent to $\mathbb{G}_n$ by $g \mapsto gx$ for each $x \in U$.

Remark. In algebraic geometry the group $\mathbb{G}_n$ is called a *torus*. This is the reason for the notion "toric variety." A "variety" is an irreducible algebraic subset of $\mathbb{C}^n$ or $\mathbb{P}^n$. It may have singularities. Here we do not allow singularities, but we allow arbitrary manifolds that do not have to be algebraic.

More information about toric varieties can be found in [Oda88], [Fu93], and [Ew96].

We wish to restrict the closures of $\mathbb{C}^n$ as much as possible. If the axes of $\mathbb{C}^n$ can be extended to the closure, then the action of $\mathbb{G}_n$ should also extend to the closure.

> **Definition.** A closure (X, U, Φ) of $\mathbb{C}^n$ is called a *toric closure* if the action of $\mathbb{G}_n$ on $\mathbb{C}^n$ extends to an action on X such that $\Phi(g \cdot \mathbf{z}) = g \cdot \Phi(\mathbf{z})$ for $g \in \mathbb{G}_n$ and $\mathbf{z} \in \mathbb{C}^n$.

Examples

1. The projective space $\mathbb{P}^n$ is a toric variety with

$$g \cdot (z_0 : \ldots : z_n) := (z_0 : gz_1 : \ldots : gz_n)$$

   and $\mathbb{G}_n = (\mathbb{C}^*)^n \subset \mathbb{C}^n = U_0 \subset \mathbb{P}^n$. Since the embedding $\mathbb{C}^n = U_0 \hookrightarrow \mathbb{P}^n$ is "equivariant," $\mathbb{P}^n$ is a toric closure of $\mathbb{C}^n$.
2. A further simple example is the Osgood space $\mathbb{O}^n$. Here the toric action is given by

$$g \cdot \big((z_0^1 : z_1^1), \ldots, (z_0^n : z_1^n)\big) := \big((z_0^1 : gz_1^1), \ldots, (z_0^n : gz_1^n)\big).$$

3. Let $A_d \subset \mathbb{N}_0^n$ be the set of multi-indices ν with $|\nu| \le d$ and define $\Phi_d : \mathbb{G}_n \to \mathbb{P}^N$ by $\Phi_d(\mathbf{z}) := (\mathbf{z}^\nu : \nu \in A_d)$. Then $X_d := \overline{\Phi_d(\mathbb{G}_n)} \subset \mathbb{P}^N$ is the image of $\mathbb{P}^n$ under the Veronese map $v_{d,n}$ with

$$v_{d,n}(z_0 : z_1 : \ldots : z_n) = \left(z_0^{d-|\nu|}\mathbf{z}^\nu : \nu \in A_d\right).$$

Obviously, X_d is a toric closure of $\mathbb{C}^n$.

7.5 Proposition. *Let (X, U, Φ) be a regular closure of $\mathbb{C}^n$ and $f : Z \to X$ a proper modification with center $S \subset X_\infty$. Then $(Z, f^{-1}(U), f^{-1} \circ \Phi)$ is again a regular closure.*

PROOF: It is clear that $(Z, f^{-1}(U), f^{-1} \circ \Phi)$ is a closure. Now let p be a polynomial. Then the holomorphic function $p \circ \Phi^{-1} : U \to \mathbb{C}$ has a meromorphic extension $\widehat{p}$ on X. But then $\widehat{p} \circ f$ is a meromorphic extension of $p \circ (f^{-1} \circ \Phi)^{-1} = (p \circ \Phi^{-1}) \circ f$. ∎

Now let (X, U, Φ) be an arbitrary regular toric closure of $\mathbb{C}^n$. The meromorphic extensions f_i of $z_i \circ \Phi^{-1}$ define a meromorphic mapping $\mathbf{f} = (f_1, \ldots, f_n) : X \rightharpoonup \mathbb{O}^n$ and therefore a generalized modification $\widehat{X}$ of X in $\mathbb{O}^n$. We know that $\widehat{X}$ is an irreducible component of the analytic set

$$A = \left\{(x; z_1, \ldots, z_n) \in X \times \mathbb{O}^n : z_0^i f_1^i(x) - z_1^i f_0^i(x) = 0 \text{ for } i = 1, \ldots, n\right\},$$

where we write $z_i \in \mathbb{P}^1$ in the form $z_i = \left(z_0^i : z_1^i\right)$ and f_i in the form $f_i = \left(f_0^i : f_1^i\right)$.

The two projections $\pi := \mathrm{pr}_1|_{\widehat{X}} : \widehat{X} \to X$ and $q := \mathrm{pr}_2|_{\widehat{X}} : \widehat{X} \to \mathrm{O}^n$ are proper holomorphic maps. It may be that $\widehat{X}$ has singularities over the indeterminacy set S of $\mathbf{f}$. Nevertheless, we consider $\widehat{X}$ as a closure of $\mathbb{C}^n$. We take $\widehat{U} := \pi^{-1}(U)$ as dense open set, and the biholomorphic map $\widehat{\Phi} := \pi^{-1} \circ \Phi : \mathbb{C}^n \to \widehat{U}$. This is possible, since $\mathbf{f}|_U = \Phi^{-1}$ is holomorphic and therefore $S \subset X - U$.

The preimage $\pi^{-1}(X - U) = \widehat{X} - \widehat{U}$ is a nowhere dense analytic subset, and the meromorphic map $\mathbf{f} \circ \pi : \widehat{X} \rightharpoonup \mathbb{O}^n$ coincides on $\widehat{U}$ with the holomorphic map q. By the identity theorem it follows that $\mathbf{f} \circ \pi = q$ is holomorphic everywhere on $\widehat{X}$. Since $(\mathbf{f} \circ \pi)|_{\widehat{U}} = \Phi^{-1} \circ \pi = \widehat{\Phi}^{-1}$, every polynomial extends to a meromorphic function (without points of indeterminacy) on $\widehat{X}$. So the new closure $\widehat{X}$ is regular. Since $q|_{\widehat{U}} = \widehat{\Phi}^{-1}$ maps $\widehat{U}$ biholomorphically onto $\mathbb{C}^n$, the map $q : \widehat{X} \to \mathbb{O}^n$ is a proper modification of $\mathbb{O}^n$ (in a generalized sense, since $\widehat{X}$ may have singularities).

Let us now consider the q-fibers. If $\mathbf{z}_0 \in \mathbb{O}^n$ is given, the fiber $q^{-1}(\mathbf{z}_0)$ is contained in the set

$$F := (\mathbf{f}|_U)^{-1}(\mathbf{z}_0) \cup \{x \in X - U \,:\, \exists \mathbf{z}_\nu \in \mathbb{C}^n \text{ with } \Phi(\mathbf{z}_\nu) \to x,\, \mathbf{z}_\nu \to \mathbf{z}_0\}.$$

For $\mathbf{z}_0 \in \mathbb{C}^n$ we obtain $q^{-1}(\mathbf{z}_0) = \{\Phi(\mathbf{z}_0)\}$, whereas for $\mathbf{z}_0 \in \mathbb{O}^n - \mathbb{C}^n$ the fiber is a compact subset of the hypersurface $X - U$.

It is easy to check that $\widehat{X}$ is even a toric closure. One uses the fact that X and $\mathbb{O}^n$ both are toric and that $\mathbf{f}|_U$ is $\mathbb{G}_n$-equivariant.

Example

Let $X = \mathbb{P}^n$ be projective space and carry out the above construction. The functions $f_i : \mathbb{P}^n \rightharpoonup \mathbb{P}^1$ are given by

$$f_i : (z_0 : z_1 : \ldots : z_n) \mapsto (z_0 : z_i).$$

These are meromorphic functions with polar set H_0 and indeterminacy set $H_0 \cap H_i$. In the inhomogeneous coordinates $w_0, \ldots, \widehat{w_j}, \ldots, w_n$ with $w_k = z_k/z_j$ the function f_i is equal to w_i/w_0 for $i \neq j$ (respectively to $1/w_0$ for $i = j$).

Now $\widehat{X}$ is given by

$$\widehat{X} = \big\{(w; z_1, \ldots, z_n) \in \mathbb{P}^n \times \mathbb{O}^n \,:\, z_0^i w_i - z_1^i w_0 = 0 \text{ for } i = 1, \ldots, n\big\}.$$

This is built from $\mathbb{P}^n$ by a finite sequence of quadratic transformations, and therefore regular. The space $\widehat{X}$ is a regular toric closure of $\mathbb{C}^n$ in the original sense.

There are many toric closures, but $\mathbb{P}^n$ and $\mathbb{O}^n$ are the most important ones. More information about this topic is given in [BrMo78] and [PS88].

Exercises

1. Calculate the strict transform of $X = \{(z, w) \,:\, w^2 = z^2(z+1)\}$ when blowing up $\mathbb{C}^2$ at the origin.
2. Show that $K_{\mathbb{P}^n} \cong \mathcal{O}(-n-1)$ and $K_Y \cong \mathcal{O}(d-n-1)$ for every regular hypersurface $Y \subset \mathbb{P}^n$ of degree d.
3. Let Σ_1 be the projective bundle associated to the vector bundle $V = \mathcal{O}_{\mathbb{P}^1} \oplus \mathcal{O}_{\mathbb{P}^1}(1)$ (cf. Exercise 5.10). Show that the bundle space of Σ_1 is biholomorphically equivalent to the blowup $Q_p(\mathbb{P}^2)$ with $p = (1:0:0)$.
4. Consider the meromorphic function $f(z, w) := w/z$ on $\mathbb{C}^2$ and show that its graph $\overline{G_f} \subset \mathbb{C}^2 \times \mathbb{P}^1$ is equal to the blowup of $\mathbb{C}^2$ at the origin.
5. Let X be a complex manifold and L a holomorphic line bundle over X. Show that any set of linearly independent global sections $s_0, \ldots, s_N \in \Gamma(X, L)$ defines a meromorphic map $\Phi : X \to \mathbb{P}^N$ by

   $$\Phi(x) := (s_0(x) : \ldots : s_N(x)).$$

   What is the exact meaning of this definition?

6. Let X be a compact connected complex manifold. A set of meromorphic functions $f_1, \dots, f_m \in \mathscr{M}(X)$ is called *analytically dependent* if $\mathbf{f}(\overline{G_\mathbf{f}}) \neq \mathbb{O}^m$, where $\mathbf{f} = (f_1, \dots, f_m)$ is the second projection $\overline{G_\mathbf{f}} \subset X \times \mathbb{O}^m \to \mathbb{O}^m$. Let P be the union of the polar sets of the f_i and show that the following properties are equivalent:
   (a) $f_1, \dots, f_m$ are analytically dependent.
   (b) The Jacobian of $\mathbf{f} : X - P \to \mathbb{C}^m$ has rank less than m.
   (c) There is a nonempty open subset $U \subset X - P$, a domain $G \subset \mathbb{C}^m$ with $\mathbf{f}(U) \subset G$, and a holomorphic function g on G such that $g(f_1(x), \dots, f_m(x)) \equiv 0$.

   The functions $f_1, \dots, f_m$ are called *algebraically dependent* if there is a polynomial $p \neq 0$ such that $p(f_1(x), \dots, f_m(x)) \equiv 0$ on $X - P$. Prove that every set of analytically dependent functions is also algebraically dependent. (Hint: The map $\overline{G_\mathbf{f}} \to \mathbb{O}^m$ is proper. Use without proof Remmert's proper mapping theorem, which says that the image of an analytic set under a proper holomorphic mapping is also an analytic set. Then use Chow's theorem.)
7. Let $\Phi : \mathbb{G}_2 \to \mathbb{P}^3$ be defined by $\Phi(z_1, z_2) := (1 : z_1 : z_2 : z_1 z_2)$. Show that $X = \overline{\Phi(\mathbb{G}_2)} \subset \mathbb{P}^3$ is a closed submanifold that coincides with the image of the *Segre map* $\sigma_{1,1} : \mathbb{P}^1 \times \mathbb{P}^1 \to \mathbb{P}^3$ with $\sigma_{1,1}((z_0 : z_1), (w_0 : w_1)) := (z_0 w_0 : z_1 w_0 : z_0 w_1 : z_1 w_1)$. Prove that X is a toric closure of $\mathbb{C}^2$.

Chapter V

Stein Theory

1. Stein Manifolds

Introduction

Definition. A complex manifold X is called *holomorphically spreadable* if for any point $x_0 \in X$ there are holomorphic functions $f_1, \ldots, f_N$ on X such that x_0 is isolated in the set

$$N(f_1, \ldots, f_N) = \{x \in X \,:\, f_1(x) = \cdots = f_N(x) = 0\}.$$

It is clear that the holomorphic map $\mathbf{f} := (f_1, \ldots, f_N) : X \to \mathbb{C}^N$ is discrete at x_0. In [Gr55] it is shown that if $n = \dim(X)$, then there is a discrete holomorphic map $\pi : X \to \mathbb{C}^n$. Thus (X, π) is a branched domain over $\mathbb{C}^n$. Furthermore, it follows that the topology of X has a countable base. Note that if X is *holomorphically separable*, i.e., for any $x, y \in X$ with $x \neq y$ there exists a holomorphic function f on X with $f(x) \neq f(y)$, then it is holomorphically spreadable.

If X is holomorphically spreadable, $A \subset X$ a compact analytic set, and $x_0 \in A$, then there are holomorphic functions $f_1, \ldots, f_N$ on X (which then must be constant on A) such that x_0 is isolated in $N(f_1, \ldots, f_N)$. So x_0 must be an isolated point of A, and it follows that every compact analytic subset of X is finite!

Definition. A complex manifold X is called *holomorphically convex* if for any compact set $K \subset X$ the *holomorphically convex hull*

$$\widehat{K} := \Big\{ x \in X \,:\, |f(x)| \le \sup_X |f| \text{ for every } f \in \mathcal{O}(X) \Big\}$$

is likewise a compact subset of X.

Every compact complex manifold is holomorphically convex. A complex manifold with a countable base is holomorphically convex if and only if for any infinite discrete subset $D \subset X$ there exists a holomorphic function f on X such that $\sup_D |f| = \infty$. The proof is the same as in $\mathbb{C}^n$.

Definition. A *Stein manifold* is a connected complex manifold that is holomorphically spreadable and holomorphically convex.

A Stein manifold of dimension $n > 0$ cannot be compact. In the converse direction, Behnke and Stein proved in 1948 that every noncompact Riemann surface is a Stein manifold (see [BeSt48]). For $n \geq 2$, a domain $G \subset \mathbb{C}^n$ is Stein if and only if it is holomorphically convex. By the Cartan–Thullen theorem this is the case exactly if G is a domain of holomorphy. This remains true for (unbranched) Riemann domains over $\mathbb{C}^n$, since Cartan–Thullen is also valid for such domains. Hence for $n \geq 2$ there are many noncompact domains in $\mathbb{C}^n$ and over $\mathbb{C}^n$ that are not Stein.

1.1 Proposition. *Let $f : X \to Y$ be a finite holomorphic map between complex manifolds. If Y is a Stein manifold, then X is also Stein.*

In particular, every closed submanifold of a Stein manifold is Stein.

PROOF: It is necessary to show only that X is holomorphically convex. For this let $K \subset X$ be a compact set and note that $f(K)$ is also compact. We consider an arbitrary point $x \in \widehat{K}$. For $g \in \mathcal{O}(Y)$ we have $g \circ f \in \mathcal{O}(X)$ and $|g \circ f(x)| \leq \sup_K |g \circ f|$, so $f(x) \in \widehat{f(K)}$ and $x \in f^{-1}(\widehat{f(K)})$. Since Y is holomorphically convex and f proper, $f^{-1}(\widehat{f(K)}$ is compact as well. As a closed set in a compact set, $\widehat{K}$ is compact.

Since $\mathbb{C}^n$ is Stein, every closed submanifold of $\mathbb{C}^n$ is also Stein. ∎

Example

Every affine-algebraic manifold is isomorphic to a closed submanifold of $\mathbb{C}^n$. Therefore, it is a Stein manifold. On the other hand, we shall see later on that there exists an algebraic surface that is Stein, but not affine-algebraic.

1.2 Proposition. *If X is a Stein manifold and $f \in \mathcal{O}(X)$, then $X - N(f)$ is Stein.*

PROOF: If X is a Stein manifold, then it is immediate that $X - N(f)$ is holomorphically spreadable. Let D be an infinite discrete set in $X - N(f)$. If it is discrete in X, nothing remains to be proved. If it has a cluster point $x_0 \in N(f)$, then $g := 1/f$ is holomorphic on $X - N(f)$ and unbounded on D. ∎

Fundamental Theorems. The following important theorems cannot be proved completely within the constraints of this book. One needs deeper tools, including sheaf theory.

1.3 Theorem A. *Let $\pi : V \to X$ be an analytic vector bundle over a Stein manifold X. Then for any $x_0 \in X$ there are global holomorphic sections $s_1, \ldots, s_N \in \Gamma(X, V)$ with the following property:*

If $U = U(x_0) \subset X$ is an open neighborhood and $s \in \Gamma(U, V)$, then there exists an open neighborhood $V = V(x_0) \subset U$ and holomorphic functions $f_1, \ldots, f_N$ on V such that

$$s|_V = f_1 \cdot s_1 + \cdots + f_N \cdot s_N.$$

1.4 Theorem B. *Let $\pi : V \to X$ be an analytic vector bundle over a Stein manifold X. Then $H^1(X, V) = 0$.*

Moreover, if $A \subset X$ is an analytic set, $\mathscr{U}$ an open covering of X and $\xi \in Z^1(\mathscr{U}, V)$ such that $\xi_{\nu\mu}|_{U_{\nu\mu} \cap A} = 0$, then we can find a cochain $\eta \in C^0(\mathscr{U}, V)$ with $\eta_\nu|_{U_\nu \cap A} = 0$ and $\delta\eta = \xi$.

The first part of this theorem will be proved at the end of this chapter.

1.5 Oka's principle. *Let X be a Stein manifold.*

1. *Every topological fiber bundle over X has an analytic structure.*
2. *If two analytic fiber bundles over X are topologically equivalent, they are also analytically equivalent.*

Cousin-I Distributions.

A *Mittag-Leffler distribution* on $\mathbb{C}$ consists of a discrete set $\{z_\nu : \nu \in \mathbb{N}\}$ in $\mathbb{C}$ together with principal parts h_ν of Laurent series at each z_ν. A solution of this distribution is a meromorphic function on $\mathbb{C}$ with the given principal parts and no other pole. We can express the situation in the language of Čech cohomology. Define $U_0 := \mathbb{C} - \{z_\nu : \nu \in \mathbb{N}\}$ and $h_0 := 0$, and let U_ν be an open neighborhood of z_ν that contains no z_μ with $\mu \neq \nu$. Then $h_\nu|_{U_\nu}$ is meromorphic, and $h_\nu - h_\mu$ is holomorphic on $U_{\nu\mu}$. A solution is a meromorphic function f such that $f - h_\nu$ is holomorphic on U_ν.

> **Definition.** Let X be a complex manifold. A *Cousin-I distribution* on X consists of an open covering $\mathscr{U} = \{U_\iota : \iota \in I\}$ together with meromorphic functions f_ι on U_ι such that $f_\iota - f_\kappa$ is holomorphic on $U_{\iota\kappa}$.
>
> A *solution* of the Cousin-I distribution is a meromorphic function f on X such that $f - f_\iota$ is holomorphic on U_ι.

If a Cousin-I distribution is given, then a cocycle $\xi \in Z^1(\mathscr{U}, \mathcal{O})$ is defined by $\xi_{\iota\kappa} := (f_\kappa - f_\iota)|_{U_{\iota\kappa}}$. If the distribution has a solution f, then $\eta_\iota := (f_\iota - f)|_{U_\iota}$ defines a cochain $\eta \in C^0(\mathscr{U}, \mathcal{O})$ with $\eta_\kappa - \eta_\iota = \xi_{\iota\kappa}$. In other words, $\delta\eta = \xi$. Conversely, if there exists an η with $\delta\eta = \xi$, then $f|_{U_\iota} := f_\iota - \eta_\iota$ defines a meromorphic function f on X with $f_\iota - f = \eta_\iota$ on U_ι. This means that f is a solution.

The following result is a consequence of Theorem B for the trivial bundle $\mathcal{O}_X = X \times \mathbb{C}$.

1.6 Proposition. *On a Stein manifold every Cousin-I distribution has a solution.*

In particular, on an open Riemann surface every Mittag-Leffler distribution has a solution.

Remark. The condition "Stein" is not necessary. If X is a complex manifold with $H^1(X, \mathcal{O}) = 0$ (for example $X = \mathbb{P}^1$), then every Cousin-I distribution has a solution. And even if $H^1(X, \mathcal{O}) \neq 0$, then the cohomology class of a given distribution may be zero. This implies that this special distribution has a solution.

Cousin-II Distributions.

Recall the Weierstrass theorem: Let $\{z_\nu : \nu \in \mathbb{N}\}$ be a discrete set in $\mathbb{C}$, and (n_ν) a sequence of positive integers. Then there exists a holomorphic function f on $\mathbb{C}$ that has zeros exactly at the z_ν, and these are of orders n_ν. The distribution $(z_\nu, n_\nu)_{\nu \in \mathbb{N}}$ is a divisor D on $\mathbb{C}$ with $D \geq 0$ and $\operatorname{div}(f) = D$.

A divisor D on a complex manifold X is called a *principal divisor* if there is a meromorphic function m on X with $\operatorname{div}(m) = D$. If $D \geq 0$, then m is holomorphic. So we are interested in conditions on X such that every divisor is principal. From the exact sequence $\mathscr{M}(X)^* \xrightarrow{\operatorname{div}} \mathscr{D}(X) \xrightarrow{\delta} \operatorname{Pic}(X)$ it follows that $\operatorname{Ker}(\delta) = \mathscr{D}(X)$ is a necessary and sufficient condition. This is, for example, the case if every analytic line bundle on X is trivial. Due to Oka's principle, if X is Stein, it is sufficient that every topological line bundle be trivial. One can prove that the topological line bundles on X are classified by elements of $H^2(X, \mathbb{Z})$. Here we will show directly that the vanishing of $H^2(X, \mathbb{Z})$ is sufficient for the solvability of the generalized Weierstrass problem.

> **Definition.** Let X be a complex manifold. A *Cousin-II distribution* on X consists of an open covering $\mathscr{U} = \{U_\iota : \iota \in I\}$ together with meromorphic functions $f_\iota \neq 0$ on U_ι such that on $U_{\iota\kappa}$ there are nowhere vanishing holomorphic functions $g_{\iota\kappa}$ with $f_\iota = g_{\iota\kappa} \cdot f_\kappa$.
>
> A *solution* of the Cousin-II distribution is a meromorphic function f on X such that $f|_{U_\iota} = h_\iota^{-1} \cdot f_\iota$ with a nowhere vanishing holomorphic function h_ι on U_ι.

A Cousin-II distribution is therefore a meromorphic section in an analytic line bundle L that is defined by the transition functions $g_{\iota\kappa}$. If L is trivial, then there are nowhere vanishing holomorphic functions h_ι on U_ι with $g_{\iota\kappa} = h_\iota h_\kappa^{-1}$, and $f|_{U_\iota} := h_\iota^{-1} \cdot f_\iota$ defines a solution.

The Cousin-II distribution uniquely determines the cocycle $\gamma = (g_{\iota\kappa}) \in Z^1(\mathscr{U}, \mathcal{O}^*)$. If f is a solution with $f|_{U_\iota} = h_\iota^{-1} \cdot f_\iota$, then

$$g_{\iota\kappa} = f_\iota f_\kappa^{-1} = h_\iota h_\kappa^{-1}, \quad \text{so } \gamma = \delta h.$$

The following shows that the converse also holds.

1.7 Proposition. *Let $\gamma \in Z^1(\mathscr{U}, \mathcal{O}^*)$ be the cocycle of a Cousin-II distribution. The distribution is solvable if and only if $\gamma \in B^1(\mathscr{U}, \mathcal{O}^*)$.*

If $H^1(X, \mathcal{O}^*) = 0$, every Cousin-II distribution on X has a solution.

Unfortunately, Theorem B cannot be applied to $H^1(X, \mathcal{O}^*)$. So even for Stein manifolds we need an additional topological condition.

Chern Class and Exponential Sequence.

Let X be an arbitrary n-dimensional connected complex manifold. We choose an open covering $\mathscr{U} = \{U_\iota : \iota \in I\}$ of X such that the U_ι and all intersections $U_{\iota\kappa}$ are simply connected and construct an exact sequence of abelian groups:

$$H^1(\mathscr{U}, \mathbb{Z}) \xrightarrow{j} H^1(\mathscr{U}, \mathcal{O}) \xrightarrow{e} H^1(\mathscr{U}, \mathcal{O}^*) \xrightarrow{c} H^2(\mathscr{U}, \mathbb{Z}).$$

1. If $\xi = (\xi_{\iota\kappa}) \in Z^1(\mathscr{U}, \mathbb{Z})$, then also $\xi \in Z^1(\mathscr{U}, \mathcal{O})$. If ξ is an integer-valued coboundary, then it is also a holomorphic one. Thus j is well defined.
2. Let $f = (f_{\iota\kappa}) \in Z^1(\mathscr{U}, \mathcal{O})$ be given. Then we define

$$e([f]) := [\exp(2\pi \mathrm{i} f_{\iota\kappa})].$$

Obviously, e is well defined, and $e \circ j([\xi]) = [1]$. On the other hand, if $e([f]) = [1]$, then there are nowhere vanishing holomorphic functions h_ι with $\exp(2\pi \mathrm{i} f_{\iota\kappa}) = h_\iota h_\kappa^{-1}$. Since the U_ι are simply connected, there are holomorphic functions g_ι with $\exp(g_\iota) = h_\iota$. Then $\exp(g_\iota - g_\kappa) = \exp(2\pi \mathrm{i} f_{\iota\kappa})$, and it follows that there are integers $\xi_{\iota\kappa}$ with

$$f_{\iota\kappa} = \frac{1}{2\pi \mathrm{i}} g_\iota - \frac{1}{2\pi \mathrm{i}} g_\kappa + \xi_{\iota\kappa}.$$

This means that $[f] = j([\xi])$.
3. Here we define $c : H^1(\mathscr{U}, \mathcal{O}^*) \to H^2(\mathscr{U}, \mathbb{Z})$. Let $h = (h_{\iota\kappa}) \in Z^1(\mathscr{U}, \mathcal{O}^*)$ be given. Since the $U_{\iota\kappa}$ are simply connected, there are holomorphic functions $g_{\iota\kappa}$ on $U_{\iota\kappa}$ with $\exp(2\pi \mathrm{i} g_{\iota\kappa}) = h_{\iota\kappa}$. They are determined only up to some $2\pi \mathrm{i} \xi_{\iota\kappa}$ with $\xi_{\iota\kappa} \in \mathbb{Z}$, but

$$\eta_{\nu\mu\varrho} := (g_{\mu\varrho} - g_{\nu\varrho} + g_{\nu\mu})|_{U_{\nu\mu\varrho}}$$

uniquely defines a cohomology class $[\eta] \in H^2(\mathscr{U}, \mathbb{Z})$, since

$$\exp(2\pi \mathrm{i} \eta_{\nu\mu\varrho}) = h_{\mu\varrho} h_{\nu\varrho}^{-1} h_{\nu\mu} = 1$$

and

$$(\xi_{\mu\varrho} - \xi_{\nu\varrho} + \xi_{\nu\mu})|_{U_{\nu\mu\varrho}} = (\delta\xi)_{\nu\mu\varrho}.$$

Let $c([h]) := [\eta]$.

If $[h] = e([f])$, then there are nowhere vanishing holomorphic functions h_ι with

$$h_{\iota\kappa} = h_\iota \cdot \exp(2\pi \mathrm{i} f_{\iota\kappa}) \cdot h_\kappa^{-1} = \exp(2\pi \mathrm{i}(f_{\iota\kappa} + g_\iota - g_\kappa)),$$

where the g_ι are suitable holomorphic functions on U_ι. We have $c \circ e([f]) = [\eta]$ with $\eta_{\nu\mu\varrho} = (f_{\mu\varrho} - f_{\nu\varrho} + f_{\nu\mu})|_{U_{\nu\mu\varrho}} = 0$, since f is a cocycle.

Now let $[h]$ be given with $c([h]) = 0$. Then there is an element $\xi = (\xi_{\iota\kappa}) \in C^1(\mathscr{U}, \mathbb{Z})$ with

$$g_{\mu\varrho} - g_{\nu\varrho} + g_{\nu\mu} = \xi_{\mu\varrho} - \xi_{\nu\varrho} + \xi_{\nu\mu} \quad \text{on} \quad U_{\nu\mu\varrho},$$

where $\exp(2\pi \mathrm{i} g_{\iota\kappa}) = h_{\iota\kappa}$. So $f_{\iota\kappa} := g_{\iota\kappa} - \xi_{\iota\kappa}$ is a cocycle with values in $\mathcal{O}$, and $\exp(2\pi \mathrm{i} f_{\iota\kappa}) = h_{\iota\kappa}$. This means that $e([f]) = [h]$.

If the covering $\mathscr{U}$ is fine enough, everything is independent of the covering. By construction it is acyclic with respect to $\mathbb{Z}$.

Definition. Let $h = (h_{\iota\kappa}) \in Z^1(\mathscr{U}, \mathcal{O}^*)$ be a cocycle defining an analytic line bundle $L \in \mathrm{Pic}(X) = H^1(X, \mathcal{O}^*)$. Then $c(h) := c([h]) = c(L) \in H^2(\mathscr{U}, \mathbb{Z}) = H^2(X, \mathbb{Z})$ is called the *Chern class* of h (or of L).

Let us return to the Cousin-II distributions.

1.8 Proposition. *Let X be a Stein manifold. Then a Cousin-II distribution on X is solvable if and only if its Chern class vanishes.*

If $H^2(X, \mathbb{Z}) = 0$, every Cousin-II distribution is solvable.

PROOF: If X is Stein, then $H^1(X, \mathcal{O}) = 0$ and $c : H^1(X, \mathcal{O}^*) \to H^2(X, \mathbb{Z})$ is injective. Let h be a Cousin-II distribution on X. Then h is solvable if and only if $[h] = 0$, and that is the case if and only if $c(h) = 0$.

If in addition $H^2(X, \mathbb{Z}) = 0$, then also $H^1(X, \mathcal{O}^*) = \{1\}$. ∎

Remarks As in the case of Cousin-I distributions, here it is sufficient that $H^1(X, \mathcal{O}) = 0$, and X need not be Stein. On the other hand, there are examples of simply connected Stein manifolds where Cousin-II is not solvable. The condition $H^2(X, \mathbb{Z}) = 0$ is essential, and for the solvability of any special distribution the vanishing of the Chern class is necessary.

In the Stein case, using higher cohomology groups one can show that $c : H^1(X, \mathcal{O}^*) \to H^2(X, \mathbb{Z})$ is even bijective.

For a noncompact Riemann surface X it is always the case that $H^2(X, \mathbb{Z}) = 0$. Therefore, in this case the Weierstrass problem is always solvable.

Extension from Submanifolds. Let X be a complex manifold and $A \subset X$ an analytic set. A continuous function $f : A \to \mathbb{C}$ is called *holomorphic on A* if for every $x \in A$ there is an open neighborhood $U = U(x) \subset X$ and a holomorphic function $\widehat{f}$ on U such that $\widehat{f}|_{U\cap A} = f|_{U\cap A}$. One sees easily that this notion is well defined, but the *local continuation* $\widehat{f}$ is not uniquely determined.

1.9 Theorem. *Let X be a Stein manifold, $A \subset X$ an analytic set, and $f : A \to \mathbb{C}$ a holomorphic function. Then there exists a holomorphic function $\widehat{f}$ on X with $\widehat{f}|_A = f$.*

PROOF: In the differentiable category such a theorem would be proved with the help of a partition of unity. Here in complex analytic geometry we use cohomology. It is a typical application of Theorem B.

We can find an open covering $\mathscr{U} = (U_\iota)_{\iota\in I}$ of X and holomorphic functions $\widehat{f}_\iota$ on U_ι with $\widehat{f}_\iota|_{U_\iota\cap A} = f|_{U_\iota\cap A}$. Then $\xi_{\iota\kappa} := \widehat{f}_\iota - \widehat{f}_\kappa$ defines a cocycle $\xi \in Z^1(\mathscr{U}, \mathcal{O})$, with $\xi_{\iota\kappa}|_{U_{\iota\kappa}\cap A} = 0$. By Theorem B there is a cochain $\eta \in C^0(\mathscr{U}, \mathcal{O})$ with $\eta_\iota|_{U_\iota\cap A} = 0$ and $\delta\eta = \xi$. Then

$$\widehat{f}_\iota - \widehat{f}_\kappa = \eta_\kappa - \eta_\iota \text{ on } U_{\iota\kappa},$$

and $\widehat{f}|_{U_\iota} := \widehat{f}_\iota + \eta_\iota$ defines a global holomorphic function $\widehat{f}$ on X. Obviously, $\widehat{f}|_A = f$.

■

Unbranched Domains of Holomorphy. Let X be an unbranched domain of holomorphy over $\mathbb{C}^n$. Obviously, X is holomorphically spreadable, and since Cartan–Thullen holds for such domains, X is holomorphically convex. It follows directly that X is a Stein manifold, and all theorems on Stein manifolds are applicable in this case.

We consider now an unbranched domain of holomorphy $p : X \to \mathbb{P}^n$. We will show that it is also a Stein manifold.

Let $\pi : \mathbb{C}^{n+1} - \{\mathbf{0}\} \to \mathbb{P}^n$ be the canonical projection. Then we obtain a new manifold

$$\widehat{X} := \{(x, \mathbf{z}) \in X \times (\mathbb{C}^{n+1} - \{\mathbf{0}\}) \,:\, p(x) = \pi(\mathbf{z})\}$$

together with two canonical projections $\widetilde{\pi} : \widehat{X} \to X$ and $\widetilde{p} : \widehat{X} \to \mathbb{C}^{n+1} - \{\mathbf{0}\}$.

Since $(p, \pi) : X \times (\mathbb{C}^{n+1} - \{\mathbf{0}\}) \to \mathbb{P}^n \times \mathbb{P}^n$ is a submersion, the fiber product $\widehat{X} = (p, \pi)^{-1}(\Delta_{\mathbb{P}^n})$ is in fact an $(n+1)$-dimensional complex manifold, and we have the following commutative diagram:

$$\begin{array}{ccc} \widehat{X} & \xrightarrow{\widetilde{p}} & \mathbb{C}^{n+1} - \{\mathbf{0}\} \\ \widetilde{\pi}\downarrow & & \downarrow\pi \\ X & \xrightarrow[p]{} & \mathbb{P}^n \end{array}$$

For a point $P_0 := (x_0, \mathbf{z}_0)$ of $\widehat{X}$ there exist open neighborhoods $U = U(x_0) \subset X$ and $V = V(\pi(\mathbf{z}_0)) \subset \mathbb{P}^n$ such that $p : U \to V$ is biholomorphic. Then $\pi^{-1}(V)$ is a neighborhood of $\mathbf{z}_0$, $\widetilde{\pi}^{-1}(U) = (U \times (\mathbb{C}^{n+1} - \{\mathbf{0}\})) \cap \widehat{X}$ a neighborhood of P_0, and $\widetilde{p} : \widetilde{\pi}^{-1}(U) \to \pi^{-1}(V)$ a biholomorphic map with $\widetilde{p}^{-1} : \mathbf{z} \mapsto ((p|_U)^{-1}(\pi(\mathbf{z})), \mathbf{z})$. So $\widetilde{p} : \widehat{X} \to M := \mathbb{C}^{n+1} - \{\mathbf{0}\}$ is an unbranched Riemann domain. It is fibered over X by $\widetilde{\pi}$ with fibers isomorphic to $\mathbb{C}^*$.

By hypothesis X is a domain of holomorphy. Therefore, it is pseudoconvex. Now we use notation and results from Section II.8. If X has only removable boundary points, then X has locally the form $U - A$, with an open set $U \subset \mathbb{C}^n$ and an analytic hypersurface A. But then $\widehat{X}$ also has such a form and is pseudoconvex. If X has at least one nonremovable boundary point, then $\widehat{X}$ (which is something like a cone over X) has only non-removable boundary points over $\mathbf{0} \in \mathbb{C}^{n+1}$. The set of these boundary points over $\mathbf{0}$ is thin in $\breve{\partial}\widehat{X}$. Outside the origin $\widehat{X}$ is pseudoconvex, since it locally looks like $X \times \mathbb{C}$. By a theorem of Grauert/Remmert (see Section II.8) it follows that $\widehat{X}$ is pseudoconvex everywhere. From Oka's theorem we know that then $\widehat{X}$ is a Stein manifold.

To show that X is also Stein, we have first to show that X is holomorphically convex. If not, we would have an infinite sequence $D = \{x_i \,:\, i \in \mathbb{N}\}$ in X without a cluster point such that every holomorphic function f on X is bounded on D. We have that $\widehat{D} := \widetilde{\pi}^{-1}(D)$ is an analytic subset in $\widehat{X}$, and $f|_{\widetilde{\pi}^{-1}(x_i)} := i$ defines a holomorphic function f on $\widehat{D}$. Since $\widehat{X}$ is Stein, there exists a holomorphic function $\widehat{f}$ on $\widehat{X}$ with $\widehat{f}|_{\widehat{D}} = f$. On every fiber $\widetilde{\pi}^{-1}(x) \cong \mathbb{C}^*$ we have a Laurent expansion

$$\widehat{f}|_{\widetilde{\pi}^{-1}(x)} = \sum_{\nu=-\infty}^{\infty} a_\nu(x) z^\nu.$$

Since $\widetilde{\pi}$ is a submersion, we can define a holomorphic function g on X by $g(x) := a_0(x)$. Then $g(x_i) = i$, and this is a contradiction.

In the same way we can show that for $x, y \in X$ with $x \neq y$ there exists a holomorphic function f on X with $f(x) \neq f(y)$. Thus X is holomorphically separable and therefore Stein.

The Embedding Theorem. By a theorem of Whitney every differentiable manifold can be embedded into a space $\mathbb{R}^N$ of sufficiently high dimension. In general, this is false for complex manifolds, since, e.g., in $\mathbb{C}^N$ there are no positive-dimensional compact complex submanifolds. However, Stein manifolds can be embedded.

1.10 Theorem. *If X is an n-dimensional Stein manifold, then there exists an embedding $j : X \hookrightarrow \mathbb{C}^{2n+1}$.*

The proof is due to Remmert (see his Habilitationsschrift and [Re56]), Narasimhan ([Nar60]), and Bishop ([Bi61]).

O. Forster ([Fo70]) proved the following theorem:

1.11 Theorem. *Any 2-dimensional Stein manifold can be embedded into $\mathbb{C}^4$, but there is a 2-dimensional Stein manifold that cannot be embedded into $\mathbb{C}^3$.*

In general, for $n \geq 2$ every n-dimensional Stein manifold can be embedded into $\mathbb{C}^{2n}$, and for $n \geq 6$ even into $\mathbb{C}^{2n-[(n-2)/3]}$. For arbitrary n there always exists a proper immersion into $\mathbb{C}^{2n-1}$. But there are examples of n-dimensional Stein manifolds X that cannot be embedded into $\mathbb{C}^N$ (and not immersed into $\mathbb{C}^{N-1}$) for $N := n + \left[\frac{n}{2}\right]$.

At the end of his paper Forster conjectured that every n-dimensional Stein manifold can be embedded into $\mathbb{C}^{N+1}$ and immersed into $\mathbb{C}^N$.

In 1992 J. Schürmann showed that Forster's conjecture is true for $n \geq 2$ (see [Schue92],[Schue97]). In 1970/71 Stehlé had given an embedding of the unit disk Δ into $\mathbb{C}^2$ (see [St72]), and in 1973 Laufer an embedding of certain annuli into $\mathbb{C}^2$. At the moment it is an open problem whether every noncompact Riemann surface can be embedded into $\mathbb{C}^2$. There has been quite a bit of progress on this by the work of J. Globevnik and B. Stensønes (see [GlSte95]), but no final result.

The Serre Problem. Above it was proved that if $f : X \to Y$ is a finite holomorphic mapping and Y a Stein manifold, then X is also Stein.

In 1953 Serre posed the following problem: Is the total space of a holomorphic fiber bundle with Stein base Y and Stein fiber F a Stein manifold? ([Se53]).

It is easy to show that the answer is positive in the case of an analytic vector bundle.

Between 1974 and 1980 it was proved that the answer is positive in the case of 1-dimensional fibers (Siu, Sibony, Hirschowitz and Mok). In 1976 Siu generalized this to the case where the fiber is a bounded domain $G \subset \mathbb{C}^n$ with $H^1(G, \mathbb{C}) = 0$, and in 1977 Diederich and Fornæss showed that the Serre conjecture is true if the fiber is a bounded domain in $\mathbb{C}^n$ with $\mathcal{C}^2$ smooth boundary.

In 1977 Skoda ([Sko77]) gave the first counterexample, a fiber bundle with $\mathbb{C}^2$ as fiber and an open (not simply connected) set in $\mathbb{C}$ as base. Demailly improved the example; he used the complex plane or a disk as base. Finally, in 1985 G. Coeuré and J.J. Loeb presented a counterexample with $\mathbb{C}^*$ as base and a bounded pseudoconvex Reinhardt domain in $\mathbb{C}^2$ as fiber. Over the years a number of positive examples have been found, for example by Matsushima/Morimoto, G. Fischer, Ancona, Siu, and Stehlé. Examples where the fiber is a so-called Banach–Stein space play an important role, but we cannot go into the details here.

Exercises

1. Show that the Cartesian product $X \times Y$ of two Stein manifolds is Stein.
2. Use Theorems A and B to prove that the total space of a vector bundle over a Stein manifold is Stein.
3. Give an example of a Cousin-I distribution on $\mathbb{C}^2 - \{\mathbf{0}\}$ that has no solution.
4. Let $U_j := \{\mathbf{z} \in \mathbb{C}^3 : z_j \neq 0\}$ and $\mathscr{U} = \{U_1, U_2, U_3\}$. Show that every Cousin-I distribution with respect to $\mathscr{U}$ has a solution.
5. Let X be a complex manifold, $\pi : L \to X$ a holomorphic line bundle, and $Z \subset L$ the zero section. Prove that if $L - Z$ is Stein, then X is a Stein manifold as well.

2. The Levi Form

Covariant Tangent Vectors. Let X be an n-dimensional complex manifold and $x \in X$ a point. The elements of the tangent space $T_x(X)$ (abbreviated by T) are the *(contravariant) tangent vectors* at x. If $z_1, \ldots, z_n$ are local coordinates at x, then every tangent vector can be written in the form

$$v = \sum_\nu v_\nu \frac{\partial}{\partial z_\nu} + \sum_\nu \overline{v}_\nu \frac{\partial}{\partial \overline{z}_\nu}.$$

Of course, T has a natural structure of an n-dimensional complex vector space.

Now we consider the space $F = F(T)$ of complex-valued real linear forms on T. For example, if f is a local (real- or complex-valued) smooth function at x, then its *differential* $(df)_x \in F$ is defined by $(df)_x(v) := v[f]$. It is uniquely determined by the germ of f at x. In local coordinates we get

$$(df)_x(v) = \sum_\nu v_\nu f_{z_\nu}(x) + \sum_\nu \overline{v}_\nu f_{\overline{z}_\nu}(x).$$

It follows that

$$(d\overline{f})_x(v) = \overline{(df)_x(v)}.$$

In particular, we have the elements $dz_\nu, d\overline{z}_\nu \in F$ defined by

$$dz_\nu(v) := v[z_\nu] \quad \text{and} \quad d\overline{z}_\nu(v) := v[\overline{z}_\nu].$$

Then $dz_\nu = dx_\nu + \mathrm{i}\, dy_\nu$ and $d\overline{z}_\nu = \overline{dz_\nu} = dx_\nu - \mathrm{i}\, dy_\nu$, for $\nu = 1, \ldots, n$.

The space F is the complexification of the $2n$-dimensional real vector space $T^* = \operatorname{Hom}_{\mathbb{R}}(T, \mathbb{R})$. Therefore, $F = T^* \oplus \mathrm{i}\, T^*$ is a $2n$-dimensional complex vector space, with basis

$$\{dz_1, \ldots, dz_n, d\overline{z}_1, \ldots, d\overline{z}_n\}.$$

We call the elements of F *complex covariant tangent vectors* or *complex 1-forms* at x.

We have $(df)_x = \sum_\nu f_{z_\nu}(x)\, dz_\nu + \sum_\nu f_{\overline{z}_\nu}(x)\, d\overline{z}_\nu$.

In general, a *complex covariant r-tensor* at x is an $\mathbb{R}$-multilinear mapping

$$\varphi : \underbrace{T \times \cdots \times T}_{r\text{-times}} \to \mathbb{C}.$$

The *tensor product* $\varphi \otimes \psi$ of an r-tensor and an s-tensor is the $(r+s)$-tensor given by

$$(\varphi \otimes \psi)(v_1, \ldots, v_r, v_{r+1}, \ldots, v_{r+s}) := \varphi(v_1, \ldots, v_r) \cdot \psi(v_{r+1}, \ldots, v_{r+s}).$$

The set of r-tensors carries the structure of a complex vector space, and the assignment $(\varphi, \psi) \mapsto \varphi \otimes \psi$ is $\mathbb{C}$-bilinear. For example, $(dz_\nu \otimes d\overline{z}_\mu)(v, w) = v_\nu \cdot \overline{w}_\mu$.

Hermitian Forms. Let X be an n-dimensional complex manifold. The notion of a plurisubharmonic function in a domain $G \subset \mathbb{C}^n$ was already defined in Chapter II. Of course, a plurisubharmonic function on a complex manifold is a real-valued function that is ($\mathscr{C}^\infty$) differentiable and plurisubharmonic with respect to all local coordinates belonging to the complex structure of X. The notion of plurisubharmonicity is invariant with respect to holomorphic coordinate transformations. So in order to prove plurisubharmonicity at some point of X it is enough to prove it with resepct to an arbitrary coordinate system. However, here we wish to express the notion of plurisubharmonicity in invariant terms. We do this by Hermitian forms.

Definition. A *Hermitian form* at $x_0 \in X$ is a Hermitian form

$$H : T_{x_0} \times T_{x_0} \to \mathbb{C}.$$

The form H is called *positive semidefinite* if $H(v, v) \geq 0$ for all v, and it is called *positive definite* if $H(v, v) > 0$ for $v \neq 0$.

A Hermitian form has a unique representation

$$H = \sum_{i,j=1}^{n} h_{ij} dz_i \otimes d\overline{z}_j,$$

where $\mathbf{H} := \left(h_{ij} \;\middle|\; i, j = 1, \ldots, n\right)$ is a Hermitian matrix; i.e., it satisfies the equation $\mathbf{H} = \overline{\mathbf{H}}^{\,t}$. In the following we suppress the symbol $\otimes$ and write $H = \sum_{i,j} h_{ij} dz_i d\overline{z}_j$.

With respect to the local coordinates we can associate to any tangent vector $v = \sum_i v_i \partial/\partial z_i + \sum_i \overline{v}_i \partial/\partial \overline{z}_i$ the corresponding vector $\mathbf{v} = (v_1, \ldots, v_n) \in \mathbb{C}^n$ and write

$$H(v, w) = \sum_{i,j=1}^{n} h_{ij} v_i \overline{w}_j = \mathbf{v} \cdot \mathbf{H} \cdot \overline{\mathbf{w}}^{\,t}.$$

Coordinate Transformations. Assume that $\mathbf{F} : G \to B$ is a holomorphic map of domains in $\mathbb{C}^n$ given by equations

$$w_k = f_k(z_1, \ldots, z_n), \text{ for } k = 1, \ldots, n.$$

If g is a differentiable function in B, by the complex chain rule it follows that

$$(g \circ \mathbf{F})_{z_i} = \sum_k (g_{w_k} \circ \mathbf{F}) \cdot (f_k)_{z_i} \quad \text{and} \quad (g \circ \mathbf{F})_{\overline{z}_i} = \sum_k (g_{\overline{w}_k} \circ \mathbf{F}) \cdot \overline{(f_k)_{z_i}}.$$

The transformation of a (contravariant) tangent vector (i.e., a derivation) ξ is given by $\mathbf{F}_*(\xi)[g] = \xi[g \circ \mathbf{F}]$. This means that

$$\begin{aligned} \mathbf{F}_*(\xi) &= \sum_{k=1}^{n} \xi[f_k] \frac{\partial}{\partial w_k} + \sum_{k=1}^{n} \xi[\overline{f_k}] \frac{\partial}{\partial \overline{w}_k} \\ &= \sum_{k=1}^{n} \left(\sum_{i=1}^{n} \xi_i \cdot (f_k)_{z_i} \right) \frac{\partial}{\partial w_k} + \sum_{k=1}^{n} \left(\sum_{i=1}^{n} \overline{\xi}_i \cdot \overline{(f_k)_{z_i}} \right) \frac{\partial}{\partial \overline{w}_k}, \end{aligned}$$

or, if $\xi \sim \boldsymbol{\xi} = (\xi_1, \ldots, \xi_n)$, then

$$\mathbf{F}_* \xi \sim \left(\sum_i \xi_i \cdot (f_1)_{z_i}, \ldots, \sum_i \xi_i \cdot (f_n)_{z_i} \right) = \boldsymbol{\xi} \cdot J_{\mathbf{F}}^t.$$

Now, a covariant tangent vector φ at $\mathbf{w} \in B$ will be transported in the opposite direction:

$$\mathbf{F}^* \varphi(\xi) := \varphi(\mathbf{F}_* \xi).$$

In particular, $\mathbf{F}^*(df)_{\mathbf{w}}(\xi) = (df)_{\mathbf{w}}(\mathbf{F}_*\xi) = (\mathbf{F}_*\xi)[f] = \xi[f \circ \mathbf{F}] = d(f \circ \mathbf{F})_{\mathbf{z}}(\xi)$ for $\mathbf{w} = \mathbf{F}(\mathbf{z})$. This gives us the formula

$$\mathbf{F}^*((df)_{\mathbf{F}(\mathbf{z})}) = d(f \circ \mathbf{F})_{\mathbf{z}}.$$

Therefore we also write $\varphi \circ \mathbf{F} := \mathbf{F}^* \varphi$ for arbitrary covariant vectors φ.

If H is a Hermitian form at $\mathbf{w} = \mathbf{F}(\mathbf{z}) \in B$, then we can define a Hermitian form $\mathbf{F}^* H$ at $\mathbf{z}$ by

$$\mathbf{F}^* H(\xi, \eta) := H(\mathbf{F}_* \xi, \mathbf{F}_* \eta).$$

If $\mathbf{J} = J_{\mathbf{F}}$ is the Jacobian of $\mathbf{F}$ at $\mathbf{z}$ and $H = \sum_{k,l} h_{k,l}\, dw_k\, d\overline{w}_l$, then

$$\mathbf{F}^* H(\xi, \eta) = (\boldsymbol{\xi} \cdot \mathbf{J}^t) \cdot \mathbf{H} \cdot \overline{(\boldsymbol{\eta} \cdot \mathbf{J}^t)}^{\,t} = \boldsymbol{\xi} \cdot (\mathbf{J}^t \cdot \mathbf{H} \cdot \overline{\mathbf{J}}) \cdot \overline{\boldsymbol{\eta}}^{\,t}.$$

So $\mathbf{F}^* H$ is given by the Hermitian matrix $\mathbf{J}^t \cdot \mathbf{H} \cdot \overline{\mathbf{J}}$.

Plurisubharmonic Functions. Assume that p is a real-valued ($\mathscr{C}^\infty$) differentiable function in a domain $B \subset \mathbb{C}^n$ and $\mathbf{w} \in B$ a point. We consider at $\mathbf{w}$ the Hermitian form

$$H_p = \sum_{k,l} \frac{\partial^2 p}{\partial w_k \partial \overline{w}_l}(\mathbf{w})\, dw_k\, d\overline{w}_l,$$

given by the Hermitian matrix

$$\mathbf{H}(p,\mathbf{w}) := \left(\frac{\partial^2 p}{\partial w_k \partial \overline{w}_l}(\mathbf{w}) \;\middle|\; k,l = 1,\ldots,n \right).$$

If $\mathbf{F} : G \to B$ is a biholomorphic transformation with $\mathbf{F}(\mathbf{z}) = \mathbf{w}$, then a direct calculation shows that

$$\mathbf{H}(p \circ \mathbf{F}, \mathbf{z}) = J_{\mathbf{F}}(\mathbf{z})^{\,t} \cdot \mathbf{H}(p,\mathbf{w}) \cdot \overline{J_{\mathbf{F}}(\mathbf{z})}.$$

Therefore, $\mathbf{F}^* H_p$ will be described by the Hermitian matrix $\mathbf{H}(p \circ \mathbf{F}, \mathbf{z})$, i.e., $\mathbf{F}^* H_p = H_{p \circ \mathbf{F}}$.

Now let X be an n-dimensional complex manifold, p a real-valued smooth function on X, and $x \in X$ an arbitrary point. Assume that $\varphi : U \to B \subset \mathbb{C}^n$ is a local coordinate system at x. Then a tangent vector ξ at x is uniquely determined by a pair $(\varphi, \boldsymbol{\xi})$, $\boldsymbol{\xi} \in \mathbb{C}^n$.

We define the Hermitian form $H_p : T_x(X) \times T_x(X) \to \mathbb{C}$ by

$$H_p(\xi, \eta) := H_{p \circ \varphi^{-1}}(\boldsymbol{\xi}, \boldsymbol{\eta}).$$

If ψ is another coordinate system at x and $\xi \sim (\psi, \widetilde{\boldsymbol{\xi}})$, then

$$\boldsymbol{\xi} = \widetilde{\boldsymbol{\xi}} \cdot J_{\varphi \circ \psi^{-1}}(\psi(x))^{\,t} = (\varphi \circ \psi^{-1})_*(\widetilde{\boldsymbol{\xi}})$$

and

$$\begin{aligned} H_{p \circ \psi^{-1}}(\widetilde{\boldsymbol{\xi}}, \widetilde{\boldsymbol{\eta}}) &= H_{(p \circ \varphi^{-1}) \circ (\varphi \circ \psi^{-1})}(\widetilde{\boldsymbol{\xi}}, \widetilde{\boldsymbol{\eta}}) \\ &= H_{p \circ \varphi^{-1}}((\varphi \circ \psi^{-1})_* \widetilde{\boldsymbol{\xi}}, (\varphi \circ \psi^{-1})_* \widetilde{\boldsymbol{\eta}}) \\ &= H_{p \circ \varphi^{-1}}(\boldsymbol{\xi}, \boldsymbol{\eta}). \end{aligned}$$

We see that the definition of H_p is independent of the local coordinates.

Definition. Assume that p is a real-valued smooth function on the complex manifold X. Then $\mathrm{Lev}(p)(x,\xi) := H_p(\xi,\xi)$ is called the *Levi form* of p at x. It is the quadratic form on $T_x(X)$ associated with H_p, and it does not depend on local coordinates.

The function p is called *plurisubharmonic* on a subset $M \subset X$ if the Levi form of p is positive semidefinite at any point $x \in M$. If the Levi form is positive definite at every point, we say that p is *strictly plurisubharmonic*.

Example

Assume that $f_1, \ldots, f_m$ are holomorphic functions on X. We show that $p := \sum_{k=1}^{m} f_k \cdot \overline{f}_k$ is plurisubharmonic in X. In fact, we have

$$\begin{aligned} H_p &= \sum_{i,j} p_{z_i,\overline{z}_j}\, dz_i\, d\overline{z}_j \\ &= \sum_{k=1}^{m} \Big(\sum_{i,j} (f_k)_{z_i} \cdot \overline{(f_k)_{z_j}}\, dz_i\, d\overline{z}_j \Big) \\ &= \sum_{k=1}^{m} \left(\Big(\sum_{i=1}^{n} (f_k)_{z_i}\, dz_i \Big) \otimes \Big(\sum_{j=1}^{n} \overline{(f_k)_{z_j}}\, d\overline{z}_j \Big) \right). \end{aligned}$$

That means that

$$\operatorname{Lev}(p)(x,\xi) = H_p(\xi,\xi) = \sum_{k=1}^{m} \Big| \sum_i (f_k)_{z_i}(x)\, \xi_i \Big|^2 \geq 0.$$

The Maximum Principle. A nonconstant plurisubharmonic function does not take on a maximum.

2.1 Theorem. *Assume that $A \subset X$ is a compact connected analytic set and that p is a plurisubharmonic function on X. Then $p|_A$ is constant.*

PROOF: We may assume that A is irreducible. If there is a point $x_0 \in A$ where $p|_A$ takes its maximum value c, we shall prove:

(∗) *The function p is identically equal to c in a small neighborhood of x_0.*

If we know (∗), then we consider the set K of all points $x \in A$ with $p(x) = c$. The open kernel K° is not empty. If $K^\circ \neq A$, there is a boundary point x_1 of K in A. The function p also takes its maximum at x_1. So by (∗) it follows again that $p(x) = c$ in a neighborhood of x_1, which is a contradiction.

So we have only to prove (∗). We may assume that A is an analytic subset of $\mathbb{C}^n$ and $x_0 = \mathbf{0}$. If the codimension of A is equal to d, then there is an $(n-d)$-dimensional domain $G' \subset \mathbb{C}^{n-d}$ with $\mathbf{0}' \in G'$ and pseudopolynomials $\omega_1(z_1; \mathbf{z}'), \ldots, \omega_d(z_d; \mathbf{z}')$ over G' such that A is an embedded-analytic subset of the joint zero set of the ω_i in a neighborhood of $\mathbf{0}$, and $\mathbf{0}$ is the only point of A over $\mathbf{0}'$.

We take a ball $B \subset G'$ around $\mathbf{0}'$ and restrict everything to an arbitrary complex line $\ell \subset B$ through $\mathbf{0}'$. The restriction $A|\ell$ is denoted by A'. Let z be a linear coordinate on ℓ with origin $0'$ such that the embedding of ℓ in B

is given by $z \mapsto (z_1, \dots, z_{n-d}) = (a_1 z, \dots, a_{n-d} z)$. The restriction $\omega_i|_\ell$ may contain multiple factors. We throw away such superfluous factors, so we can assume that every $\omega_i|_\ell$ is free of multiple factors.

Let us first assume that the union of the discriminant sets of the pseudopolynomials over ℓ consists only of the point $\mathbf{0}'$ and that A' is irreducible and has s sheets over ℓ. Then A' is the Riemann surface of $\sqrt[s]{z}$. We write A' in the form

$$A' \approx \{(t; a_1 z, \dots, a_{n-d} z) \, : \, z = t^s\}.$$

By $\mathbf{F}(t) := (t; a_1 t^s, \dots, a_{n-d} t^s)$ we have a local parametrization of A'. Then $(p \circ \mathbf{F})_{t\bar{t}}(t) = \mathbf{H}(p \circ \mathbf{F}, t)$ is given again by $\mathbf{J}^t \cdot \mathbf{H}(p, \mathbf{F}(t)) \cdot \overline{\mathbf{J}}$, where $\mathbf{J} = (1; a_1, \dots, a_{n-d})^t$ denotes the Jacobian of the holomorphic map $\mathbf{F}$. The proof is the same as in the case of a biholomorphic map $\mathbf{F}$. So $(p \circ \mathbf{F})_{t\bar{t}} \geq 0$, and $p \circ \mathbf{F}$ is a subharmonic function of t. We get $p' := p|_{A'} \equiv c$ on A'.

The same is true if A' is not irreducible but has $\mathbf{0}'$ as the joint discriminant set, since $\mathbf{0}$ is the only point of A' over $\mathbf{0}'$.

Now assume that the union of the discriminant sets is general. Every point in A' can be connected with $\mathbf{0}$. We introduce the subset K of all points $x \in A'$ with $p'(x) = c$. If $K^\circ \neq A'$, there is a boundary point x_1 of K in A'. We know that A' is an embedded-analytic set. Then there is a neighborhood $U(x_1) \subset A'$ that is embedded-analytic over a disk $B'(z_1) \subset \ell$ around a point $z_1 \in B$ such that over z_1 the only point of U is x_1 and the union discriminant set consists of z_1 only. Then we get $p'|_U \equiv c$ (by the same argument as above), which is a contradiction to the property "boundary point." So $p' \equiv c$ follows.

This holds for all ℓ, and therefore $p \equiv c$ over the whole ball B ,which is in a full open neighborhood of x_0 in A. So we have $(*)$. ∎

Exercises

1. Assume that p is a real-valued smooth function on the complex manifold X. If $\xi \sim \boldsymbol{\xi}$ is a tangent vector and φ a complex coordinate system at $x_0 \in X$, then define

$$(\partial p)_{x_0}(\xi) := \sum_{\nu=1}^{n} (p \circ \varphi^{-1})_{z_\nu}(\varphi(x_0)) \cdot \xi_\nu \, .$$

   Show that $(\partial p)_{x_0} : T_{x_0} \to \mathbb{C}$ is a complex-valued real linear form that does not depend on the local coordinates. Prove the following formulas:
   (a)

$$\begin{aligned} \operatorname{Lev}(p \cdot q)(x, \xi) &= p(x) \cdot \operatorname{Lev}(q)(x, \xi) + q(x) \cdot \operatorname{Lev}(p)(x, \xi) \\ &\quad + 2 \operatorname{Re}\left((\partial p)_x(\xi) \cdot \overline{(\partial q)_x(\xi)} \right). \end{aligned}$$

(b)
$$\mathrm{Lev}(h \circ p)(x, \xi) = h''(p(x)) \cdot |(\partial p)_x(\xi)|^2 + h'(p(x)) \cdot \mathrm{Lev}(p)(x, \xi).$$

2. Let $G \subset\subset X$ be a relatively compact domain with smooth boundary. Show that there is an open neighborhood U of ∂G in X and a real-valued smooth function φ on U such that $G \cap U = \{x \in U : \varphi(x) < 0\}$ and $(\partial\varphi)_x \neq 0$ for $x \in \partial G$. Show that
$$H_x(\partial G) = \{\xi \in T_x(X) : (\partial\varphi)_x(\xi) = 0\}$$
is a well-defined subspace of $T_x(X)$ that does not depend on the boundary function φ.

   Show that if for every $x \in \partial G$ there is a local boundary function ψ such that $\mathrm{Lev}(\psi)$ is positive definite on $H_x(\partial G)$, then φ can be chosen as a strictly plurisubharmonic function.
3. Let $G \subset\subset X$ be a relatively compact domain with smooth boundary, and $\varphi : U(\partial G) \to \mathbb{R}$ a global boundary function. If $\mathrm{Lev}(\varphi)$ has for every $x \in \partial G$ at least one negative eigenvalue on $H_x(\partial G)$, G is called *pseudoconcave*.

   Show that if X is connected and there is a nonempty pseudoconcave domain in X, then every global holomorphic function on X is constant.

3. Pseudoconvexity

Pseudoconvex Complex Manifolds. If X is an arbitrary complex manifold, then there exists a sequence of compact subsets $K_i \Subset X$ with the following properties:

1. The set K_{i-1} is always contained in the open kernel $(K_i)^\circ$ of K_i.
2. $\displaystyle\bigcup_{i=1}^{\infty} K_i = X$.

If X is *holomorphically convex*, then the K_i can be chosen in such a way that the holomorphically convex hull
$$\widehat{K}_i = \Big\{x \in X : |f(x)| \le \sup_{K_i} |f| \text{ for all } f \in \mathcal{O}(X)\Big\}$$
always equals K_i. (One uses the same proof as for domains in $\mathbb{C}^n$.)

Therefore, for every point $x \in X - K_i$ there is a holomorphic function f in X such that $|f(x)| > 1$ and $|f| < 1$ on K_i. By passing over to a multiple and a power of f, we can make $|f|$ arbitrarily small on K_i and arbitrarily big in a fixed neighborhood of x.

Since $K_{i+2} - (K_{i+1})^\circ$ is compact, there are finitely many holomorphic functions $f_i^{(1)}, \ldots, f_i^{(N_i)}$ in X such that for $p_i := \sum_{\nu=1}^{N_i} |f_i^{(\nu)}|^2$ the following hold:

1. $\sup_{K_i} p_i < 2^{-i}$.
2. $p_i > i$ on $K_{i+2} - (K_{i+1})^\circ$.

Of course, p_i is a nonnegative plurisubharmonic function in X.

The sum $\sum_{i=1}^{\infty} p_i$ converges compactly on X to a nonnegative function p, with $\{x \in X : p(x) < c\} \subset\subset X$ for every $c > 0$.

If f is a holomorphic function on a domain $G \subset \mathbb{C}^n$, and $G' \subset\subset G$ a subdomain, then we have the Cauchy estimate

$$|D^\nu f(\mathbf{z})| \leq \frac{\nu!}{\delta_G(\mathbf{z})^{|\nu|}} \cdot \sup_{\overline{G'}} |f|, \text{ for } \mathbf{z} \in G'.$$

Using this estimate in the intersection of a local coordinate system for X with $(K_i)^\circ$, one shows that all derivatives of $\sum_i p_i$ converge compactly in X to the corresponding derivatives of p. So p is $\mathscr{C}^\infty$ and again a plurisubharmonic function. One can even show that p is real-analytic (see [DoGr60]).

Definition. A complex manifold X is called *pseudoconvex* if there exists a nonnegative smooth plurisubharmonic exhaustion function p on X (i.e., a $\mathscr{C}^\infty$ function p with $\{x \in X : p(x) \leq r\} \subset\subset X$ for all $r > 0$ such that the Levi form of p is everywhere positive semidefinite).

If we can find for p a strictly plurisubharmonic function in X, then X is called *stricly pseudoconvex* or *1-complete*[1]. If p is strictly plurisubharmonic only outside a compact set $K \subset X$, then X is called *1-convex* or *strongly pseudoconvex (at infinity)*.

In the literature a pseudoconvex manifold is often called *weakly 1-complete.*

Above it was shown that

every holomorphically convex complex manifold is pseudoconvex.

We shall prove later on that every 1-complete complex manifold is holomorphically convex, and even Stein (solution of the Levi problem). Also, strongly pseudoconvex manifolds are holomorphically convex. But in general this is not true for weakly 1-complete (i.e., pseudoconvex) manifolds.

Examples. Strict pseudoconvexity (1-completeness) is one of the most important notions in the analysis of complex manifolds. Many constructions can be carried out only in the strict pseudoconvex case. Let us consider some examples.

Example 1: The theory was inaugurated by the following result:

[1] An n-dimensional complex manifold X is called *q-complete* if it has an exhaustion function p such that at every point of X the Levi form of p has at least $n - q + 1$ positive eigenvalues.

An unbranched domain G over $\mathbb{C}^n$ is strictly pseudoconvex if and only if it is a domain of holomorphy.

For proofs see [Oka53], [Br54], [No54].

Example 2: Every unbranched Hartogs convex domain over $\mathbb{C}^n$ is strictly pseudoconvex. See [Ri68], where a smoothing procedure for strictly pseudoconvex functions is introduced.

Example 3: The proofs of the following statements are elementary.

Every compact complex manifold is pseudoconvex. The Cartesian product of finitely many (strictly) pseudoconvex complex manifolds is (strictly) pseudoconvex. Any submanifold of a (strictly) pseudoconvex complex manifold is (strictly) pseudovonvex.

Example 4: Assume that $G \subset \mathbb{C}^n$ is a domain of holomorphy and that $A \subset G$ is an analytic hypersurface. Then $G - A$ is Hartogs convex (and therefore strictly pseudoconvex). For the proof we just take a Hartogs figure

$$\mathsf{H} = \{\mathbf{t} : \varepsilon < |t_1| < 1, |t_i| < 1 \text{ for } i \geq 2\} \cup \{\mathbf{t} : |t_1| < 1, |t_i| < \varepsilon \text{ for } i \geq 2\}$$

in the unit polydisk $\mathsf{P}^n = \{\mathbf{t} : |t_i| < 1 \text{ for } i = 1, \ldots, n\}$ and a biholomorphic mapping $\mathbf{F} : \mathsf{H} \to G - A$. The mapping $\mathbf{F}$ extends to a holomorphic mapping $\mathsf{P}^n \to G$. If $\widehat{A} = \mathbf{F}^{-1}(A) \subset \mathsf{P}^n$ is not empty, then it is an analytic hypersurface in $\mathsf{P}^n - \mathsf{H}$. Some lines $L(\mathbf{t}') = \{\mathbf{t} = (t_1, \mathbf{t}') : t_1 \in \mathbb{C}\}$ will intersect it in a compact subset of $\mathsf{P}^n - \mathsf{H}$; for other $\mathbf{t}'$ the intersection is empty (see Figure V.1).

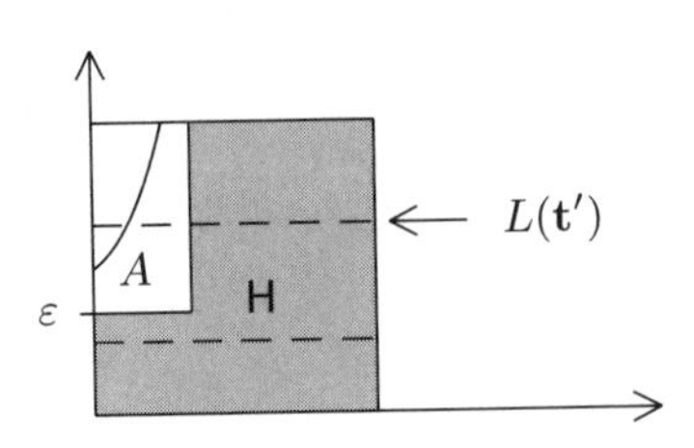

Figure V.1. Hartogs convexity of $G - A$

There is a limit $\mathbf{t}'_0$ with $\varepsilon < |\mathbf{t}'_0| < 1$ such that the intersection of $L(\mathbf{t}'_0)$ with $\widehat{A}$ is not empty but in arbitrarily small neighborhoods of $\mathbf{t}'_0$ there are points $\mathbf{t}'$ for which the intersection is empty. If $\mathbf{t} \in \widehat{A} \cap L(\mathbf{t}'_0)$, then there is a neighborhood U where $\widehat{A}$ is given by a holomorphic equation $f = 0$. The function $t_1 \mapsto f(t_1, \mathbf{t}'_0)$ has isolated zeros in $U \cap L(\mathbf{t}'_0)$, and by the theorem of Rouché $t_1 \mapsto f(t_1, \mathbf{t}')$ also has to vanish at some points of $U \cap L(\mathbf{t}')$ for $\mathbf{t}'$ near to $\mathbf{t}'_0$. This is a contradiction.

Thus $\mathbf{F}(\mathsf{P}^n) \subset G - A$, and the domain $G - A$ is Hartogs pseudoconvex.

Example 5: Let $A \subset \mathbb{P}^n$ be an analytic hypersurface. We know that there is a homogeneous polynomial ω such that

$$A = \{(z_0 : z_1 : \ldots : z_n) \in \mathbb{P}^n \, : \, \omega(z_0, \ldots, z_n) = 0\}.$$

Let ω be homogeneous of order s. Then

$$p(z_0 : \ldots : z_n) := \log \Big\{ \Big(\sum_{i=0}^{n} z_i \overline{z}_i \Big)^s \Big/ |\omega(z_0, \ldots, z_n)|^2 \Big\}$$

is a well-defined smooth exhaustion function for the affine algebraic manifold $\mathbb{P}^n - A$. We calculate the Levi form in local coordinates $t_\nu = z_\nu / z_0$, $\nu = 1, \ldots, n$ using the properties of the logarithm. Since the Levi form of $f + \overline{f}$ vanishes for any holomorphic function f, it follows that

$$\begin{aligned} \operatorname{Lev}(p)(\mathbf{t}_0, \boldsymbol{\xi}) &= s \cdot \operatorname{Lev}(\log(1 + \|\mathbf{t}\|^2))(\mathbf{t}_0, \boldsymbol{\xi}) \\ &= s \cdot \left(\frac{-1}{(1 + \|\mathbf{t}_0\|^2)^2} \cdot |\langle \mathbf{t}_0 \, , \, \boldsymbol{\xi} \rangle|^2 + \frac{1}{1 + \|\mathbf{t}_0\|^2} \cdot \|\boldsymbol{\xi}\|^2 \right) \\ &= \frac{s}{(1 + \|\mathbf{t}_0\|^2)^2} \cdot \left(\|\boldsymbol{\xi}\|^2 + (\|\mathbf{t}_0\|^2 \cdot \|\boldsymbol{\xi}\|^2 - |\langle \mathbf{t}_0 \, , \, \boldsymbol{\xi} \rangle|^2) \right), \end{aligned}$$

and this expression is positive for $\boldsymbol{\xi} \neq \mathbf{0}$. So p is strictly plurisubharmonic everywhere, and $X = \mathbb{P}^n - A$ is 1-complete. In this case we can show directly that X is holomorphically convex:

Every function

$$f(z_0 : \ldots : z_n) := \frac{z_0^{s_0} \cdots z_n^{s_n}}{\omega(z_0, \ldots, z_n)}, \qquad \text{with } \sum_{i=0}^{n} s_i = s,$$

is holomorphic in X, and the maximum of the absolute values of all these functions tends to infinity as $(z_0 : \ldots : z_n)$ approaches A. So $\widehat{K} \subset\subset X$ for any subset $K \subset\subset X$. Consequently, X is holomorphically convex, and it is even Stein, as one can see from the following theorem.

3.1 Theorem. *Let X be a holomorphically convex connected complex manifold that contains no compact analytic subset of positive dimension. Then X is a Stein manifold.*

PROOF: Let $x_0 \in X$ be an arbitrary point. Then the set

$$A := \{x \in X \, : \, f(x) = f(x_0) \text{ for every } f \in \mathcal{O}(X)\}$$

is a closed analytic subset of X. Clearly, it is contained in the holomorphically convex hull $\widehat{\{x_0\}} = \{x \in X : |f(x)| \le |f(x_0)| \text{ for all } f \in \mathcal{O}(X)\}$. Since $\widehat{\{x_0\}}$ has to be compact, A is likewise compact. This is possible only if A consists of isolated points. Then there exists an open neighborhood $U = U(x_0)$ and holomorphic functions $f_1, \ldots, f_N$ in U such that

$$\{x_0\} = A \cap U = \{x \in U : f_1(x) = \cdots = f_N(x) = 0\}.$$

So X is holomorphically spreadable and therefore Stein. ∎

3.2 Proposition. *A 1-complete complex manifold cannot contain compact analytic subsets of positive dimension.*

PROOF: Let p be a strictly plurisubharmonic exhaustion function in the manifold X. Then p is plurisubharmonic, and by the maximum principle it must be constant on any compact connected analytic subset $A \subset X$. If A has positive dimension, then there is a point $x \in A$ and an open neighborhood $U = U(x) \subset X$ such that $A \cap U$ is a submanifold of U of positive dimension. The function $p|_{A \cap U}$ is strictly plurisubharmonic and constant. That is impossible. ∎

3.3 Corollary. *Let X be a 1-complete manifold that is holomorphically convex. Then X is a Stein manifold.*

At the end of this chapter we will see that the condition "holomorphically convex" is not necessary.

3.4 Corollary. *Let $A \subset \mathbb{P}^n$ be an analytic hypersurface. Then $\mathbb{P}^n - A$ is a Stein manifold. Every analytic subset $B \subset \mathbb{P}^n$ of positive dimension meets A in at least one point.*

Example 6: There is a famous theorem by H. Cartan:

> *A domain $G \subset \mathbb{C}^2$ is a domain of holomorphy if and only if the first Cousin problem is always solvable.*

The solvability is also true for higher-dimensional domains of holomorphy, but there is a greater class of domains with this property. Take, e.g., the domain $G \subset \mathsf{P}^n \subset \mathbb{C}^n$ (with $n \ge 3$) that is the union of the three open sets

$$\begin{aligned} U_1 &:= \{\mathbf{z} \in \mathsf{P}^n : |z_1| > \varepsilon\}, \\ U_2 &:= \{\mathbf{z} \in \mathsf{P}^n : |z_2| > \varepsilon\}, \\ U_3 &:= \{\mathbf{z} \in \mathsf{P}^n : |(z_3, \ldots, z_n)| < \varepsilon\}. \end{aligned}$$

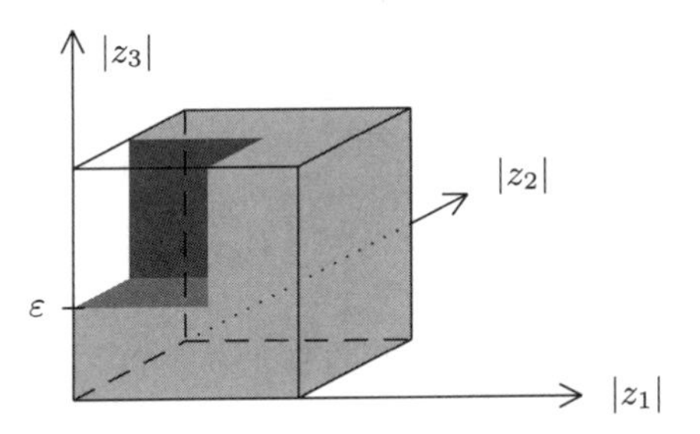

Figure V.2. Cousin-I in a domain which is not Hartogs convex

The domain $G = U_1 \cup U_2 \cup U_3$ is not Hartogs pseudoconvex (see Figure V.2). But all three covering elements U_i are domains of holomorphy. Therefore, we have $H^1(U_i, \mathcal{O}) = 0$ for $i = 1, 2, 3$, as will be proved in Section V.5. So $\mathscr{U} = \{U_1, U_2, U_3\}$ is an acyclic covering for $\mathcal{O}$, and every Cousin-I distribution in G can be given by a cocycle $f \in H^1(\mathscr{U}, \mathcal{O})$, up to a coboundary. Such an f consists of holomorphic functions f_{ij} in U_{ij} with $f_{12} + f_{23} + f_{31} = 0$ on U_{123}. This implies that the Laurent series of f_{12} (around the origin) in U_{123} contains no powers $z_1^i z_2^j$ with $i < 0, j < 0$. By the identity theorem this is true on the whole set U_{12}. Therefore, we can subtract a coboundary $\delta\{g_1, g_2, 0\}$ from our cocycle such that thereafter $f_{12} \equiv 0$.

Then the new f_{23}, f_{13} coincide on U_{123}. Together they give a holomorphic function h in $(U_1 \cup U_2) \cap U_3$ that extends holomorphically to U_3. Then the new cocycle f is equal to the coboundary $\delta(0, 0, h)$, and hence the old cocycle also cobounds (i.e., is a coboundary). Consequently the Cousin-I problem is solvable.

Example 7: Assume that X is an n-dimensional complex manifold and that $x_0 \in X$ is a point. We can blow up X in x_0. Then we obtain an n-dimensional complex manifold $\widehat{X}$, an $(n-1)$-dimensional complex submanifold $A \subset \widehat{X}$ that is isomorphic to $\mathbb{P}^{n-1}$, and a proper holomorphic map $\pi : \widehat{X} \to X$ that maps A to x_0 and $\widehat{X} - A$ biholomorphically onto $X - \{x_0\}$. We have a strictly pseudoconvex neighborhood U around x_0. We can lift the strictly plurisubharmonic exhaustion function p on U by π to $\pi^{-1}(U)$. This is a strongly pseudoconvex neighborhood of A that is not, however, strictly pseudoconvex.

Example 8: For a similar example consider the analytic set

$$A := \{\mathbf{z} \in \mathbb{C}^{2n} \, : \, z_1 z_{n+j} - z_j z_{n+1} = 0 \text{ for } j = 2, \dots, n\}.$$

Outside of the origin A is regular of dimension $n+1$. For example, if $z_i \neq 0$, then $z_{n+j} = z_j \cdot z_{n+i}/z_i$ for $j = 1, \dots, n$. So A is there parameterized by $z_1, \dots, z_n$ and z_{i+1}. It follows that $\dim(A) = n + 1$.

We have a meromorphic map from $\mathbb{C}^{2n}$ to the Osgood space $\mathbb{O}^n$ given by $w_i = z_i/z_{n+i}$, $i = 1, \dots, n$. Its graph is the set

$$\widehat{X} = \{(\mathbf{z}, w) \in \mathbb{C}^{2n} \times \mathbb{O}^n \,:\, w_i z_{n+i} - z_i = 0 \text{ for } i = 1, \dots, n\}.$$

We denote by X the closure of the part of $\widehat{X}$ that lies over $A - \{\mathbf{0}\}$. Let $\pi : X \to A$ be the restriction of the canonical projection from $A \times \mathbb{O}^n$ to A. Then π maps a 1-dimensional projective space onto $\mathbf{0}$ and the rest biholomorphically onto $A - \{\mathbf{0}\}$.

The space X is an $(n+1)$-dimensional complex manifold, locally given by the equations

$$\begin{aligned} z_i &= w_i z_{n+i} && \text{for } i = 1, \dots, n, \\ w_i &= w_1 && \text{for } i = 2, \dots, n. \end{aligned}$$

The set $E(s) := \{(z_1, \dots, 2n) \in A \,:\, z_1/z_{n+1} = \cdots = z_n/z_{2n} = s\}$ is an n-dimensional plane for every $s \in \mathbb{P}^1$, and A is the union of all these planes. Since $E(s_1) \cap E(s_2) = \{\mathbf{0}\}$ for $s_1 \neq s_2$, A is singular at the origin. It follows that X is a vector bundle of rank n over $\pi^{-1}(\mathbf{0}) \cong \mathbb{P}^1$.

Now we use the function $p(\mathbf{z}) = \sum_i z_i \bar{z}_i$ on $\mathbb{C}^{2n}$. It induces a strictly plurisubharmonic function on the complex manifold $A - \{\mathbf{0}\}$. The $(n+1)$-dimensional complex manifold X is strongly pseudoconvex by $p \circ \pi$, but not 1-complete, since it contains the compact analytic subset $\pi^{-1}(\mathbf{0})$.

Example 9: Consider the covariant tangent bundle T' of $\mathbb{P}^n$. If ξ is a global holomorphic vector field on $\mathbb{P}^n$, then a plurisubharmonic function p_ξ on T' is defined by

$$p_\xi(\omega_x) := \omega_x(\xi_x) \cdot \overline{\omega_x(\xi_x)}.$$

We consider local coordinates in a set $U_i \subset \mathbb{P}^n$, for example $t_\nu := z_\nu/z_0$ for $\nu = 1, \dots, n$ in the case $i = 0$. Then every ω over U_i can be written uniquely in the form $\omega = \sum_\nu \omega_\nu dt_\nu$. So $t_1, \dots, t_n, \omega_1, \dots, \omega_n$ are local coordinates in T' over U_i. If $\xi = \sum_\nu \xi_\nu \partial/\partial z_\nu$, then

$$p_\xi(\mathbf{t}, \boldsymbol{\omega}) = \Big(\sum_\nu \omega_\nu \xi_\nu\Big) \cdot \Big(\overline{\sum_\nu \omega_\nu \xi_\nu}\Big).$$

We have the following $n + n^2$ holomorphic vector fields over U_i that extend to $\mathbb{P}^n$:

$$\frac{\partial}{\partial t_\nu} = q(0, \dots, \underbrace{z_0}_{\nu\text{th place}}, \dots, 0) \text{ and } t_\mu \frac{\partial}{\partial t_\nu} = q(0, \dots, \underbrace{z_\mu}_{\nu\text{th place}}, \dots, 0),$$

where $q : \mathcal{O}(1) \oplus \cdots \oplus \mathcal{O}(1) \to T\mathbb{P}^n$ is the canonical bundle epimorphism in the Euler sequence. Plurisubharmonic functions p_ν^i, respectively $p_{\nu\mu}^i$, are defined in local coordinates by $\omega_\nu \overline{\omega}_\nu$, respectively $\omega_\nu \overline{\omega}_\nu \cdot t_\mu \bar{t}_\mu$. By adding them all we obtain the plurisubharmonic function

$$p_i = \left(\sum_{\nu=1}^{n} \omega_\nu \overline{\omega}_\nu\right) \cdot \left(1 + \sum_{\mu=1}^{n} t_\mu \bar{t}_\mu\right).$$

Finally, if we add the p_i constructed for the various U_i, $i = 0, \dots, n$, we obtain a function p. If Z denotes the zero section in T', then p vanishes on Z and is strictly plurisubharmonic and positive outside of Z. It tends to ∞ for $\|\boldsymbol{\omega}\| \to \infty$. So the complex manifold T' is strongly pseudovonvex, but not 1-complete.

Example 10: This example is probably due to J.P. Serre.

Assume that E is an elliptic curve (i.e., a compact Riemann surface of genus 1). Then E is a 1-dimensional torus, given by a lattice of periods $(1, \xi)$, where the imaginary part of ξ is positive. We may write the elements of E as real linear combinations $z = s \cdot 1 + t \cdot \xi$, with $s, t \in [0, 1]$. The first cohomology group $H^1(E, \mathcal{O})$ is equal to $\mathbb{C}$ (this will follow from results of the next chapter, but it is also a very classical result in the theory of Riemann surfaces).

We have a covering $\mathscr{U}$ of E consisting of the two elements

$$\begin{aligned} U_1 &= \{z = s + t\xi \,:\, 0 \le s \le 1/2, \quad 0 \le t \le 1\}, \\ U_2 &= \{z = s + t\xi \,:\, 1/2 \le s \le 1, \quad 0 \le t \le 1\}. \end{aligned}$$

Denote by $C \subset E$ the circle $\{z = 1 + t \cdot \xi \,:\, 0 \le t \le 1\}$ and define there the function $f \equiv 1$ (and 0 on the other component of $U_1 \cap U_2$). Then f is a cocycle in $Z^1(\mathscr{U}, \mathcal{O})$. It yields a nonvanishing cohomology class, since otherwise we would obtain a nonconstant bounded holomorphic function on $\mathbb{C}$. We construct a fiber bundle A above E with fibers $A_x = \mathbb{C}$ by gluing above C the point $(z, w) \in U_2 \times \mathbb{C}$ with $(z, w - 1) \in U_1 \times \mathbb{C}$. We denote this point in the bundle by $[z, w]$.

Now, A is topologically trivial (since we can find a continuous function g on U_2 that coincides with f on C such that $\delta\{0, g\} = f$), but it is not analytically trivial, because it defines a nontrivial cohomology class. It follows from the construction that the notions of real lines, planes, and convexity are well defined in A.

Let $\widehat{A}$ be the bundle with typical fiber $\mathbb{P}^1$ that is obtained from A by adding the point at infinity to each fiber. Then $\widehat{A}$ is compact and has the infinite cross section $D = D_\infty$ over E. We put $X := A = \widehat{A} - D$. Then X does not contain any compact analytic set of dimension 1. Otherwise, there would be a number b such that the analytic set meets every fiber in exactly b points. We could pass over to their barycenters and would obtain an ordinary holomorphic cross section in A. This would imply the triviality of A.

We consider the real 3-dimensional surface

$$S = S_r = \{[z, w] \in A \,:\, z = s + t\xi,\ w = s + r\, exp(2\pi \mathrm{i} \cdot \theta) \text{ and } 0 \le \theta \le 1\}.$$

Here $r > 0$ is a very big number. The surface S bounds a tube G, which converges to A as r tends to ∞.

We look at the holomorphic tangent H in any point of S. We can consider H as a real plane in A. By subtracting s in the fiber above $s + t\xi$ we pass over from A to $E \times \mathbb{C}$. The hypersurface S is transformed diffeomorphically into a convex cylinder over E and H into a contacting plane that does not enter the interior of the cylinder. So H does not enter the interior of the tube G. If H were contained in S, then the transformed H would lie in the boundary of the cylinder and therefore be compact. So H itself would be a compact analytic set. We saw that such a set does not exist. So there remains only the possibility that H contacts S of first order; i.e., the intersection $H \cap S$ is a real line. We can choose a complex coordinate $\zeta = x + \mathrm{i}y$ in H such that our real line is exactly the x-axis.

If ϱ is a smooth defining function for S, then ϱ behaves on H like the function y^2. It follows that $(\varrho|_H)_{\zeta\overline{\zeta}} > 0$. This means that the Levi form of ϱ is always positive definite on the holomorphic tangent H. Now we can construct a strictly plurisubharmonic function $\widehat{\varrho}$ in $U - D$ (where U is a neighborhood of D in $\widehat{A}$) whose level sets are the manifolds S_r such that $\widehat{\varrho}$ converges to ∞ when approaching D.[2] It follows that X is strongly pseudoconvex.

Later on we shall prove that a strongly pseudoconvex manifold is holomorphically convex, and since X contains no positive-dimensional compact analytic subset, it is a Stein manifold. So there are many holomorphic functions in X. Assume that there is a meromorphic function on $\widehat{A}$ that has poles of order m on D only. Then the coefficient of the highest polar part of f is a holomorphic cross section η in the mth tensor power of the normal bundle of D. Because this normal bundle is topologically trivial, η cannot have zeros on D. So f tends to infinity approaching D, and no analytic set $\{f = \text{const}\}$ meets D. This is a contradiction, since A is not analytically trivial. Every holomorphic function on X must have essential singularities on D.

Analytic Tangents. Let G be a domain in $\mathbb{C}^n$ with $n \geq 2$ and p a strictly plurisubharmonic function in G. Denote by X the set $\{\mathbf{z} \in G : p(\mathbf{z}) < 0\}$. Let $B \subset\subset G$ be an open subset and $\mathbf{w} \in \partial X \cap \overline{B}$ an arbitrary point.

The expansion of p in $\mathbf{w}$ is given by

$$p(\mathbf{z}) = p(\mathbf{w}) + 2\,\mathrm{Re}\, Q(\mathbf{w}, \mathbf{z} - \mathbf{w}) + \mathrm{Lev}(p)(\mathbf{w}, \mathbf{z} - \mathbf{w}) + R(\mathbf{w}, \mathbf{z} - \mathbf{w}),$$

where

$$Q(\mathbf{w}, \mathbf{h}) = \mathbf{h} \cdot \nabla p(\mathbf{w})^{\,t} + \frac{1}{2}\mathbf{h} \cdot \mathrm{Hess}(p)(\mathbf{w}) \cdot \mathbf{h}^{\,t},$$

[2] We may assume that ϱ is globally defined and $\varrho = r$ on S_r. Then we choose an unbounded strictly monotonic smooth function $h : \mathbb{R} \to \mathbb{R}$ such that h''/h' is very large. The function $\widehat{\varrho} := h \circ \varrho$ will do the job.

with the complex Hessian $\text{Hess}(p)(\mathbf{w}) = \Big(p_{z_i z_j}(\mathbf{w}) \;\Big|\; i,j = 1,\dots,n\Big)$, and

$$\lim_{\mathbf{h}\to\mathbf{0}} \frac{R(\mathbf{w},\mathbf{h})}{\|\mathbf{h}\|^2} = 0.$$

The map $\mathbf{h} \mapsto Q(\mathbf{w},\mathbf{h})$ is a holomorphic polynomial of degree 2, for every $\mathbf{w}$. The assignment $\mathbf{w} \mapsto Q(\mathbf{w},\mathbf{h})$ is smooth.

Let $G' \subset\subset G$ be an open neighborhood of $\overline{B}$. We can find constants $c, k > 0$ such that

$$\text{Lev}(p)(\mathbf{w}, \mathbf{z}-\mathbf{w}) + R(\mathbf{w}, \mathbf{z}-\mathbf{w}) \ge k\|\mathbf{z}-\mathbf{w}\|^2 - c\|\mathbf{z}-\mathbf{w}\|^3$$

for $\mathbf{z} \in G'$ and arbitrary $\mathbf{w}$.

We say that a real number $\varepsilon > 0$ is *sufficiently small* with respect to B if every ball $U(\mathbf{w})$ with center $\mathbf{w} \in \partial X \cap \overline{B}$ and radius ε belongs to G', and

$$k\|\mathbf{z}-\mathbf{w}\|^2 - c\|\mathbf{z}-\mathbf{w}\|^3 \ge 0$$

is valid on these $U(\mathbf{w})$.

Assuming that this is the case, for $\mathbf{w} \in \partial X \cap \overline{B}$ we define the analytic set

$$A(\mathbf{w}) := \{\mathbf{z} \in U(\mathbf{w}) \,:\, Q(\mathbf{w}, \mathbf{z}-\mathbf{w}) = 0\}$$

in $U(\mathbf{w})$. On $A(\mathbf{w})$ we have $p(\mathbf{z}) \ge 0$. Therefore, $A(\mathbf{w}) - \{\mathbf{w}\}$ is outside of X. Since $\mathbf{w}$ belongs to the boundary of X, this implies that $A = A(\mathbf{w})$ has codimension 1. We call A an *analytic tangent* (or, in German, a *Stützfläche*) for X at $\mathbf{w}$.

3.5 Proposition. *Let G, p, X, B, G' be as above. There exists a differentiable family $A(\mathbf{w})$ of analytic tangents to ∂X at the points $\mathbf{w} \in \overline{B} \cap \partial X$ with $A(\mathbf{w}) \cap \overline{X} = \{\mathbf{w}\}$.*

Here "differentiable" means that the defining quadratic polynomials for $A(\mathbf{w})$ depend smoothly on $\mathbf{w}$.

Exercises

1. Let X be a complex manifold and $Y \subset X$ a closed complex submanifold. Construct a nonnegative smooth function $f : X \to \mathbb{R}$ with the following properties:
   (a) $\text{Lev}(f)(x,\xi) \ge 0$ for every $x \in X$, $\xi \in T_x(X)$.
   (b) For every $x \in Y$ there is a linear subspace $P_x \subset T_x(X)$ such that $P_x + T_x(Y) = T_x(X)$ and $\text{Lev}(f)(x,\xi) > 0$ for $\xi \in P_x$, $\xi \ne 0$.
2. Let X be a 1-complete complex manifold and $f_1, \dots, f_q \in \mathcal{O}(X)$. Show that $X - N(f_1, \dots, f_q)$ is q-complete.
3. Let $G \subset\subset \mathbb{C}^n$ be a strictly convex domain with smooth boundary. Construct the differentiable family $A(\mathbf{w})$ of analytic tangents to ∂G.

4. Let φ be a strictly plurisubharmonic smooth function defined in a neighborhood $U = U(\mathbf{0}) \subset \mathbb{C}^n$ with $\varphi(\mathbf{0}) = 0$, and ψ an arbitrary smooth function in U. Prove that there are $r > 0$ and $\varepsilon > 0$ sufficiently small such that $\varphi + \varepsilon \cdot \psi$ is strictly plurisubharmonic in $\mathsf{B}_r(\mathbf{0})$ and

$$\{\mathbf{z} \in \mathsf{B}_r(\mathbf{0}) : (\varphi + \varepsilon \cdot \psi)(\mathbf{z}) < 0\}$$

is a Stein manifold.

4. Cuboids

Distinguished Cuboids. If Q is a closed subset in $\mathbb{C}^n$, then in this section we say that something *is defined on* Q if it is defined in a small neighborhood of Q. Two objects are called *equal on* Q if they coincide in a small neighborhood of Q. So actually we consider "germs" along Q.

Definition. A *cuboid* is a closed domain

$$Q = \{(z_1, \dots, z_n) = (x_1 + \mathrm{i}x_{n+1}, \dots, x_n + \mathrm{i}x_{2n}) \in \mathbb{C}^n : a_i \le x_i \le b_i\},$$

where $a_i < b_i$ are real numbers for $i = 1, \dots, 2n$.

If there are partitions

$$a_i = a_i^0 < a_i^1 < \cdots < a_i^{m_i} = b_i, \qquad \text{for } i = 1, \dots, 2n,$$

then we denote by $\mathcal{A}$ the system of sequences

$$a_i^j,\ i = 1, \dots, 2n, \quad j = 0, \dots, m_i.$$

A closed covering $\mathscr{U}_{\mathcal{A}}$ of Q is defined by the system of cuboids

$$Q_{j_1,\dots,j_{2n}} = \{\mathbf{z} : a_i^{j_i - 1} \le x_i \le a_i^{j_i}, \text{ for } i = 1, \dots, 2n\}.$$

For any open covering of Q there is a closed cuboid covering which is finer.

In the following it will be not enough to have a covering of Q. Additionally, we need a system of complex submanifolds

$$Q = X_0 \supset X_1 \supset \cdots \supset X_{i-1} \supset X_i \supset \cdots \supset X_s, \quad s \le n,$$

where X_i has dimension $n - i$, such that there is a holomorphic function f_i in X_{i-1} vanishing everywhere on X_i to first order (and maybe also vanishing at points of $X_{i-1} - X_i$).

Definition. A *distinguished cuboid* is a cuboid that is equipped with a system $\{(X_i, f_i) : i = 1, \dots, s\}$. The number $n - s$ is called the *manifold dimension* of the distinguished cuboid.

Vanishing of Cohomology. As described above, everything on a closed cuboid is assumed to be defined on an open neighborhood of the cuboid. Therefore, we can consider cocycles and coboundaries with respect to closed cuboid coverings.

4.1 Proposition. *Assume that Q is a cuboid and $\mathscr{U}_{\mathcal{A}}$ a closed cuboid covering of Q. Then every cocycle $\xi \in Z^1(\mathscr{U}_{\mathcal{A}}, \mathcal{O})$ cobounds.*

PROOF: First we consider a very simple system $\mathcal{A}$. We just take the case where $m_1 = 2$ and all other $m_i = 1$. Then

$$\mathscr{U}_{\mathcal{A}} = \{Q_0 = Q_{1,1,\ldots,1}, Q_1 = Q_{2,1,\ldots,1}\}$$

has the minimal possible number of elements. The cocycle ξ is given by one holomorphic function ξ_{01} on $Q_0 \cap Q_1 = \{\mathbf{z} \in Q : z_1 = a_1^1\}$. Let ε be a small positive number. Then in the z_1-plane we can choose two continuous paths α_j from $a_1^1 + \mathrm{i}(a_{n+1}^0 - \varepsilon)$ to $a_1^1 + \mathrm{i}(a_{n+1}^1 + \varepsilon)$ (for $j = 0$ on the left side and for $j = 1$ on the right side of the line $x_1 = a_1^1$, see Figure V.3) and define $\eta = \{\eta_0, \eta_1\}$ by the Cauchy integral

$$\eta_j(z_1, \ldots, z_n) := \frac{1}{2\pi \mathrm{i}} \int_{\alpha_j} \frac{\xi_{01}(w, z_2, \ldots, z_n)}{w - z_1}\, dw.$$

Then we get $\delta\eta = \xi$.

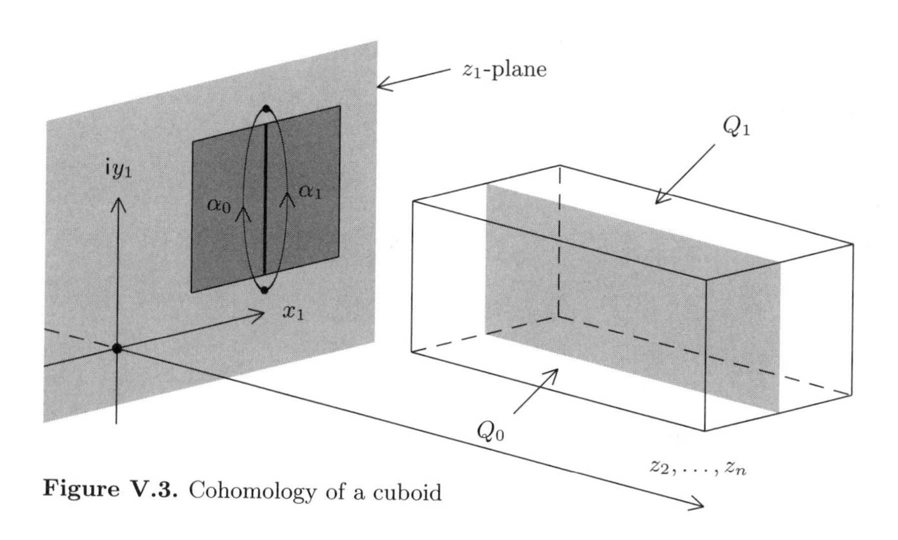

Figure V.3. Cohomology of a cuboid

The next step is an induction on the number $m = \sum_i m_i$. We just handled the case $m = 2n + 1$. Now we assume that $m > 2n + 1$ and that the proposition already has been proved for any number smaller than m. We put $i_0 := \max\{i : 1 \le i \le 2n \text{ and } m_i \ge 2\}$ and define

$$Q' := \{\mathbf{z} \in Q : x_{i_0} \le a_{i_0}^{m_{i_0}-1}\},\ Q'' := \{\mathbf{z} \in Q : x_{i_0} \ge a_{i_0}^{m_{i_0}-1}\}.$$

By the induction hypothesis, we have $\xi|_{\mathscr{U}'} = \delta(\eta')$ and $\xi|_{\mathscr{U}''} = \delta(\eta'')$ with cuboid coverings $\mathscr{U}'$ for Q' and $\mathscr{U}''$ for Q''. Hence, $\xi - \delta\{\eta', \eta''\}$ is a cocycle with respect to the covering $\mathscr{U}$ that is given by a holomorphic function g on the surface

$$Q' \cap Q'' = \{\mathbf{z} \in Q : x_{i_0} = a_{i_0}^{m_{i_0}-1}\}.$$

As above, by integration we obtain holomorphic functions η_0 on Q' and η_1 on Q'' with $g = \eta_1 - \eta_0$. So we have a cochain $\sigma = \{\eta' + \eta_0, \eta'' + \eta_1\}$ on Q with $\delta\sigma = \xi$ (since the functions η_j are already cocycles). This completes the induction. ■

Vanishing on the Embedded Manifolds. We consider a distinguished cuboid Q in $\mathbb{C}^n$ with manifold dimension $n - s$, and for $i = 0, \ldots, s$ we prove the following:

4.2 Proposition. *Every cocycle of $Z^1(\mathscr{U}_\mathcal{A} \cap X_i, \mathcal{O})$ cobounds, and every holomorphic function on X_{i+1} can be extended to a holomorphic function on X_i.*

PROOF: We carry out an induction on i. In the case $i = 0$ the proposition has been proved already. The induction hypothesis now states that it holds for some i, $0 \le i \le s - 1$, and we prove the extension property for $i + 1$. Assume that g is the holomorphic function on X_{i+1}. Define $g(\mathbf{z}) = 0$ on the other connected components of $N(f_{i+1}) \subset X_i$. If the covering $\mathscr{U}_\mathcal{A}$ is sufficiently small, we can extend this g to a cochain $\eta \in C^0(\mathscr{U}_\mathcal{A} \cap X_i, \mathcal{O})$. The coboundary $\delta(\eta)$ vanishes on $N(f_{i+1})$. Therefore, there is a cochain $\alpha \in C^1(\mathscr{U}_\mathcal{A} \cap X_i, \mathcal{O})$ with $\delta(\eta) = \alpha \cdot f_{i+1}$. It is clear that α is a cocycle, and by the induction hypothesis there is a $\gamma \in C^0(\mathscr{U}_\mathcal{A} \cap X_i, \mathcal{O})$ with $\alpha = \delta(\gamma)$. Since $\delta(\eta - f_{i+1} \cdot \gamma) = 0$, we get $\widehat{g} = \eta - f_{i+1} \cdot \gamma$ as a holomorphic extension of g.

Now we prove that any $\xi \in Z^1(\mathscr{U}_\mathcal{A} \cap X_{i+1}, \mathcal{O})$ cobounds. As in the proof of the preceeding theorem we have to show this only in the case where $m_1 = 2$ and $m_i = 1$ for $i > 1$. So ξ is simply a holomorphic function g on $(Q_{1,1,\ldots,1} \cap Q_{2,1,\ldots,1}) \cap X_{i+1}$. We have to find holomorphic functions f' on $Q_{1,1,\ldots,1} \cap X_{i+1}$ and f'' on $Q_{2,1,\ldots,1} \cap X_{i+1}$ with $f'' - f' = g$. For that we first extend g to $(Q_{1,1,\ldots,1} \cap Q_{2,1,\ldots,1}) \cap X_i$, construct f', f'' for X_i (induction hypothesis), and then restrict them to X_{i+1}. That completes the proof. ■

Cuboids in a Complex Manifold. We assume that X is an n-dimensional complex manifold and p a smooth real-valued function on X with $p(x) \to 2$ when $x \to \partial X$.[3] We assume further that p is strictly plurisub-

[3] This means that for every $\varepsilon > 0$ there is a compact subset $K \subset X$ such that $p(x) > 2 - \varepsilon$ for $\in X - K$.

harmonic on $\{p(x) = 1\}$ and define $Y := \{x \in X : p(x) < 1\}$. Then Y is a relatively compact strongly pseudoconvex open subset of X.

For an n-tuple $\mathbf{z} = (z_1, \dots, z_n)$ we use the norm

$$|\mathbf{z}| := \sup_i \max(|\mathrm{Re}(z_i)|, |\mathrm{Im}(z_i|).$$

For an open neighborhood $G \subset X$ of ∂Y we want to apply the results from the end of V.3. (It does not matter that there are no global coordinates on G.) We consider a relatively compact open neighborhood $B = B(\partial Y) \subset\subset G$ and choose a G' with $B \subset\subset G' \subset\subset G$.

For any point $x_0 \in \partial Y$ there is a compact cuboid U^* with center x_0 in a coordinate neighborhood S around x_0. If a real number $\varepsilon > 0$ is given, we can choose every U^* so small that $U^* \subset U(\mathbf{w}) \subset\subset S$ for every point $\mathbf{w} \in U^*$, where $U(\mathbf{w})$ is the ball with center $\mathbf{w}$ and radius ε (with respect to the local coordinates). Then for every $\mathbf{w} \in U^* \cap \partial Y$ we have the analytic tangent $A(\mathbf{w})$ given by a quadratic polynomial $f_{\mathbf{w}}$ in $U(\mathbf{w})$. It follows that $A(\mathbf{w}) \cap U^* \cap \overline{Y} = \mathbf{w}$, and the function $p|U^* \cap (A(\mathbf{w}) - \mathbf{z})$ is positive. This is illustrated by Figure V.4.

For an open subset $Y' \subset\subset Y$ we have the following proposition.

4.3 Proposition. *There is a distinguished cuboid $Q^* = U^* \times Q \subset \mathbb{C}^N$ with the following property:*

For $s = N - n$ the submanifold X_s is projected biholomorphically onto a compact set $U' \subset U^*$ with $Y' \cap U^* \subset U' \subset Y$.

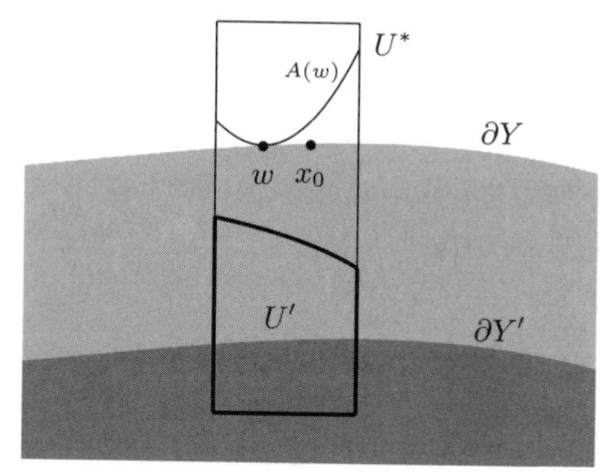

Figure V.4. Projection of the distinguished cuboid

PROOF: We may work in $\mathbb{C}^n$ and assume that x_0 is the origin. We use the differentiable family of analytic tangents $A(\mathbf{w}) = N(f_{\mathbf{w}})$ with $\mathbf{w} \in \partial Y \cap U^*$.

Take a positive number r so big that we always have $|1/f_{\mathbf{w}}(\mathbf{z})| < r$ for $\mathbf{w} \in \partial Y \cap U^*$ and $\mathbf{z} \in Y' \cap U^*$. In a small neighborhood of $\mathbf{w}$ we have $|1/f_{\mathbf{w}}(\mathbf{z})| > r$. Therefore, by compactness there are finitely many of these $f_{\mathbf{w}}$, say $f_1, \ldots, f_s$, with $\max_i |1/f_i(\mathbf{z})| > r$ for all $\mathbf{z} \in \partial Y \cap U^*$. For a fixed i we denote by Q_i' the cuboid $\{\mathbf{z}' = (z_{n+1}, \ldots, z_{n+i}) : |\mathbf{z}'| \le r\}$ and by Q_i'' the cuboid $\{\mathbf{z}'' = (z_{n+i+1}, \ldots, z_{n+s}) : |\mathbf{z}''| \le r\}$. We put $Q^* = U^* \times Q_i' \times Q_i''$.

The submanifolds $X_i \subset Q^*$ are obtained in the following way: Consider the graph of the i-tuple $(1/f_1, \ldots, 1/f_i)$ in $U^* \times Q_i'$, take the union of those connected components that contain points over Y, and multiply this union by Q_i''. The manifold X_i has dimension $N - i$. In X_{i-1} we have the holomorphic function $z_{n+i} \cdot f_i(\mathbf{z}) - 1$, which vanishes on X_i to order 1.

Finally, the projection U' of X_s contains no point of ∂Y. Hence, it is contained in Y. Since $U^* \cap Y'$ is contained in U', the proposition is proved. ∎

4.4 Corollary. *In the above notation,* $H^1(U', \mathcal{O}) = 0$.

Enlarging U'. We use the same notation as before and construct a set $\widehat{U}' \subset U^* \cap Y$ that is bigger than U' where the vanishing theorem still holds.

For that purpose we take an open set $\widehat{Y}'$ instead of Y' with $Y' \subset\subset \widehat{Y}' \subset\subset Y$ and $U' \subset \widehat{Y}'$. We need a bigger $\widehat{r} > r$ such that we still have $|1/f_{\mathbf{w}}(\mathbf{z})| < \widehat{r}$ on $\widehat{Y}'$ for $\mathbf{w} \in \partial Y \cap U^*$. But now $\max_i |1/f_i(\mathbf{z})| > \widehat{r}$ no longer is true on $\partial Y \cap U^*$. So we add some functions to the old ones, $f_{s+1}, \ldots, f_{\widehat{s}}$, such that we get the old situation again.

We get the Q_i' and the Q_i'', for $i = 1, \ldots, \widehat{s}$. But we write $\widehat{Q}_i'$ instead of Q_i', $\widehat{Q}_i''$ for Q_i'', and $\widehat{X}_i$ for X_i. Then we have $\widehat{Q}^* = U^* \times \widehat{Q}_i' \times \widehat{Q}_i'' \subset \mathbb{C}^{\widehat{N}}$ with $N < \widehat{N}$. The projection of $X_{\widehat{s}}$ is a compact set $\widehat{U}' \subset Y \cap U^*$ with $\widehat{Y}' \cap U^* \subset \widehat{U}'$. So again we have the vanishing of the cohomology.

The following statement is proved in the next paragraph.

> (∗) *Every holomorphic function on* U' *can be approximated arbitrarily well by holomorphic functions on* $\widehat{U}'$.

Suppose all cochains are given with respect to a covering $\mathcal{U}$ of $\widehat{U}' \cong \widehat{X}_{\widehat{s}}$. If $\xi \in Z^1(\mathcal{U}, \mathcal{O})$ is a cocycle over $\widehat{U}'$, we have $\xi = \delta(\eta)$ over U' and $\xi = \delta(\widehat{\eta})$ over $\widehat{U}'$, with cochains η and $\widehat{\eta}$ with respect to the coverings $\mathcal{U} \cap U'$ and $\mathcal{U} \cap \widehat{U}'$. Then $\widehat{\eta} - \eta$ is a holomorphic function over U', which we approximate by a sequence of holomorphic functions $\widehat{g}_j \in \mathcal{O}(\widehat{U}')$. We may replace the cochain $\widehat{\eta}$ by $\widehat{\eta}_j := \widehat{\eta} - \widehat{g}_j$ and obtain that $\widehat{\eta}_j$ approximates η over U'.

Now we can prove the following theorem.

4.5 Theorem. *If $x_0 \in \partial Y$ and U^* is a small cuboid in a local coordinate system around x_0, then $H^1(Y \cap U^*, \mathcal{O}) = 0$.*

PROOF: Take a fixed open covering $\mathscr{U}$ of $Y \cap U^*$ and exhaust Y by a sequence of domains Y_i with $Y_i \subset\subset Y_{i+1} \subset\subset Y$. We get a sequence of compact sets $U_i' \subset U^* \cap Y$ with $Y_i \cap U^* \subset U_i'$.

If $\xi \in Z^1(\mathscr{U}, \mathcal{O})$ is a cocycle, for every i we get a cochain η_i on U_i' such that $\xi = \delta(\eta_i)$ on U_i'. As we have seen above, we may assume that the difference of η_{i+1} and η_i on U_i' is as small as we want. Then the sequence η_i converges to some $\eta \in C^0(\mathscr{U}, \mathcal{O})$ with $\delta(\eta) = \xi$. ∎

Approximation. We first prove the following simple result.

4.6 Proposition. *Assume that Q is a cuboid in $\mathbb{C}^N$. Then any holomorphic function f on Q can be approximated arbitrarily well by polynomials.*

PROOF: We make the induction hypothesis that f is already a polynomial in the variables $\mathbf{z}' = (z_1, \dots, z_s)$. Let γ_ε be the boundary of the rectangle with corners $(a_{s+1} - \varepsilon, a_{N+s+1} - \varepsilon)$, $(b_{s+1} + \varepsilon, a_{N+s+1} - \varepsilon)$, $(b_{s+1} + \varepsilon, b_{N+s+1} + \varepsilon)$, and $(a_{s+1} - \varepsilon, b_{N+s+1} + \varepsilon)$ in the (x_{s+1}, x_{N+s+1})-plane. We have the integral representation

$$f(\mathbf{z}', z_{s+1}, \mathbf{z}'') = \frac{1}{2\pi \mathbf{i}} \int_{\gamma_\varepsilon} \frac{f(\mathbf{z}', \xi, \mathbf{z}'')}{\xi - z_{s+1}} \, d\xi.$$

It is necessary to approximate $1/(\xi - z_{s+1})$ with a fixed $\xi \in |\gamma_\varepsilon|$ by a polynomial. We consider only the case where $\operatorname{Re}(\xi) = b_{s+1} + \varepsilon$. In this case there exists a disk D (with a center far to the left) such that $\xi \in \partial D$ and the rectangle with corners (a_{s+1}, a_{N+s+1}), (b_{s+1}, a_{N+s+1}), (b_{s+1}, b_{N+s+1}), and (a_{s+1}, b_{N+s+1}) is contained in D. Then we expand $1/(\xi - z_{s+1})$ into a power series on the disk and get the desired approximation. So the induction step is complete. Since nothing remains to be shown for $s = 0$, the proof is finished. ∎

We have to consider the cuboids $Q = U^* \times Q_s'$ and $\widehat{Q} = U^* \times \widehat{Q}'_{\widehat{s}}$. For this denote by M the Cartesian product of the last $\widehat{s} - s$ factors of the Cartesian product $\widehat{Q}'_{\widehat{s}}$. It follows that $Q_s' \times M \subset \widehat{Q}'_{\widehat{s}}$, and $S := (U^* \times Q_s' \times M) \cap \widehat{X}_{\widehat{s}}$ is projected biholomorphically onto X_s and onto U'. Every holomorphic function on S is the restriction of a holomorphic function in $U^* \times Q_s' \times M$ and can be approximated arbitrarily well by polynomials, in particular by holomorphic functions on the larger set $U^* \times \widehat{Q}_{\widehat{s}}$. Therefore, every holomorphic function on U' can be approximated by holomorphic functions on $\widehat{U}'$, and we have proved $(*)$.

Finally, we wish to replace U^* by its open kernel $(U^*)^\circ$. For this we exhaust U^* by relatively compact cuboids U_j^* with $U_j^* \subset\subset U_{j+1}^*$, and choose the functions f_i, $i = 1, \ldots, \widehat{s}$ (from the "enlarging" process), independently of j.

Then we can approximate cochains on $U_j' \subset U_j^*$ by those on $\widehat{U}_j' \subset U_j^*$, and we can approximate these again by cochains on $\widehat{U}_{j+1}'$. So for any 1-cocycle ξ on $(U^*)^\circ \cap Y$ we obtain (as a limit) a cochain η on $(U^*)^\circ \cap Y$ with $\delta\eta = \xi$. Then we have the following theorem.

4.7 Theorem. *If $x_0 \in \partial Y$ and U^* is a small cuboid in a local coordinate system around x_0, then $H^1(Y \cap (U^*)^\circ, \mathcal{O}) = 0$.*

Exercises

1. Let $G \subset\subset \mathbb{C}^n$ be a strictly convex subdomain with smooth boundary. Use coverings with distinguished cuboids to prove that $H^1(G, \mathcal{O}) = 0$.
2. Prove for the same domain G that every holomorphic function in G can be approximated by polynomials.

5. Special Coverings

Cuboid Coverings. Again let X be an n-dimensional complex manifold and p a smooth real-valued function on X with $p(x) \to 2$ for $x \to \partial X$. Assume that p is strictly plurisubharmonic on $\{p(x) = 1\}$, and let $Y := \{x \in X : p(x) < 1\}$. We also consider open sets $Y_\alpha = \{x \in X : p(x) < 1 + \alpha\}$, where $0 \le \alpha < \varepsilon$ and ε is a very small positive number such that also Y_α is strongly pseudoconvex.

Since ∂Y is compact, it can be covered by finitely many cuboids U_i^* around points $x_i \in \partial Y$ which are always contained relatively compactly in a local coordinate neighborhood such that x_i is the origin of the coordinate system. If $U_i^* = \{\mathbf{z} : |\mathbf{z}| \le r_i\}$ and $0 < t_i < r_i$, then we may also assume that the open sets $U_i = \{\mathbf{z} : |\mathbf{z}| < t_i\}$ cover ∂Y. By adding more cuboids, the t_i can be chosen as small as necessary.

We also need these properties for the sets ∂Y_α, and for sets $\partial\widehat{Y}$, where $\widehat{Y}$ is an open set between Y and Y_α. For that we have to move the centers of the cuboids a little bit. If α is very small, the U_i still cover ∂Y.

There may exist a compact set $K \subset Y$ that is not yet covered by the U_i. Then we construct a covering of K with additional cuboids $U_i \subset\subset U_i^* \subset\subset Y$. Finally, we have the following result.

5.1 Proposition. *For a small $\varepsilon > 0$, for any α with $0 \le \alpha < \varepsilon$ and any open set $\widehat{Y}$ with $Y \subset \widehat{Y} \subset Y_\alpha$ there exist coverings $\mathscr{U}^* = \{U_i^* : i = 1, \ldots, m\}$*

and $\mathscr{U} = \{U_i : i = 1, \dots, m\}$ *of* $\widehat{Y}$ *with* $U_i \subset\subset U_i^*$ *such that the following hold:*

1. *Every* U_i^* *is a compact cuboid in a local coordinate system around a point* x_i *that corresponds to the origin in this coordinate system. The set* U_i^* *is small enough in the sense of Sections V.3 and V.4 to have families of analytic tangents in* U_i^*.
2. *There are numbers* $0 < t_i < r_i$ *such that*

$$U_i = \{\mathbf{z} : |z| < t_i\} \quad \textit{and} \quad U_i^* = \{\mathbf{z} : |z| \le r_i\}$$

in the chosen coordinates.

3. *For each* i, *either* $x_i \in \widehat{Y}$ *or* $U_i^* \subset Y$.
4. *For each* i, *the "star"* $S_i = \bigcup_{j:\, U_j \cap U_i \neq \varnothing} U_j$ *is contained in* U_i^*.

We call $(\mathscr{U}, \mathscr{U}^*)$ a *special pair of coverings* for $(Y, \widehat{Y})$ (cf. Figure V.5).

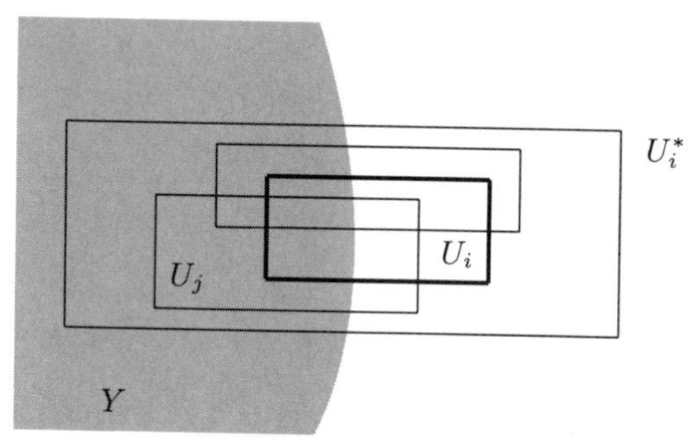

Figure V.5. Special pairs of coverings

The Bubble Method. Let a small $\varepsilon > 0$ be fixed in the sense of the preceding paragraph. We choose a special pair $(\mathscr{U}, \mathscr{U}^*)$ of coverings for $(Y, \widehat{Y})$. Since $H^1(U_i \cap Y, \mathcal{O}) = 0$, it follows that $\mathscr{U} \cap Y$ is acyclic, and every cohomology class of $H^1(Y, \mathcal{O})$ can be given by a cocycle $\xi \in Z^1(\mathscr{U} \cap Y, \mathcal{O})$.

Let $x_0 \in \partial Y$ be an arbitrary point and i an index such that $x_0 \in U_i$. Then we have a cochain $\eta = \{\eta_j\} \in C^0(\mathscr{U} \cap U_i^* \cap Y, \mathcal{O})$ with $\delta(\eta) = \xi | (U_i^* \cap Y)$. We change η_j to 0 if $U_j \not\subset S_i$. Then η is a cochain over Y, and the cocycle $\xi - \delta(\eta)$ represents the same cohomology class as ξ. But $\xi - \delta(\eta)$ vanishes on $U_i \cap Y$.

Now we use the *bubble method*. By altering the strongly plurisubharmonic function p a little bit on some compact subset of U_i (let us say by subtracting a smooth function with compact support) we enlarge Y a little bit to a strongly pseudoconvex open subset $\widehat{Y} \supset Y$ with $\widehat{Y} - Y \subset\subset U_i$. Then every cohomology class represented by some $\xi \in Z^1(\mathscr{U} \cap Y, \mathcal{O})$ extends to $\widehat{Y}$. So the restriction $H^1(\widehat{Y}, \mathcal{O}) \to H^1(Y, \mathcal{O})$ is surjective. We assume that for $\widehat{Y}$ (after a small movement) the conditions for U_i, U_i^* are still satisfied.

In the next step we go to another $x_0 \in \widehat{Y}$ and apply the procedure to $\widehat{Y}$. Continuing in this way, after finitely many steps we have the following result.

5.2 Proposition. *There is a fundamental system of strongly pseudoconvex neighborhoods Y_α of $\overline{Y}$ such that the restriction $H^1(Y_\alpha, \mathcal{O}) \to H^1(Y, \mathcal{O})$ is surjective.*

Fréchet Spaces. The following theory can be found in the famous multigraphed notes [CS53]. See also [GuRo65].

Assume that V is a complex vector space.

Definition. A *seminorm* in V is a map $p : V \to \{x \in \mathbb{R} : x \geq 0\}$ with the following properties:

1. $p(a \cdot f) = |a| \cdot p(f)$ for $a \in \mathbb{C}$, $f \in V$.
2. $p(f_1 + f_2) \leq p(f_1) + p(f_2)$, if f_1, $f_2 \in V$.

If a sequence of seminorms p_i, $i \in \mathbb{N}$, is given in V, we can define a topology in V. If $\varepsilon > 0$ and $m \in \mathbb{N}$, we call the set

$$U_\varepsilon^{(m)}(0) = \{f \in V : p_i(f) < \varepsilon \text{ for } i = 1, \dots, m\}$$

an (ε, m)-neighborhood of 0. As m increases the (ε, m)-neighborhoods, of course, become smaller.

A subset $W \subset V$ with $0 \in W$ is called a *neighborhood* of 0 if there is an (ε, m)-neighborhood $U_\varepsilon^{(m)}(0) \subset W$. If $f_0 \in V$ is an arbitrary element and W a neighborhood of 0, then

$$f_0 + W := \{f_0 + f : f \in W\}$$

is a neighborhood of f_0. Finally, a set $M \subset V$ is called *open* if it is a neighborhood of any $f \in M$.

Definition. A sequence (f_k) in V is called *convergent* to f if for any neighborhood $W(f)$ there is a number k_0 such that $f_k \in W$ for $k \geq k_0$.

The sequence (f_k) is called a *Cauchy sequence* if for all $\varepsilon > 0$ and $m \in \mathbb{N}$ there is a number k_0 such that $p_i(f_l - f_k) < \varepsilon$ for $i = 1, \dots, m$ and $k, l \geq k_0$.

Definition. A *Fréchet topology* on a complex vector space V is a topology defined by a sequence (p_i) of seminorms, with the following properties:

1. For every $f \in V - \{0\}$ there is an i with $p_i(f) > 0$; i.e., the topology is Hausdorff.
2. The space V is complete in this topology: Every Cauchy sequence converges.

A *Fréchet space* is a complex vector space with a Fréchet topology.

Every Fréchet space possesses a metric that defines its topology.

The vector space $\mathcal{O}(X)$ of holomorphic functions on a complex manifold X is a Fréchet space. We just take an increasing sequence of compact subsets K_i with $K_i \subset (K_{i+1})^\circ$ and $\bigcup_i K_i = X$, and we define $p_i(f) := \sup|f(K_i)|$. If (f_k) is a Cauchy sequence of holomorphic functions on X, then $f_k|(K_i)^\circ$ converges uniformely to a holomorphic function on $(K_i)^\circ$. Hence (f_k) converges compactly to a holomorphic function f on X. This means that (f_k) converges to f in the topology of $\mathcal{O}(X)$.

Since every Fréchet space V is a metric space, a subset $M \subset V$ is compact if and only if every sequence (f_k) in M has a convergent subsequence with limit in M.

Definition. A linear map $v : E \to F$ between two Fréchet spaces is called *compact* or *completely continuous* if there is a neighborhood $U(0) \subset E$ such that $v(U)$ is relatively compact in F.

Obviously, every completely continuous map is continuous.

We have the following famous theorem of Schwartz.

5.3 Theorem of L. Schwartz. *Assume that E, F are two Fréchet spaces and that $u, v : E \to F$ are two continuous linear maps with the following properties:*

1. *u is surjective.*
2. *v is completely continuous.*

Then the quotient $F/(u+v)(E)$ has finite dimension.

A typical example of a completely continuous map is given as follows:

Let X be a complex manifold and $Y \subset\subset X$ an open subset. Using the increasing sequence of compact subsets from above, we have $Y \subset K_i$ for some i. Now, $U := \{f \in \mathcal{O}(X) : p_i(f) < 1\}$ is a neighborhood of 0 in $E := \mathcal{O}(X)$. Let $v : E \to F := \mathcal{O}(Y)$ be defined by $v(f) := f|_Y$. Then

$$v(U) \subset \mathscr{B} := \{g \in \mathcal{O}(Y) : \sup|g| \le 1\}.$$

Since $\mathscr{B}$ is closed and bounded, it follows from Montel's theorem that it is compact. This shows that the restriction map v is completely continuous.

Remark. It is clear that we can replace the space of holomorphic functions by the space of holomorphic cross sections in an analytic vector bundle.

Finiteness of Cohomology. We wish to apply Fréchet space theory to our standard situation.

Let X be an n-dimensional complex manifold and $Y = \{x \in X : p(x) < 1\}$, where p is a smooth exhaustion function on X with $\sup(p) = 2$. Assume, as usual, that p is strictly plurisubharmonic on a neighborhood of ∂Y. Let a small real number $\varepsilon > 0$ be fixed such that Y_α is strongly pseudoconvex for $0 \le \alpha < \varepsilon$. We have shown that the restriction map $H^1(Y_\alpha, \mathcal{O}) \to H^1(Y, \mathcal{O})$ is surjective for sufficiently small α.

We start with a special pair $(\mathscr{U}, \mathscr{U}^*)$ of cuboid coverings for (Y, Y_α), choose numbers $\widehat{t}_i$ with $t_i < \widehat{t}_i < r_i$ and define $\widehat{U}_i = \{\mathbf{z} : |\mathbf{z}| < \widehat{t}_i\}$. This gives a covering $\widehat{\mathscr{U}} = \{\widehat{U}_i : i = 1, \ldots, m\}$ of $\overline{Y}_\alpha$.

The spaces $C^0(\mathscr{U} \cap Y, \mathcal{O})$, $Z^1(\mathscr{U} \cap Y, \mathcal{O})$, and $Z^1(\widehat{\mathscr{U}} \cap Y_\alpha, \mathcal{O})$ are Fréchet spaces, as is $Z^1(\widehat{\mathscr{U}} \cap Y_\alpha, \mathcal{O}) \oplus C^0(\mathscr{U} \cap Y, \mathcal{O})$. Since the cohomology can be extended from Y to Y_α, it follows that the map

$$u : E = Z^1(\widehat{\mathscr{U}} \cap Y_\alpha; \mathcal{O}) \oplus C^0(\mathscr{U} \cap Y, \mathcal{O}) \to F = Z^1(\mathscr{U} \cap Y, \mathcal{O}),$$

with $u(\xi, \eta) := \xi|_{\mathscr{U} \cap Y} + \delta(\eta)$, is surjective. The map $v : E \to F$ is defined as the negative of the restriction map from $Z^1(\widehat{\mathscr{U}} \cap Y_\alpha, \mathcal{O})$ to $Z^1(\mathscr{U} \cap Y, \mathcal{O})$, i.e., by $v(\xi, \eta) := -\xi|_{\mathscr{U} \cap Y}$. Then v is completely continuous.

The map $u + v$ is the coboundary map $\delta : C^0(\mathscr{U} \cap Y, \mathcal{O}) \to Z^1(\mathscr{U} \cap Y, \mathcal{O})$, since the first summand goes to 0. So the quotient $F/(u+v)(E)$ is the first cohomology group of Y. Its complex dimension is finite by Schwartz's theorem. Thus we have proved the following result.

5.4 Theorem. *If Y is a strongly pseudoconvex relatively compact subset of a complex manifold X, then* $\dim_{\mathbb{C}} H^1(Y, \mathcal{O}) < \infty$.

Holomorphic Convexity. We consider the same situation as above and choose an $\alpha > 0$ such that $Y \subset\subset Y_\alpha$ and:

1. $\dim_{\mathbb{C}} H^1(Y_\alpha, \mathcal{O}) < \infty$.
2. For every $x \in \partial Y$ we have an analytic tangent $A(x) \subset Y_\alpha$ to ∂Y, with $A(x) \cap \overline{Y} = \{x\}$, given by a holomorphic function f_x in a neighborhood U of $A(x)$.

Let $x \in \partial Y$ be an arbitrary point, $f = f_x$, $A = A(x)$, and U a suitable neighborhood of A. Then $\mathscr{U}_x = \{U \cap Y_\alpha, Y_\alpha - A\}$ is a covering of Y_α consisting of only two elements. So $Z^1(\mathscr{U}_x, \mathcal{O}) = \mathcal{O}(U \cap Y_\alpha - A)$, and every function f^{-k} belongs to $Z^1(\mathscr{U}_x, \mathcal{O})$. Since $H^1(Y_\alpha, \mathcal{O})$ is finite-dimensional, for $m \gg 0$ there are complex numbers $a_1, \ldots, a_m$, not all zero, such that the cocycle given by the function

$$g = (a_1 \cdot f^{-1} + \cdots + a_m \cdot f^{-m})|_{(U \cap Y_\alpha - A)}$$

is cohomologous to 0. Therefore, we find holomorphic functions h_1 in $Y_\alpha - A$ and h_2 in $U \cap Y_\alpha$ such that $h_1 - h_2 = g$. The function

$$h := \begin{cases} h_1 & \text{on } Y_\alpha - A, \\ h_2 + g & \text{on } U \cap Y_\alpha, \end{cases}$$

is meromorphic in Y_α with poles in A. So $h|_Y$ is a holomorphic function, and for $y \to x$ it tends to ∞.

If K is compact in Y, then the holomorphic convex hull $\widehat{K}$ does not approach x. This is true for every $x \in \partial Y$. So $\widehat{K}$ is compact. As a consequence we have the following result.

5.5 Theorem. *If Y is a strongly pseudoconvex relatively compact subset of a complex manifold X, then Y is holomorphically convex.*

Moreover, we have a solution of the Levi problem in this special case.

5.6 Theorem. *Assume that $Y = \{x \in X : p(x) < 1\}$, where $p(x)$ is a smooth function in X with $p(x) \to 2$ for $x \to \partial X$ that is strictly plurisubharmonic on ∂Y. If Y contains no higher-dimensional compact analytic subsets, then Y is a Stein manifold.*

PROOF: In Section V.3 we showed that every holomorphically convex manifold that contains no compact analytic subset of positive dimension is a Stein manifold.
■

5.7 Corollary. *If p is a strictly plurisubharmonic exhaustion function of X, then $Y = \{x \in X : p(x) < 1\}$ is a Stein manifold.*

This is clear, since under the assumption there are no compact analytic subsets of positive dimension in X.

Negative Line Bundles. Assume now that Z is a compact n-dimensional complex manifold and that F is a holomorphic line bundle on Z. Let $\pi : F \to Z$ be the canonical projection.

Definition. The line bundle F is called *negative* if there is a strongly pseudoconvex neighborhood $Y \subset\subset F$ of the zero section of F.

We identify Z with the zero section in F, and we call Y a *tube* around Z.

For an arbitrary holomorphic vector bundle V of rank s on Z we consider the pullback bundle π^*V on F. If $U \subset Z$ is an open set, then the holomorphic cross sections in π^*V over $\pi^{-1}(U)$ are holomorphic maps $f : \pi^{-1}(U) \to V$ such that for any $z \in U$ the restriction of f to the fiber F_z has values in V_z.

The trivial line bundle on F is denoted by $\mathscr{O}_F$. Its local holomorphic cross sections are the local holomorphic functions on F. If $V|_U \cong U \times \mathbb{C}^s$, then the holomorphic cross sections of π^*V over $\pi^{-1}(U)$ are s-tuples of local holomorphic functions. Since the cocycles in our theory are given by local holomorphic functions, we may replace $\mathscr{O}_F$ by π^*V on F and get the same results for $X = F$ and the strongly pseudoconvex set Y. So we have $\dim_{\mathbb{C}} H^1(Y, \pi^*V) < \infty$.

Among the holomorphic maps $f : \pi^{-1}(U) \to V$ that correspond to sections in π^*V we have the maps f that on every fiber F_z are homogeneous of degree m (with values in V_z). We denote the space of these maps by $\mathscr{O}_m(\pi^{-1}(U), V)$. It can be identified with the space of holomorphic cross sections in the tensor product $V \otimes F^{-m}$ (see Section IV.2).

We choose a finite open covering $\mathscr{U} = \{U_i : i = 1, 2, \ldots, l\}$ of Z. Let $\widehat{\mathscr{U}}$ be the covering of F given by the sets $\widehat{U}_i = \pi^{-1}(U_i)$ We denote by $Z^1_m(\widehat{\mathscr{U}}, \pi^*V)$ the vector space of cocycles $\xi_m = \{\xi^m_{ij}\}$ with $\xi^m_{ij} \in \mathscr{O}_m(\pi^{-1}(U_{ij}), V)$. These are homogeneous of degree m on the fibers of F. We also call them *cocycles of degree m*. Every finite sum $\xi = \sum_i \xi_i$ of cocycles of distinct degrees m_i is contained in $Z^1(F, \pi^*V)$ and can be restricted to $Z^1(Y, \pi^*V)$. It is a coboundary if and only if all ξ_i are coboundaries. Since the cohomology $H^1(Y, \pi^*V)$ is finite, we obtain the following theorem:

5.8 Theorem. *Assume that F is a negative line bundle and V an arbitrary holomorphic vector bundle on the compact manifold Z. Then there is an integer m_0 such that $H^1(Z, V \otimes F^{-m}) = 0$ for all $m \geq m_0$.*

Remark. A line bundle F on Z is called *positive* if its complex dual F' is negative. Since $(F')' = F$, the above result can be reformulated as follows:

If F is a positive line bundle on Z, then $H^1(Z, V \otimes F^m) = 0$ for $m \geq m_0$.

Bundles over Stein Manifolds. Now we again consider a general n-dimensional complex manifold X and assume that we have a strictly plurisubharmonic function $\widetilde{p}$ in X with $\widetilde{p}(x) \to 2$ when $x \to \partial X$. Then $Z = \{x \in X : \widetilde{p}(x) < 1\}$ is called a *special Stein manifold.*

We use similar methods to those in the last paragraph. Now let F be the trivial line bundle $X \times \mathbb{C}$ on X and the strictly plurisubharmonic function p

on F be defined by

$$p(x, w) := \widetilde{p}(x) + w\overline{w}.$$

The manifold $Y = \{(x, w) \in F : x \in Z \text{ and } p(x, w) < 1\}$ is a strictly pseudoconvex tube around the zero section in $F|_Z$. However, Z is not compact now; it is only relatively compact in X. The tube extends to a neighborhood of $\overline{Z}$. The boundary of the restriction to $\overline{Z}$ has an "edge" $\partial Y \cap (\partial Z \times \mathbb{C})$.

But the proof of finiteness of cohomology goes through in the same way as if there were no edge. We have to apply the method of analytic tangents to ∂Y and ∂Z simultaneously. If V is a holomorphic vector bundle on X, we denote the pullback of V to F by $\widehat{\mathcal{O}}$. It then follows that $\dim_{\mathbb{C}} H^1(Y|_Z, \widehat{\mathcal{O}}) < \infty$. But in this situation all the tensor powers of F are again the trivial line bundle. We have $V \otimes F^{-m} \cong V$ for every m. From this we conclude the following:

5.9 Theorem. *If Z is a special Stein manifold (as described above) and V a holomorphic vector bundle on Z, then $H^1(Z, V) = 0$.*

Exercises

1. Let E be a Fréchet space and let $U = U(0) \subset E$ be a neighborhood such that $\overline{U}$ is compact. Prove that E is finite-dimensional.
2. Let $u : E \to F$ be a continuous linear map between Fréchet spaces. Prove that if $u(E)$ is finite-dimensional, then u is compact.
3. Let E be a Fréchet space. A subset $M \subset E$ is called *bounded* if for every neighborhood $U = U(0) \subset E$ there is an $r_0 \in \mathbb{R}$ such that $M \subset r \cdot U$ for every $r \geq r_0$. Show that a compact set $K \subset E$ is closed and bounded. Let X be a complex manifold. Show that every closed and bounded subset $K \subset \mathcal{O}(X)$ is compact.
4. Let X be an n-dimensional Stein manifold and $Y \subset X$ a closed submanifold of dimension $n - 1$. Prove that the line bundle $N_X(Y)$ is positive.
5. Let $F = \mathcal{O}(-1)$ be the tautological bundle over $\mathbb{P}^2$. Prove that F is negative and that

$$H^1(\mathbb{P}^2, \mathcal{O}(k)) = H^1(\mathbb{P}^2, F^{-k}) = 0 \quad \text{for every } k \in \mathbb{Z}.$$

6. Let $X \subset \mathbb{P}^3$ be a regular hypersurface and $i : X \hookrightarrow \mathbb{P}^3$ the canonical injection. Prove that $H^1(X, i^*\mathcal{O}(k)) = 0$ for $k \in \mathbb{Z}$.

6. The Levi Problem

Enlarging: The Idea of the Proof. We wish to show that every strongly pseudoconvex (i.e., 1-convex) manifold is holomorphically convex, and that every strictly pseudoconvex (i.e., 1-complete) manifold is Stein. This solves the Levi problem for complex manifolds.

Recall that an n-dimensional manifold X is called *strongly pseudoconvex* if there is a smooth exhaustion function p on X that is strictly plurisubharmonic on $\{x \in X : p(x) \geq 1\}$.

For $s \geq 1$ the open sublevel set $X_s = \{x \in X : p(x) < s\}$ is holomorphically convex, and its cohomology with coefficients in $\mathcal{O}$ is finite. We will construct an open set $\widehat{X}$ with $X_s \subset\subset \widehat{X}$ by extending X_s in several steps such that every holomorphic function on X_s can be approximated by holomorphic functions on $\widehat{X}$.

In the first step we choose a point $x_1 \in \{x \in X : p(x) = s\}$, a local coordinate system around x_1, a sufficiently small compact cuboid U_1^* around x_1 in that coordinate system, and an open concentric cuboid $U_1 \subset\subset (U_1^*)^\circ$ such that every analytic tangent $A(x_0)$ with $x_0 \in \bar{U}_1 - X_s$ is defined in U_1^*.

By disturbing p a little bit in a neighborhood of x_1, we get a new function p_2 and a bigger set $X^2 = \{p_2 < s\}$ with $X_s \subset X^2$ and $x_1 \in X^2$. We do this carefully enough to get X^2 strongly pseudoconvex again. Since everything happens in a cuboid, we can apply the theory of Section V.4 and obtain the desired approximation property for functions in $X^1 = X_s$ and X^2. Then we repeat this process and construct a bigger set X^2, and so on.

Since $\overline{X}_s$ is compact, finitely many set $X^1, X^2, X^3, \ldots, X^N$ already cover $\overline{X}_s$, and we order them such that X^{i+1} is obtained by enlarging X^i in a neighborhood of some point $x_i \in \overline{X}_s$. Finally, $X_s \subset\subset \widehat{X} := X^N$, and every holomorphic function on X_s can be approximated by holomorphic functions on $\widehat{X}$.

Enlarging: The First Step. We choose an $s \geq 1$. Then p is strictly plurisubharmonic in a neighborhood of ∂X_s, and also on $X - X_s$. We put $\overline{\partial} X_s = \{x \in X : p(x) = s\}$ and get $\partial X_s \subset \overline{\partial} X_s$.

Assume that $x_1 \in \overline{\partial} X_s$ is a point that is the origin in a local coordinate system, and that $U_1^* = \{\mathbf{z} : |\mathbf{z}| \leq r\}$ is a compact cuboid in that coordinate system. Furthermore, let $U_1 \subset\subset U_1^*$ be a concentric open cuboid such that every analytic tangent $A(x_0)$ with $x_0 \in \overline{U}_1 - X_s$ is defined in U_1^*. On $A(x_0)$ we have $p(x) > s$ outside x_0, and there is an $\varepsilon > 0$ with $p(x) > s + \varepsilon$ on $\partial U_1^* \cap A(x_0)$. We assume that p is strictly plurisubharmonic on U_1^*.

6.1 Proposition. *We can add to p a function $h \leq 0$ with support $\overline{U}_1$ such that $h(x_1) < 0$, $p + h$ is still strictly plurisubharmonic, and the following approximation property is satisfied:*

If $\widehat{X} := \{x \in X : p(x) + h(x) < s\}$, then every holomorphic function (or cross section in a holomorphic vectorbundle) over X_s can be approximated over X_s (i.e., on every compact subset of X_s) arbitrarily well by holomorphic functions (respectively holomorphic cross sections) over $\widehat{X}$.

PROOF: Assume that $K \subset X_s$ is compact. We choose an $R \gg 0$ and consider the family $A(x)$ of analytic tangents to the points of $\overline{U}_1 \cap \partial X_s$. There is a finite set of points $y_1^*, \dots, y_{m_1}^* \in \overline{U}_1 \cap \partial X_s$ such that for the defining functions f_i^* of $A(y_i^*)$ and the corresponding vector $g^* := (1/f_1^*, \dots, 1/f_{m_1}^*)$ we have the estimate

$$|g^*(x)| := \max_i |1/f_i^*(x)| > R \text{ on } \partial X_s \cap \overline{U}_1.$$

Now we replace the points y_i^* by points y_i very near to y_i^*, with $p(y_i) > s$, and the functions f_i^* by f_i (the defining function for $A(y_i)$) such that we still have $|g(x)| > R$ on $\partial X_s \cap \overline{U}_1$, for $g := (1/f_1, \dots, 1/f_{m_1})$.

We choose R and m_1 so big that $|g(x)| < R$ in $K \cap U_1^*$.

Now we add a smooth function h with support $\overline{U}_1$ that is strictly negative on U_1. We let $\widehat{X} := \{x \in X : p(x) + h(x) < s\}$ and may assume that the points y_i are not contained in $\widehat{X}$ and that $|g(x)| > R$ on $\widehat{X} - X_s$. We can make h and its derivatives so small that $p + h$ is still strictly plurisubharmonic in U_1^* (see Figure V.6).

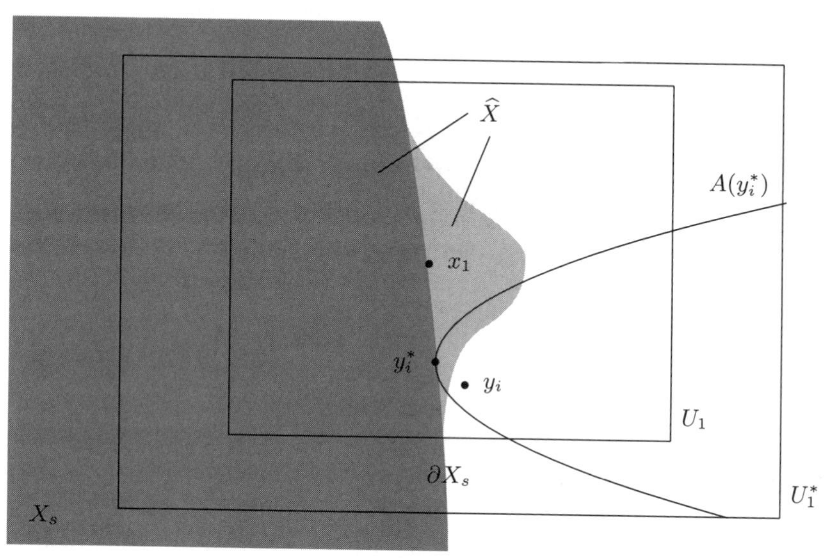

Figure V.6. The enlarging process

Next we take an $\widehat{R} > R$ and add further functions f_x with $x \in \partial\widehat{X}$, say $f_{m_1+1}, \dots, f_m$, such that for $g = (1/f_1, \dots, 1/f_m)$ it follows that $|g(x)| > \widehat{R}$ at every point of $\partial\widehat{X} \cap U_1^*$, and $|g(x)| < \widehat{R}$ in $K \cap U_1^*$. Then we define

$$\widehat{U} = \{x \in \widehat{X} \cap U_1^* : |g(x)| \leq \widehat{R}\}.$$

We define U' by replacing the number $\widehat{R}$ by R in the first m_1 inequalities.

We use the enlarging method of Section V.4 and prove in the same way that every holomorphic function in U' can be approximated by those in $\widehat{U}$.

Now we let $\widehat{U}$ and U' increase by making the number m bigger (i.e., adding new functions) and, simultaneously, by also making R and $\widehat{R}$ bigger with $R < \widehat{R}$. So we can construct sequences U_j', $\widehat{U}_j$ where the $\widehat{U}_j$ converge to $U_1^* \cap \widehat{X}$ and the U_j' to $U_1^* \cap X_s$. Hence, we have the approximation of the holomorphic functions in $X_s \cap U_1^*$ by those in $\widehat{X} \cap U_1^*$.

Let $\mathscr{W}$ be the open covering of $\widehat{X}$ consisting of the two elements $W_1 = X_s$ and $W_2 = U_1^* \cap \widehat{X}$. On the intersection W_{12} we can approximate every holomorphic function by those on W_2. Since the cohomology of $\widehat{X}$ is finite, there are cocycles $\xi_1, \ldots, \xi_k \in Z^1(\mathscr{W}, \mathcal{O})$ such that the corresponding cohomology classes $\overline{\xi}_1, \ldots, \overline{\xi}_k$ form a basis of the image of $Z^1(\mathscr{W}, \mathcal{O})$ in $H^1(\widehat{X}, \mathcal{O})$. The mapping $\varphi : \mathbb{C}^k \times C^0(\mathscr{W}, \mathcal{O}) \to Z^1(\mathscr{W}, \mathcal{O})$ defined by

$$\varphi(a_1, \ldots, a_k, \eta) := a_1\xi_1 + \cdots + a_k\xi_k + \delta(\eta)$$

is a surjective mapping of Fréchet spaces. By the open mapping theorem (see [Ru74]) φ is an open map.

If g is a holomorphic function in X_s, then g can be approximated on W_{12} by a sequence of holomorphic functions $f_j \in \mathcal{O}(W_2)$. Since φ is open, the cocycle $g - f_j$ (which is a small function on W_{12}) can be written as a sum $f_j' + \delta(\eta_j)$, with a small function $f_j' \in \mathcal{O}(W_{12})$ and a (small) cochain η_j. Replacing $f_j - f_j'$ by a suitable approximation f, and η_j by a suitable (and also small) cochain $\{\varrho_1, \varrho_2\}$, we get $g - f = \varrho_2 - \varrho_1$, with $\varrho_1 \in \mathcal{O}(W_1)$ and $\varrho_2 \in \mathcal{O}(W_2)$. The function

$$\widehat{g}(x) := \begin{cases} g(x) + \varrho_1(x) & \text{on } W_1, \\ f(x) + \varrho_2(x) & \text{on } W_2, \end{cases}$$

is holomorphic on $\widehat{X}$ and approximates g on X_s as well as we want. ∎

Enlarging: The Whole Process. Let $X^1 = X_s$, $p_1 = p$, and $h_1 = h$. Then we define $X^2 := \widehat{X}$ and $p_2 := p + h$. Repeating the process from above with a suitable point $x_2 \in \overline{\partial} X^2$, cuboids $U_2 \subset\subset U_2^*$, and a suitable function h_2 we obtain a new set $\widehat{X} = X^3$ that is defined by a new function $p_3 = p_2 + h_2$, and every holomorphic function on X^2 can be approximated arbitrarily well by holomorphic functions on X^3.

Continuing in this way we can always choose the correction functions h_i so small that the strict plurisubharmonicity is never disturbed. After finitely many steps we obtain the following:

6.2 Proposition. *There is a strictly pseudoconvex complex manifold $\widehat{X}$ with $X_s \subset\subset \widehat{X}$ such that every holomorphic function on X_s can be approximated over X_s arbitrarily well by holomorphic functions on $\widehat{X}$.*

We have the analoguous result for cochains.

6.3 Proposition. *Assume that $\mathscr{W}$ is an open covering of $\widehat{X}$ and that $\xi \in Z^1(\mathscr{W}, \mathcal{O})$ is a cocycle with $\xi | X_s = \delta(\eta_s)$, $\xi = \delta(\eta)$. Then, by changing η, the difference $\eta - \eta_s$ can be made arbitrarily small on X_s.*

Remark. We can replace $\widehat{X}$ by a set $X_{s+\varepsilon}$, with an $\varepsilon > 0$, since such a manifold is contained in $\widehat{X}$.

Solution of the Levi Problem.

Let X be an n-dimensional strongly pseudoconvex manifold that as above is endowed with a smooth exhaustion function p that is strictly plurisubharmonic on $\{p \geq 1\}$.

The solution of the Levi problem will be given in three steps. First we prove an approximation theorem for X_s and X.

6.4 Theorem. *Every holomorphic function in X_s, $s \geq 1$, can be approximated by holomorphic functions in X.*

PROOF: There is a maximal $\widehat{s}$ with $s < \widehat{s} \leq \infty$ such that every holomorphic function on X_s can be approximated by those on $X_{\widehat{s}}$. We have to prove $\widehat{s} = \infty$. If this were not the case, we could approximate each holomorphic function on $X_{\widehat{s}}$ by those on $X_{\widehat{s}+\varepsilon}$ with $\varepsilon > 0$. This would imply the approximation of the holomorphic functions on X_s by those on $X_{\widehat{s}+\varepsilon}$ and contradict maximality. ∎

Assume now that K is a compact set in X_s, $s \geq 1$. Then K is also contained in X_t for $t > s$. We denote the holomorphically convex hull of K in X_t by $\widehat{K}_t$. In the second step of our proof we show that $\widehat{K}_t = \widehat{K}_s$ for every $t \geq s$.

It is always the case that $\widehat{K}_s \subset \widehat{K}_t$, and, due to the approximation results, $\widehat{K}_t \cap X_s = \widehat{K}_s$. Suppose that there exists a $t > s$ such that $\widehat{K}_t$ is bigger than $\widehat{K}_s$. Then we can find a minimal $t' \geq s$ with $\widehat{K}_t \subset \overline{X}_{t'}$. It follows that $t' < t$ (since X_t is holomorphically convex), and there is at least one point $x_0 \in \widehat{K}_t \cap \partial X_{t'}$. We choose a very small number $\varepsilon > 0$ and consider the set $X_{t'+\varepsilon}$.

Let $A(x_0) \subset X_{t'+\varepsilon}$ be the analytic tangent to $X_{t'}$ at x_0. We may assume that there is a neighborhood $U = U(x_0)$ and a holomorphic function g in U such that $A(x_0) = \{x \in U \cap X_{t'+\varepsilon} : g(x) = 0\}$, and we find a meromorphic function f on $X_{t'+\varepsilon}$ with poles only on $A(x_0)$.

Then we construct a sequence of points x_ν outside of $\overline{X}_{t'}$ converging to x_0 and meromorphic functions f_ν on $X_{t'+\varepsilon}$ with poles only on $A(x_\nu)$. This can be done in such a way that f_ν converges to f with respect to the Fubini metric in $\overline{\mathbb{C}}$ (which will be described in Section VI.3). Then f_ν is holomorphic in a neighborhood of $\overline{X}_{t'}$, $|f_\nu(x_0)|$ takes arbitrarily large values, and $\sup_K|f_\nu|$ stays bounded. By approximation we can find a holomorphic function f^* on X_t such that $|f^*(x_0)| > \sup_K|f^*|$, i.e., $x_0 \notin \widehat{K}_t$. This is a contradiction.

Let us give some more details for the construction of f and f_ν. To begin with f, we use the open covering $\mathscr{U} = \{U_1, U_2\}$ of $X_{t'+\varepsilon}$ given by $U_1 = U \cap X_{t'+\varepsilon}$ and $U_2 = X_{t'+\varepsilon} - A(x_0)$. Since $\dim H^1(X_{t'+\varepsilon}, \mathcal{O}) < \infty$, there are complex numbers $a_1, \ldots, a_m$ such that the function $a := a_1 g^{-1} + \cdots + a_m g^{-m}$ is a cocycle on U_{12} that is cohomologous to zero. Then $a = f_2 - f_1$, with $f_1 \in \mathcal{O}(U_1)$ and $f_2 \in \mathcal{O}(U_2)$. So we get the meromorphic function f that is equal to $a + f_1$ on U_1, and equal to f_2 on U_2. It has $A(x_0)$ as polar set.

We can find a holomorphic family of analytic tangents $A(z)$, always given by one holomorphic equation $g_z = 0$, with z arbitrarily near to x_0. If m is big enough, then for every z we have a vector space P_z of principal parts $p = a_1 g_z^{-1} + \cdots + a_m g_z^{-m}$ and a linear map $\varphi_z : P_z \to H^1(X_{t'+\varepsilon}, \mathcal{O})$. Let $E_z \subset P_z$ be the kernel of φ_z. Then $\dim(E_z) \geq 1$, and there is a minimal dimension $m_0 \leq m$ for E_z that will occur at generic points. Taking E_z only at generic points and forming the closure for $z \to x_0$, we get a regular holomorphic family of vector spaces. If we always take meromorphic functions f_z with principal part in E_z, we get a continuous family (f_z) and the desired convergence $f_z \to f$ for $z \to x_0$.

The last step of our proof is simple:

Let $K \subset X$ be any compact set. Then K is contained in some K_s, $s \geq 1$, and from above we know that $\widehat{K}_t = \widehat{K}_s$ for every $t > s$. Let x be an arbitrary point of the holomorphically convex hull $\widehat{K}$. Then $x \in X_t$, for some $t > s$. If f is a holomorphic function in X_t, then there is a holomorphic approximation $\widehat{f} \in \mathcal{O}(X)$. Since we have $|\widehat{f}(x)| \leq \sup_K|\widehat{f}|$, we can conclude that $|f(x)| \leq \sup_K|f|$ as well. So $x \in \widehat{K}_t$, and $\widehat{K} = \widehat{K}_s$. This proves the following theorem.

6.5 Theorem. *If X is strongly pseudoconvex, then X is also holomorphically convex. It is a Stein manifold if and only if there are no compact analytic subsets in X of positive dimension.*

In addition, we have the following result.

6.6 Proposition. *If X contains a compact analytic subset A of positive dimension, then already $A \subset X_1$.*

The PROOF is the same as that for $\widehat{K} = \widehat{K}_1$.

We also obtain the following theorem.

6.7 Theorem. *If X is a strictly pseudoconvex (i.e., 1-complete) manifold and V a holomorphic vector bundle on X, then $H^1(X, V) = 0$.*

PROOF: This has already been proved for every distinguished Stein manifold X_s, $s \geq 1$. By approximation of coboundaries we get the result for X. ∎

The Compact Case. Let X be a **compact** complex manifold, and V a holomorphic vector bundle on X. We can find a pair of finite open coverings $(\mathscr{U}, \mathscr{U}^*)$ for X consisting of concentric cuboids $U_i \subset\subset U_i^*$ such that $H^1(U_i, V) = H^1(U_i^*, V) = 0$ for all i. Then $C^0(\mathscr{U}, V)$, $Z^1(\mathscr{U}, V)$ and $Z^1(\mathscr{U}^*, V)$ are Fréchet spaces. The finiteness of the coverings is essential here!

Again we consider the linear mappings

$$u, v : Z^1(\mathscr{U}^*, V) \oplus C^0(\mathscr{U}, V) \to Z^1(\mathscr{U}, V)$$

such that u is surjective, v is completely continuous, and $\text{Im}(u + v) = B^1(\mathscr{U}, V)$. Since the coverings are acyclic, it follows from the theorem of Schwartz that $\dim_{\mathbb{C}} H^1(X, V) < \infty$.

All restriction maps $\Gamma(U_i^*, V) \to \Gamma(U, V)$ are completely continuous. From this it also follows that $\text{id} : \Gamma(X, V) \to \Gamma(X, V)$ is completely continuous, and $\Gamma(X, V)$ is finite-dimensional. So we have the following theorem.

6.8 Theorem. *If X is a compact complex manifold, then*

$$\dim_{\mathbb{C}} \Gamma(X, V) < \infty \quad \text{and} \quad \dim_{\mathbb{C}} H^1(X, V) < \infty.$$

Exercises

1. Apply the results of this section to domains in $\mathbb{C}^n$. What do strong and strict pseudoconvexity mean in this case?
2. Let X be a strongly pseudoconvex complex manifold. Prove that there is a compact subset $K \subset X$ such that every irreducible compact analytic subset $A \subset X$ of positive dimension is contained in K.
3. Let X be a Stein manifold and V a holomorphic vector bundle over X. Prove that for every $x_0 \in X$ there exists an open neighborhood $U = U(x_0) \subset X$ and holomorphic sections $\xi_1, \ldots, \xi_N \in \Gamma(X, V)$ such that for every $x \in U$, V_x is generated by $\xi_1(x), \ldots, \xi_N(x)$.

Chapter VI

Kähler Manifolds

1. Differential Forms

The Exterior Algebra. Let X be an n-dimensional complex manifold and $x \in X$ a point. We consider complex-valued alternating multilinear forms on the tangent space $T_x(X)$ (abbreviated by T).

We assume that the reader is familiar with real multilinear algebra!

Definition. A complex *r-form* (or *r-dimensional differential form*) at x is an alternating $\mathbb{R}$-multilinear mapping

$$\varphi : \underbrace{T \times \cdots \times T}_{r\text{-times}} \to \mathbb{C}\,.$$

The set of all complex r-forms at x is denoted by F^r.

Remarks

1. By convention, $F^0 = \mathbb{C}$. $F^1 = F(T)$ is the complexification of the $2n$-dimensional real vector space $T_x^* = \operatorname{Hom}_{\mathbb{R}}(T, \mathbb{R})$.
2. Since T is $(2n)$-dimensional over $\mathbb{R}$, every alternating multilinear form on T with more than $2n$ arguments must be zero. So $F^r = 0$ for $r > 2n$.
3. In general, F^r is a complex vector space. We can represent an element $\varphi \in F^r$ uniquely in the form $\varphi = \operatorname{Re}(\varphi) + \mathrm{i}\operatorname{Im}(\varphi)$, where $\operatorname{Re}(\varphi)$ and $\operatorname{Im}(\varphi)$ are real-valued r-forms at x. Then it follows directly that
$$\dim_{\mathbb{C}} F^r = \binom{2n}{r}.$$
4. We associate with each element $\varphi \in F^r$ a complex-conjugate element $\overline{\varphi} \in F^r$ by setting $\overline{\varphi}(v_1, \dots, v_r) := \overline{\varphi(v_1, \dots, v_r)}$. We have:
   (a) $\overline{\varphi} = \operatorname{Re}(\varphi) - \mathrm{i}\operatorname{Im}(\varphi)$.
   (b) $\overline{\overline{\varphi}} = \varphi$.
   (c) $\overline{(\varphi + \psi)} = \overline{\varphi} + \overline{\psi}$.
   (d) φ is real if and only if $\overline{\varphi} = \varphi$.

Now let $\varphi \in F^r$ and $\psi \in F^s$ be given. The *wedge product* $\varphi \wedge \psi \in F^{r+s}$ is defined by

$$\begin{aligned}\varphi \wedge \psi(v_1, \dots, v_r, v_{r+1}, \dots, v_{r+s}) \\ := \frac{1}{r!s!} \sum_{\sigma \in S_{r+s}} (\operatorname{sgn} \sigma) \varphi(v_{\sigma(1)}, \dots, v_{\sigma(r)}) \cdot \psi(v_{\sigma(r+1)}, \dots, v_{\sigma(r+s)}) .\end{aligned}$$

Then $\varphi \wedge \psi(v, w) = \varphi(v) \cdot \psi(w) - \varphi(w) \cdot \psi(v)$ for 1-forms φ, ψ, and in general:

1. $\varphi \wedge \psi = (-1)^{rs} \psi \wedge \varphi$ (anticommutativity).
2. $(\varphi \wedge \psi) \wedge \omega = \varphi \wedge (\psi \wedge \omega)$ (associativity).

In particular, $\varphi \wedge \varphi = 0$ for every 1-form φ.

We also write $\bigwedge^r F$ instead of F^r. With the multiplication "$\wedge$" the vector space

$$\bigwedge F := \bigoplus_{r=0}^{\infty} \bigwedge^r F = \bigoplus_{r=0}^{2n} F^r$$

becomes a graded associative (noncommutative) $\mathbb{C}$-algebra with 1. It is called the *exterior algebra* at x.

For the moment let $\omega_j := dz_j$ and $\omega_{n+j} := d\overline{z}_j$ for $j = 1, \dots, n$. Then F^r is generated by the elements

$$\omega_{\nu_1} \wedge \cdots \wedge \omega_{\nu_r}, \text{ with } 1 \le \nu_1 < \cdots < \nu_r \le 2n.$$

The number of these elements is exactly $\binom{2n}{r}$. So they form a basis, and every $\varphi \in F^r$ has a uniquely determined representation

$$\varphi = \sum_{1 \le \nu_1 < \cdots < \nu_r \le 2n} a_{\nu_1 \dots \nu_r} \omega_{\nu_1} \wedge \cdots \wedge \omega_{\nu_r},$$

with complex coefficients $a_{\nu_1 \dots \nu_r}$.

Forms of Type (p, q). Now we consider the influence of the complex structure.

> **Definition.** Let $p, q \in \mathbb{N}_0$ and $p + q = r$. A form $\varphi \in F^r$ is called a *form of type* (p, q) if
>
> $$\varphi(cv_1, \dots, cv_r) = c^p \overline{c}^q \cdot \varphi(v_1, \dots, v_r) \text{ for all } c \in \mathbb{C}.$$

1.1 Proposition. *If $\varphi \in F^r$ is a nonzero form of type (p, q), then p and q are uniquely determined.*

PROOF: Suppose φ is of type (p, q) and of type (p', q'). Since $\varphi \neq 0$ there exist tangent vectors $v_1, \dots, v_r$ such that $\varphi(v_1, \dots, v_r) \neq 0$. Then

$$\varphi(cv_1, \ldots, cv_r) = \begin{cases} c^p \overline{c}^q \cdot \varphi(v_1, \ldots, v_r), \\ c^{p'} \overline{c}^{q'} \cdot \varphi(v_1, \ldots, v_r). \end{cases}$$

Therefore, $c^p \overline{c}^q = c^{p'} \overline{c}^{q'}$ for each $c \in \mathbb{C}$. If $c = e^{\mathrm{i}t}$, then $e^{\mathrm{i}t(p-q)} = e^{\mathrm{i}t(p'-q')}$, for arbitrary $t \in \mathbb{R}$. That can hold only when $p - q = p' - q'$. Since $p + q = p' + q' = r$, it follows that $p = p'$ and $q = q'$. ∎

1.2 Proposition.

1. *If φ is of type (p, q), then $\overline{\varphi}$ is of type (q, p).*
2. *If φ, ψ are both of type (p, q), then $\varphi + \psi$ and $\lambda \cdot \varphi$ (with $\lambda \in \mathbb{C}$) are also of type (p, q).*
3. *If φ is a form of type (p, q) and ψ of type (p', q'), then $\varphi \wedge \psi$ is of type $(p + p', q + q')$.*

We leave the PROOF to the reader.

Example

We have $dz_\nu(cv) = (cv)[z_\nu] = c \cdot (v[z_\nu]) = c \cdot dz_\nu(v)$, since z_ν is holomorphic. So dz_ν is a form of type $(1, 0)$.

From $d\overline{z}_\nu = \overline{dz_\nu}$ it follows that $d\overline{z}_\nu$ is a form of type $(0, 1)$.

Then $dz_{i_1} \wedge \cdots \wedge dz_{i_p} \wedge d\overline{z}_{j_1} \wedge \cdots \wedge d\overline{z}_{j_q}$ (with $1 \le i_1 < \cdots < i_p \le n$ and $1 \le j_1 < \cdots < j_q \le n$) is a form of type (p, q).

1.3 Theorem. *Any r-form φ has a uniquely determined representation*

$$\varphi = \sum_{p+q=r} \varphi^{(p,q)},$$

where $\varphi^{(p,q)} \in F^r$ are forms of type (p, q).

PROOF: The existence of the desired representation follows from the fact that the forms $dz_{i_1} \wedge \cdots \wedge dz_{i_p} \wedge d\overline{z}_{j_1} \wedge \cdots \wedge d\overline{z}_{j_q}$ constitute a basis of F^r.

For the uniqueness let

$$\varphi = \sum_{p+q=r} \varphi^{(p,q)} = \sum_{p+q=r} \widetilde{\varphi}^{(p,q)}.$$

Then

$$\sum_{p+q=r} \psi^{(p,q)} = 0 \quad \text{for } \psi^{(p,q)} := \varphi^{(p,q)} - \widetilde{\varphi}^{(p,q)}.$$

It follows that

$$0 = \sum_{p+q=r} \psi^{(p,q)}(cv_1, \ldots, cv_r) = \sum_{p+q=r} c^p \overline{c}^q \cdot \psi^{(p,q)}(v_1, \ldots, v_r).$$

For fixed $(v_1, \ldots, v_r)$ we obtain a polynomial equation in the ring $\mathbb{C}[c, \overline{c}]$, and all coefficients $\psi^{(p,q)}(v_1, \ldots, v_r)$ must vanish. Since we can choose $v_1, \ldots, v_r$ arbitrarily, we have $\varphi^{(p,q)} = \widetilde{\varphi}^{(p,q)}$ for all p, q. ∎

Bundles of Differential Forms. Let X be an n-dimensional complex manifold. We denote by $F(X)$ the complexified cotangent bundle $T^*(X) \otimes \mathbb{C}$. It has the spaces $F(T_x(X)) = \mathrm{Hom}_{\mathbb{R}}(T_x(X), \mathbb{C}) = T_x(X)^* \otimes \mathbb{C}$ of complex covariant tangent vectors as fibers, so it is a (topological) complex vector bundle of rank $2n$.[1] It even has a real-analytic structure, but not a complex-analytic structure.

If E is a (topological) complex vector bundle of rank m over X, then for $0 \le r \le m$ we can construct a bundle $\bigwedge^r E$ of rank $\binom{m}{r}$ over X such that $(\bigwedge^r E)_x = \bigwedge^r (E_x)$ for every $x \in X$. If E is given by transition functions g_{ij}, then $\bigwedge^r E$ is given by the matrices $g_{ij}^{(r)}$ whose entries are the $(r \times r)$ minors of g_{ij}.

In particular we have the vector bundles $\bigwedge^r F(X)$ of rank $\binom{2n}{r}$, with a real-analytic structure.

> **Definition.** An *r-form* (or an *r-dimensional differential form*) on an open set $U \subset X$ is a smooth section $\omega \in \Gamma(U, \bigwedge^r F(X))$.

So an r-form ω on U assigns to every $x \in U$ an r-form ω_x at x. If $z_1, \ldots, z_n$ are local coordinates in a neighborhood of x, then $\omega_j := dz_j$ and $\omega_{n+j} := d\overline{z}_j$ form a basis of the 1-forms on this neighborhood, and there is a representation

$$\omega_x = \sum_{1 \le i_1 < \cdots < i_r \le 2n} a_{i_1 \ldots i_r}(x) \omega_{i_1} \wedge \cdots \wedge \omega_{i_r},$$

where $x \mapsto a_{i_1 \ldots i_r}(x)$ are smooth functions.

Henceforth, the set of all (smooth) r-forms on U will be denoted by $\mathscr{A}^r(U)$, and the subset of all forms of type (p, q) by $\mathscr{A}^{(p,q)}(U)$.

If f is a smooth function on U, then its *differential* $df \in \mathscr{A}^1(U)$ is given by $x \mapsto (df)_x$. In local coordinates we have

$$df = \sum_{\nu=1}^{n} f_{z_\nu}\, dz_\nu + \sum_{\nu=1}^{n} f_{\overline{z}_\nu}\, d\overline{z}_\nu.$$

A *(smooth) vector field* is a smooth section of the tangent bundle $T(X)$. So in local coordinates it can be written in the form

[1] The tangent bundle $T(X)$ is a complex-analytic vector bundle of rank n. Here we denote its **real** dual bundle by $T^*(X)$. It is a real-analytic bundle of rank $2n$ (over $\mathbb{R}$). The **complex** dual bundle $T'(X)$ of $T(X)$ (with fiber $T'_x(X) = \mathrm{Hom}_{\mathbb{C}}(T_x(X), \mathbb{C})$) is a complex-analytic bundle of rank n over $\mathbb{C}$.

$$\xi = \sum_{\nu=1}^{n} \xi_\nu \frac{\partial}{\partial z_\nu} + \sum_{\nu=1}^{n} \overline{\xi}_\nu \frac{\partial}{\partial \overline{z}_\nu},$$

where the coefficients ξ_ν are smooth functions. Then we can apply df to such a vector field and obtain

$$df(\xi) = \sum_{\nu=1}^{n} \xi_\nu f_{z_\nu} + \sum_{\nu=1}^{n} \overline{\xi}_\nu f_{\overline{z}_\nu}.$$

For any open set U the differential can be generalized to the *Poincaré map* $d = d_U : \mathscr{A}^r(U) \to \mathscr{A}^{r+1}(U)$ in the following way:

If $\omega = \sum_{1 \le i_1 < \cdots < i_r \le 2n} a_{i_1 \dots i_r} \omega_{i_1} \wedge \cdots \wedge \omega_{i_r}$ is the basis representation in a coordinate neighborhood U, then

$$d_U(\omega) := \sum_{1 \le i_1 < \cdots < i_r \le 2n} da_{i_1 \dots i_r} \wedge \omega_{i_1} \wedge \cdots \wedge \omega_{i_r}.$$

One can show that this definition is independent of the choice of the local coordinates and that d has the following properties:

1. If f is a smooth function, then df is the differential of f.
2. d is $\mathbb{C}$-linear.
3. $d \circ d = 0$.
4. $d_V(\omega|_V) = (d_U\omega)|_V$, for $\omega \in \mathscr{A}^r(U)$.
5. If $\varphi \in \mathscr{A}^r(U)$ and $\psi \in \mathscr{A}^s(U)$, then $d(\varphi \wedge \psi) = d\varphi \wedge \psi + (-1)^r \varphi \wedge d\psi$.
6. d is a real operator; that is, $d\overline{\varphi} = \overline{d\varphi}$. In particular, $d\varphi = d(\operatorname{Re}\varphi) + \mathrm{i}\, d(\operatorname{Im}\varphi)$.

The differential $d\omega$ is called the *total derivative* or *exterior derivative* of ω.

Now we consider the decomposition of an r-form into a sum of forms of type (p, q). We use some abbreviations. If $I = (i_1, \dots, i_p)$ and $J = (j_1, \dots, j_q)$ are multi-indices in increasing order, we write

$$a_{IJ}\, dz_I \wedge d\overline{z}_J$$

instead of

$$a_{i_1 \dots i_p, j_1 \dots j_q} dz_{i_1} \wedge \cdots \wedge dz_{i_p} \wedge d\overline{z}_{j_1} \wedge \cdots \wedge d\overline{z}_{j_q}.$$

So a general r-dimensional differential form ω has the unique representation

$$\omega = \sum_{p+q=r} \sum_{\substack{|I|=p \\ |J|=q}} a_{IJ}\, dz_I \wedge d\overline{z}_J,$$

and the Poincaré differential of ω is given by

$$d\omega = \sum_{p+q=r} \sum_{\substack{|I|=p \\ |J|=q}} da_{IJ} \wedge dz_I \wedge d\overline{z}_J.$$

If f is a smooth function, then $df = \partial f + \overline{\partial} f$, with

$$\partial f = \sum_{\nu=1}^{n} f_{z_\nu}\, dz_\nu \quad \text{and} \quad \overline{\partial} f = \sum_{\nu=1}^{n} f_{\overline{z}_\nu}\, d\overline{z}_\nu.$$

Here ∂f has type $(1,0)$, $\overline{\partial} f$ has type $(0,1)$, and df type $(1,1)$.

1.4 Proposition. *If φ is a form of type (p,q), then $d\varphi$ has a unique decomposition $d\varphi = \partial\varphi + \overline{\partial}\varphi$ with a $(p+1,q)$-form $\partial\varphi$ and a $(p,q+1)$-form $\overline{\partial}\varphi$.*

PROOF: If $\varphi = \sum_{I,J} a_{IJ} dz_I \wedge d\overline{z}_j$, define

$$\partial\varphi := \sum_{I,J} \partial a_{IJ} \wedge dz_I \wedge d\overline{z}_J \quad \text{and} \quad \overline{\partial}\varphi := \sum_{I,J} \overline{\partial} a_{IJ} \wedge dz_I \wedge d\overline{z}_J.$$

Then $d\varphi = \partial\varphi + \overline{\partial}\varphi$ is the unique decomposition of the $(p+q+1)$-form $d\varphi$ into forms of pure type. ∎

For general r-forms the derivatives with respect to z and $\overline{z}$ are defined in the obvious way. (In the French literature one writes $d'\varphi$ instead of $\partial\varphi$ and $d''\varphi$ instead of $\overline{\partial}\varphi$.)

1.5 Theorem.

1. *∂ and $\overline{\partial}$ are $\mathbb{C}$-linear operators with $d = \partial + \overline{\partial}$.*
2. *$\partial\partial = 0$, $\overline{\partial}\overline{\partial} = 0$, and $\partial\overline{\partial} + \overline{\partial}\partial = 0$.*
3. *$\partial, \overline{\partial}$ are **not** real. We have*
   $\overline{\partial\varphi} = \overline{\partial}\overline{\varphi}$ *and* $\overline{\overline{\partial}\varphi} = \partial\overline{\varphi}$.
4. *If φ is an r-form and ψ is arbitrary, then*

$$\begin{aligned} \partial(\varphi \wedge \psi) &= \partial\varphi \wedge \psi + (-1)^r \varphi \wedge \psi, \\ \overline{\partial}(\varphi \wedge \psi) &= \overline{\partial}\varphi \wedge \psi + (-1)^r \varphi \wedge \overline{\partial}\psi. \end{aligned}$$

PROOF: It suffices to prove this for forms of pure type. Then the formulas can be easily derived from the corresponding formulas for d and the uniqueness of the decomposition into forms of type (p,q). ∎

Remark. Sometimes the operator $d^c := \mathrm{i}(\overline{\partial} - \partial)$ will be used. Then $\overline{d^c\varphi} = d^c\overline{\varphi}$, so d^c is a real operator with $d^c d^c = 0$. We have $dd^c = 2\mathrm{i}\,\partial\overline{\partial}$.

A smooth function f is holomorphic if and only if $\overline{\partial} f = 0$. Correspondingly, it follows for a $(p,0)$-form $\varphi = \sum_{|I|=p} a_I\, dz_I$ that $\overline{\partial}\varphi = 0$ if and only if all coefficients a_I are holomorphic. Hence we make the following definition:

Definition. Let φ be a p-form on the open set $U \subset X$.

1. φ is called *holomorphic* if φ is of type $(p, 0)$ and $\overline{\partial}\varphi = 0$.
2. φ is called *antiholomorphic* if and only if φ is of type $(0, p)$ and $\partial\varphi = 0$.

The set of holomorphic p-forms on U is denoted by $\Omega^p(U)$.

Clearly, φ is antiholomorphic if and only if $\overline{\varphi}$ is holomorphic.

Exercises

1. Let X be an n-dimensional complex manifold, f a smooth function on X, and ω a smooth form of type $(n-1, n-1)$ on X. Prove that
$$df \wedge d^c\omega = d\omega \wedge d^c f\,.$$
2. A real differential form of type $(1, 1)$ is called *positive* if $\mathrm{i}\omega(v, v) > 0$ for every tangent vector $v \neq 0$. Prove that $\mathrm{i}\partial\overline{\partial}\|\mathbf{z}\|^2$ is a positive form on $\mathbb{C}^n$.
3. Consider the $(n, n-1)$-form η_0 on $\mathbb{C}^n$ defined by
$$\eta_0 := (-1)^{n(n-1)/2} \sum_{k=1}^{n} (-1)^k \overline{z}_k\, dz_1 \wedge \cdots \wedge dz_n \wedge d\overline{z}_1 \wedge \cdots \wedge \widehat{d\overline{z}_k} \wedge \cdots \wedge d\overline{z}_n$$
and calculate $\overline{\partial}\eta_0$. Let f be a holomorphic function on an open neighborhood of the closed ball $\overline{\mathsf{B}_r(\mathbf{0})}$, and $\omega := \big(f(\mathbf{z})/\|\mathbf{z}\|^{2n}\big) \cdot \eta_0$. Show that $d\omega = 0$.
4. Let $X := \mathbb{P}^n$. On $U_0 := \{(z_0 : z_1 : \ldots : z_n) \in \mathbb{P}^n : z_0 \neq 0\}$ we use the holomorphic coordinates $t_\nu = z_\nu / z_0$. Then $\omega_0 = dt_1 \wedge \cdots \wedge dt_n$ is a holomorphic n-form on U_0. Show that a holomorphic n-form on an open subset $U \subset X$ is the same as a holomorphic section over U in the canonical bundle K_X. Prove that there is a meromorphic section s in K_X with $s|_{U_0} = \omega_0$ and $\operatorname{div}(s) = -(n+1)H_0$, where $H_0 = \{z_0 = 0\}$ is the hyperplane at infinity.

2. Dolbeault Theory

Integration of Differential Forms. We recall some results from real analysis.

Let $G \subset \mathbb{C}^n \cong \mathbb{R}^{2n}$ be a domain. A closed subset $M \subset G$ is called an r-dimensional submanifold of class $\mathscr{C}^k$ if for each $\mathbf{z} \in M$ there exists a neighborhood U of $\mathbf{z}$ in G and a $\mathscr{C}^k$ map $\mathbf{f} : U \to \mathbb{R}^{2n-r}$ such that:

1. $U \cap M = \mathbf{f}^{-1}(\mathbf{0})$.
2. The rank of the (real) Jacobi matrix of $\mathbf{f}$ is equal to $2n - r$.

It follows from the (real) implicit function theorem that for any $\mathbf{z} \in M$ there is a neighborhood U of $\mathbf{z}$, an open set $W \subset \mathbb{R}^r$, and a $\mathscr{C}^k$ map $\varphi : W \to G$ such that:

1. φ maps W homeomorphically onto U.
2. The rank of the Jacobian matrix of φ is equal to r.

Then φ is called a *local parametrization* of M.

An r-dimensional differential form $\omega = \sum_{p+q=r} \sum_{I,J} a_{IJ}\, dz_I \wedge d\overline{z}_J$ is called an r-form of class $\mathscr{C}^k$ if all coefficients are functions of class k. Then $d\omega$ is an $(r+1)$-form of class $\mathscr{C}^{k-1}$. We always assume that $k \geq 1$.

If $\Phi : B_1 \to B_2$ is a $\mathscr{C}^k$ map between domains $B_1 \subset \mathbb{C}^m$ and $B_2 \subset \mathbb{C}^n$, then every r-form ω of class $\mathscr{C}^k$ on B_2 can be lifted back to B_1 by

$$(\Phi^*\omega)_{\mathbf{z}}(v_1, \dots, v_r) := \omega_{\Phi(\mathbf{z})}(\Phi_* v_1, \dots, \Phi_* v_r).$$

If $u_1, \dots, u_{2n}$ are real coordinates in B_2 then

$$\Phi^*\Big(\sum a_{i_1 \dots i_r}\, du_{i_1} \wedge \dots \wedge du_{i_r}\Big) = \sum (a_{i_1 \dots i_r} \circ \Phi)\, d(u_{i_1} \circ \Phi) \wedge \dots \wedge d(u_{i_r} \circ \Phi).$$

Therefore, $\Phi^*\omega$ is again an r-form of class $\mathscr{C}^k$. We have:

1. $\Phi^*(\omega_1 \wedge \omega_2) = \Phi^*\omega_1 \wedge \Phi^*\omega_2$.
2. $\Phi^*(d\omega) = d(\Phi^*\omega)$, in particular $\Phi^*(df) = d(f \circ \Phi)$.

Now let $\omega = a\, dz_1 \wedge \dots \wedge dz_n \wedge d\overline{z}_1 \wedge \dots \wedge d\overline{z}_n$ be an arbitrary $(2n)$-form. Since $dz_i \wedge d\overline{z}_i = -2\mathrm{i}\, dx_i \wedge dy_i$, it follows that

$$\begin{aligned} \omega &= a \cdot (-1)^{n(n-1)/2} dz_1 \wedge d\overline{z}_1 \wedge \dots \wedge dz_n \wedge d\overline{z}_n \\ &= a \cdot (-1)^{n(n-1)/2} \cdot (-2)^n \mathrm{i}^n dx_1 \wedge dy_1 \wedge \dots \wedge dx_n \wedge dy_n \\ &= a \cdot (-1)^{n(n+1)/2} (2\mathrm{i})^n dx_1 \wedge dy_1 \wedge \dots \wedge dx_n \wedge dy_n. \end{aligned}$$

We write $\widetilde{a} := a \cdot (-1)^{n(n+1)/2}(2\mathrm{i})^n$. If $B \subset \mathbb{C}^n$ is a bounded domain and $\widetilde{a}$ continuous in a neighborhood of $\overline{B}$, then

$$\int_B \omega := \int_B \widetilde{a}(x_1 + \mathrm{i}y_1, \dots, x_n + \mathrm{i}y_n)\, dx_1 dy_1 \cdots dx_n dy_n.$$

If $M \subset G$ is an r-dimensional submanifold of class $\mathscr{C}^k$, ω an r-form of class $\mathscr{C}^k$, and $K \subset M$ a compact subset that is contained in the range of a local parametrization $\varphi : W \to M$, then we define

$$\int_K \omega := \int_{\varphi^{-1}(K)} \varphi^*\omega.$$

The integral does not depend on the parametrization if we allow only changes of parameters with positive Jacobi determinant. If K cannot be covered by a single parametrization, then one uses a partition of unity.

Finally we have the famous Stokes's theorem.

2.1 Stokes's theorem. *Let $G \subset \mathbb{C}^n$ be a bounded domain with smooth boundary and ω a smooth $(n-1)$-form in a neighborhood of $\overline{G}$. Then*

$$\int_{\partial G} \omega = \int_G d\omega.$$

The Inhomogeneous Cauchy Formula. We consider a generalization of Cauchy's integral formula in the 1-dimensional case. Let $G \subset\subset \mathbb{C}$ be a domain with smooth boundary, and $U = U(\overline{G})$ an open neighborhood of the closed domain.

2.2 Theorem. *Let f be a continuously differentiable (complex-valued) function on U. Then, for $z \in G$, we have*

$$f(z) = \frac{1}{2\pi \mathrm{i}} \int_{\partial G} \frac{f(\zeta)}{\zeta - z}\, d\zeta + \frac{1}{2\pi \mathrm{i}} \int_G \frac{(\partial f/\partial \overline{\zeta})(\zeta)}{\zeta - z}\, d\zeta \wedge d\overline{\zeta}.$$

PROOF: Let $z \in G$ be fixed, choose an $r > 0$ such that the disk $\mathsf{D}_r(z)$ lies relatively compactly in G, and define $G_r := G - \overline{\mathsf{D}_r(z)}$. The differential form

$$\omega(z) := \frac{1}{2\pi \mathrm{i}} \cdot \frac{f(\zeta)}{\zeta - z}\, d\zeta$$

is of class $\mathscr{C}^1$ in an open neighborhood $V = V(\overline{G_r})$, and its Poincaré differential is

$$d\omega(z) = -\frac{1}{2\pi \mathrm{i}} \cdot \frac{(\partial f/\partial \overline{\zeta})(\zeta)}{\zeta - z}\, d\zeta \wedge d\overline{\zeta}.$$

By the theorem of Stokes we have

$$\int_{G_r} d\omega(z) = \int_{\partial G_r} \omega(z) = \int_{\partial G} \omega(z) - \int_{\partial \mathsf{D}_r(z)} \omega(z).$$

If $0 < \varepsilon < r$ and $A_{\varepsilon,r} := \mathsf{D}_r(z) - \mathsf{D}_\varepsilon(z)$, then for any continuous function g on $\overline{\mathsf{D}_r(z)}$,

$$\begin{aligned}\int_{A_{\varepsilon,r}} \frac{g(\zeta)}{\zeta - z}\, d\zeta \wedge d\overline{\zeta} &= \int_\varepsilon^r \int_0^{2\pi} \frac{g(z + \varrho \cdot e^{\mathrm{i}t})}{\varrho \cdot e^{\mathrm{i}t}} \cdot 2\mathrm{i}\varrho\, dt\, d\varrho \\ &= 2\mathrm{i} \int_\varepsilon^r \int_0^{2\pi} g(z + \varrho \cdot e^{\mathrm{i}t}) \cdot e^{-\mathrm{i}t}\, dt\, d\varrho.\end{aligned}$$

This integral exists and stays bounded if ε tends to zero. So

$$\int_G d\omega(z) = \int_{\partial G} \omega(z) - \lim_{r \to 0} \int_{\partial \mathsf{D}_r(z)} \omega(z).$$

In order to show that the limit on the right side is equal to $f(z)$ we estimate

$$\begin{aligned}\Big|\int_{\partial \mathsf{D}_r(z)} \omega(z) - f(z)\Big| &= \Big|\frac{1}{2\pi \mathrm{i}}\int_0^{2\pi} f(z+re^{\mathrm{i}t})\cdot \mathrm{i}\,dt - \frac{1}{2\pi}\int f(z)\,dt\Big| \\ &= \Big|\frac{1}{2\pi}\int_0^{2\pi} \left(f(z+re^{\mathrm{i}t}) - f(z)\right)\,dt\Big| \\ &\leq \sup_{[0,2\pi]} |f(z+re^{\mathrm{i}t}) - f(z)| \to 0 \text{ for } r\to 0.\end{aligned}$$

It follows that $f(z) = \int_{\partial G}\omega(z) - \int_G d\omega(z)$. ∎

The $\overline{\partial}$-Equation in One Variable. The Poincaré lemma from real analysis can be formulated as follows:

> *Let $G \subset \mathbb{C}^n$ be a star-shaped (e.g., a convex) region, $\varphi \in \mathscr{A}^r(G)$, $r>0$, and $d\varphi = 0$. Then there exists a $\psi \in \mathscr{A}^{r-1}(G)$ with $d\psi = \varphi$.*

We want to prove a similar theorem for the $\overline{\partial}$ operator. For this we start in one variable using the integral

$$\mathrm{Ch}_f(z) := \frac{1}{2\pi \mathrm{i}}\int_G \frac{f(\zeta)}{\zeta - z}\,d\zeta\wedge d\overline{\zeta},$$

for bounded functions $f\in\mathscr{C}^1(G)$.

2.3 Theorem. *Let $f\in\mathscr{C}^1(\mathbb{C})$ be a function with compact support. Then there is a continuously differentiable function u on $\mathbb{C}$ with $u_{\overline{z}} = f$.*

PROOF: Choose an $R>0$ such that $\operatorname{supp}(f) \subset\subset D := \mathsf{D}_R(0)$. Then $u := \mathrm{Ch}_f$ is a continuously differentiable function on D, and we can assume that the integral is taken over the whole plane. It follows that

$$u(z) = \frac{1}{2\pi \mathrm{i}}\int_{\mathbb{C}} \frac{f(\zeta + z)}{\zeta}\,d\zeta\wedge d\overline{\zeta},$$

and therefore (applying formulas for the derivative of parametric integrals)

$$\begin{aligned} u_{\overline{z}}(z) &= \frac{1}{2\pi \mathrm{i}}\int_{\mathbb{C}} \frac{\partial f/\partial\overline{\zeta}(\zeta+z)}{\zeta}\,d\zeta\wedge d\overline{\zeta} \\ &= \frac{1}{2\pi \mathrm{i}}\int_{\mathbb{C}} \frac{\partial f/\partial\overline{\zeta}(\zeta)}{\zeta - z}\,d\zeta\wedge d\overline{\zeta}\end{aligned}$$

$$\begin{aligned} &= \frac{1}{2\pi \mathrm{i}}\int_{D} \frac{\partial f/\partial\overline{\zeta}(\zeta)}{\zeta - z}\,d\zeta\wedge d\overline{\zeta} \\ &= f(z) - \frac{1}{2\pi \mathrm{i}}\int_{\partial D} \frac{f(\zeta)}{\zeta - z}\,d\zeta \;=\; f(z),\end{aligned}$$

since $f|_{\partial D} \equiv 0$. ■

2.4 Corollary. *Let $G \subset\subset \mathbb{C}$ be a bounded domain, and $f \in \mathscr{C}^1(G)$ a bounded function. Then there exists a continuously differentiable function u on G with $u_{\overline{z}} = f$.*

PROOF: Again we define $u := \mathrm{Ch}_f$.

Let $z_0 \in G$ be an arbitrary point and $D := \mathsf{D}_r(z_0)$ a disk with $D \subset\subset G$. There exists a smooth function ϱ on $\mathbb{C}$ with

1. $0 \le \varrho \le 1$,
2. $\varrho|_D \equiv 1$,
3. $\mathrm{supp}(\varrho) \subset\subset G$.

We define $f_1 := \varrho \cdot f$ and $f_2 := f - f_1$. Then f_1 is a continuously differentiable function on $\mathbb{C}$ with $f_1|_D = f|_D$, and f_2 a function on G with $f_2|_D \equiv 0$.

There exists a $\mathscr{C}^1$ function u_1 on $\mathbb{C}$ with $(u_1)_{\overline{z}} = f_1$. The function $u_2 := \mathrm{Ch}_{f_2}$ is defined on G, and given by

$$u_2(z) = \frac{1}{2\pi \mathrm{i}} \int_{G-D} \frac{f_2(\zeta)}{\zeta - z}\, d\zeta \wedge d\overline{\zeta}.$$

The integrand is continuous and bounded on $G - D$, and holomorphic in z, for $z \in D$.

Obviously, $u = u_1 + u_2$ on G, and on D we have $u_{\overline{z}} = (u_1)_{\overline{z}} = f_1 = f$. ■

A Theorem of Hartogs. Now we are able to prove the so-called *Kugelsatz*, also known as *Hartogs' theorem*:

2.5 Theorem. *Let $G \subset \mathbb{C}^n$ be a domain, $n \ge 2$, $K \subset G$ a compact subset. If $G - K$ is connected, any holomorphic function f on $G - K$ has a holomorphic extension $\widehat{f}$ on G.*

PROOF: Choose an open neighborhood $U = U(K) \subset\subset G$ and a smooth function ϱ on $\mathbb{C}^n$ with $\mathrm{supp}(\varrho) \subset\subset G$ and $\varrho|_U \equiv 1$. Define

$$g_k := f \cdot \frac{\partial \varrho}{\partial \overline{z}_k}, \qquad \text{for } k = 1, \ldots, n.$$

We have

$$\frac{\partial g_k}{\partial \overline{z}_l} = f \cdot \frac{\partial^2 \varrho}{\partial \overline{z}_k \partial \overline{z}_l} = \frac{\partial g_l}{\partial \overline{z}_k}.$$

For fixed $\mathbf{z}'' = (z_2, \ldots, z_n)$ we define

$$u(z_1, z_2, \ldots, z_n) := \mathrm{Ch}_{g_1(z_1, \mathbf{z}'')}(z_1) = \frac{1}{2\pi \mathrm{i}} \int_{\mathbb{C}} \frac{g_1(\zeta, z_2, \ldots, z_n)}{\zeta - z_1}\, d\zeta \wedge d\overline{\zeta}.$$

Then $u_{\overline{z}_1} = g_1$ (we can apply the previous corollary, since g_1 has compact support), and for $k \geq 2$ we get

$$\begin{aligned} u_{\overline{z}_k}(z_1, \ldots, z_n) &= \frac{1}{2\pi \mathrm{i}} \int_{\mathbb{C}} \frac{(g_1)_{\overline{z}_k}(\zeta, z_2, \ldots, z_n)}{\zeta - z_1} \, d\zeta \wedge d\overline{\zeta} \\ &= \frac{1}{2\pi \mathrm{i}} \int_{\mathbb{C}} \frac{(g_k)_{\overline{z}_1}(\zeta, z_2, \ldots, z_n)}{\zeta - z_1} \, d\zeta \wedge d\overline{\zeta} \\ &= g_k(z_1, z_2, \ldots, z_n). \end{aligned}$$

For $\mathbf{z} \notin \operatorname{supp}(\varrho)$ we have $u_{\overline{z}_k}(\mathbf{z}) = 0$, $k = 1, \ldots, n$. Therefore, u is a holomorphic function outside $\operatorname{supp}(\varrho)$. Since $g_1(\zeta, \mathbf{z}') = 0$ for arbitrary ζ and large $\|\mathbf{z}'\|$, it follows that $u(z_1, \mathbf{z}') = 0$ for large $\|\mathbf{z}'\|$, and so u is defined everywhere and is identically zero on the unbounded connected component Z of $G - K$.

There exists a nonempty open subset V of $Z \cap (G - \operatorname{supp}(\varrho)) \subset G - K$. The function $\widehat{f} := (1 - \varrho) \cdot f + u$ is defined on G and coincides with f on V. In addition, we have

$$\widehat{f}_{\overline{z}_k} = \begin{cases} -f \cdot \varrho_{\overline{z}_k} + u_{\overline{z}_k} = -g_k + g_k = 0 & \text{in } G - K, \\ u_{\overline{z}_k} = g_k = 0 & \text{in } U. \end{cases}$$

So $\widehat{f}$ is holomorphic in G. ∎

Dolbeault's Lemma. Before proving the $\overline{\partial}$ analogue of Poincaré's lemma we need the following result.

2.6 Proposition. *Let $P \subset\subset Q$ be polydisks around the origin in $\mathbb{C}^n$. Let g be a smooth function in Q that is holomorphic in $z_{k+1}, \ldots, z_n$, for some $k \geq 1$.*

Then there exists a smooth function u in Q that is again holomorphic in $z_{k+1}, \ldots, z_n$ such that $u_{\overline{z}_k} = g$ on P.

PROOF: There are disks $E \subset\subset D$ around 0 in the z_k-plane such that $P = P' \times E \times P''$ and $Q = Q' \times D \times Q''$. Let ϱ be a smooth nonnegative function on $\mathbb{C}$ with $\varrho|_E \equiv 1$ and $\varrho|_{\mathbb{C}-D} \equiv 0$. Then we define

$$u(z_1, \ldots, z_n) := \frac{1}{2\pi \mathrm{i}} \int_{\mathbb{C}} \frac{\varrho(\zeta) \cdot g(z_1, \ldots, z_{k-1}, \zeta, z_{k+1}, \ldots, z_n)}{\zeta - z_k} \, d\zeta \wedge d\overline{\zeta}.$$

The function u is defined on Q, and the integrand has compact support. Therefore,

$$u_{\overline{z}_k}(\mathbf{z}) = (\varrho \cdot g)(\mathbf{z}) = g(\mathbf{z}) \text{ on } P,$$

and by the rules for parametric integration we get $u_{\overline{z}_i} = 0$ for $i = k+1, \ldots, n$. So u is holomorphic in these variables. ∎

2.7 Dolbeault's lemma. *Let $P \subset\subset Q$ be polydisks around the origin in $\mathbb{C}^n$. For each $\varphi \in \mathscr{A}^{p,q}(Q)$ with $q > 0$ and $\overline{\partial}\varphi = 0$ there exists a $\psi \in \mathscr{A}^{p,q-1}(P)$ with $\overline{\partial}\psi = \varphi|_P$.*

PROOF: Without loss of generality we may assume that $p = 0$. Let $k \le n$ be the smallest number such that φ does not involve the differentials $d\overline{z}_{k+1}, \dots, d\overline{z}_n$. We carry out the proof by induction on k.

For $k = 0$ there is nothing to prove, since $q > 0$.

Now we assume that $k \geq 1$ and the theorem has already been proved for $k-1$. Then we write

$$\varphi = d\overline{z}_k \wedge \alpha + \beta,$$

where $\alpha \in \mathscr{A}^{0,q-1}(Q)$ and $\beta \in \mathscr{A}^{0,q}(Q)$ are forms not involving $d\overline{z}_k, \dots, d\overline{z}_n$.

If $\alpha = \sum^*_{|J|=q-1} a_J d\overline{z}_J$ (where $\sum^*$ means that $d\overline{z}_k, \dots, d\overline{z}_n$ do not occur), then

$$\begin{aligned} 0 = \overline{\partial}\varphi &= -d\overline{z}_k \wedge \overline{\partial}\alpha + \overline{\partial}\beta \\ &= -\sum_{\nu \neq k} d\overline{z}_k \wedge d\overline{z}_\nu \wedge \left(\sum^*_{|J|=q-1} (a_J)_{\overline{z}_\nu}\, d\overline{z}_J \right) + \overline{\partial}\beta. \end{aligned}$$

Then it follows that $(a_J)_{\overline{z}_\nu} \equiv 0$ for $\nu > k$. By the above proposition there are smooth functions A_J in Q that are holomorphic in $\overline{z}_{k+1}, \dots, \overline{z}_n$ such that $(A_J)_{\overline{z}_k} = a_J$ on P. We define $\gamma := \sum^*_{|J|=q-1} A_J\, d\overline{z}_J$. Then

$$\begin{aligned} \overline{\partial}\gamma &= \sum^*_{|J|=q-1} \sum_{\nu \notin J} (A_J)_{\overline{z}_\nu}\, d\overline{z}_\nu \wedge d\overline{z}_J \\ &= \sum^*_{|J|=q-1} \left(a_j\, d\overline{z}_k \wedge d\overline{z}_J + \sum_{\substack{\nu<k \\ \nu \notin J}} (A_J)_{\overline{z}_\nu}\, d\overline{z}_\nu \wedge d\overline{z}_J \right) \\ &= d\overline{z}_k \wedge \sum^*_{|J|=q-1} a_J\, d\overline{z}_J + \cdots \\ &= d\overline{z}_k \wedge \alpha + \eta, \end{aligned}$$

where η does not involve $d\overline{z}_k, \dots, d\overline{z}_n$. We see that also $\beta - \eta$ does not involve $d\overline{z}_k, \dots, d\overline{z}_n$.

Since $\beta - \eta = (\varphi - d\overline{z}_k \wedge \alpha) - (\overline{\partial}\gamma - d\overline{z}_k \wedge \alpha) = \varphi - \overline{\partial}\gamma$, we have $\overline{\partial}(\beta - \eta) = \overline{\partial}\varphi = 0$, and the induction hypothesis implies that there is a form $\sigma \in \mathscr{A}^{0,q-1}(P)$ with $\overline{\partial}\sigma = (\beta - \eta)|_P$. Setting $\psi := \sigma + \gamma$, it follows that

$$\overline{\partial}\psi = (\beta - \eta) + (d\overline{z}_k \wedge \alpha + \eta) = \varphi \quad \text{on } P.$$

This completes the proof. ∎

We immediately obtain the following result for manifolds X:

If $U \subset X$ is an open set and $\varphi \in \mathscr{A}^{0,q}(U)$ a smooth form with $q \geq 1$ and $\overline{\partial}\varphi = 0$, then for every $x \in U$ there exists an open neighborhood $V = V(x) \subset U$ and a form $\psi \in \mathscr{A}^{0,q-1}(V)$ with $\overline{\partial}\psi = \varphi|_V$.

Without proof we mention the following result, which provides more precise information (see [GrLi70] and [Li70]).

2.8 Theorem (I. Lieb). *Let $G \subset\subset \mathbb{C}^n$ be a Levi convex domain with smooth boundary and $\omega = \sum_J a_J \, d\overline{z}_J$ a smooth $(0,q)$-form on G with $\overline{\partial}\omega = 0$. Moreover, suppose that there is a real constant $M > 0$ with*

$$\|\omega\| := \max_J \sup_G |a_J| \leq M.$$

Then there exists a constant k independent of ω and a form ψ of type $(0, q-1)$ on G with $\overline{\partial}\psi = \omega$ and $\|\psi\| \leq k \cdot M$.

Siu and Range generalized this theorem to domains with piecewise smooth boundary (see [RaSi]). L. Hörmander has given a quite different proof for the solution of the $\overline{\partial}$-equation with L^2-estimates on pseudoconvex domains using methods of functional analysis (see [Hoe66]). In the meantime there exists a big industry of solving $\overline{\partial}$-equations, but we do not want to go here into detail.

Dolbeault Groups. Let X be an n-dimensional complex manifold and $U \subset X$ an open subset. Then we have natural inclusions[2]

$$\varepsilon : \mathbb{C} \hookrightarrow \mathscr{A}^0(U) = \mathscr{C}^\infty(U) \quad \text{and} \quad \varepsilon : \Omega^p(U) \hookrightarrow \mathscr{A}^{p,0}(U).$$

Together with the Poincaré differential d and the $\overline{\partial}$-operator we get the following sequences of $\mathbb{C}$-linear maps:

1. The *de Rham sequence*

$$0 \to \mathbb{C} \xrightarrow{\varepsilon} \mathscr{A}^0(U) \xrightarrow{d} \mathscr{A}^1(U) \xrightarrow{d} \cdots \xrightarrow{d} \mathscr{A}^n(U) \to 0.$$

   If U is diffeomorphic to a star-shaped domain, then the sequence is exact, i.e., $\{f \in \mathscr{A}^0(U) : df = 0\} = \mathbb{C}$ and $Z^r(U) := \operatorname{Ker}(d : \mathscr{A}^r(U) \to \mathscr{A}^{r+1}(U))$ is equal to $B^r(U) := \operatorname{Im}(d : \mathscr{A}^{r-1}(U) \to \mathscr{A}^r(U))$. In general, we have only $B^r(U) \subset Z^r(U)$ (because $d \circ d = 0$).
2. The *Dolbeault sequence*

$$0 \to \Omega^p(U) \xrightarrow{\varepsilon} \mathscr{A}^{p,0}(U) \xrightarrow{\overline{\partial}} \mathscr{A}^{p,1}(U) \xrightarrow{\overline{\partial}} \cdots \xrightarrow{\overline{\partial}} \mathscr{A}^{p,n}(U) \to 0.$$

   We define

$$\begin{aligned} Z^{p,q}(U) &:= \{\varphi \in \mathscr{A}^{p,q}(U) : \overline{\partial}\varphi = 0\} \quad \text{for } 0 \leq q \leq n, \\ B^{p,q}(U) &:= \overline{\partial}(\mathscr{A}^{p,q-1}(U)) \quad \text{for } 1 \leq q \leq n \quad \text{and} \quad B^{p,0}(U) := 0. \end{aligned}$$

   Then $B^{p,q} \subset Z^{p,q}$, because $\overline{\partial} \circ \overline{\partial} = 0$ (respectively $0 \subset \Omega^p(U)$).

[2] Here $\mathbb{C}$ stands for the set of locally constant functions.

Definition. The complex vector space $H^{p,q}(U) := Z^{p,q}(U)/B^{p,q}(U)$ is called the *Dolbeault (cohomology) group of type* (p,q) of the set U, and the complex vector space $H^r(U) := Z^r(U)/B^r(U)$ is called the rth *de Rham (cohomology) group* of U.

We have

$$H^{p,0}(U) = \Omega^p(U) \quad \text{and} \quad H^{p,n}(U) = \mathscr{A}^{p,n}(U)/\overline{\partial}(\mathscr{A}^{p,q-1}(U)).$$

For $q \geq 1$ the condition $H^{p,q}(U) = 0$ means that for every $\varphi \in \mathscr{A}^{p,q}(U)$ with $\overline{\partial}\varphi = 0$ there is a $\psi \in \mathscr{A}^{p,q-1}(U)$ with $\overline{\partial}\psi = \varphi$. From the solution of the $\overline{\partial}$-equation (using, for example, [Hoe66], chapter IV, or Dolbeault's lemma and an approximation theorem due to Oka) we obtain that $H^{p,q}(G) = 0$ for every strictly pseudoconvex domain $G \subset \mathbb{C}^n$ and $q \geq 1$.

If U is connected, then $H^0(U) = \mathbb{C}$. If U is star-shaped, then $H^r(U) = 0$ for $r > 0$.

Now let $\Omega_X^p := \bigwedge^p T'(X)$ be the holomorphic vector bundle of holomorphic p-forms on X (in particular, $\Omega_X^0 = \mathscr{O}_X$, $\Omega_X^1 = T'(X)$ and $\Omega_X^n = K_X$). Then $\Omega^p(U) = \Gamma(U, \Omega_X^p)$ for every open set $U \subset X$. We construct a linear map

$$D : H^1(X, \Omega_X^p) \to H^{p,1}(X).$$

For every $\xi \in Z^1(X, \Omega_X^p)$ there is an open covering $\mathscr{U} = \{U_i : i \in I\}$ of X such that ξ is given with respect to this covering. Then we can pass to any refinement. So we may assume that $\mathscr{U}$ is as fine as we want.

Let $\xi = (\xi_{ij}) \in Z^1(\mathscr{U}, \Omega_X^p)$ be given. Then $\xi_{ij} \in \Omega^p(U_{ij})$ and $\xi_{ij} + \xi_{jk} = \xi_{ik}$ on U_{ijk}. Let (e_k) be a partition of unity for the covering $\mathscr{U}$. We define differential forms

$$\varphi_i := \sum_{k \in I} e_k \xi_{ki} \in \mathscr{A}^{p,0}(U_i), \text{ for } i \in I.$$

Here $\varphi = (\varphi_i)$ is an element of $C^0(\mathscr{U}, \bigwedge^{p,0} F(X))$ with $\delta\varphi = \xi$, since

$$\varphi_j - \varphi_i = \sum_k e_k(\xi_{kj} - \xi_{ki}) = \sum_k e_k \xi_{ij} = \xi_{ij}.$$

In U_{ij} we have

$$\overline{\partial}\varphi_j - \overline{\partial}\varphi_i = \overline{\partial}\xi_{ij} = 0.$$

Therefore, a global $(p,1)$-form ω on X can be defined by $\omega|_{U_i} := \overline{\partial}\varphi_i$. Now, $\overline{\partial}\omega = 0$. So $\omega \in Z^{p,1}(X)$.

If ξ is a coboundary, $\xi = \delta\eta$ for some $\eta \in C^0(\mathscr{U}, \Omega_X^p)$, then $\eta_j - \eta_i = \xi_{ij} = \varphi_j - \varphi_i$ on U_{ij}, and a global $(p,0)$-form ψ on X can be defined by $\psi|_{U_i} := \eta_i - \varphi_i$. We get

$$\overline{\partial}(-\psi)|_{U_i} = \overline{\partial}\varphi_i - \overline{\partial}\eta_i = \omega|_{U_i}, \text{ since } \eta_i \text{ is holomorphic.}$$

So the assignment $\xi \mapsto \omega$ induces a map $D : H^1(X, \Omega_X^p) \to H^{p,1}(X)$ by $[\xi] \mapsto [\omega]$, where $[\dots]$ denotes the residue class. One can show that this definition is independent of the covering $\mathscr{U}$.

2.9 Dolbeault's theorem. *The map $D : H^1(X, \Omega_X^p) \to H^{p,1}(X)$ is an isomorphism.*

PROOF: We use the notation from above. To show that D is injective, we suppose that there is a $(p,0)$-form ϱ on X with $\overline{\partial}\varrho = \omega$. Then $\overline{\partial}\varphi_i = \overline{\partial}\varrho|_{U_i}$, for every $i \in I$. It follows that $\overline{\partial}(\varphi_i - \varrho) = 0$ and therefore $\tau_i := \varphi_i - \varrho$ is holomorphic on U_i.

Then $\tau = (\tau_i)$ is a cochain of holomorphic p-forms, with

$$(\delta\tau)_{ij} = (\varphi_j - \varrho) - (\varphi_i - \varrho) = \xi_{ij}.$$

So the class of ξ vanishes.

Now let an element $\omega \in Z^{p,1}(X)$ be given. Since $\overline{\partial}\omega = 0$, we can find a covering $\mathscr{U} = \{U_i : i \in I\}$ of small polydisks (in local coordinates) with $(p,0)$-forms φ_i such that $\overline{\partial}\varphi_i = \omega|_{U_i}$. Then $\overline{\partial}(\varphi_j - \varphi_i) = 0$ for every pair (i,j), and $\xi_{ij} := \varphi_j - \varphi_i$ lies in $\Omega^p(U_{ij})$. Clearly, $\xi = (\xi_{ij})$ is an element of $Z^1(\mathscr{U}, \Omega_X^p)$ whose cohomology class is mapped by D onto the class of ω. This shows that D is surjective. ∎

Dolbeault's theorem has an interesting consequence:

2.10 Theorem. *Let X be a Stein manifold. If φ is a $(p,1)$-form on X with $\overline{\partial}\varphi = 0$, then there exists a form ψ on X of type $(p,0)$ with $\overline{\partial}\psi = \varphi$.*

PROOF: By Theorem B for Stein manifolds we have $H^1(X, \Omega_X^p) = 0$ for every p, and the result follows from Dolbeault's theorem. ∎

Remark. Dolbeault's theorem is also true for forms of type (p,q) with $q > 1$. For the proof one has to introduce higher cohomology groups $H^q(X, \Omega_X^p)$, to generalize Theorem B for those cohomology groups and to show that $H^q(X, \Omega_X^p) \cong H^{p,q}(X)$.

In an analogous way one can prove **de Rham's theorem**:

$$H^r(X) \cong H^r(X, \mathbb{C}),$$

where $H^r(X, \mathbb{C})$ denotes the cohomology group with values in the set of locally constant functions (isomorphic to the singular cohomology group with values in $\mathbb{C}$).

Let $\mathscr{U}$ be an open covering of X such that all intersections $U_{\iota_0 \dots \iota_k}$ are contractible. In the case $r = 2$ the de Rham isomorphism

$$\varphi : H^2(X) \to H^2(\mathscr{U}, \mathbb{C}) \cong H^2(X, \mathbb{C})$$

is given as follows:

Let $\omega \in \mathscr{A}^2(X)$ be given, with $d\omega = 0$. On every U_ν there exists an element $\beta_\nu \in \mathscr{A}^1(U_\nu)$ with

$$\omega|_{U_\nu} = d\,\beta_\nu.$$

Then $d(\beta_\mu - \beta_\nu) = 0$ on $U_{\nu\mu}$. Therefore, on every $U_{\nu\mu}$ there exists a smooth (complex-valued) function $f_{\nu\mu}$ with

$$d\,f_{\nu\mu} = \beta_\mu - \beta_\nu.$$

Now, $d(f_{\mu\lambda} - f_{\nu\lambda} + f_{\nu\mu}) = 0$ on $U_{\nu\mu\lambda}$. It follows that

$$a_{\nu\mu\lambda} := (f_{\mu\lambda} - f_{\nu\lambda} + f_{\nu\mu})|_{U_{\nu\mu\lambda}}$$

is constant.

So we have $a = (a_{\nu\mu\lambda}) \in Z^2(\mathscr{U}, \mathbb{C})$, and the de Rham class $[\omega] \in H^2(X)$ will be mapped onto the cohomology class $[a] \in H^2(\mathscr{U}, \mathbb{C})$.

Exercises

1. Let η_0 be the $(n, n-1)$-form defined in Exercise 1.3, and f a holomorphic function on an open neighborhood of the closed ball $\overline{\mathsf{B}_r(\mathbf{0})}$. Prove that

$$f(\mathbf{0}) = \frac{(n-1)!}{(2\pi \mathrm{i})^n} \int_{\partial \mathsf{B}_r(\mathbf{0})} \frac{f(\mathbf{z})}{\|\mathbf{z}\|^{2n}} \cdot \eta_0$$

(integral formula of Bochner–Martinelli).

2. Let (r_ν) be a monotone increasing sequence tending to $r > 0$, and let φ be a smooth $(0, q)$-form on $\mathsf{P}^n(\mathbf{0}, r)$ with $q > 0$ and $\overline{\partial}\varphi = 0$. Construct a sequence (ψ_ν) of smooth $(0, q-1)$-forms on $\mathsf{P}^n(\mathbf{0}, r)$ such that
   (a) $\overline{\partial}\psi_\nu = \varphi$ on $\mathsf{P}^n(\mathbf{0}, r_\nu)$.
   (b) $|\psi_{\nu+1} - \psi_\nu| < 2^{-\nu}$ on $\mathsf{P}^n(\mathbf{0}, r_{\nu-1})$ in the case $q = 1$, and $\psi_{\nu+1} = \psi_\nu$ on $\mathsf{P}^n(\mathbf{0}, r_{\nu-1})$ for $q \geq 2$.

   Show that $H^{p,q}(P) = 0$ for $q \geq 1$ and every polydisk P in $\mathbb{C}^n$.
3. Let X be a complex manifold and $\pi : E \to X$ a holomorphic vector bundle of rank q over X. If $\varphi : E|_U \to U \times \mathbb{C}^q$ is a trivialization, there are sections $\xi_\nu : U \to E$ defined by $\varphi \circ \xi_\nu(x) = (x, \mathbf{e}_\nu)$. The q-tuple $\xi = (\xi_1, \ldots, \xi_q)$ is called a *frame* for E. Use the concept of frames to define differential forms of type (p, q) with values in E locally as q-tuples of forms. Let $\mathscr{A}^{p,q}(U, E)$ be the space of (p, q)-forms with values in E over U. Show that there is an analogue of the Dolbeault sequence for forms with values in E.

3. Kähler Metrics

Hermitian metrics. Assume that X is an n-dimensional complex manifold.

> **Definition.** A *Hermitian metric* H on X is an assignment of a positive definite Hermitian form H_x to each tangent space $T_x(X)$ such that locally in complex coordinates it can be written in the form
>
> $$H_x = \sum_{i,j=1}^{n} g_{ij}(x)\, dz_i d\overline{z}_j,$$
>
> with smooth coefficients g_{ij}.

If $\alpha : [0,1] \to X$ is a differentiable arc, we can assign to it a length

$$L_H(\alpha) := \int_0^1 \sqrt{H_{\alpha(t)}(\dot{\alpha}(t), \dot{\alpha}(t))}\, dt,$$

where $\dot{\alpha}(t) = \alpha_* \partial/\partial t = \sum_\nu \alpha'_\nu(t)\partial/\partial z_\nu + \sum_\nu \overline{\alpha}'_\nu(t)\partial/\partial \overline{z}_\nu$ and $\alpha = (\alpha_1, \ldots, \alpha_n)$ in local coordinates. So

$$L_H(\alpha) = \int_0^1 \sqrt{\sum_{i,j} g_{ij}(\alpha(t))\alpha'_i(t)\overline{\alpha}'_j(t)}\, dt.$$

The length is positive if α is not constant.

Now, for a tangent vector $\xi \in T_x(X)$ a norm can be defined by

$$\|\xi\|_H := \sqrt{H_x(\xi, \xi)}.$$

Then $L_H(\alpha) = \int_0^1 \|\alpha_* \partial/\partial t\|_H\, dt$. There is a uniquely determined differentiable 1-form σ along α defined by

$$\sigma_{\alpha(t)}(\xi) = H_{\alpha(t)}\left(\xi, \frac{\alpha_* \partial/\partial t}{\|\alpha_* \partial/\partial t\|_H}\right).$$

It follows that $\alpha^*\sigma = a\, dt$, with $a = \sigma(\alpha_*\partial/\partial t) = \|\alpha_*\partial/\partial t\|_H$. Traditionally, σ is called the *line element* and is denoted by ds, though it is not the differential of a function. We have

$$L_H(\alpha) = \int_0^1 \alpha^* ds = \int_\alpha ds,$$

and $\alpha^* H_{\alpha(t)} = \|\alpha_*\partial/\partial t\|_H^2\, dt^2 = (ds)^2$. Therefore, we also denote the Hermitian metric H by ds^2.

Let $\mathscr{U} = (U_\iota)_{\iota \in I}$ be a (countable) open covering of X by local coordinate charts with coordinates $(z_1^\iota, \dots, z_n^\iota)$. Then there exists a partition of unity (ϱ_ι) for $\mathscr{U}$, and

$$ds^2 := \sum_{\iota \in I} \varrho_\iota \cdot \sum_{i=1}^{n} dz_i^\iota \, d\overline{z}_i^\iota$$

is a Hermitian metric on X. Thus we have the following result.

3.1 Proposition. *On every complex manifold there exists a Hermitian metric.*

The Fundamental Form. In a local coordinate system we associate to our Hermitian form $H = ds^2 = \sum_{i,j} g_{ij} dz_i d\overline{z}_j$ the differential form

$$\omega = \omega_H := \mathrm{i} \cdot \sum_{i,j} g_{ij} dz_i \wedge d\overline{z}_j.$$

(In the literature one often uses $\omega = \mathrm{i}/2 \cdot \sum g_{ij} dz_i \wedge d\overline{z}_j$.) We have $\overline{\omega} = -\mathrm{i} \cdot \sum_{i,j} \overline{g}_{ij} d\overline{z}_i \wedge dz_j = \omega$. So ω is a real $(1,1)$-form.

Let $\mathbf{G} = \big(g_{ij} \;\big|\; i,j = 1, \dots, n\big)$ be the (Hermitian) matrix of the coefficients of H, and $\boldsymbol{\xi}, \boldsymbol{\eta}$ the coordinate vectors of two tangent vectors ξ, η. Then $H(\xi, \eta) = \boldsymbol{\xi} \cdot \mathbf{G} \cdot \overline{\boldsymbol{\eta}}^t$.

If we apply a holomorphic coordinate transformation

$$\mathbf{z} = \mathbf{F}(\mathbf{w}) = (f_1(\mathbf{w}), \dots, f_n(\mathbf{w})),$$

we obtain

$$\mathbf{F}^*\omega = \mathrm{i} \cdot \sum_{i,j} (g_{ij} \circ \mathbf{F}) \, df_i \wedge d\overline{f}_j = \mathrm{i} \cdot \sum_{\nu,\mu} \widehat{g}_{\nu\mu} \, dw_\nu \wedge d\overline{w}_\mu.$$

This can be written in matrix form as

$$\widehat{\mathbf{G}}(\mathbf{w}) = J_{\mathbf{F}}(\mathbf{w})^t \cdot \mathbf{G}(\mathbf{F}(\mathbf{w})) \cdot \overline{J_{\mathbf{F}}(\mathbf{w})},$$

which is the transformation formula for Hermitian forms. Thus $\mathbf{F}^*\omega_H = \omega_{\mathbf{F}^*H}$. So the correspondence $H \leftrightarrow \omega$ is independent of the coordinates. The form ω is globally defined. In fact,

$$\begin{aligned} \omega(\xi, \eta) &= \mathrm{i} \cdot \sum_{i,j} g_{ij} (\xi_i \overline{\eta}_j - \overline{\xi}_j \eta_i) \\ &= \mathrm{i} \cdot (\boldsymbol{\xi} \cdot \mathbf{G} \cdot \overline{\boldsymbol{\eta}}^t - \boldsymbol{\eta} \cdot \mathbf{G} \cdot \overline{\boldsymbol{\xi}}^t) \\ &= -2 \operatorname{Im}(H(\xi, \eta)). \end{aligned}$$

The form ω_H is called the *fundamental form* of the metric H.

Definition. A *Kähler metric* on X is a Hermitian metric H with $d\omega_H = 0$. If X possesses a Kähler metric H, the pair (X, ω_H) is called a *Kähler manifold*. In this case we also say that X is a Kähler manifold.

In general, a Hermitian metric H does not satisfy the condition $d\omega_H = 0$. There are many complex manifolds that cannot have a Kähler metric. However, on a Riemann surface X every 3-form vanishes. So every Riemann surface is a Kähler manifold.

Geodesic Coordinates. Assume that H is a Hermitian metric on the manifold X and that $\mathbf{z} = (z_1, \dots, z_n)$ are holomorphic coordinates around some point $x_0 \in X$.

Definition. The coordinates $\mathbf{z}$ are called *geodesic* in x_0 (with respect to H) if $H = \sum_{i,j} g_{ij}\, dz_i d\overline{z}_j$ and

1. $g_{ij}(x_0) = \delta_{ij}$, i.e. $\mathbf{G}(x_0)$ is the unit matrix,
2. $(g_{ij})_{z_\nu}(x_0) = 0$ and $(g_{ij})_{\overline{z}_\nu}(x_0) = 0$ for $\nu = 1, \dots, n$, i.e., $(dg_{ij})_{x_0} = 0$.

We will show that the existence of geodesic coordinates is a characteristic property of Kähler metrics. For that we need a lemma.

3.2 Lemma. *$H = \sum_{i,j} g_{ij} dz_i d\overline{z}_j$ is a Kähler metric if and only if*

$$(g_{ij})_{z_\nu} = (g_{\nu j})_{z_i} \quad \text{and} \quad (g_{ij})_{\overline{z}_\nu} = (g_{i\nu})_{\overline{z}_j}, \text{ for all } i, j, \nu.$$

(The first equation implies the second one.)

PROOF: Let $\omega = \omega_H$ be the fundamental form of H. Then we have

$$\begin{aligned} d\omega &= \mathrm{i} \cdot \Big(\sum_{i,j,\nu} (g_{ij})_{z_\nu}\, dz_\nu \wedge dz_i \wedge d\overline{z}_j + \sum_{i,j,\nu} (g_{ij})_{\overline{z}_\nu}\, d\overline{z}_\nu \wedge dz_i \wedge d\overline{z}_j \Big) \\ &= \mathrm{i} \cdot \Big(\sum_j \big(\sum_{\nu<i} ((g_{ij})_{z_\nu} - (g_{\nu j})_{z_i})\, dz_\nu \wedge dz_i \big) \wedge d\overline{z}_j \\ &\qquad + \sum_i \big(\sum_{\nu<j} ((g_{i\nu})_{\overline{z}_j} - (g_{ij})_{\overline{z}_\nu})\, d\overline{z}_\nu \wedge d\overline{z}_j \big) \wedge dz_i \Big). \end{aligned}$$

Thus $d\omega = 0$ if and only if $(g_{ij})_{z_\nu} - (g_{\nu j})_{z_i} = 0$ and $(g_{i\nu})_{\overline{z}_j} - (g_{ij})_{\overline{z}_\nu} = 0$. ∎

3.3 Proposition. *H is a Kähler metric if and only if there are geodesic coordinates around every point in X.*

PROOF: We consider a fixed point $x_0 \in X$. If geodesic coordinates exist around x_0, then all first derivatives of the g_{ij} are zero there, and $d\omega = 0$ at x_0.

Assume now that H is a Kähler metric. We take a local coordinate system around x_0 such that x_0 is the origin of this coordinate system and apply a unitary transformation $\mathbf{z} = \mathbf{w} \cdot \mathbf{U}^t$ such that $\mathbf{U}^t \cdot \mathbf{G}(\mathbf{0}) \cdot \overline{\mathbf{U}} = \mathbf{E}_n$. We denote the new coordinates by $\mathbf{z}$ again and the new coefficient matrix of H also by $\mathbf{G}$ such that now $\mathbf{G}(\mathbf{0}) = \mathbf{E}_n$.

Finally, we apply a transformation

$$z_i = w_i + \frac{1}{2}\mathbf{w} \cdot \mathbf{A}_i \cdot \mathbf{w}^t, \quad i = 1, \dots, n,$$

where the $\mathbf{A}_i = \left(a^i_{k,l} \mid k, l = 1, \dots, n\right)$ are constant symmetric matrices. Then

$$(z_i)_{w_j} = \delta_{ij} + \frac{1}{2}\sum_{k,l} a^i_{kl}(\delta_{kj} w_l + w_k \delta_{lj}) = \delta_{ij} + \sum_k a^i_{jk} w_k.$$

So the Jacobian $\mathbf{J}$ of the transformation is given by $\mathbf{J}(\mathbf{w}) = \mathbf{E}_n + \sum_k \underline{\mathbf{A}}_k \cdot w_k$, with

$$\underline{\mathbf{A}}_k = \left(a^i_{jk} \;\middle|\; \begin{matrix} i = 1, \dots, n \\ j = 1, \dots, n \end{matrix}\right).$$

The coefficient matrix $\widehat{\mathbf{G}}$ of H in the new coordinates is therefore given up to first order by

$$\begin{aligned}
\widehat{\mathbf{G}}(\mathbf{w}) &= \mathbf{J}(\mathbf{w})^t \cdot \mathbf{G}(\mathbf{z}) \cdot \overline{\mathbf{J}(\mathbf{w})} \\
&= \Big(\mathbf{E}_n + \sum_k \underline{\mathbf{A}}_k w_k\Big)^t \cdot \mathbf{G}(\mathbf{z}) \cdot \overline{\Big(\mathbf{E}_n + \sum_l \underline{\mathbf{A}}_l w_l\Big)} \\
&\approx \mathbf{G}(\mathbf{z}) + \sum_k \underline{\mathbf{A}}_k^t \cdot \mathbf{G}(\mathbf{z}) w_k + \sum_l \mathbf{G}(\mathbf{z}) \cdot \overline{\underline{\mathbf{A}}_l}\, \overline{w}_l.
\end{aligned}$$

Since $(z_i)_{w_j}(\mathbf{0}) = \delta_{ij}$, we get $(\mathbf{G} \circ \mathbf{z})_{w_\lambda}(\mathbf{0}) = \mathbf{G}_{z_\lambda}(\mathbf{0})$ for all λ. So $\widehat{\mathbf{G}}_{w_\lambda}(\mathbf{0}) = \mathbf{G}_{z_\lambda}(\mathbf{0}) + \underline{\mathbf{A}}_\lambda^t$. The first derivatives of the functions $\widehat{g}_{ij}$ have to be zero at $\mathbf{0}$. We can achieve this by setting

$$a^i_{j\lambda} := -(g_{ji})_{z_\lambda}(\mathbf{0}).$$

Since $d\omega = 0$, it follows that the matrices $\mathbf{A}_i$ are in fact symmetric (use the equations of the lemma). So geodesic coordinates exist at x_0. ∎

Local Potentials. We assume that H is a Kähler metric on X and that U is a polydisk in a local coordinate system. We have the fundamental form $\omega = \mathrm{i} \cdot \sum_{i,j} g_{ij} dz_i \wedge d\overline{z}_j$ of H with $d\omega = 0$. Then by the Poincaré lemma we can construct a real 1-form φ with $d\varphi = \omega$ in U. We have the decomposition $\varphi = \varphi_0 + \varphi_1 = \sum_i a_i\, dz_i + \sum_i \bar{a}_i\, d\overline{z}_i$. Comparing types, we get $\partial\varphi_0 = \overline{\partial}\varphi_1 = 0$

and $\overline{\partial}\varphi_0 + \partial\varphi_1 = \omega$. Then, by Dolbeault's lemma, there are functions h_0, h_1 in U with $\partial h_0 = \varphi_0$ and $\overline{\partial} h_1 = \varphi_1$. It follows that

$$\mathrm{i}\partial\overline{\partial}(\mathrm{i}(h_0 - h_1)) = \overline{\partial}\partial h_0 + \partial\overline{\partial} h_1 = \overline{\partial}\varphi_0 + \partial\varphi_1 = \omega.$$

The operator $\mathrm{i}\partial\overline{\partial} = \frac{1}{2} d\, d^c$ is real. So we can replace $h^* := \mathrm{i}(h_0 - h_1)$ by the arithmetic mean h of h^* and its conjugate. Therefore, $\mathrm{i}\partial\overline{\partial} h = \omega$. The real-valued function h is called a *local potential* of the Kähler metric H. It is strictly plurisubharmonic, because

$$\mathrm{i}\partial\overline{\partial} h = \mathrm{i} \sum_{i,j} h_{z_i \overline{z}_j}\, dz_i \wedge d\overline{z}_j$$

is the fundamental form of a Hermitian metric; i.e., the Levi form of h is positive definite.

In the other direction, if local potentials h exist for a Hermitian metric H, then $\omega = \mathrm{i}\partial\overline{\partial} h$ is the fundamental form to H, and $d\omega = 0$. So H is a Kähler metric.

Pluriharmonic Functions. A real-valued function h on an open subset $U \subset X$ is called *pluriharmonic* if $\mathrm{i}\partial\overline{\partial} = 0$. In this case the 1-form $\varphi = \partial h - \overline{\partial} h$ is closed (i.e., $d\varphi = 0$) and purely imaginary (i.e., of the form $\varphi = \mathrm{i}\psi$, with a real form ψ). In fact, we have $\overline{\varphi} = -\varphi$.

If U is simply connected and x_0 a fixed point of U, we can define a function g by $g(x) := \int_{x_0}^{x} \psi$. Then $dg = \psi = -\mathrm{i}\varphi$, so $\partial g = -\mathrm{i}\partial h$ and $\overline{\partial} g = \mathrm{i}\overline{\partial} h$. Let $f := h + \mathrm{i}g$. It follows that

$$\overline{\partial} f = \overline{\partial} h + \mathrm{i}\overline{\partial} g = \overline{\partial} h - \overline{\partial} h = 0,$$

so f is holomorphic. Furthermore, $df = \partial f = 2\partial h$.

The Fubini Metric. The complex projective space $\mathbb{P}^n$ is one of the most important examples of Kähler manifolds. Let

$$\pi : (z_0, \dots, z_n) \mapsto (z_0 : \dots : z_n)$$

be the canonical projection from $\mathbb{C}^{n+1} - \{\mathbf{0}\}$ onto $\mathbb{P}^n$. The unitary linear transformations $L_{\mathbf{A}}(\mathbf{z}) = \mathbf{z} \cdot \mathbf{A}^t$ of $\mathbb{C}^{n+1}$ (with $\mathbf{A} \in \mathrm{U}(n+1)$, i.e., $\mathbf{A} \in \mathrm{GL}_n(\mathbb{C})$ and $\mathbf{A}^t \cdot \overline{\mathbf{A}} = \mathbf{E}_n$) give biholomorphic transformations $\varphi = \varphi_{\mathbf{A}} : \mathbb{P}^n \to \mathbb{P}^n$, by $\varphi_{\mathbf{A}}(\pi(\mathbf{z})) = \pi(L_{\mathbf{A}}(\mathbf{z}))$. The set of these is the group $\mathrm{PU}(n+1)$ of *projective unitary transformations.* If x_0, x_1 are arbitrary points in the projective space, then there exists a projective unitary transformation φ with $\varphi(x_0) = x_1$.

On $\mathbb{C}^{n+1}$ we have the canonical strictly plurisubharmonic function

$$p(z_0, \dots, z_n) := z_0\overline{z}_0 + \dots + z_n\overline{z}_n,$$

which is positive outside of $\mathbf{0}$. It is invariant by unitary linear transformations of $\mathbb{C}^{n+1}$.

If $x_0 \in \mathbb{P}^n$ is a point, we can find a simply connected neighborhood $U(x_0)$ and a holomorphic cross section $f : U \to \mathbb{C}^{n+1} - \{\mathbf{0}\}$ for the submersion π. The cross section is determined up to multiplication by a nowhere vanishing holomorphic function h. We use $\log(p \circ f)$ (or equivalently any function $\log(p \circ (h \cdot f))$) as local potential for a Hermitian metric. Since

$$\mathrm{i}\partial\overline{\partial}\log(p \circ (h \cdot f)) = \mathrm{i}\partial\overline{\partial}[\log(p \circ f) + \log(h) + \log(\bar{h})] = \mathrm{i}\partial\overline{\partial}\log(p \circ f),$$

we obtain a well-defined Hermitian form H on $\mathbb{P}^n$. Its fundamental form is

$$\omega_H = \mathrm{i}\partial\overline{\partial}\log(p \circ f).$$

We now show that H is positive definite.

Everything is invariant under the group $\mathrm{PU}(n+1)$. If $\mathbf{A}$ is a unitary matrix, $\varphi_{\mathbf{A}}(x_0) = x_1$, and f a holomorphic cross section in a neighborhood of x_1, then $\widetilde{f} := L_{\mathbf{A}}^{-1} \circ f \circ \varphi_A$ is a section in a neighborhood of x_0. So $p \circ \widetilde{f} = (p \circ f) \circ \varphi_{\mathbf{A}}$ is a local potential of H near x_0. It follows that $H_{x_0}(\xi, \eta) = H_{x_1}(\varphi_* \xi, \varphi_* \eta)$. Therefore, it is sufficient to prove only the positive definiteness at the point $x_0 = (1 : 0 : \cdots : 0)$.

At this point a holomorphic cross section is given by

$$f : (z_0 : \ldots : z_n) \mapsto \Big(1, \frac{z_1}{z_0}, \ldots, \frac{z_n}{z_0}\Big).$$

Therefore, we may take as potential the function

$$\begin{aligned} \log(p \circ f(z_0, \ldots, z_n)) &= \log(1 + t_1\bar{t}_1 + \cdots + t_n\bar{t}_n) \\ &= t_1\bar{t}_1 + \cdots + t_n\bar{t}_n + \text{terms of higher order}, \end{aligned}$$

where $t_i = z_i/z_0$ are the (inhomogeneous) local coordinates at x_0. Therefore, $H(\xi, \eta) = \xi_1\overline{\eta}_1 + \cdots + \xi_n\overline{\eta}_n$, and this is positive definite. The metric H constructed in this way is called the *Fubini metric* (or *Fubini–Study metric*) on $\mathbb{P}^n$. Since H is given by local potentials, it is a Kähler metric.

3.4 Proposition. *If X is a Kähler manifold, then any closed submanifold $Y \subset X$ is also a Kähler manifold.*

PROOF: We can restrict the given Kähler metric H on X to the submanifold Y. Then the restricted fundamental form $\omega_H|_Y$ corresponds to the restricted Hermitian form $H|_Y$, which again is positive definite, and $\omega_H|_Y$ is also closed. So Y is a Kähler manifold again. ∎

The following is an immediate consequence.

3.5 Theorem. *Every projective algebraic manifold is a Kähler manifold.*

Definition. Let X be an arbitrary n-dimensional complex manifold. We say that meromorphic functions $f_1, \ldots, f_k$ on X are *analytically dependent* if

$$\mathrm{rk}_x(f_1, \ldots, f_k) < k$$

at every point where each of the functions is holomorphic. They are called *algebraically dependent* if there exists a polynomial $p(w_1, \ldots, w_k)$ that is not the zero polynomial such that $p(f_1, \ldots, f_n)$ vanishes identically wherever it is defined.

X is called a *Moishezon manifold* if it is compact and has n analytically independent meromorphic functions.

On compact manifolds algebraic and analytic dependence are equivalent notions. In this case one can show that the field of all meromorphic functions on X has transcendence degree at most the dimension n of X. For projective manifolds there is equality. But a Moishezon manifold is in general not a Kähler manifold. Moishezon's theorem states that

A Moishezon manifold X is projective algebraic if and only if it is a Kähler manifold (see [Moi67], cf. also the article "Modifications" of Th. Peternell, VII.6.6 in [GrPeRe94]).

Deformations. There are many compact Kähler manifolds that are not projective algebraic. There are even examples of Kähler manifolds where every meromorphic function on them is constant.

The Kähler property is invariant under small holomorphic deformations, but not under large ones (see [Hi62], and also [GrPeRe94], VII.6.5).

Definition. Assume that X is a connected $(n+m)$-dimensional complex manifold, Y an m-dimensional complex manifold, and $\pi : X \to Y$ a proper surjective holomorphic map that has rank m at every point. Then (X, π, Y) is called a *holomorphic family of compact n-dimensional complex manifolds.*

Indeed, since π is a proper submersion, the fibers $X_y = \pi^{-1}(y)$, $y \in Y$, are compact submanifolds of X. When y varies, the fibers may have different complex structures. However, they are isomorphic as differentiable manifolds.

3.6 Theorem. *If X_0 is a Kähler manifold and $\pi : X \to Y$ is a deformation of X_0 (i.e., a holomorphic family with $X_0 = \pi^{-1}(y_0)$ for some $y_0 \in Y$), then there exists a neighborhood $U(y_0) \subset Y$ such that every fiber over U is again a Kähler manifold.*

On the other hand, it may happen that all fibers X_y except for X_{y_0} are Kähler manifolds.

The first statement can be found in many books on Kähler manifolds (for example in [MoKo71]). The original proof of the second statement was given by H. Hironaka.

Exercises

1. Let X be a complex manifold and E a holomorphic vector bundle over X. Use local frames as described in Exercise 2.3. A *Hermitian form* H on E is an assignment of a Hermitian form H_x to each fiber E_x such that $\langle \xi_i \,|\, \xi_j \rangle := H(\xi_i, \xi_j)$ is smooth for every local frame $\xi = (\xi_1, \ldots, \xi_q)$. Let h_ξ be the matrix $h_\xi := \big(\langle \xi_i \,|\, \xi_j \rangle\big)$. Calculate a transformation formula of h_ξ for changing from ξ to another frame ξ'. We call H a Hermitian scalar product (or a fiber metric) on E if H_x is positive definite for every $x \in X$. In this case, prove that if $x_0 \in X$ is fixed, then there is a local frame at x_0 such that $h_\xi(x_0) = \mathbf{E}_q$ and $(\partial h_\xi)_{x_0} = 0$.
2. For $\mathbf{z} \in \mathbb{C}^{n+1} - \{\mathbf{0}\}$ consider the linear map

$$\varphi_{\mathbf{z}} : \mathbb{C}^{n+1} \to T_{\pi(\mathbf{z})}(\mathbb{P}^n),$$

which has been defined in Section IV.5. Show that the Fubini–Study metric H is given by $H(\varphi_{\mathbf{z}}(\mathbf{v}), \varphi_{\mathbf{z}}(\mathbf{w})) = \langle \mathbf{v} \,|\, \mathbf{w} \rangle$.

For $\mathbf{z} \in \mathbb{C}^{n+1}$ with $\|\mathbf{z}\| = 1$ let $H_{\mathbf{z}} := \{\mathbf{w} \in \mathbb{C}^{n+1} : \langle \mathbf{z} \,|\, \mathbf{w} \rangle = 0\}$. For $\mathbf{w} \in H_{\mathbf{z}}$ with $\|\mathbf{w}\| = 1$ define

$$\alpha(t) := \pi((\cos t)\mathbf{z} + (\sin t)\mathbf{w}).$$

Prove that α is a closed path of length π. Using the fact that every geodesic on $\mathbb{P}^n$ is of this form, prove the formula

$$\operatorname{dist}(\pi(\mathbf{z}), \pi(\mathbf{w})) = \arctan \left| \frac{1}{\langle \mathbf{z} \,|\, \mathbf{w} \rangle} \mathbf{w} - \mathbf{z} \right| .$$

3. For decomposable k-vectors $\mathsf{A} = \mathbf{a}_1 \wedge \cdots \wedge \mathbf{a}_k$ and $\mathsf{B} = \mathbf{b}_1 \wedge \cdots \wedge \mathbf{b}_k$ define

$$\langle \mathsf{A} \,|\, \mathsf{B} \rangle := \det\Big(\langle \mathbf{a}_i \,|\, \mathbf{b}_j \rangle\Big)$$

and $|\mathsf{A}| = \langle \mathsf{A} \,|\, \mathsf{A} \rangle^{1/2}$. Prove that there is a Kähler form ω on $G_{k,n}$ locally given by $\omega = \mathrm{i}\partial\overline{\partial} \log|\mathsf{A}|$.
4. Assume that X is a connected complex manifold and that $A \subset X$ is a nowhere dense analytic set. Prove that if h is a bounded pluriharmonic function on $X - A$, then it can be extended as a pluriharmonic function to X.
5. Let $B \subset \mathbb{C}^n$ be a closed ball. Is there a plurisubharmonic function h on $\mathbb{C}^n - B$ with $\lim_{\mathbf{z} \to B} h(\mathbf{z}) = -\infty$?
6. A Kähler metric H on a complex manifold X is called *complete* if X is a complete metric space with respect to the associated distance function. Construct a complete Kähler metric on $\mathbb{C}^n - \{\mathbf{0}\}$.

4. The Inner Product

The Volume Element. Let X be an n-dimensional complex manifold with a Hermitian metric H, and $\omega = \omega_H$ the associated fundamental form.

For $x_0 \in X$ it is possible to choose coordinates $z_1, \ldots, z_n$ such that x_0 is the origin and $g_{ij}(x_0) = \delta_{ij}$. Therefore, $H_{x_0} = \sum_i dz_i d\overline{z}_i$. These are called *Euclidean coordinates.* We calculate the $2n$-form ω^n in these coordinates at x_0.

We have $\omega = \mathrm{i} \cdot \sum_\nu dz_\nu \wedge d\overline{z}_\nu$, and so

$$\begin{aligned} \omega^n &= \Bigl(\mathrm{i} \cdot \sum_{\nu=1}^{n} (dx_\nu + \mathrm{i} dy_\nu) \wedge (dx_\nu - \mathrm{i} dy_\nu)\Bigr)^n \\ &= \Bigl(\mathrm{i} \cdot \sum_{\nu=1}^{n} (-2\mathrm{i}) dx_\nu \wedge dy_\nu\Bigr)^n \\ &= 2^n \cdot \Bigl(\sum_{\nu=1}^{n} dx_\nu \wedge dy_\nu\Bigr)^n. \end{aligned}$$

A simple induction on k shows that

$$\Bigl(\sum_{\nu=1}^{n} dx_\nu \wedge dy_\nu\Bigr)^k = k! \sum_{1 \le \nu_1 < \cdots < \nu_k \le n} dx_{\nu_1} \wedge dy_{\nu_1} \wedge \cdots \wedge dx_{\nu_k} \wedge dy_{\nu_k}.$$

It follows that

$$\omega^n = 2^n \, n! \, dx_1 \wedge dy_1 \wedge \cdots \wedge dx_n \wedge dy_n$$

and

$$dx_1 \wedge dy_1 \wedge \cdots \wedge dx_n \wedge dy_n = \frac{1}{2^n \, n!} \omega^n = \frac{1}{2^n \, n!} (-2 \operatorname{Im} H_{x_0})^n.$$

Remarks

1. In general coordinates one would get

   $$\omega^n = 2^n \, n! \, g \, dx_1 \wedge dy_1 \wedge \cdots \wedge dx_n \wedge dy_n,$$

   where $g = \det(g_{ij} \mid i, j = 1, \ldots, n)$.

2. In the literature one finds other formulas:

   If ω is defined by $\omega = \frac{\mathrm{i}}{2} \sum_\nu dz_\nu \wedge d\overline{z}_\nu$, then

   $$\omega^n = n! \, dx_1 \wedge dy_1 \wedge \cdots \wedge dx_n \wedge dy_n.$$

   If, in addition, the wedge product is defined in such a way that $\varphi \wedge \psi = \frac{1}{2}(\varphi \otimes \psi - \psi \otimes \varphi)$ for 1-forms φ and ψ, then $\omega = -\frac{1}{2} \operatorname{Im} H$ and

   $$dx_1 \wedge dy_1 \wedge \cdots \wedge dx_n \wedge dy_n = \frac{1}{n!}\Bigl(-\frac{1}{2} \operatorname{Im} H\Bigr)^n.$$

Definition. The $2n$-form

$$dV := \frac{1}{2^n\, n!}\omega^n = \frac{1}{2^n\, n!}(-2\operatorname{Im} H)^n$$

is called the *volume element* associated to H.

The traditional notation "dV" does not mean that the volume element is the differential of some $(2n-1)$-form. It is easy to see that this is impossible, e.g., on compact manifolds. Moreover, in the case of a compact Kähler manifold (X, ω) it follows that $\int_X \omega^n = 2^n n! \int_X dV > 0$.

4.1 Proposition. *If (X, ω) is a compact Kähler manifold, then ω^k defines a nonzero class in the de Rham group $H^{2k}(X)$ for $1 \le k \le n = \dim(X)$.*

PROOF: It is clear that $d(\omega^k) = 0$. Suppose that there is some $(2k-1)$-form φ with $d\varphi = \omega^k$. Then $d(\varphi \wedge \omega^{n-k}) = \omega^n$, and by Stokes's theorem

$$\int_X \omega^n = \int_X d(\varphi \wedge \omega^{n-k}) = \int_{\partial X} \varphi \wedge \omega^{n-k} = 0.$$

This is a contradiction. ∎

The Star Operator

The Star Operator. Let $z_1, \ldots, z_n$ be Euclidean coordinates at $x_0 \in X$. We use real coordinates in the following form:

$$z_1 = u_1 + \mathrm{i} u_2, \ldots, z_n = u_{2n-1} + \mathrm{i} u_{2n}.$$

Then any r-form φ can be written as $\varphi = \sum_I a_I\, du_I$, where the summation is over all ascending sequences $I = (i_1, \ldots, i_r)$, $1 \le i_1 < \cdots < i_r \le 2n$.

Definition. The *(Hodge) star operator* $* : \mathscr{A}^r(X) \to \mathscr{A}^{2n-r}(X)$ is the $\mathbb{C}$-linear map defined by

$$*\, du_I := \varepsilon_{I,I'} \cdot du_{I'},$$

where $I \cup I' = \{1, \ldots, 2n\}$, $I \cap I' = \varnothing$, $\varepsilon_{I,I'} = \pm 1$, and

$$\varepsilon_{I,I'} \cdot du_I \wedge du_{I'} = dV.$$

The star operator depends on the metric (which determines Euclidean coordinates) and on the orientation (which is determined by the order of the real coordinates), but on nothing else. So it is globally defined and invariant under unitary coordinate transformations (in the sense that $*\,(\varphi \circ F) = (*\,\varphi) \circ F$ for these transformations).

4.2 Proposition.

1. *If φ is an r-form, then* $**\varphi = (-1)^r \cdot \varphi$.
2. $du_I \wedge * du_J = \begin{cases} dV & \text{if } I = J, \\ 0 & \text{otherwise.} \end{cases}$
3. $*$ *is real; i.e.,* $*\overline{\varphi} = \overline{*\varphi}$.

PROOF: (1) We have $**du_I = \varepsilon_{I',I}\varepsilon_{I,I'}\,du_I$, and

$$\begin{aligned}\varepsilon_{I',I}\varepsilon_{I,I'}\,du_{I'} \wedge du_I &= \varepsilon_{I',I}\varepsilon_{I,I'}(-1)^{r(2n-r)}\,du_I \wedge du_{I'} \\ &= \varepsilon_{I',I}(-1)^r\,dV \\ &= (-1)^r\,du_{I'} \wedge du_I.\end{aligned}$$

(2) $du_I \wedge * du_J = \varepsilon_{J,J'} du_I \wedge du_{J'} = 0$ if $I \cap J' \neq \varnothing$, and $= dV$ if $J' = I'$.

(3) We have $*(\operatorname{Re}\varphi + \mathrm{i}\operatorname{Im}\varphi) = *(\operatorname{Re}\varphi) + \mathrm{i}*(\operatorname{Im}\varphi)$, by definition. ∎

If $\varphi = \sum_I a_I du_I$ and $\psi = \sum_J b_J du_J$, then

$$\varphi \wedge *\psi = \sum_{I,J} a_I b_J du_I \wedge * du_J = \Bigl(\sum_I a_I b_I\Bigr) dV.$$

In particular, this expression is symmetric in φ and ψ.

Now we define $\overline{*}(\varphi) := *\overline{\varphi}$. This is a complex antilinear isomorphism between $\mathscr{A}^r(X)$ and $\mathscr{A}^{2n-r}(X)$.

We get $\varphi \wedge \overline{*}\psi = (\sum_I a_I \overline{b}_I)\,dV$ and $\varphi \wedge \overline{*}\varphi = (\sum_I |a_I|^2)\,dV$.

The Effect on (p,q)-Forms.

We calculate $\overline{*}\varphi$ for (p,q)-forms φ.

4.3 Lemma. *If $\varphi = dz_I \wedge d\overline{z}_J$ with $|I| = p$ and $|J| = q$, then*

$$\varphi \wedge \overline{*}\varphi = 2^{p+q}\,dV.$$

PROOF: We write $I = M \cup A$ and $J = M \cup B$, with pairwise disjoint sets $A, B, M \subset N := \{1, \dots, n\}$ such that $dz_I \wedge d\overline{z}_J = c_0\,dz_A \wedge d\overline{z}_B \wedge \omega_M$, where $\omega_M = \prod_{\mu \in M} dz_\mu \wedge d\overline{z}_\mu$ and $|c_0| = 1$.

Let $m := |M|$. Then $\omega_M = (-2\mathrm{i})^m \prod_{\mu\in M} du_{2\mu-1} \wedge du_{2\mu}$. The remainder $dz_A \wedge d\overline{z}_B$ is a form of type $(p-m, q-m)$.

If, for example, $A = \{\alpha_1, \dots, \alpha_s\}$ (with $s = p - m$), then

$$\begin{aligned}dz_A &= (du_{2\alpha_1-1} + \mathrm{i}du_{2\alpha_1}) \wedge \cdots \wedge (du_{2\alpha_s-1} + \mathrm{i}du_{2\alpha_s}) \\ &= \sum_{\nu=0}^{s} \mathrm{i}^{s-\nu} \sum_{|C|=\nu} \sigma_C du_{\alpha_C} \wedge du_{\alpha'_C},\end{aligned}$$

where $C \subset \{1, \ldots, s\}$ is an ascending sequence of length $|C|$, α_C the sequence of the $2\alpha_i - 1$, $i \in C$, and α'_C the sequence of the $2\alpha_i$, $i \in C$; σ_C is a number equal to ± 1. So we have $\sum_{\nu=0}^{s} \binom{s}{\nu} = 2^s$ summands. The analogue will be obtained for $d\overline{z}_B$.

Altogether we see that $\varphi = dz_I \wedge d\overline{z}_J$ is equal to 2^m times 2^{p+q-2m} different monomials $c_K\, du_K$ with $|c_K| = 1$. So

$$\varphi \wedge \overline{*}\,\varphi = \left(\sum_K |2^m \cdot c_K|^2\right) dV = 2^{p+q}\, dV.$$

■

Again let $\varphi = dz_I \wedge d\overline{z}_J$ with $|I| = p$, $|J| = q$ and $p + q = r$. Then $*\,\varphi$ is a $(2n - r)$-form, and it can be written as

$$*\,\varphi = \sum_{\nu+\mu=r} \sum_{\substack{|K|=n-\nu \\ |L|=n-\mu}} a_{KL}\, d\overline{z}_K \wedge dz_L,$$

with certain constants a_{KL}.

Now we apply the unitary transformation $\mathbf{F}(\mathbf{z}) = \mathbf{z} \cdot \mathbf{C}^t$, with

$$\mathbf{C} = \begin{pmatrix} e^{\mathrm{i}t_1} & & 0 \\ & \ddots & \\ 0 & & e^{\mathrm{i}t_n} \end{pmatrix}, \quad t_i \in \mathbb{R} \text{ for } i = 1, \ldots, n.$$

For $I = \{i_1, \ldots, i_p\} \subset N = \{1, \ldots, n\}$ we set $t_I := t_{i_1} + \cdots + t_{i_p}$. Then

$$\mathbf{F}^*\varphi = \exp(\mathrm{i}(t_I - t_J)) \cdot \varphi$$

and

$$(\mathbf{F}^{-1})^*(a_{KL}\, d\overline{z}_K \wedge dz_L) = a_{KL} \cdot \exp(\mathrm{i}(t_K - t_L)) \cdot d\overline{z}_K \wedge dz_L.$$

Since the star operator is invariant under unitary transformations, we have

$$(\mathbf{F}^{-1})^*(*\,(\mathbf{F}^*\varphi)) = *\,\varphi.$$

It follows that

$$*\,\varphi = \exp(\mathrm{i}(t_I - t_J)) \cdot \sum_{\nu+\mu=r} \sum_{K,L} a_{KL} \cdot \exp(\mathrm{i}(t_K - t_L))\, d\overline{z}_K \wedge dz_L.$$

The uniqueness of the representation implies that

$$\exp(\mathrm{i}(t_I + t_K - t_J - t_L)) = 1 \quad \text{for all } K, L.$$

Arbitrary t_i are allowed. If we choose them all very small, then $t_I + t_K = t_J + t_L$.

If there is a pair (K, L) with $I \cup K \neq J \cup L$, then for example there is an $i \in I$ that is not contained in $J \cup L$. Setting $t_i \neq 0$ and $t_j = 0$ for $j \neq i$ gives a contradiction. Therefore, $p + (n - \nu) = q + (n - \mu)$. On the other hand, we have $\nu + \mu = p + q$. This is possible if and only if $q = \mu$ and $p = \nu$. $*\varphi$ is a form of type $(n - q, n - p)$, and we can write

$$*\varphi = \sum_{\substack{|K|=n-p \\ |L|=n-q}} a_{KL}\, d\overline{z}_K \wedge dz_L.$$

We consider an index $i \in I \cup J$. There are three possibilities:

1. If $i \in I$ and $i \notin J$, then i must lie in L, since we always have $I \cup K = J \cup L$. We choose $t_i = \varepsilon$ (positive, but very small) and $t_j = 0$ for $j \neq i$, and we consider the equation $t_I + t_K = t_J + t_L$. Since $t_I = \varepsilon$, $t_K \geq 0$ and $t_J = 0$, it follows that $t_L \geq \varepsilon$, and therefore even $t_L = \varepsilon$. But then $t_K = 0$. This implies that $i \in L$ and $i \notin K$.
2. If $i \in J$ and $i \notin I$, then analogously $i \in K$ and $i \notin L$.
3. If $i \in I \cap J$, then φ contains the term $dz_i \wedge d\overline{z}_i$ and therefore the term $du_{2i-1} \wedge du_{2i}$. This implies that $du_{2i-1} \wedge du_{2i}$ (and therefore $dz_i \wedge d\overline{z}_i$) cannot occur in $*\varphi$.

So $I \cup K = J \cup L = \{1, \dots, n\}$ and $*\varphi = a\, d\overline{z}_{I'} \wedge dz_{J'}$, with some constant factor a. We may assume that $\varepsilon_{I,I'} = \varepsilon_{J,J'} = 1$.

4.4 Proposition. *In the above notation*

$$\overline{*}(dz_I \wedge d\overline{z}_J) = \overline{a}\, dz_{I'} \wedge d\overline{z}_{J'} \qquad \textit{with } a = 2^{p+q-n}\mathrm{i}^n(-1)^{q(n-p)+n(n+1)/2}.$$

PROOF: For $\varphi = dz_I \wedge d\overline{z}_J$ we have

$$\begin{aligned}
\varphi \wedge \overline{*}\varphi &= dz_I \wedge d\overline{z}_J \wedge (\overline{a}\, dz_{I'} \wedge d\overline{z}_{J'}) \\
&= \overline{a}(-1)^{q(n-p)}\, dz_I \wedge dz_{I'} \wedge d\overline{z}_J \wedge d\overline{z}_{J'} \\
&= \overline{a}(-1)^{q(n-p)+n(n-1)/2}\, (dz_1 \wedge d\overline{z}_1) \wedge \cdots \wedge (dz_n \wedge d\overline{z}_n) \\
&= \overline{a}(-1)^{q(n-p)+n(n-1)/2}(-2\mathrm{i})^n\, dV \\
&= \overline{a}(-1)^{q(n-p)+n(n-1)/2+n}2^n\mathrm{i}^n\, dV.
\end{aligned}$$

On the other hand, we know from the lemma that $\varphi \wedge \overline{*}\varphi = 2^{p+q}\, dV$. Comparing the coefficients, it follows that $\overline{a} = 2^{p+q-n}\mathrm{i}^{-n}(-1)^{q(n-p)+n(n+1)/2}$.
■

Remark. The formula is valid only with respect to Euclidean coordinates. And one has to observe the rule that $\varepsilon_{I,I'} = \varepsilon_{J,J'} = 1$.

The Global Inner Product. We work on a compact Hermitian manifold X. Let $\mathscr{U} = \{U_1, \ldots, U_N\}$ be a finite open covering of X with coordinates, and (ϱ_i) a partition of unity for $\mathscr{U}$. If ψ is any differentiable $2n$-form on X, we define the integral

$$\int_X \psi := \sum_{\nu=1}^{N} \int_{U_\nu} \varrho_\nu \psi.$$

(In local coordinates $\varrho_\nu \psi$ has the form $c(z_1, \ldots, z_n)\, dV$ with a differentiable function c with compact support. We already know the meaning of the integral over such a $2n$-form.) One can show that this definition is independent of the choice of the partition of unity.

If φ is an r-form, then $\psi = \varphi \wedge \bar{*}\varphi$ is a $2n$-form that locally is equal to a form $c\,dV$, with $c \geq 0$ and $c(x) = 0$ if and only if $\varphi_x = 0$.

Definition. The *inner product* of two r-forms φ, ψ on X is defined by

$$(\varphi\,,\,\psi) := \int_X \varphi \wedge \bar{*}\psi.$$

4.5 Proposition.

1. $(\varphi_1 + \varphi_2\,,\,\psi) = (\varphi_1\,,\,\psi) + (\varphi_2\,,\,\psi)$.
2. $(c \cdot \varphi\,,\,\psi) = c \cdot (\varphi\,,\,\psi)$.
3. $(\psi\,,\,\varphi) = \overline{(\varphi\,,\,\psi)}$.
4. $(\varphi\,,\,\varphi) \geq 0$, *and* $(\varphi\,,\,\varphi) = 0$ *if and only if* $\varphi = 0$.

PROOF: (1) and (2) are trivial.

(3) We have $\psi \wedge \bar{*}\varphi = \overline{\overline{\psi} \wedge *\varphi} = \overline{\varphi \wedge \bar{*}\psi}$.

(4) It is clear that $(\varphi\,,\,\varphi) \geq 0$ always. If $\varphi \neq 0$, then there is a point $x_0 \in X$ with $\varphi_{x_0} \neq 0$. If $x_0 \in U_\nu$ and $\varrho_\nu(x_0) > 0$, then we can find a small neighborhood $V = V(x_0) \subset\subset U_\nu$ such that $\varrho_\nu \varphi = c\,dV$ on V, with $c > 0$ everywhere in V. It follows easily that $(\varphi\,,\,\varphi) > 0$. ■

Since $(\ldots\,,\,\ldots)$ is a Hermitian scalar product on $\mathscr{A}^r(X)$, a norm on this space is defined by $\|\varphi\| := (\varphi\,,\,\varphi)^{1/2}$.

4.6 Proposition. *Two forms of different type are orthogonal to each other.*

PROOF: Let φ, ψ be two forms of type (p, q), respectively (s, t), such that $p + q = s + t = r$. Then $\varphi \wedge \bar{*}\psi$ is a form of type $(p + n - s, q + n - t)$. If $s > p$, then $q > t$ and therefore $q + n - t > n$. If $s < p$, then $p + n - s > n$. So $\varphi \wedge \bar{*}\psi = 0$ in these cases. ■

Unfortunately, $\mathscr{A}^r(X)$ is not complete in the induced topology, so that we cannot apply Hilbert space methods directly. We will show how to construct a closure of $\mathscr{A}^r(X)$.

Currents. Let X be a compact Hermitian manifold. A sequence $\varphi_\nu = \sum_I a_{I,\nu}\, du_I$ of (smooth) r-forms on X is said to be *convergent to zero* in $\mathscr{A}^r(X)$ if for every α all sequences $D^\alpha a_{I,\nu}$ are uniformly convergent to zero.

> **Definition.** A *current of degree* $2n-r$ is an $\mathbb{R}$-linear map $T : \mathscr{A}^r(X) \to \mathbb{C}$ such that if (φ_ν) is a sequence of r-forms converging to zero, then $T(\varphi_\nu)$ converges to zero in $\mathbb{C}$.

The set of all currents of degree $2n-r$ is denoted by $\mathscr{A}^r(X)'$.

Examples

1. A current of degree $2n$ is a distribution in the sense of L. Schwartz.
2. Let ψ be a differential form of degree $2n-r$. Then ψ defines a current T_ψ of degree $2n-r$ by
$$T_\psi[\varphi] := \int_X \psi \wedge \varphi.$$
It is easy to see that T_ψ is, in fact, a current. So $\mathscr{A}^{2n-r}(X) \subset \mathscr{A}^r(X)'$.
3. Let $M \subset X$ be an r-dimensional differential submanifold. Then a current T_M is defined by $T_M[\varphi] := \int_M \varphi$. Therefore, we say that a current of degree $2n-r$ has *dimension* r.

Let (φ_ν) be a sequence in $\mathscr{A}^r(X)$ converging to some r-form φ (i.e., $\varphi_\nu - \varphi \to 0$). Then also $\bar{*}(\varphi_\nu - \varphi) \to 0$, and therefore $T_\psi[\bar{*}(\varphi_\nu - \varphi)] \to 0$ in $\mathbb{C}$ for every $\psi \in \mathscr{A}^r(X)$. It follows that

$$\begin{aligned} \lim_{\nu\to\infty} (\psi\,,\varphi_\nu) &= \lim_{\nu\to\infty} \int_X \psi \wedge \bar{*}\varphi_\nu = \lim_{\nu\to\infty} T_\psi[\bar{*}\varphi_\nu] \\ &= T_\psi[\bar{*}\varphi] = \int_X \psi \wedge \bar{*}\varphi = (\psi\,,\varphi). \end{aligned}$$

> **Definition.** For $T \in \mathscr{A}^r(X)'$ and $\varphi \in \mathscr{A}^{2n-r}(X)$ we define
> $$(T\,,\varphi) := T[\bar{*}\varphi].$$

In particular, we have $(T_\psi\,,\varphi) = (\psi\,,\varphi)$. If $\psi_\nu \to \psi$, then $(T_\psi\,,\varphi) = \lim_{\nu\to\infty}(\psi_\nu\,,\varphi)$. This motivates the following.

Definition. If (ψ_ν) is a Cauchy sequence in $\mathscr{A}^{2n-r}(X)$, then $T = \lim_{\nu\to\infty} \psi_\nu \in \mathscr{A}^r(X)'$ is defined by

$$(T\,,\,\varphi) := \lim_{\nu\to\infty} (\psi_\nu\,,\,\varphi)\,.$$

The limit T is called a *square integrable current.*

We omit the proof that T is actually a current.

Before defining the notion of a limit of a sequence of currents we note that

$$\begin{aligned} T = \lim_{\nu\to\infty} \psi_\nu \quad &\iff \quad (T\,,\,\varphi) - (\psi_\nu\,,\,\varphi) \to 0 \text{ for all } \varphi \in \mathscr{A}^{2n-r}(X) \\ &\iff \quad T[\varphi] - T_{\psi_\nu}[\varphi] \to 0 \text{ for all } \varphi \in \mathscr{A}^r(X). \end{aligned}$$

Definition. A sequence (T_ν) of currents in $\mathscr{A}^r(X)'$ is called *convergent* to a current $T \in \mathscr{A}^r(X)'$ (written as $T = \lim_{\nu\to\infty} T_\nu$) if

$$\lim_{\nu\to\infty} (T[\varphi] - T_\nu[\varphi]) = 0 \text{ for all } \varphi \in \mathscr{A}^r(X).$$

We state the following without proof.

4.7 Theorem. *The space $\mathscr{A}^r(X)'$ of currents of degree $2n-r$ is complete; i.e., every Cauchy sequence converges. The set of square integrable currents forms a complete subspace of $\mathscr{A}^r(X)'$, which we call the* closure $\overline{\mathscr{A}}^{2n-r}(X)$ *of the space of $(2n-r)$-forms.*

Exercises

1. Compute the volume of the complex projective space $\mathbb{P}^n$ with respect to the Fubini metric, and the volume of a p-dimensional linear subspace.
2. Let (X, ω) be a compact Kähler manifold and $Y \subset X$ a closed submanifold with $\dim(Y) = m$. Then $\text{vol}(Y) = c \cdot \int_Y \omega^m$ for some constant c. Calculate this constant!
3. Try to define the product of a current and a differential form!
4. For $T \in \mathscr{A}^r(X)'$ define $dT \in \mathscr{A}^{r+1}(X)'$ by $dT[\varphi] := (-1)^{r+1}T[d\varphi]$. Show that dT is in fact a current with $d(dT) = 0$. Calculate dT for the case that T is a form or a submanifold.

5. Hodge Decomposition

Adjoint Operators. A complex vector space with a Hermitian scalar product is called a *unitary vector space.* Let V_1, V_2 be two unitary vector spaces, and $T : V_1 \to V_2$ an operator (i.e., a $\mathbb{C}$-linear mapping). If there exists an operator $T^* : V_2 \to V_1$ with

$$(T(v)\,,\,w) = (v\,,\,T^*(w)) \qquad \text{for } v \in V_1,\ w \in V_2,$$

then T^* is called an *adjoint operator* for T. It is clear that T^* is uniquely determined, but in general it may not exist.

If V_1, V_2 are unitary vector spaces and $T : V_1 \to V_2$ is a continuous operator, then it follows from the Cauchy–Schwarz inequality that $v \mapsto (T(v)\,,\,w)$ is continuous for any fixed $w \in V_2$. If V_1, V_2 are Hilbert spaces, then it follows by the Riesz representation theorem that there exists an element $T^*(w)$ with $(T(v)\,,\,w) = (v\,,\,T^*(w))$, and that $w \mapsto T^*(w)$ is continuous.

Now let X be an n-dimensional complex manifold.

5.1 Lemma. *If X is compact and φ a $(2n-1)$-form on X, then $\int_X d\varphi = 0$.*

PROOF: Let $\mathscr{U} = \{U_\nu : \nu = 1, \dots, N\}$ be a finite open covering of X by local coordinates and (ϱ_ν) a subordinate partition of unity. We choose open subsets $U_\nu^* \subset\subset U_\nu$ with piecewise smooth boundary and $\operatorname{supp}(\varrho_\nu) \subset U_\nu^*$. Then

$$\int_X d\varphi = \sum_\nu \int_{U_\nu} d(\varrho_\nu \varphi) = \sum_\nu \int_{U_\nu^*} d(\varrho_\nu \varphi) = \sum_\nu \int_{\partial U_\nu^*} \varrho_\nu \varphi = 0,$$

by Stokes's theorem. ∎

5.2 Proposition. *The operator $\delta := -\bar{*}\, d \bar{*} : \mathscr{A}^{r+1}(X) \to \mathscr{A}^r(X)$ is the adjoint of $d : \mathscr{A}^r(X) \to \mathscr{A}^{r+1}(X)$.*

PROOF: Let $\varphi \in \mathscr{A}^r(X)$, $\psi \in \mathscr{A}^{r+1}(X)$ be two arbitrary forms. Then $\varphi \wedge \bar{*}\psi$ is a $(2n-1)$-form, and

$$d(\varphi \wedge \bar{*}\psi) = d\varphi \wedge \bar{*}\psi + (-1)^r \varphi \wedge d\bar{*}\,\psi.$$

Therefore,

$$\begin{aligned}(d\varphi\,,\,\psi) &= \int_X d\varphi \wedge \bar{*}\psi = \int_X d(\varphi \wedge \bar{*}\psi) - (-1)^r \int_X \varphi \wedge d\bar{*}\,\psi \\ &= (-1)^{r+1+2n-r} \int_X \varphi \wedge \bar{*}\bar{*}\, d\bar{*}\,\psi = -\int_X \varphi \wedge \bar{*}(\bar{*}\, d\bar{*}\,\psi) \\ &= (\varphi\,,\,-\bar{*}\, d\bar{*}\,\psi)\,.\end{aligned}$$

∎

5.3 Proposition. *The operator* $\vartheta := -\bar{*}\,\partial\bar{*} : \mathscr{A}^{p+1,q}(X) \to \mathscr{A}^{p,q}(X)$, *respectively* $\overline{\vartheta} := -\bar{*}\,\overline{\partial}\bar{*} : \mathscr{A}^{p,q+1}(X) \to \mathscr{A}^{p,q}(X)$, *is the adjoint of* ∂, *respectively* $\overline{\partial}$.

PROOF: We consider only the first case. Let φ be a form of type (p, q) with $p + q = r$ and ψ a form of type $(p + 1, q)$. Then $\varphi \wedge \bar{*}\,\psi$ is a form of type $(n - 1, n)$. Therefore, $\overline{\partial}(\varphi \wedge \bar{*}\,\psi) = 0$ and

$$d(\varphi \wedge \bar{*}\,\psi) = \partial(\varphi \wedge \bar{*}\,\psi) = \partial\varphi \wedge \bar{*}\,\psi + (-1)^r \varphi \wedge \partial(\bar{*}\,\psi).$$

It follows that

$$\begin{aligned}(\partial\varphi\,,\,\psi) &= \int_X \partial\varphi \wedge \bar{*}\,\psi = \int_X d(\varphi \wedge \bar{*}\,\psi) - (-1)^r \int_X \varphi \wedge \partial(\bar{*}\,\psi)\\ &= (-1)^{r+1+2n-r} \int_X \varphi \wedge \bar{*}\bar{*}\,\partial(\bar{*}\,\psi) \;=\; (\varphi\,,\, -\bar{*}\,\partial\bar{*}\,\psi)\,.\end{aligned}$$

∎

5.4 Proposition. *We have*

1. $\delta = \vartheta + \overline{\vartheta}$,
2. $\delta\delta = 0$, $\vartheta\vartheta = 0$ *and* $\overline{\vartheta\vartheta} = 0$,
3. $\vartheta\overline{\vartheta} + \overline{\vartheta}\vartheta = 0$,
4. $\overline{\vartheta\varphi} = \overline{\vartheta}(\overline{\varphi})$ *and* $\overline{\overline{\vartheta}\varphi} = \vartheta(\overline{\varphi})$.

The proofs are trivial.

The Kählerian Case. Now let X be an n-dimensional Kähler manifold. For the moment X is not assumed to be compact. A canonical operator $L : \mathscr{A}^r(X) \to \mathscr{A}^{r+2}(X)$ is defined by

$$L\varphi := \frac{1}{2}\,\varphi \wedge \omega,$$

where ω is the fundamental form. If φ is a (p, q)-form, then $L\varphi$ is a $(p+1, q+1)$-form. Since $d\omega = 0$ and ω is of type $(1, 1)$, we also have $\partial\omega = \overline{\partial}\omega = 0$. It follows that

$$\partial L\varphi = \frac{1}{2}\,\partial(\varphi \wedge \omega) = \frac{1}{2}\,\partial\varphi \wedge \omega = L\partial\varphi$$

and

$$\overline{\partial} L\varphi = \frac{1}{2}\,\overline{\partial}(\varphi \wedge \omega) = \frac{1}{2}\,\overline{\partial}\varphi \wedge \omega = L\overline{\partial}\varphi.$$

In addition, we have

$$\overline{L\varphi} = L\overline{\varphi}.$$

5.5 Proposition. *If X is a compact (Kähler) manifold, then the adjoint operator $\Lambda : \mathscr{A}^{r+2}(X) \to \mathscr{A}^r(X)$ of L is given by*

$$\Lambda = (-1)^r \,\bar{*}\, L \,\bar{*}.$$

PROOF: If ψ is an $(r+2)$-form, then $\omega \wedge \bar{*}\psi$ is a $(2n-r)$-form, and

$$\begin{aligned}(L\varphi\,,\psi) &= \frac{1}{2}\int_X \varphi \wedge \omega \wedge \bar{*}\psi = \frac{1}{2}\int_X (-1)^r \varphi \wedge \overline{**}\omega \wedge \bar{*}\psi \\ &= (\varphi\,, (-1)^r \,\bar{*}\, L \,\bar{*}\,\psi)\,.\end{aligned}$$

∎

Note also that Λ is a real operator: $\overline{\Lambda\varphi} = \Lambda\overline{\varphi}$.

5.6 Proposition. *The commutativity relations $\vartheta\Lambda = \Lambda\vartheta$ and $\overline{\vartheta}\Lambda = \Lambda\overline{\vartheta}$ hold.*

We leave the proof to the reader.

Bracket Relations. Let k, l be integers, and $S : \mathscr{A}^r(X) \to \mathscr{A}^{r+k}(X)$, $T : \mathscr{A}^r(X) \to \mathscr{A}^{r+l}(X)$ linear operators that are defined for every r. Then

$$[S,T] := S \circ T - T \circ S : \mathscr{A}^r(X) \to \mathscr{A}^{r+k+l}$$

is called the *bracket* or the *commutator* of S and T.

Example

As we have seen above, $\vartheta, \overline{\vartheta} : \mathscr{A}^r \to \mathscr{A}^{r-1}$ and $\Lambda : \mathscr{A}^r \to \mathscr{A}^{r-2}$ have the brackets $[\vartheta, \Lambda] = [\overline{\vartheta}, \Lambda] = 0$ (as operators from $\mathscr{A}^r$ to $\mathscr{A}^{r-3}$). Before we calculate more brackets, we have to calculate $\vartheta\varphi$ very explicitely for a form φ of type (p,q). For that we need geodesic coordinates. So everything that follows is valid only on Kähler manifolds.

Let $\varphi = \sum_{I,J} a_{IJ}\, dz_I \wedge d\overline{z}_J$ be an arbitrary (p,q)-form. We consider geodesic coordinates at a point $x_0 \in X$ such that $\omega = \mathrm{i}\sum_\nu dz_\nu \wedge d\overline{z}_\nu$. Then

$$\bar{*}\varphi = \sum_{I,J} \overline{a}_{IJ}\, \bar{*}(dz_I \wedge d\overline{z}_J) = \overline{a(p,q)} \cdot \sum_{I,J} \overline{a}_{IJ}\, dz_{I'} \wedge d\overline{z}_{J'},$$

where $a(p,q) = 2^{p+q-n}\,\mathrm{i}^n(-1)^{q(n-p)+n(n+1)/2}$, and $\varepsilon_{I,I'} = \varepsilon_{I',J'} = 1$. In a neighborhood of x_0 the formula is more complicated, because the coefficients g_{ij} of the metric are involved. We have to apply $\overline{\partial}$ in this neighborhood, but because we are working with geodesic coordinates, we obtain the exact value at x_0 by simply applying $\overline{\partial}$ to our representation of $\bar{*}\varphi$ at x_0. It follows that

$$\overline{\partial}\,\bar{*}\varphi = \overline{a(p,q)} \cdot (-1)^{n-p} \sum_{I,J}\sum_{\iota \in J} \overline{(a_{IJ})_{z_\iota}}\, dz_{I'} \wedge d\overline{z}_\nu \wedge d\overline{z}_{J'}.$$

Now we need some notation.

If $J = (j_1, \ldots, j_q)$, then $J(\nu) := (j_1, \ldots, \widehat{j_\nu}, \ldots, j_q)$ for $\nu = 1, \ldots, q$. Furthermore, let I^* and $J^*(\nu)$ be arrangements of I, respectively $J(\nu)$, such that $\varepsilon_{I',I^*} = \varepsilon_{(j_\nu, J'),J^*(\nu)} = 1$.

Then $dz_{I^*} = (-1)^{p(n-p)}\, dz_I$ and $d\overline{z}_{J^*(\nu)} = (-1)^{n-q+q(n-q)+\nu-1} d\overline{z}_{J(\nu)}$, and we get

$$\begin{aligned}
\overline{\vartheta}\varphi &= -\overline{*}\,\overline{\partial}\,\overline{*}\varphi \\
&= -a(p,q)\cdot \overline{a(n-p,n-q+1)}\cdot(-1)^{n-p}\sum_{I,J}\sum_{\nu=1}^{q}(a_{IJ})_{z_{j_\nu}}\, dz_{I^*}\wedge d\overline{z}_{J^*(\nu)} \\
&= a(p,q)\cdot \overline{a(n-p,n-q+1)}\cdot(-1)^{n-p+p(n-p)+n-q+q(n-q)} \\
&\quad\times \sum_{I,J}\sum_{\nu=1}^{q}(-1)^{\nu}(a_{IJ})_{z_{j_\nu}}\, dz_I\wedge d\overline{z}_{J(\nu)} \\
&= 2\cdot(-1)^{p}\sum_{I,J}\sum_{\nu=1}^{q}(-1)^{\nu}(a_{IJ})_{z_{j_\nu}}\, dz_I\wedge d\overline{z}_{J(\nu)}.
\end{aligned}$$

5.7 Theorem. *On a compact Kähler manifold* $[\overline{\vartheta}, L] = \mathrm{i}\,\partial$.

PROOF: In order to calculate $\overline{\vartheta}L\varphi$ and $L\overline{\vartheta}\varphi$ for a (p,q)-form φ we begin with

$$\begin{aligned}
L\varphi &= \frac{1}{2}\omega\wedge\varphi \quad (\text{since } \varphi\wedge\omega = \omega\wedge\varphi) \\
&= \left(\frac{\mathrm{i}}{2}\sum_{\lambda=1}^{n} dz_\lambda\wedge d\overline{z}_\lambda\right)\wedge\left(\sum_{I,J} a_{IJ}\, dz_I\wedge d\overline{z}_J\right) \\
&= \frac{\mathrm{i}}{2}(-1)^{p}\sum_{I,J}\sum_{\lambda\in I'\cap J'} a_{IJ}\, dz_\lambda\wedge dz_I\wedge d\overline{z}_\lambda\wedge d\overline{z}_J.
\end{aligned}$$

From that we get

$$\begin{aligned}
\overline{\vartheta}\,L\varphi &= \\
&= \mathrm{i}\,(-1)^{p+1}\sum_{I,J}\sum_{\lambda\in I'\cap J'}(-1)^{p+1}(a_{IJ})_{z_\lambda}\, dz_\lambda\wedge dz_I\wedge d\overline{z}_J \\
&\quad + \mathrm{i}\,(-1)^{p+1}\sum_{I,J}\sum_{\lambda\in I'\cap J'}\sum_{\nu=1}^{q}(-1)^{p+\nu+1}(a_{IJ})_{z_{j_\nu}}\, dz_\lambda\wedge dz_I\wedge d\overline{z}_\lambda\wedge d\overline{z}_{J(\nu)} \\
&= \mathrm{i}\sum_{I,J}\sum_{\lambda\in I'\cap J'}(a_{IJ})_{z_\lambda}\, dz_\lambda\wedge dz_I\wedge d\overline{z}_J + \varphi_0,
\end{aligned}$$

with

$$\varphi_0 := \mathrm{i} \cdot \sum_{I,J} \sum_{\lambda \in I' \cap J'} \sum_{\nu=1}^{q} (-1)^{\nu} (a_{IJ})_{z_{j_\nu}} \, dz_\lambda \wedge dz_I \wedge d\overline{z}_\lambda \wedge d\overline{z}_{J(\nu)}.$$

On the other hand,

$$\begin{aligned}
L\overline{\vartheta}\varphi &= \frac{1}{2}\omega \wedge \overline{\vartheta}\varphi \\
&= \left(\mathrm{i} \sum_{\lambda=1}^{n} dz_\lambda \wedge d\overline{z}_\lambda \right) \wedge (-1)^p \sum_{I,J} \sum_{\nu=1}^{q} (-1)^{\nu} (a_{IJ})_{z_{j_\nu}} \, dz_I \wedge d\overline{z}_{J(\nu)} \\
&= \mathrm{i} \sum_{I,J} \sum_{\nu=1}^{q} \sum_{\lambda \in I' \cap J(\nu)'} (-1)^{\nu} (a_{IJ})_{z_{j_\nu}} \, dz_\lambda \wedge dz_I \wedge d\overline{z}_\lambda \wedge d\overline{z}_{J(\nu)} \\
&= \mathrm{i} \sum_{I,J} \sum_{\nu=1}^{q} (-1)^{\nu} (-1)^{\nu-1} (a_{IJ})_{z_{j_\nu}} dz_{j_\nu} \wedge dz_I \wedge d\overline{z}_J \\
&\quad + \mathrm{i} \sum_{I,J} \sum_{\lambda \in I' \cap J'} \sum_{\nu=1}^{q} (-1)^{\nu} (a_{IJ})_{z_{j_\nu}} \, dz_\lambda \wedge dz_I \wedge d\overline{z}_\lambda \wedge d\overline{z}_{J(\nu)} \\
&= -\mathrm{i} \sum_{I,J} \sum_{\lambda \in I' \cap J} (a_{IJ})_{z_\lambda} \, dz_\lambda \wedge dz_I \wedge d\overline{z}_J + \varphi_0.
\end{aligned}$$

In total we obtain

$$\overline{\vartheta} L\varphi - L\overline{\vartheta}\varphi = \mathrm{i} \sum_{I,J} \sum_{\lambda \in I'} (a_{IJ})_{z_\lambda} \, dz_\lambda \wedge dz_I \wedge d\overline{z}_J = \mathrm{i}\partial\varphi.$$

■

5.8 Corollary. $[\vartheta, L] = -\mathrm{i}\,\overline{\partial}$, $[\partial, \Lambda] = -\mathrm{i}\,\overline{\vartheta}$ *and* $[\overline{\partial}, \Lambda] = \mathrm{i}\,\vartheta$.

The proof is an easy exercise.

The Laplacian.
Let X be an n-dimensional compact complex manifold.

Definition. The (*real*) *Laplacian* $\Delta : \mathscr{A}^r(X) \to \mathscr{A}^r(X)$ is defined by

$$\Delta\varphi := (d\,\delta + \delta\,d)\varphi.$$

The *complex Laplacian* $\Box : \mathscr{A}^{p,q}(X) \to \mathscr{A}^{p,q}(X)$ is defined by

$$\Box\,\varphi := (\overline{\partial}\,\overline{\vartheta} + \overline{\vartheta}\,\overline{\partial})\varphi.$$

Since d and δ are real operators, Δ is also real.

5.9 Proposition. *On a compact Kähler manifold,* $\Box = \overline{\Box}$.

PROOF: We have $\partial\overline{\vartheta} = -\overline{\vartheta}\,\partial$ and $\overline{\partial}\,\vartheta = -\vartheta\,\overline{\partial}$, as can be seen from the bracket relations. Therefore,

$$\begin{aligned} -\mathsf{i}\,\square &= -\mathsf{i}(\overline{\partial}\,\overline{\vartheta} + \overline{\vartheta}\,\overline{\partial}) \\ &= \overline{\partial}(-\mathsf{i}\,\overline{\vartheta}) + (-\mathsf{i}\,\overline{\vartheta})\overline{\partial} \\ &= \overline{\partial}(\partial\Lambda - \Lambda\partial) + (\partial\Lambda - \Lambda\partial)\overline{\partial}. \end{aligned}$$

From this one easily derives that $\overline{-\mathsf{i}\,\square} = \mathsf{i}\,\square$. ∎

A further consequence is

$$\mathsf{i}\,\square = \partial(\overline{\partial}\Lambda - \Lambda\overline{\partial}) + (\overline{\partial}\Lambda - \Lambda\overline{\partial})\partial = \mathsf{i}(\partial\vartheta + \vartheta\partial).$$

Therefore, $\square = \partial\vartheta + \vartheta\partial$ and $\overline{\square\varphi} = (\partial\vartheta + \vartheta\partial)\overline{\varphi} = \square\overline{\varphi}$; i.e., $\square$ is a real operator.

5.10 Proposition. *On a compact Kähler manifold,* $\square = \frac{1}{2}\Delta$.

PROOF: We have

$$\begin{aligned} \Delta &= d\delta + \delta d \\ &= (\partial + \overline{\partial})(\vartheta + \overline{\vartheta}) + (\vartheta + \overline{\vartheta})(\partial + \overline{\partial}) \\ &= (\partial\,\vartheta + \vartheta\,\partial) + (\overline{\partial}\,\overline{\vartheta} + \overline{\vartheta}\,\overline{\partial}) \\ &= 2\square. \end{aligned}$$

∎

This formula is not valid on general compact manifolds!

Harmonic Forms. Let X be a compact Hermitian (not necessarily Kähler) complex manifold. We define

$$\begin{aligned} \mathscr{B}^r(X) &:= d\mathscr{A}^{r-1}(X), \\ \mathscr{D}^r(X) &:= \delta\mathscr{A}^{r+1}(X), \\ \mathscr{H}^r(X) &:= \{\varphi \in \mathscr{A}^r(X) \,:\, \Delta\varphi = 0\}. \end{aligned}$$

The elements of $\mathscr{H}^r(X)$ are called *harmonic forms.*

Example

For calculating the star operator on (p, q)-forms one needs the number $a(p, q) = 2^{p+q-n}\mathsf{i}^n(-1)^{q(n-p)+n(n+1)/2}$. If $n = 1$, then $a(1, 0) = -\mathsf{i}$, $a(0, 1) = \mathsf{i}$ and $a(1, 1) = -2\mathsf{i}$. It follows that in geodesic coordinates

$$\begin{aligned} \Delta f = \delta df &= -\overline{*}\,d\,\overline{*}(f_z\,dz + f_{\overline{z}}\,d\overline{z}) \\ &= -\overline{*}\,d(\mathsf{i}\,f_{\overline{z}}\,d\overline{z} - \mathsf{i}\,f_z\,dz) \\ &= -\overline{*}(2\mathsf{i}\,f_{z\overline{z}}\,dz \wedge d\overline{z}) \\ &= -4f_{z\overline{z}} = -f_{xx} - f_{yy}. \end{aligned}$$

This motivates the name "harmonic."

5.11 Proposition.

1. $\varphi \in \mathscr{H}^r \iff d\varphi = \delta\varphi = 0$.
2. *The vector spaces* $\mathscr{B}^r$, $\mathscr{D}^r$, *and* $\mathscr{H}^r$ *are mutually orthogonal.*

PROOF: (1) If $\Delta\varphi = 0$, then $(\varphi\,,\, d\delta\varphi) + (\varphi\,,\, \delta d\varphi) = 0$. But $(\varphi\,,\, d\delta\varphi) = (\delta\varphi\,,\, \delta\varphi)$ and $(\varphi\,,\, \delta d\varphi) = (d\varphi\,,\, d\varphi)$. Therefore, $d\varphi = \delta\varphi = 0$.

(2) We have $(d\varphi\,,\, \delta\psi) = (\varphi\,,\, \delta\delta\psi) = 0$ for all φ, ψ. And if $\Delta\psi = 0$, then $(d\varphi\,,\, \psi) = (\varphi\,,\, \delta\psi) = 0$, and analogously, $(\delta\varrho\,,\, \psi) = (\varrho\,,\, d\psi) = 0$ for all φ, ϱ. So the spaces are mutually orthogonal. ■

5.12 Hodge theorem. *If X is a compact Hermitian manifold, then*

$$\mathscr{A}^r(X) = \mathscr{B}^r(X) \oplus \mathscr{D}^r(X) \oplus \mathscr{H}^r(X).$$

PROOF: We give only a sketch of the proof. For more details see [dRh84], [Schw50], or [GriHa78], and for a very explicit and rather elementary proof see [War71].

(1) If we assume that $\mathscr{A}^r$ is finite-dimensional, then the proof is very easy. We let V be the orthogonal complement of $\mathscr{B}^r + \mathscr{D}^r$. Then $\mathscr{A}^r = \mathscr{B}^r \oplus \mathscr{D}^r \oplus V$, and it follows immediately that $\Delta\varphi = 0$ for every $\varphi \in V$.

(2) If $\mathscr{A}^r$ is infinite-dimensional, then we consider its closure, the Hilbert space $\mathscr{L}^2$ of square integrable currents. We can define derivatives of currents by

$$\frac{\partial T}{\partial u_\nu}[\varphi] := -T\Big[\frac{\partial \varphi}{\partial u_\nu}\Big].$$

So it is possible to speak of *harmonic currents* (i.e., currents T with $\Delta T = 0$). Now, the Laplacian is an *elliptic operator*, the definition of which we now recall.

A linear differential operator P has locally (in real coordinates $u_1, \ldots, u_n$) the form

$$P = \sum_{|\alpha| \le k} a_\alpha(\mathbf{u}) D^\alpha : \mathscr{C}^\infty(U, \mathbb{R}^M) \to \mathscr{C}^\infty(U, \mathbb{R}^N),$$

with $(N \times M)$ matrices a_α and $D^\alpha = \partial^{|\alpha|}/(\partial u_1^{\alpha_1} \cdots \partial u_n^{\alpha_n})$. Then the *symbol* of P at $\mathbf{u}$ is the linear map

$$\sigma_P(\mathbf{u}, \boldsymbol{\xi}) := \sum_{|\alpha| = k} a_\alpha(\mathbf{u}) \boldsymbol{\xi}^\alpha : \mathbb{R}^M \to \mathbb{R}^N,$$

s on $\boldsymbol{\xi} \in \mathbb{R}^n - \{\mathbf{0}\}$. The operator P is called *elliptic* if $\sigma_P(\mathbf{u}, \boldsymbol{\xi})$ every $\mathbf{u}$ and every $\boldsymbol{\xi} \neq \mathbf{0}$. Since $\sigma_\Delta(\mathbf{u}, \boldsymbol{\xi}) = -\|\boldsymbol{\xi}\|^2 \cdot \mathrm{id}$, it follows ptic.

We state the **regularity theorem**:

Let $P : \mathscr{A}^r(X) \to \mathscr{A}^r(X)$ be an elliptic operator, and φ an element of $\mathscr{A}^r(X)$. If there is a square integrable current T with $P(T) = \varphi$, then there is an r-form ψ with $T = T_\psi$.

If $\mathscr{V}$ is the orthogonal complement to $\overline{\mathscr{B}^r} \oplus \overline{\mathscr{D}^r}$ in $\mathscr{L}^2$, then simple Hilbert space theory implies that $\mathscr{L}^2 = \overline{\mathscr{B}^r} \oplus \overline{\mathscr{D}^r} \oplus \mathscr{V}$, and the sum is orthogonal.

Just as in the finite-dimensional case it follows that $\Delta(\mathscr{V}) = 0$, and using approximation of currents by forms one sees that every harmonic current belongs to $\mathscr{V}$. From the regularity theorem it follows that $\mathscr{V} = \mathscr{H}^r$.

Finally, one shows that if ψ is an r-form and $T_\psi = T + S + T_\varphi$, with a harmonic form φ and currents $T \in \overline{\mathscr{B}^r}$, $S \in \overline{\mathscr{D}^r}$, then there are also forms $\tau \in \mathscr{B}^r$, $\sigma \in \mathscr{D}^r$ such that $\psi = \tau + \sigma + \varphi$. ∎

5.13 Theorem. *If X is a compact Hermitian manifold, then $H^r(X) = \mathscr{H}^r(X)$.*

PROOF: The de Rham group $H^r(X)$ is the quotient of the closed forms modulo the exact forms. Let $\varphi \in \mathscr{A}^r(X)$ be given with $d\varphi = 0$. Then we have a unique decomposition $\varphi = d\psi + \delta\varrho + H\varphi$, with $\Delta(H\varphi) = 0$. Now,

$$\begin{aligned} (\delta\varrho\,,\,\delta\varrho) &= (\varphi\,,\,\delta\varrho) \quad \text{(because of the orthogonality)} \\ &= (d\varphi\,,\,\varrho) = 0. \end{aligned}$$

So $\varphi = d\psi + H\varphi$, and we define $h : H^r(X) \to \mathscr{H}^r(X)$ by $h : [\varphi] \mapsto H\varphi$. This is well defined, and since $\varphi = d\psi \iff H\varphi = 0$, it is injective. If α is harmonic, then $d\alpha = 0$ and $H\alpha = \alpha$. So h is surjective. ∎

Remark. By the theorem of de Rham we have $H^r(X) \cong H^r(X, \mathbb{C})$. Therefore, $\beta_r(X) := \dim_{\mathbb{C}}(\mathscr{H}^r(X))$ is a topological invariant. We have

$$\beta_r(X) = b_r(X) = \mathrm{rk}(H^r(X, \mathbb{Z})),$$

the rth Betti number. In Chapter IV only b_1 was introduced, since we did not work with higher cohomology groups.

5.14 Theorem (Poincaré duality). *On every n-dimensional compact Hermitian manifold we have $H^r(X) \cong H^{2n-r}(X)$.*

PROOF: It is easy to see that $\bar{*}\Delta = \Delta\bar{*}$. Therefore, $\bar{*}: \mathscr{H}^r(X) \to \mathscr{H}^{2n-r}(X)$ is an isomorphism (depending on the metric). Then we apply the previous theorem. ∎

Remark. The original Poincaré duality states that there is an isomorphism $H^r(X,\mathbb{Z}) \to H_{2n-r}(X,\mathbb{Z})$. Our theorem here is a very weak version of this topological result.

Consequences. Let X be an n-dimensional compact Hermitian manifold. We define subspaces of $\mathscr{A}^{p,q}(X)$ by

$$\begin{aligned} \mathscr{B}^{p,q} &:= \bar{\partial}\mathscr{A}^{p,q-1}, \\ \mathscr{D}^{p,q} &:= \bar{\vartheta}\mathscr{A}^{p,q+1}, \\ \mathscr{H}^{p,q} &:= \{\varphi \in \mathscr{A}^{p,q} : \Box\varphi = 0\}. \end{aligned}$$

We have $\Box\varphi = 0 \iff \bar{\partial}\varphi = \bar{\vartheta}\varphi = 0$, and therefore the three spaces are mutually orthogonal.

5.15 Decomposition theorem for (p,q)-forms. *We have*

$$\mathscr{A}^{p,q} = \mathscr{B}^{p,q} \oplus \mathscr{D}^{p,q} \oplus \mathscr{H}^{p,q}.$$

The proof is the same as in the case of r-forms. One needs the ellipticity of $\Box$, which can be shown directly (and is trivially given on Kähler manifolds, since then $\Box = \frac{1}{2}\Delta$).

5.16 Theorem. $H^{p,q}(X) \cong \mathscr{H}^{p,q}(X)$ *for all* p, q.

The proof is the same as for $H^r(X)$.

5.17 Finiteness theorem. $\dim_{\mathbb{C}} H^{p,q}(X) < \infty$ *for all* p, q.

Using the theory of elliptic operators one can prove that $\mathscr{H}^{p,q}$ is finite-dimensional. This implies our result. On the other hand, we know already from Dolbeault's theorem that $H^{p,0}(X) \cong \Omega^p(X) = H^0(X, \Omega^p_X)$ and $H^{p,1}(X) \cong H^1(X, \Omega^p_X)$, where Ω^p_X is the complex analytic bundle of holomorphic p-forms. By the results of Chapter V these cohomology groups are finite-dimensional. This can be generalized to higher cohomology. So we have two different ways to prove the finiteness theorem.

5.18 Serre–Kodaira duality theorem. *On an n-dimensional compact Hermitian manifold* $H^{p,q}(X) \cong H^{n-p,n-q}(X)$ *for all* p, q.

PROOF: It is easy to see that $\bar{*}\Box = \Box\bar{*}$. Therefore, $\bar{*}$ defines an isomorphism from $\mathscr{H}^{p,q}$ onto $\mathscr{H}^{n-p,n-q}$. ∎

As mentioned above, Poincaré duality can be proved by topological means. This is not possible for the Serre–Kodaira duality.

Now we consider applications to Kähler manifolds.

5.19 Theorem. *If X is an n-dimensional compact Kähler manifold, then*

1. $\mathscr{H}^r(X) = \bigoplus_{p+q=r} \mathscr{H}^{p,q}(X)$.
2. $\overline{\mathscr{H}^{p,q}(X)} = \mathscr{H}^{q,p}(X)$.

PROOF: (1) If $\varphi \in \mathscr{H}^r$, then there is a unique decomposition

$$\varphi = \sum_{p+q=r} \varphi^{(p,q)},$$

with $\varphi^{(p,q)} \in \mathscr{A}^{p,q}$. It follows that

$$0 = \Delta\varphi = \sum_{p+q=r} \Delta\varphi^{(p,q)}.$$

If X is a Kähler manifold, then $\Delta = 2\Box$. Therefore, $\Delta\varphi^{(p,q)}$ is a form of type (p,q). The uniqueness of the decomposition then implies that $\Delta\varphi^{(p,q)} = 0$ for all p, q. So $\varphi^{(p,q)} \in \mathscr{H}^{p,q}(X)$.

(2) Since $\Box$ is a real operator on Kähler manifolds, complex conjugation defines an isomorphism from $\mathscr{H}^{p,q}$ to $\mathscr{H}^{q,p}$. ∎

5.20 Decomposition theorem of Hodge–Kodaira. *For every compact Kähler manifold and every r,*

$$H^r(X) \cong \bigoplus_{p+q=r} H^{p,q}(X).$$

In particular,

$$H^1(X, \mathbb{C}) = H^1(X, \mathcal{O}_X) \oplus H^0(X, \Omega^1_X).$$

The theorem of Hodge–Kodaira follows immediately from the decomposition theorem on harmonic forms.

We define $\beta_{p,q}(X) := \dim_{\mathbb{C}} H^{p,q}(X)$. Then we have the equations

1. $\beta_r = \sum_{p+q=r} \beta_{q,r}$,
2. $\beta_{p,q} = \beta_{q,p}$,
3. $\beta_{p,q} = \beta_{n-p,n-q}$,
4. $\beta_r = \beta_{2n-r}$.

5.21 Corollary. *If r is odd, then $\beta_r(X)$ is even.*

PROOF: We have

$$\beta_r = \sum_{p+q=r} \beta_{p,q} = \sum_{\substack{p+q=r \\ p<r/2}} \beta_{p,q} + \sum_{\substack{p+q=r \\ p>r/2}} = 2 \cdot \sum_{\substack{p+q=r \\ p<r/2}} \beta_{p,q}.$$

■

5.22 Proposition. *If $L : \mathscr{A}^r(X) \to \mathscr{A}^{r+2}(X)$ is the map given by $L(\varphi) = \frac{1}{2}\varphi \wedge \omega$, then $L^r(1) \in \mathscr{H}^{r,r}$.*

PROOF: We have $\overline{\partial}(L^r(1)) = L^r\overline{\partial}(1) = 0$, and

$$\overline{\vartheta}(L^r(1)) = (L\overline{\vartheta} + \mathrm{i}\,\partial)L^{r-1}(1) = L\overline{\vartheta}L^{r-1}(1) + \mathrm{i}\,L^{r-1}\partial(1) = L\overline{\vartheta}L^{r-1}(1).$$

Repeating this, we finally have $\overline{\vartheta}(L^r(1)) = L^r\overline{\vartheta}(1) = 0$. Thus $L^r(1)$ is harmonic. ■

5.23 Corollary. *If r is even, then $\beta_r(X) \neq 0$.*

PROOF: We have

$$L^r(1) = \left(\frac{1}{2}\right)^r \omega^r = c\, dz_1 \wedge d\overline{z}_1 \wedge \cdots \wedge dz_r \wedge d\overline{z}_r,$$

with $c \neq 0$ for $0 \leq r \leq n$. Therefore, $\mathscr{H}^{r,r} \neq 0$. Since $\mathscr{H}^{r,r} \subset \mathscr{H}^{2r}$, it follows that $\beta_{2r}(X) \neq 0$. ■

This gives a topological condition for Kähler manifolds.

Examples

1. Every Riemann surface X is a Kähler manifold. If X is compact, then $g := \frac{1}{2}b_1(X)$ is called the *genus* of X. For example, an elliptic curve (a 1-dimensional torus) has genus 1, since its homology is generated by two independent cycles.

   Now we have $\beta_{1,0} + \beta_{0,1} = b_1 = 2g$ and $\beta_{1,0} = \beta_{0,1}$. So $\beta_{1,0} = \beta_{0,1} = g$, i.e.,

   $$g = \dim_{\mathbb{C}} H^{1,0}(X) = \dim_{\mathbb{C}} H^0(X, K_X).$$

   The genus is the number of independent holomorphic 1-forms on X.
2. Let H be an n-dimensional Hopf manifold, $n \geq 2$. Then $b_1 = 1$, as we have shown in Chapter IV. So H cannot be a Kähler manifold!

Exercises

1. Prove the commutativity relations $\vartheta\Lambda = \Lambda\vartheta$ and $\overline{\vartheta}\Lambda = \Lambda\overline{\vartheta}$.
2. Prove the bracket relations $[\vartheta, L] = -\mathrm{i}\,\overline{\partial}$, $[\partial, \Lambda] = -\mathrm{i}\,\overline{\vartheta}$, and $[\overline{\partial}, \Lambda] = \mathrm{i}\,\vartheta$.

3. Let X be an n-dimensional compact complex manifold. Use the norm $\|\varphi\| = (\varphi, \varphi)^{1/2}$ on $\mathscr{A}^r(X)$ and prove for a given cohomology class $u \in H^r(X)$ that among all representatives $\varphi \in \mathscr{A}^r(X)$ of u a representative φ_0 has minimal norm if and only if $\delta\varphi_0 = 0$.
4. Calculate $H^{p,q}(X)$ for $X = \mathbb{P}^n$ and all p, q.

6. Hodge Manifolds

Negative Line Bundles. Assume that X is an n-dimensional compact complex manifold and that F is a holomorphic line bundle on X. We define a new notion of negativity for F. The new definition is formally more restrictive, and if we would generalize it to arbitrary vector bundles, it would, in fact, be different from our old notion. But it turns out that for line bundles the new negativity is equivalent to the old one.

Let a system of trivializations $\varphi_\iota : F|_{U_\iota} \to U_\iota \times \mathbb{C}$ (with transition functions $g_{\iota\kappa}$) be given. A *fiber metric* on F is given by a system h of positive smooth functions h_ι such that

$$h_\kappa(x) = h_\iota(x) \cdot |g_{\iota\kappa}(x)|^2 \text{ on } U_{\iota\kappa}.$$

Then a smooth function $\chi_h : F \to \mathbb{R}$ can be defined by

$$\chi_h \circ \varphi_\iota^{-1}(x, w) = h_\iota(x) \cdot |w|^2.$$

Definition. A line bundle F over a compact manifold X is called *Griffiths negative* if there exists a system of trivializations $\varphi_\iota : F|_{U_\iota} \to U_\iota \times \mathbb{C}$ (with transition functions $g_{\iota\kappa}$), and a fiber metric $h = (h_\iota)$ on F such that χ_h is strictly plurisubharmonic on $F - Z_F$ (where Z_F denotes the zero section in F).

6.1 Proposition. *The line bundle F is Griffiths negative if and only if there exists a system of positive smooth functions ϱ_ι on U_ι such that:*

1. *$-\log \varrho_\iota$ is strictly plurisubharmonic on U_ι.*
2. *$\varrho_\iota = |g_{\iota\kappa}| \cdot \varrho_\kappa$ on $U_{\iota\kappa}$.*

PROOF: Let h be any fiber metric on F and consider a point $x_0 \in X$. There is a trivialization φ at x_0 such that $h_\varphi(x_0) = 1$ and all first derivatives of h_φ vanish at x_0. Then $h_\varphi(x) \cdot |w|^2$ is strictly plurisubharmonic at every (x_0, w) with $w \neq 0$ if and only if h_φ is strictly plurisubharmonic at x_0. Taking such trivializations φ_ι we have only to set $\varrho_\iota := (h_\iota)^{-1/2}$. ∎

The *tube* $T := \{v \in F : \chi_h(v) < 1\}$ is a strongly pseudoconvex neighborhood of the zero section in F. Locally, for $T_\iota := T \cap \pi^{-1}(U_\iota)$, we have

$$T_\iota = \{(x, w) \in U_\iota \times \mathbb{C} : h_\iota(x)|w|^2 < 1\} = \{(x, w) : |w| < \varrho_\iota(x)\}.$$

The boundary of T is given by the equation $w\bar{w}/\varrho_\iota^2(x) = 1$.

6.2 Proposition. *The line bundle F is Griffiths negative if and only if F is negative (in the old sense).*

PROOF: One direction is trivial. To show that every negative line bundle is Griffiths negative one needs rotations in the fibers and a smoothing procedure. For details see [Gr62]. ∎

Definition. The line bundle F on X is called *Griffiths positive* if the complex dual F' is negative.

It is clear that F is Griffiths positive if and only it is positive in the old sense.

Special Holomorphic Cross Sections. We assume that F is a positive line bundle on a compact manifold X. Then F' is negative. We always denote the coordinate on the fibers of F' by w. Of course, the variable w depends on the local trivialization.

The strongly pseudoconvex tube around the zero section of F' is locally given by

$$T = \{(\mathbf{z}, w) \in U \times \mathbb{C} : |w| < \varrho(\mathbf{z})\},$$

where U is a domain in $\mathbb{C}^n$. From the last chapter we know that $\dim_{\mathbb{C}} H^1(T, \mathcal{O}) < \infty$. In this situation we prove the following result.

6.3 Proposition. *If m is sufficiently large, the following two statements are valid:*

1. *If $x_1, x_2 \in X$ are two different points, then there are holomorphic cross sections $s_0, s_1 \in \Gamma(X, F^m)$ with $s_0(x_1) \neq 0$ and $s_0(x_2) \neq 0$ such that the quotient s_1/s_0 has different values at these points.*
2. *If $x_0 \in X$ is a point, then there are holomorphic cross sections*

$$s_0, s_1, \ldots, s_n \in \Gamma(X, F^m)$$

with $s_0(x_0) \neq 0$ and $\det J_{(s_1/s_0, \ldots, s_n/s_0)}(x_0) \neq 0$.

PROOF: If s_0, s_1 are two sections in a line bundle with $s_0(x_0) \neq 0$, then there is a holomorphic function f in a neighborhood of x_0 such that $s_1(x) = f(x) \cdot s_0(x)$ in that neighborhood. The quotient s_1/s_0 is defined to be equal to f near x_0.

We operate in several steps:

(1) Let $x_0 \in X$ be a point. We assume that in local coordinates x_0 corresponds to the origin. We call a trivialization of F' around x_0 *usable* if

$$\log(\varrho(\mathbf{z})) = \sum_{i,j} g_{ij}(\mathbf{z}) z_i \bar{z}_j + \|\mathbf{z}\|^3 \cdot a(\mathbf{z})$$

in a neighborhood of $\mathbf{0}$. Here a is a smooth function and $\sum_{i,j} g_{ij} z_i \bar{z}_j$ is negative definite.

The existence of a usable trivialization follows easily. We take an arbitrary trivialization and write

$$\log(\varrho(\mathbf{z})) = a_0 + \operatorname{Re}(Q(\mathbf{z})) + \sum_{i,j} g_{ij}(\mathbf{z}) z_i \bar{z}_j + \|\mathbf{z}\|^3 \cdot a(\mathbf{z}),$$

with $a_0 \in \mathbb{R}$ and a quadratic polynomial $Q(\mathbf{z})$ with $Q(\mathbf{z}) = 0$. Then we change the trivialization by dividing w by $\exp(a_0 + Q(\mathbf{z}))$.

(2) Now let x_1, x_2 be two different points in X. We choose usable trivializations simultaneously over neighborhoods $U_1(x_1)$ and $U_2(x_2)$, and, moreover, strictly increasing sequences $0 < r_1 < r_2 < \cdots < 1$ and $0 < s_1 < s_2 < \cdots < 1$ such that $\left| s_i/r_i \right| \neq 1$ and $\{w = r_i\} \cap T$, $\{w = s_i\} \cap T$ are analytic sets in T, contained in $F'|U_1$ (respectively in $F'|U_2$, cf. Figure VI.1). We get

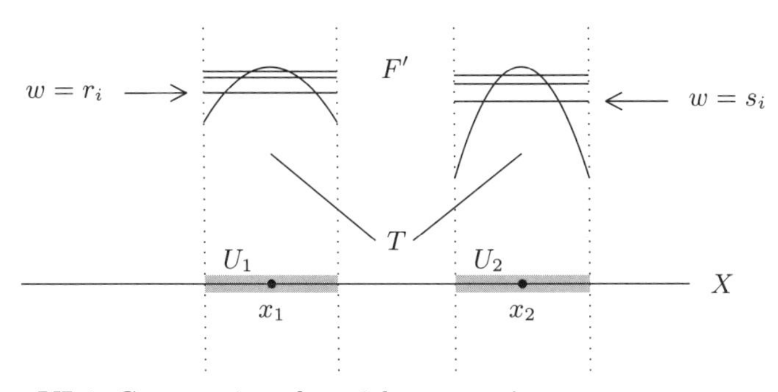

Figure VI.1. Construction of special cross sections

principal parts of meromorphic functions $g_i = 1/(1 - w/r_i)$ in $T \cap (F'|U_1)$ and $g_i^* = 1/(1 - w/r_i)$ in $T \cap (F'|U_2)$, and $h_i = 1/(1 - w/s_i)$ in $T \cap (F'|U_2)$ as well. Because of the finiteness of cohomology there are complex numbers a_i that are not all zero such that there are meromorphic functions g, h in T with principal parts $\sum_i a_i(g_i + g_i^*)$ and $\sum_i a_i(g_i + h_i)$. We may assume that the first nonvanishing coefficient a_i is equal to 1.

The functions g, h are holomorphic in a neighborhood of the zero section, and locally they have power series expansions

$$g(x,w)=\sum_{m=0}^{\infty} g_m(x)w^m \quad \text{and} \quad h(x,w)=\sum_{m=0}^{\infty} h_m(x)w^m.$$

The coefficients g_m (respectively h_m) define global holomorphic sections $s_{0,m}$ (respectively $s_{1,m}$) in the tensor bundle F^m.

If over x_1 the first nonvanishing part of g is $a_i g_i = 1/(1-w/r_i)$, then we have

$$g(x_1,w)=\sum_{m=0}^{\infty}\left(\frac{1}{r_i^m}+c_m\right)w^m, \text{ with } c_m \to 0 \text{ for } m \to \infty.$$

So for large m we have $g_m(x_1) \approx 1/(r_i^m)$, and also $g_m(x_2) \approx 1/(r_i^m)$. Analogously, $h_m(x_1) \approx 1/(r_i^m)$ and $h_m(x_2) \approx 1/(s_i^m)$. We take $s_0 = s_{0,m}$ and $s_1 = s_{1,m}$ for such a large m. Then the quotient s_1/s_0 at x_1 has a value that is nearly 1, and at x_2 nearly $(r_i/s_i)^m$. These values are different for large m.

(3) Now we work at some point x_0, and over a neighborhood U of x_0 we define $g_i = 1/(1-w/r_i)$ and $h_{i,j}(\mathbf{z},w) = 1/[1-(w/r_i)\cdot(1+\varepsilon z_j)]$, for $j=1,\ldots,n$. If U and ε are sufficiently small, the polar sets of g_i and $h_{i,j}$ are always analytic sets in T. As above, we obtain meromorphic functions g and h_j in T, for $j=1,\ldots,n$, which have power series expansions $g(x,w)=\sum_m g_m(x)w^m$ and $h_j(x,w)=\sum_m h_{j,m}(x)w^m$ in a neighborhood of the zero section. Again let g_i be the first principal part with nonvanishing coefficient a_i. For large m we have $g_m(x_0) \approx (1/r_i)^m$ and $h_{j,m}(x_0) \approx (1/r_i)^m(1+m\cdot\varepsilon z_j)$. As above, g_m and $h_{j,m}$ define sections s_0 and s_j in F^m with $(s_j/s_0)(x_0) \approx 1+m\varepsilon z_j$. So the Jacobian of $(s_1/s_0,\ldots,s_n/s_0)$ at x_0 is approximately $m^n\varepsilon^n$ and hence is not zero for m large enough. ∎

Projective Embeddings. Let X be an n-dimensional compact complex manifold and F a positive line bundle on X. Then for large m we have many global holomorphic cross sections in F^m. If there are sections $s_0,\ldots,s_N$ that do not vanish simultaneously, then

$$\Phi(x) := (s_0(x):\ldots:s_N(x))$$

defines a holomorphic map $\Phi: X \to \mathbb{P}^n$.

6.4 Kodaira's embedding theorem. *For sufficiently large m, there are holomorphic cross sections $s_0,\ldots,s_N$ in F^m such that the map*

$$\Phi: x \to (s_0(x):\cdots:s_N(x))$$

is an embedding of X in $\mathbb{P}^N$.

PROOF: In the preceding paragraph it was shown that for any point $x_0 \in X$ there is a holomorphic cross section s_0 in some tensor power of F that does not vanish at x_0.

The set $X_1 = \{x \in X : s_0(x) = 0\}$ is a proper analytic set. We decompose it into (finitely many) irreducible components, and for every irreducible component we choose some point in that component and a holomorphic cross section that does not vanish at this point. In this way we obtain finitely many holomorphic cross sections $s_0, \dots, s_l$ whose joint zero set X_2 has dimension $n-2$ at most. We decompose this again into irreducible components and continue the procedure. After finitely many steps we obtain holomorphic cross sections $s_0, \dots, s_q$ in some tensor power F^{m_0} that do not vanish simultaneously.

For any two points $x_1 \ne x_2 \in X$ there are two holomorphic cross sections t_0, t_1 such that $(t_0(x_i) : t_1(x_i))$ are homogeneous coordinates of two different points. The set $Y_1 = \{(x,y) \in X \times X : (t_0(x) : t_1(x)) = (t_0(y) : t_1(y))\}$ has lower dimension. We decompose it into irreducible components and proceed as above. After finitely many steps we get holomorphic cross sections $t_0, \dots, t_p$ such that $x \mapsto (t_0(x) : \dots : t_p(x))$ is injective.

For any point $x_0 \in X$ there are holomorphic cross sections $u_0, u_1, \dots, u_r$ such that $u_0(x_0) \ne 0$ and the map $x \to (u_0(x) : \dots : u_r(x))$ has a nonvanishing Jacobian at x_0. Altogether we have holomorphic cross sections $f_0, \dots, f_N$ that do not vanish simultaneously such that the map $\Phi : x \to (f_0(x) : \dots : f_N(x))$ is an injective immersion. Since X is compact, Φ is an embedding. ∎

It follows from Chow's theorem that X has a projective algebraic structure and a genuine Zariski topology.

Hodge Metrics. We consider an n-dimensional complex Kähler manifold X. Let H be the Kähler metric on X, and

$$\omega = \omega_H = \mathrm{i} \sum_{i,j} g_{ij}\, dz_i \wedge d\bar{z}_j$$

the associated fundamental form. It is a closed real form of type $(1,1)$.

We choose an open covering $\mathscr{U} = \{U_\nu : \nu \in N\}$ of X such that all intersections $U_{\nu_1,\dots,\nu_k}$ are contractible. If $\mathscr{V}$ is of the same kind and a refinement of $\mathscr{U}$, and G is an abelian group, then the induced homomorphism $H^i(\mathscr{U}, G) \to H^i(\mathscr{V}, G)$ is an isomorhism, for $i = 0, 1, 2$. We choose $\mathscr{U}$ to be fine enough so that H has a potential h_ν in every U_ν:

$$\omega|_{U_\nu} = \mathrm{i}\, \partial\overline{\partial} h_\nu.$$

Here the h_ν are smooth real-valued functions, and the differences $h_{\nu\mu} = h_\mu - h_\nu$ are pluriharmonic in $U_{\nu\mu}$.

If we set $\beta_\nu := -\mathrm{i}\, \partial h_\nu$, then $d\beta_\nu = \mathrm{i}\, \partial\overline{\partial} h_\nu = \omega|_{U_\nu}$. The closed form $dh_{\nu\mu} = \partial h_{\nu\mu} + \overline{\partial} h_{\nu\mu}$ (respectively $\mathrm{i}\, d^c h_{\nu\mu} = \partial h_{\nu\mu} - \overline{\partial} h_{\nu\mu}$) is real (respectively purely imaginary). The sum

$$dh_{\nu\mu} + \mathrm{i}\, d^c h_{\nu\mu} = 2\partial h_{\nu\mu}$$

is a closed holomorphic form of type $(1,0)$. If $x_0 \in U_{\nu\mu}$ is a fixed point, then

$$g_{\nu\mu}(x) := \int_{x_0}^{x} \mathrm{i}\, d^c h_{\nu\mu}$$

defines a purely imaginary function on $U_{\nu\mu}$. It is determined up to a purely imaginary constant. Then $f_{\nu\mu} := h_{\nu\mu} + g_{\nu\mu}$ is a holomorphic function (since $df_{\nu\mu} = 2\partial h_{\nu\mu}$).

We have $d(f_{\mu\lambda} - f_{\nu\lambda} + f_{\nu\mu}) = 0$, and therefore $f_{\mu\lambda} - f_{\nu\lambda} + f_{\nu\mu} \in \mathrm{i}\mathbb{R}$. Then $f = (f_{ij}) \in C^1(\mathscr{U}, \mathcal{O})$ and $\delta f \in Z^2(\mathscr{U}, \mathrm{i}\mathbb{R})$. On the other hand,

$$\beta_\mu - \beta_\nu = \frac{1}{\mathrm{i}} \partial h_{\nu\mu} = \frac{1}{2\mathrm{i}}\, df_{\nu\mu}.$$

This shows that $(1/2\,\mathrm{i}\,)\delta f$ defines a cohomology class $c = c(H, \mathscr{U}) \in H^2(\mathscr{U}, \mathbb{R}) \cong H^2(X, \mathbb{R})$ that corresponds to ω under the de Rham isomorphism (cf. the description at the end of Section 2).

In the following we do not distinguish between $H^2(\mathscr{U}, \mathbb{Z})$ and its image in $H^2(\mathscr{U}, \mathbb{R})$. If a cohomology class of $H^2(\mathscr{U}, \mathbb{R})$ lies in $H^2(\mathscr{U}, \mathbb{Z})$, then we call such a class an *integral class*.

Definition. A Kähler metric H is called a *Hodge metric* if it is integral. A *Hodge manifold* is a compact complex manifold that carries a Hodge metric.

Clearly, every closed complex submanifold of a Hodge manifold is again a Hodge manifold.

6.5 Theorem. *Let X be a compact complex manifold. Every Hodge metric on X defines via the de Rham isomorphism a cohomology class in $H^2(X, \mathbb{Z})$ that is the Chern class of a negative line bundle F on X.*

PROOF: Let ω be the fundamental form of the Hodge metric. We take a finite contractible covering $\mathscr{U} = \{U_k : k = 1, \dots, q\}$ that is so fine that in every U_k we have a potential h_k for ω. As above, we construct the holomorphic functions f_{kl} with $df_{kl} = 2\partial h_{kl}$. Then $\delta\{f_{kl}\} \in Z^2(\mathscr{U}, 2\,\mathrm{i}\,\mathbb{Z})$, and $\gamma_{kl} := \exp(\pi f_{kl})$ defines a cocycle $\gamma = (\gamma_{kl}) \in Z^1(\mathscr{U}, \mathcal{O}^*)$ that determines a holomorphic line bundle F on X. If $\varphi_k : F|_{U_k} \to U_k \times \mathbb{C}$ are the trivializations, then

$$\varphi_k \circ \varphi_l^{-1}(x, w) = (x, \exp(\pi f_{kl}(x)) \cdot w).$$

We define

$$T_k := \{(x, w) \in U_k \times \mathbb{C} : |w| < \exp(-\pi h_k(x))\},$$

and get $\varphi_k \circ \varphi_l^{-1}(T_l|_{U_{kl}}) = T_k|_{U_{kl}}$. So we have a tube T around the zero section in F with $T|_{U_k} = T_k$. Since $-\log(\exp(-\pi h_k)) = \pi h_k$ is strictly plurisubharmonic, it follows that F is negative. ∎

The above procedure can be reversed.

6.6 Theorem. *Assume that X is a compact complex manifold, and F a line bundle on X with a tube T around the zero section such that*

$$T|_{U_k} = \{(x, w) : |w| < \exp(-\pi h_k)\}$$

with strictly plurisubharmonic functions h_k. Then the h_k are local potentials of a Hodge metric on X.

All this finally implies that

Hodge manifolds are just projective algebraic manifolds.

This theorem was first proved by K. KODAIRA. We shall see in the next section that it contains a generalization of the classical period relations for the n-dimensional complex torus. These were originally derived by completely different methods (see, e.g., [Ko54]).

Example

On $\mathbb{P}^n$ we have the Fubini metric H and consider the metric $(1/(2\pi))H$ with

$$h_k = \frac{1}{2\pi} \log\left(\left| \frac{z_0}{z_k} \right|^2 + \cdots + \left| \frac{z_n}{z_k} \right|^2 \right)$$

as local potentials, for $k = 0, 1, \ldots, n$. It follows that

$$h_{kl} = h_l - h_k = \frac{1}{\pi}(\ln(|z_k/z_l|)),$$

and $f_{kl} = \frac{1}{\pi} \log(z_k/z_l)$ is a holomorphic function with $\mathrm{Re}(f_{kl}) = h_{kl}$. Therefore, the transition functions of the associated line bundle F are the functions $\gamma_{kl} = \exp(\pi f_{kl}) = z_k/z_l$, and F is the (negative) tautological bundle.

Exercises

1. Assume that X is a compact Riemann surface and that $x_0 \in X$ is a point. Show that the bundle $[-x_0]$ (associated to the divisor $(-1) \cdot x_0$) is negative.
2. Compute the embedding of $\mathbb{P}^n$ into $\mathbb{P}^N$ defined by a basis of the vector space $\Gamma(\mathbb{P}^n, \mathcal{O}(2))$ and determine the image.
3. Let X be a compact complex manifold and $L \to X$ a negative line bundle. Prove that $\Gamma(X, L) = \{0\}$.
4. Let X be a compact complex manifold and $\pi : E \to X$ a holomorphic vector bundle. Assume that there are trivializations $\varphi_\iota : E|_{U_\iota} \to U_\iota \times \mathbb{C}^q$ with transition functions $g_{\iota\kappa}$.
   (a) A fiber metric on E is given by a system of smooth functions $h_\iota : U_\iota \to M_{n,n}(\mathbb{C})$ such that $h_\iota(x)$ is a positive definite Hermitian matrix

for every $x \in U_\iota$, and $h_\kappa = g^t_{\iota\kappa} \cdot h_\iota \cdot \overline{g}_{\iota\kappa}$ on $U_{\iota\kappa}$. Prove that there is a smooth nonnegative function $\chi_h : E \to \mathbb{R}$ with

$$\chi_h \circ \varphi_\iota^{-1}(x, \mathbf{v}) = \mathbf{v} \cdot h_\iota(x) \cdot \overline{\mathbf{v}}^{\,t}.$$

Prove that if $x_0 \in U_\iota$, then φ_ι can be replaced by a trivialization ψ such that $h_\psi(x_0) = \mathbf{E}_q$ (= identity matrix) and $(\partial h_\psi)_{x_0} = 0$. Such a trivialization is called *normal* at x_0.

(b) The bundle E is called *Griffiths negative* if h can be chosen in such a way that χ_h is strictly plurisubharmonic on $E - Z_E$ (where Z_E is the zero section in E). As in the case of line bundles, E is called *Griffiths positive* if the complex dual E' is Griffiths negative.

If $\varphi : E|_U \to U \times \mathbb{C}^q$ is a trivialization and $\varphi(e) = (x, \mathbf{v})$, then the tangential map $\varphi_* : T_e(E) \to T_x(X) \oplus \mathbb{C}^q$ is an isomorphism. Denote the space $(\varphi_*)^{-1}(T_x(X))$ by $H_e(\varphi)$ (space of "horizontal" tangent vectors at e with respect to φ). Prove that E is Griffiths positive if and only if there is a fiber metric h on E such that for every $e \in E - Z_E$ with $\pi(e) = x$ there is a normal trivialization φ at x such that

$$\operatorname{Lev}(\chi_h)(e, \xi) < 0 \text{ for } \xi \in H_e(\varphi),\ \xi \neq 0\,.$$

(c) Prove that $T(\mathbb{P}^n)$ is Griffiths positive with the Fubini metric.

(d) Let E be a Griffiths positive vector bundle and $F \subset E$ a subbundle. Show that E/F is Griffiths positive. What does this mean for the normal bundle of a submanifold of $\mathbb{P}^n$?

5. Let X be a compact complex manifold and $L \to X$ a positive line bundle. Show that there is an open neighborhood $U = U(Z_E) \subset\subset E$ such that $E - U$ is strongly pseudoconvex.

7. Applications

Period Relations. Recall the definition of the n-dimensional torus. We take $2n$ vectors $\mathbf{w}_1, \dots, \mathbf{w}_{2n} \in \mathbb{C}^n$ that are linearly independent over $\mathbb{R}$. The additive subgroup

$$\Gamma = \{\mathbf{z} = m_1\mathbf{w}_1 + \cdots + m_{2n}\mathbf{w}_{2n} \,:\, m_i \in \mathbb{Z}\} \subset \mathbb{C}^n$$

acts freely on $\mathbb{C}^n$ by translations. The quotient space $T^n = \mathbb{C}^n/\Gamma$ is an n-dimensional compact complex manifold and is called a *torus*. We say that $\mathbf{W} = \left(\mathbf{w}_1^t, \dots, \mathbf{w}_{2n}^t\right) \in M_{n,2n}(\mathbb{C})$ is the *period matrix* of T^n.

For $\mathbf{z} \in \mathbb{C}^n$ we denote its equivalence class in T^n by $[\mathbf{z}]$. By the quotient map $\pi : \mathbb{C}^n \to T^n$, the affine space $\mathbb{C}^n$ becomes the universal covering space of T^n. Since π is locally biholomorphic, we can use the affine coordinates $z_1, \dots, z_n$ of $\mathbb{C}^n$ as local coordinates on T^n.

Every torus has the structure of an abelian complex Lie group:

$$[\mathbf{v}] + [\mathbf{w}] := [\mathbf{v} + \mathbf{w}].$$

It is clear that this definition does not depend on the representatives of the equivalence classes.

If $\mathbf{w} \in \mathbb{C}^n$ is a fixed vector, the *translation* $\tau_{\mathbf{w}} : \mathbb{C}^n \to \mathbb{C}^n$ is defined by $\tau_{\mathbf{w}}(\mathbf{z}) := \mathbf{z} + \mathbf{w}$. It induces a biholomorphic map $T^n \to T^n$ by $[\mathbf{z}] \mapsto [\mathbf{z} + \mathbf{w}]$, which we also denote by $\tau_{\mathbf{w}}$. It depends only on the equivalence class of $\mathbf{w}$.

From topology it is known that $H_2(T^n, \mathbb{Z})$ is a free group spanned by the $\binom{2n}{2}$ cycles

$$c_{ij}(s,t) := s\mathbf{w}_i + t\mathbf{w}_j, \quad 0 \le s \le 1,\ 0 \le t \le 1,$$

with $i < j$. Since T^n is compact, every de Rham cohomology class $[\omega] \in H^2(T^n)$ is uniquely determined by the *periods*

$$v_{ij} := \int_{c_{ij}} \omega, \quad i < j.$$

On $\mathbb{C}^n$ we have the Euclidean metric $ds^2 = dz_1 d\overline{z}_1 + \cdots + dz_n d\overline{z}_n$, which is invariant under translations. It induces a Kähler metric H_0 on T^n with fundamental form

$$\omega_0 = \mathrm{i} \sum_{\nu=1}^{n} dz_\nu \wedge d\overline{z}_\nu.$$

It follows that every torus T^n is a Kähler manifold. Using the Euclidean volume element, for every differentiable function f on T^n we define the *mean value* $M[f]$ by

$$M[f](\mathbf{z}) := \frac{1}{I} \int_{\mathbf{w} \in T^n} f \circ \tau_{\mathbf{w}}(\mathbf{z})\, dV \quad \text{with } I := \int_{T^n} dV.$$

It is clear that $M[f]$ is invariant under arbitrary translations and therefore a constant function.

7.1 Lemma. *If $\widetilde{\omega}$ is the fundamental form of some Kähler metric $\widetilde{H}$ on T^n, then there is another Kähler metric H on T^n such that the associated fundamental form $\omega = \omega_H$ has constant coefficients and the same periods as $\widetilde{\omega}$. If $\widetilde{H}$ is a Hodge metric, then H is also a Hodge metric.*

PROOF: If $\widetilde{\omega} = \mathrm{i} \sum_{\nu,\mu} \widetilde{g}_{\nu\mu}\, dz_\nu \wedge d\overline{z}_\mu$, then we define $g_{\nu\mu} := M[\widetilde{g}_{\nu\mu}]$ and

$$\omega := \mathrm{i} \sum_{\nu,\mu} g_{\nu\mu}\, dz_\nu \wedge d\overline{z}_\mu.$$

The form ω has constant coefficients. Since the $\widetilde{g}_{\nu\mu}$ are periodic (i.e., invariant under translations by elements of Γ), we have

$$\int_{c_{ij}} (\widetilde{g}_{\nu\mu} \circ \tau_{\mathbf{w}})\, dz_\nu \wedge d\overline{z}_\mu = \int_{c_{ij}} \widetilde{g}_{\nu\mu}\, dz_\nu \wedge d\overline{z}_\mu, \text{ for every } \mathbf{w} \in \mathbb{C}^n.$$

Using Fubini we compute

$$\begin{aligned} v_{ij} = \int_{c_{ij}} \omega &= \mathrm{i}\sum_{\nu,\mu} \int_{c_{ij}} \left(\frac{1}{I} \int_{\mathbf{w}\in T^n} \widetilde{g}_{\nu\mu} \circ \tau_{\mathbf{w}}\, dV \right) dz_\nu \wedge d\overline{z}_\mu \\ &= \mathrm{i}\sum_{\nu,\mu} \frac{1}{I} \int_{\mathbf{w}\in T^n} \left(\int_{c_{ij}} \widetilde{g}_{\nu\mu} \circ \tau_{\mathbf{w}}(\mathbf{z})\, dz_\nu \wedge d\overline{z}_\mu \right) dV \\ &= \frac{1}{I} \int_{T^n} \left(\mathrm{i}\sum_{\nu\mu} \int_{c_{ij}} \widetilde{g}_{\nu\mu}\, dz_\nu \wedge d\overline{z}_\mu \right) dV \\ &= \frac{1}{I} \int_{T^n} \widetilde{v}_{ij}\, dV = \widetilde{v}_{ij}. \end{aligned}$$

Now H is a Hodge metric if and only if ω_H defines an integral class. Since a cohomology class in $H^2(T^n)$ is uniquely determined by the periods, the second assertion also follows. ∎

With $\mathbf{w}_i = \big(w_1^{(i)}, \ldots, w_n^{(i)}\big)$ for $i = 1, \ldots, 2n$, we have

$$\begin{aligned} \int_{c_{ij}} dz_\nu \wedge d\overline{z}_\mu &= \int_0^1 \int_0^1 \big(w_\nu^{(i)}\, ds + w_\nu^{(j)}\, dt\big) \wedge \big(\overline{w}_\mu^{(i)}\, ds + \overline{w}_\mu^{(j)}\, dt\big) \\ &= w_\nu^{(i)} \overline{w}_\mu^{(j)} - w_\nu^{(j)} \overline{w}_\mu^{(i)}, \end{aligned}$$

and therefore

$$\begin{aligned} v_{ij} &= \mathrm{i}\sum_{\nu,\mu} g_{\nu\mu} \big(w_\nu^{(i)} \overline{w}_\mu^{(j)} - w_\nu^{(j)} \overline{w}_\mu^{(i)}\big) \\ &= -2\,\mathrm{Im}\Big(\sum_{\nu,\mu} g_{\nu\mu} w_\nu^{(i)} \overline{w}_\mu^{(j)}\Big) = -2\,\mathrm{Im}\big(\mathbf{w}_i^t \cdot \mathbf{G} \cdot \overline{\mathbf{w}}_j\big), \end{aligned}$$

for $\mathbf{G} := (g_{\nu\mu})$.

7.2 Proposition. *The torus T^n is a Hodge manifold if and only if all periods v_{ij} are integral.*

PROOF: Since every element of T^n has a unique representative

$$\mathbf{z} = u_1\mathbf{w}_1 + \cdots + u_{2n}\mathbf{w}_{2n}, \text{ with } u_i \in [0, 1),$$

we can use $\mathbf{u} = (u_1, \ldots, u_{2n})$ as real coordinates on T^n.

We have a special contractible covering $\mathscr{U} = \{U_{\mathbf{k}}\}$ of T^n. For $\mathbf{k} = (k_1, \ldots, k_{2n})$ with $k_i \in \{0,\ 1\}$ we define the open set

$$U_{\mathbf{k}} = \Big\{ \mathbf{u} \in T^n : \frac{1}{2} \cdot k_i \le u_i \le \frac{1}{2} \cdot (k_i + 1) \Big\} \text{ (see Figure VI.2).}$$

The boundaries of the $U_{\mathbf{k}}$ consist of parts of the "walls" S_j and S_j^*, which are defined by

$$S_j^* = \Big\{\mathbf{u} \in T^n \,:\, u_j = \frac{1}{2}\Big\} \quad \text{and} \quad S_j = \big\{\mathbf{u} \in T^n \,:\, u_j = 1\big\}.$$

Since $\mathscr{U}$ is contractible, the singular cohomology group $H^2(T^n)$ is isomorphic

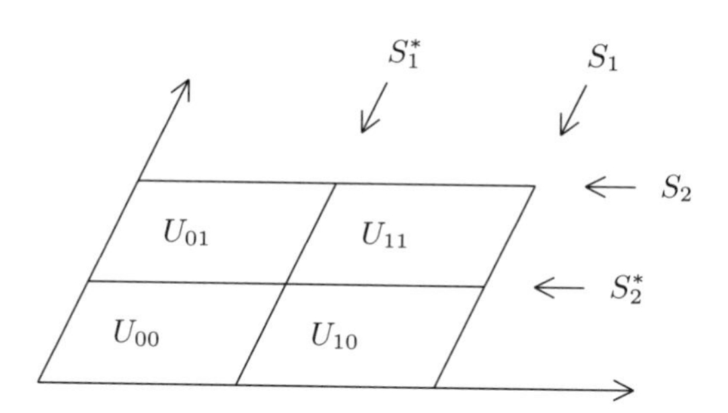

Figure VI.2. A special covering for the torus

to the cohomology group $H^2(\mathscr{U},\mathbb{Z})$. In particular, $H^2(\mathscr{U},\mathbb{Z}) = \mathbb{Z}$ for the torus T^1, which has real dimension 2. The class $[du_1 \wedge du_2]$ is integral, and every integral class is an integral multiple of $[du_1 \wedge du_2]$. If we denote by $\mathscr{U}^{(n)}$ the covering of T^n, then an easy induction argument shows that a cohomology class in $H^2(\mathscr{U}^{(n)},\mathbb{Z})$ is integral (respectively real) if and only if it has the form $\sum_{\nu<\mu} a_{\nu\mu}\, du_\nu \wedge du_\mu$, with $a_{\nu\mu}$ in $\mathbb{Z}$ (respectively $\mathbb{R}$):

For any $\nu < \mu$ the class $[du_\nu \wedge du_\mu]$ is integral. Assume that the case $n-1$ has already been proved. Then $T^n = T^1 \times T^{n-1}$, and after subtracting an integral sum of $[du_i \wedge du_j]$ with $i < j < n$ the cohomology class is given by a covering $\mathscr{U}^{(1)} \times T^{n-1}$ and is therefore a lifting of some integral multiple of $du_{2n-1} \wedge du_{2n}$. So

$$[\omega] = \sum_{i<j} v_{ij}[du_i \wedge du_j] \text{ with } v_{ij} = \int_{c_{ij}} \omega \in \mathbb{R},$$

and $[\omega]$ is integral if and only if all periods v_{ij} are integral. ∎

We call T^n an *abelian variety* if T^n is projective algebraic.

7.3 Proposition. *A torus T^n is an abelian variety if and only if all periods are integral (with respect to some Kähler form ω).*

More precisely, we have the following result on period relations.

7.4 Theorem. *Let T^n be given by the period matrix $\mathbf{W} = \big(\mathbf{w}_1^t, \ldots, \mathbf{w}_{2n}^t\big)$. Then T^n is an abelian variety if and only if there is a positive definite Hermitian matrix $\mathbf{G}$ such that $\operatorname{Im}\big(\mathbf{W}^t \cdot \mathbf{G} \cdot \overline{\mathbf{W}}\big)$ is integral.*

The Siegel Upper Halfplane

Definition. The *(generalized) Siegel upper halfplane* (of degree n) is the $n(n+1)/2$-dimensional domain

$$\mathcal{H}_n := \{\mathbf{Z} = \mathbf{X} + \mathrm{i}\mathbf{Y} \in M_n(\mathbb{C}) \,:\, \mathbf{Z}^t = \mathbf{Z}, \text{ and } \mathbf{Y} \text{ positive definite}\}.$$

7.5 Theorem. *If $\mathbf{Z} \in \mathcal{H}_n$, then $\mathbf{W} = (\mathbf{E}_n, \mathbf{Z}) \in M_{n,2n}(\mathbb{C})$ is the period matrix of an abelian variety T^n.*

PROOF: We take for $\mathbf{G}$ the positive definite symmetric real matrix $\mathbf{Y}^{-1}$. Then

$$\begin{aligned}
\mathbf{W}^t \cdot \mathbf{G} \cdot \overline{\mathbf{W}} &= \begin{pmatrix} \mathbf{E}_n \\ \mathbf{X}^t + \mathrm{i}\mathbf{Y}^t \end{pmatrix} \cdot \mathbf{Y}^{-1} \cdot (\mathbf{E}_n, \mathbf{X} - \mathrm{i}\mathbf{Y}) \\
&= \begin{pmatrix} \mathbf{E}_n \\ \mathbf{X}^t + \mathrm{i}\mathbf{Y}^t \end{pmatrix} \cdot (\mathbf{Y}^{-1}, \mathbf{Y}^{-1} \cdot \mathbf{X} - \mathrm{i}\mathbf{E}_n) \\
&= \begin{pmatrix} \mathbf{Y}^{-1} & \mathbf{Y}^{-1} \cdot \mathbf{X} - \mathrm{i}\mathbf{E}_n \\ \mathbf{X}^t \cdot \mathbf{Y}^{-1} + \mathrm{i}\mathbf{E}_n & \mathbf{X}^t \cdot \mathbf{Y}^{-1} \cdot \mathbf{X} + \mathbf{Y} \end{pmatrix}.
\end{aligned}$$

It follows that

$$\operatorname{Im}(\mathbf{W}^t \cdot \mathbf{G} \cdot \overline{\mathbf{W}}) = \begin{pmatrix} \mathbf{O} & -\mathbf{E}_n \\ \mathbf{E}_n & \mathbf{O} \end{pmatrix}$$

is integral, and the period relations are satisfied. ∎

A converse of the theorem is also true, but we do not give the proof here.

Semipositive Line Bundles. We consider an n-dimensional compact complex manifold X with a holomorphic line bundle F on X.

Definition. The bundle F is called *seminegative* if there is a tube T around the zero cross section such that $T = \{(\mathbf{z}, w) \,:\, |w| < \varrho(\mathbf{z})\}$ in local coordinates and $-\ln(\varrho)$ is everywhere plurisubharmonic and strictly plurisubharmonic at at least one point of X. The bundle F is called *semipositive* if the dual bundle F' is seminegative.

We give an example of a seminegative bundle F with $\dim H^1(T, \mathcal{O}) = \infty$. The boundary function $-\ln(\varrho)$ can be chosen to be strictly plurisubharmonic outside a nowhere dense analytic set in X.

Here is the construction. We take a compact Riemann surface R of genus $g \geq 1$ and a negative line bundle L on R with $H^1(R, L') \neq 0$. The complex dual L' is positive, and the cross sections in L^{-m}, $m > 0$, are holomorphic functions on L that are homogeneous polynomials of order m on the fibers on L.

We add a point at infinity to each fiber of L and obtain a compact algebraic surface X. We denote by D the divisor of all points at infinity, and by $F(m)$ the line bundle over X, associated to the divisor mD. Then all cross sections in L^{-m} are sections in $F(m)$ over X that vanish on the zero section $Z = Z_L \subset L \subset X$ of order m. These sections are holomorphic functions on the dual bundle $F(-m) := F(m)'$ over X (associated to the divisor $-mD$) that are linear on the fibers of $F(-m)$, but vanish over Z. Let $\mathscr{F}$ be the set of all these functions. The function $f(x) \equiv 1$ on X is also a cross section in $F(m)$ that vanishes on D to mth order, but it does not vanish on Z. It gives a holomorphic function on $F(-m)$ that is linear on the fibers, vanishes over D, but does not vanish over Z. We denote it by h and add it to the system $\mathscr{F}$. Now, $\mathscr{F}$ is a set of holomorphic functions on $F(-m)$ that are linear on the fibers and whose joint zero set is the zero cross section of $F(-m)$.

Forming the sum of several products $f\bar{f}$, $f \in \mathscr{F}$, we construct a plurisubharmonic function ψ on $F(-m)$ with the following properties:

1. The zero set of ψ is the zero cross section in $F(-m)$.
2. ψ is quadratic on the fibers.
3. $\psi(z)$ tends to ∞ as z approaches ∂F.
4. ψ is strictly plurisubharmonic outside of $F(-m)|_Z$, but plurisubharmonic everywhere.

So our tube $T = \{(w, x) : \psi(w, x) < 1\}$ is pseudoconvex, and strongly pseudoconvex outside of $F(-m)|_Z$.

The restriction $F(-m)|_Z$ is trivial. Therefore, the holomorphic functions on $(F(-m)|_Z) \cap T$ consist of certain convergent power series $\sum_i a_i(x)w^i$, where w is a coordinate on the fibers. All powers i appear. The first cohomology of Z with coefficients in these power series is infinite-dimensional, since it contains the direct sum of the cohomology with coefficients in those holomorphic functions that are polynomials on the fibers with fixed degree i. We can extend these polynomials to $F(-m)$ by replacing w by h. If we restrict to these polynomials, we get an injection of the cohomology of Z with coefficients in the polynomials on the fibers into $H^1(T, \mathcal{O})$. So the cohomology $H^1(T, \mathcal{O})$ is likewise infinite-dimensional.

Moishezon Manifolds. We previously mentioned Moishezon manifolds (see Section VI.3). An n-dimensional compact complex manifold is called a *Moishezon manifold* if it has n analytically independent meromorphic functions. Without proof we state here a result that was originally called the Grauert–Riemenschneider conjecture (cf. [GrRie70]).

7.6 Theorem. *If a compact complex manifold X has a semipositive line bundle, then it is a Moishezon manifold.*

The conjecture was proved by Y.T. Siu (see [Siu84] and the summary in [GrPeRe94], VII.6).

To obtain a result in the other direction we have to blow up along a d-codimensional submanifold $Y \subset X$ with $d \geq 2$ (cf. Section IV.7, "Monoidal Transformations"). This leads to a proper modification $\widehat{X}$ of X, where Y is replaced by a 1-codimensional submanifold $\widehat{Y}$ and the map $\widehat{Y} \to Y$ is a fiber bundle with typical fiber $\mathbb{P}^{d-1}$.

7.7 Theorem. *If X is a Moishezon manifold, there is a finite sequence of monoidal transformations with d-codimensional centers with various d that lead to a proper modification $\widehat{X}$ of X such that $\widehat{X}$ is a projective algebraic variety.*

This statement is proved in [Moi67]. We know that $\widehat{X}$ carries a negative line bundle. By blowing down to X we get a "meromorphic" negative line bundle on X. Now over certain points of X there is a family of lines instead of one line.

In the case that X is a surface every Moishezon manifold is projective algebraic (see, e.g., [ChoKo52]). But already in dimension three there are Moishezon manifolds that are not projective algebraic. The first example was given in Russia (the "Russian counterexample" or "the complex Sputnik," as it was called at Princeton in 1958), but see also [Nag58] or better just the classical example of Hironaka ([Hi60]), or [Pe93].

For a proof of the following important result see [Moi67] or [Pe86].

7.8 Theorem. *Assume that X is a Moishezon manifold which has a Kähler metric. Then X is a projective algebraic manifold.*

Exercises

1. Prove that the torus T^2 given by the period matrix
$$\mathbf{W} = \begin{pmatrix} 1 & 0 & \sqrt{-2} & \sqrt{-5} \\ 0 & 1 & \sqrt{-3} & \sqrt{-7} \end{pmatrix}$$
has only constant meromorphic functions.
2. Prove that the Siegel upper halfplane is contractible.
3. For $\mathbf{W} \in M_{n,2n}(\mathbb{C})$ let $\widehat{\mathbf{W}} := \binom{\mathbf{W}}{\overline{\mathbf{W}}}$. Recall the results of Exercise 5.3 in chapter IV and prove that for every n-dimensional torus T there is a matrix $\mathbf{Z} \in M_n(\mathbb{C})$ with $\det \operatorname{Im}(\mathbf{Z}) > 0$ and $T \cong T_{(\mathbf{E}_n, \mathbf{Z})}$.
4. Find a compact complex surface X with a seminegative line bundle F such that F is not holomorphically convex.

Chapter VII

Boundary Behavior

1. Strongly Pseudoconvex Manifolds

The Hilbert Space. Assume that X is an n-dimensional complex manifold that carries a (real analytic) Kähler metric $ds^2 = \sum g_{ij} dz_i d\overline{z}_j$, where the g_{ij} are real analytic functions of the local coordinates $z_1, \ldots, z_n$. We consider a domain $\Omega \subset\subset X$ whose boundary $\partial\Omega$ is $\mathscr{C}^\infty$-smooth and strictly Levi convex. In this chapter such a domain is called strongly pseudoconvex.

If $x_0 \in X$ is a point, we can introduce local coordinates in a neighborhood of x_0 such that $\mathbf{G} = (g_{ij}(x_0))$ is the identity matrix. We call them *Euclidean coordinates* in x_0. They are determined up to holomorphic coordinate transformations that at x_0 are given to first order by a unitary matrix. We have a Euclidean volume element dV in x_0, and also the complex star operator $\overline{*}$ given by

$$\varphi \wedge \overline{*}\psi = 2^{p+q} \cdot \left(\sum_{I,J} a_{IJ} \cdot \overline{b}_{IJ} \right) dV,$$

for forms $\varphi = \sum a_{IJ}\, dz_I \wedge d\overline{z}_J$ and $\psi = \sum b_{IJ}\, dz_I \wedge d\overline{z}_J$ of type (p,q). Using the Euclidean coordinates we also have a scalar product at x_0 that is given by

$$\langle \varphi(x_0) \,|\, \psi(x_0) \rangle := 2^{p+q} \sum a_{IJ} \cdot \overline{b}_{IJ}.$$

This is invariant by unitary transformations and therefore independent of the choice of the Euclidean coordinates.

Assume now that φ and ψ are continuous on $\overline{\Omega}$. Then we have the inner product

$$(\varphi, \psi) = \int_\Omega \langle \varphi(x) \,|\, \psi(x) \rangle \, dV = \int_\Omega \varphi \wedge \overline{*}\psi.$$

This is a pre-Hilbert space that completes to the Hilbert space $L^2_{p,q} = L^2_{p,q}(\Omega)$ of (p,q)-forms with square integrable coefficients. It contains the space $\mathscr{A}^{p,q}(\overline{\Omega})$ of $\mathscr{C}^\infty$-(p,q)-forms over $\overline{\Omega}$ as a dense subspace, and also the space $\mathscr{A}_0^{p,q}(\Omega)$ of such forms with compact support in Ω. Every element of $L^2_{p,q}$ is a square integrable current of degree $p+q$ (or *bidegree* (p,q)).

Operators. First, we shall extend the operator $\overline{\partial} : \mathscr{A}_0^{p,q}(\Omega) \to \mathscr{A}_0^{p,q+1}(\Omega)$ to a bigger domain $\mathrm{dom}(\overline{\partial}) \subset L^2_{p,q}$. For this we use the fact that there is a uniquely determined adjoint operator $\overline{\vartheta} : \mathscr{A}_0^{p,q+1}(\Omega) \to \mathscr{A}_0^{p,q}(\Omega)$ such that

$$(\overline{\partial}\varphi, \psi) = (\varphi, \overline{\vartheta}\psi), \text{ for all } \varphi \in \mathscr{A}_0^{p,q}(\Omega),\ \psi \in \mathscr{A}_0^{p,q+1}(\Omega).$$

The proof for this statement is the same as in the case of compact Kähler manifolds.

Now the existence of the extension follows easily. If $\varphi \in L^2_{p,q}$ is given, we define the current $\overline{\partial}\varphi$ by $(\overline{\partial}\varphi, \psi) = (\varphi, \overline{\vartheta}\psi)$ for all $\psi \in \mathscr{A}_0^{p,q+1}(\Omega)$. We say that φ is an element of $\operatorname{dom}(\overline{\partial})$ if the current $\overline{\partial}\varphi$ is contained in $L^2_{p,q+1}$. The (unbounded) operator $\overline{\partial}$ is now a map from $\operatorname{dom}(\overline{\partial})$ to $\operatorname{dom}(\overline{\partial})$, with $\overline{\partial}\circ\overline{\partial} = 0$ (since $\overline{\vartheta}\circ\overline{\vartheta} = 0$).

Next we shall extend $\overline{\vartheta}$ to the conjugate operator $\overline{\partial}^*$ on a dense subset $\operatorname{dom}(\overline{\partial}^*) \subset L^2_{p,q+1}$. Assume that $\psi \in L^2_{p,q+1}$. If $\varphi \in \mathscr{A}_0^{p,q}(\Omega)$, we define $(\varphi, \overline{\partial}^*\psi) = (\overline{\partial}\varphi, \psi)$ and get a current $\overline{\partial}^*\psi$ of bidegree (p,q) over Ω. For ψ an element of $\operatorname{dom}(\overline{\partial}^*)$ we require that the current $\overline{\partial}^*\psi$ be contained in $L^2_{p,q}$. But this is not yet enough. We wish to have $(\varphi, \overline{\partial}^*\psi) = (\overline{\partial}\varphi, \psi)$ for every $\varphi \in \operatorname{dom}(\overline{\partial})$. At the moment we have this only for $\varphi \in \mathscr{A}_0^{p,q}(\Omega)$. But a general $\varphi \in \operatorname{dom}(\overline{\partial})$ can be approximated by a sequence $\varphi_i \in \mathscr{A}_0^{p,q}(\Omega)$. Assume now the validity of

$$(*) \qquad |(\overline{\partial}\varphi, \psi)| \le c \cdot \|\varphi\|, \text{ for all } \varphi \in \operatorname{dom}(\overline{\partial}),$$

where $\|\varphi\| = \sqrt{(\varphi, \varphi)}$ and $c > 0$ is a constant that depends only on ψ. Then we have also that $(\overline{\partial}\varphi_i, \psi)$ converges to $(\overline{\partial}\varphi, \psi)$, and $(\varphi_i, \overline{\partial}^*\psi)$ to $(\varphi, \overline{\partial}^*\psi)$. This gives the required statement.

If, on the other hand, $(\overline{\partial}\varphi, \psi) = (\varphi, \overline{\partial}^*\psi)$ is always valid, then $(*)$ follows. So we define

$$\operatorname{dom}(\overline{\partial}^*) := \{\psi \in L^2_{p,q+1} : \overline{\partial}^*\psi \in L^2_{p,q} \text{ and } (*) \text{ is valid}\}.$$

Next we consider the operator $\square = \overline{\partial}^*\overline{\partial} + \overline{\partial}\,\overline{\partial}^*$ that maps a form in $L^2_{p,q}$ onto a current of bidegree (p,q) over Ω. We define

$$\operatorname{dom}(\square) := \{\varphi \in L^2_{p,q} : \varphi \in \operatorname{dom}(\overline{\partial}) \cap \operatorname{dom}(\overline{\partial}^*),\ \overline{\partial}\varphi \in \operatorname{dom}(\overline{\partial}^*),\ \overline{\partial}^*\varphi \in \operatorname{dom}(\overline{\partial})\}.$$

Then $\square$ maps $\operatorname{dom}(\square)$ to $L^2_{p,q}$. For $\varphi, \psi \in \operatorname{dom}(\square)$ we have the equality

$$(\square\varphi, \psi) = (\overline{\partial}\varphi, \overline{\partial}\psi) + (\overline{\partial}^*\varphi, \overline{\partial}^*\psi) = (\varphi, \square\psi).$$

The *space of harmonic forms* is defined to be the space

$$\mathscr{H}^{p,q} := \{\varphi \in \operatorname{dom}(\square) : \square(\varphi) = 0\}.$$

We get very simply that $\mathscr{H}^{p,q}$ is closed in $\operatorname{dom}(\square)$ and that $\square(\operatorname{dom}(\square))$ is orthogonal to $\mathscr{H}^{p,q}$.

Boundary Conditions. We now consider forms $\varphi \in \mathscr{A}^{p,q}(\overline{\Omega})$. In this case all derivatives $\overline{\partial}\varphi, \overline{\partial}^* \varphi$ and also those of higher order are square integrable. So $\mathrm{dom}(\square) \cap \mathscr{A}^{p,q}(\overline{\Omega})$ consists of those forms φ for which the equation $(*)$ is valid for φ and $\overline{\partial}\varphi$. We assume that in a neigborhood $U(\partial\Omega)$ there is a smooth real function r such that $U \cap \Omega = \{x \in U : r(x) < 0\}$ and that $dr(x) \neq 0$ for $x \in U$. Then by a computation (partial integration) one has the following:

> *A form ψ belongs to $\mathrm{dom}(\overline{\partial}^*)$ if and only if $\partial r \wedge \overline{*}\psi \equiv 0$ on $\partial\Omega$. This holds for arbitrary ψ, not only for elements in the inverse image of the space of smooth differential forms.*

The following is then immediate.

1.1 Proposition. *A form φ belongs to $\mathrm{dom}(\square)$ if and only if the following two boundary conditions are satisfied on $\partial\Omega$:*

1. $\partial r \wedge \overline{*}\varphi \equiv 0$.
2. $\partial r \wedge \overline{*}\,\overline{\partial}\varphi \equiv 0$.

These conditions are called the *$\overline{\partial}$-Neumann boundary conditions*. For literature see [DL81] and [FoKo72]. The main result of the next section is the following:

> *Assume that $q > 0$. Under the hypotheses given above, the harmonic forms of type (p, q) (i.e., elements of $\mathscr{H}^{p,q}$) represent the elements of the cohomology group $H^{p,q}(\Omega)$.*

The first proof of this theorem was given in [Ko63].

Exercises

1. Give an explicit description of the operator $\overline{\partial}^*$.
2. Prove that the boundary conditions for the operator $\overline{\partial}$ and $\mathscr{C}^\infty$ functions are always satisfied.
3. Prove that the boundary conditions are never satisfied if $q = n$ and the form is not identically zero on $\partial\Omega$.
4. Build a theory of harmonic forms for a "cohomology with compact support" $H_0^1(\Omega, \mathcal{O})$. What are the boundary conditions?

2. Subelliptic Estimates

Sobolev Spaces. We look for an operator

$$N : \mathscr{A}^{p,q}(\Omega) \to \mathscr{A}^{p,q}(\Omega) \cap \mathrm{dom}(\square)$$

such that $\square \circ N = \mathrm{id}$. Finding such an operator is the so-called *Neumann problem*. To solve this problem we need Fourier transforms and the so-called *subelliptic estimates*.

We begin by introducing the notion of a Sobolev space. We start with the vector space $\mathscr{S}$ of *rapidly decreasing* $\mathscr{C}^\infty$-functions $f : \mathbb{R}^N \to \mathbb{C}$ with

$$|\mathbf{x}^\alpha D^\beta f(\mathbf{x})| \to 0 \text{ for } \|\mathbf{x}\| \to \infty \text{ and for all } \alpha, \beta \in \mathbb{N}^N.$$

Here

$$\mathbf{x}^\alpha = x_1^{\alpha_1} \cdots x_N^{\alpha_N}, \quad D^\beta f = \frac{\partial^{|\beta|}}{\partial x_1^{\beta_1} \cdots \partial x_N^{\beta_N}}, \text{ and } |\beta| = \beta_1 + \cdots + \beta_N.$$

We have the Fourier transfomation $\mathcal{F} : \mathscr{S} \to \mathscr{S}$ given by

$$\mathcal{F}[f](\mathbf{x}) = \widehat{f}(\mathbf{x}) := \frac{1}{(2\pi)^{N/2}} \int f(\mathbf{y}) e^{-\mathrm{i}\langle \mathbf{x}, \mathbf{y} \rangle} \, d\mathbf{y}.$$

For every $s \in \mathbb{R}$ we define the norm $\|\ldots\|_s$ by

$$\|f\|_s^2 = \frac{1}{(2\pi)^N} \int \left(1 + \|\mathbf{x}\|^2\right)^s \left| \widehat{f}(\mathbf{x}) \right|^2 d\mathbf{x}.$$

The *Sobolev space* of index s is defined as the completion $H_s(\mathbb{R}^N)$ of $\mathscr{S}$ with respect to this norm.

Now we work in the space $\mathbb{R}^{N+1}$ with coordinates $(\mathbf{x}, r) = (x_1, \ldots, x_N, r)$, and consider the half-space $\mathbb{R}_-^{N+1} = \{(\mathbf{x}, r) : r \le 0\}$ and $\mathscr{C}^\infty$ functions $f(\mathbf{x}, r)$ with compact support in $\mathbb{R}_-^{N+1}$. Then for fixed r these functions are in $\mathscr{S}$. We take the Fourier transforms with respect to $\mathbf{x}$, for fixed r, and define

$$|||f|||_s^2 = \int_{-\infty}^0 \|f(\ldots, r)\|_s^2 \, dr.$$

This norm is called the *tangential Sobolev norm.*

Now let us return to our strongly pseudoconvex domain $\Omega \subset\subset X$. We put $N = 2n - 1$, and take any $x_0 \in \partial\Omega$ and a neighborhood $U(x_0) \subset X$ with local coordinates

$$z_1 = x_1 + \mathrm{i}x_2, \ldots, z_{n-1} = x_{2n-3} + \mathrm{i}x_{2n-2}, z_n = x_{2n-1} + \mathrm{i}r.$$

Here r is the boundary function for $\partial\Omega$ as in the preceding section. In the following we will work only with respect to these local coordinates. If $f \in \mathscr{A}_0^{0,0}(U \cap \overline{\Omega})$, then f is $\mathscr{C}^\infty$-smooth and has compact support in $\mathbb{R}_-^{N+1}$. Hence, we can define the norm $|||f|||_s^2$. For $\varphi \in \mathscr{A}_0^{p,q}(U \cap \overline{\Omega})$ we put $|||\varphi|||_s^2 = \sum_f |||f|||_s^2$, where f runs through the coefficients of φ.

Definition. The Neumann problem for (p, q)-forms is called *subelliptic* on Ω if there is an $\varepsilon > 0$ such that for every $x_0 \in \partial\Omega$ there is a coordinate neighborhood $U(x_0)$ and a constant $c > 0$ such that

$$\|\varphi\|_\varepsilon^2 \le c \cdot Q(\varphi, \varphi)$$

for all $\varphi \in \mathscr{A}_0^{p,q}(U \cap \overline{\Omega}) \cap \operatorname{dom}(\overline{\partial}^*)$.

Here $Q(\varphi, \psi) = (\varphi, \psi) + (\overline{\partial}\varphi, \overline{\partial}\psi) + (\overline{\partial}^*\varphi, \overline{\partial}^*\psi)$ is the so-called *energy form* (or the *Dirichlet inner product*).

2.1 Theorem. *If Ω is strongly pseudoconvex, the Neumann problem is subelliptic for all (p, q)-forms with $q > 0$.*

The proof of this theorem is rather difficult, of course. The so-called Kohn–Morrey inequality is involved. See [DL81], for instance.

The Neumann Operator. There is an old theory on elliptic boundary problems, beginning with Dirichlet's problem (see the old papers of L. Bers and L. Nirenberg, for instance). Here the Gårding inequality plays an important role. If it holds for a problem, existence and regularity of solutions of the problem can be proved. In our terminology it can be expressed in the form

$$\|\varphi\|_1^2 \le c \cdot Q(\varphi, \varphi).$$

If $\varepsilon < 1$, then the inequality for subellipticity is much weaker than the Gårding inequality. But nevertheless, in the papers of Kohn and Nirenberg it was proved that there are results on the smoothness of the solution of the Neumann problem that are similar to those in the old theory and that follow from subellipticity alone. We state a first important result (see [DL81]):

2.2 Proposition. *Assume that Ω is strongly pseudoconvex and that the Neumann problem is subelliptic for (p, q)-forms on Ω. Then there is a unique operator $N : L^2_{p,q} \to \operatorname{dom}(\Box)$ with the following properties:*

1. *$N(\mathscr{H}^{p,q}) = 0$.*
2. *$\Box \circ N = \operatorname{id}$.*
3. *$N(L^2_{p,q})$ is orthogonal to $\mathscr{H}^{p,q}$.*

This operator N is called the *Neumann operator*. Under the above conditions the following theorem is true:

2.3 Theorem.

1. *The harmonic space $\mathscr{H}^{p,q}$ is finite-dimensional. It is contained in $\mathscr{A}^{p,q}(\overline{\Omega})$, and for $\varphi \in \mathscr{H}^{p,q}$ we have $\partial r \wedge \overline{*}\varphi \equiv 0$ on $\partial\Omega$.*
2. *The Neumann operator N is compact with respect to the $L_2^{p,q}$-norm.*
3. *We have $\mathscr{A}^{p,q}(\overline{\Omega}) = \mathscr{H}^{p,q} \oplus \Box N(\mathscr{A}^{p,q}(\overline{\Omega}))$.*
4. *If $\varphi \in \mathscr{A}^{p,q}(\overline{\Omega})$, $\overline{\partial}\varphi = 0$, then φ is orthogonal to $\overline{\partial}^*\overline{\partial}N(\mathscr{A}^{p,q}(\overline{\Omega}))$.*

For a proof see [KoNi65] and [FoKo72]. Both papers are difficult to read.

In the case $X = \mathbb{C}^n$ there is a simpler proof. A domain $\Omega \subset\subset \mathbb{C}^n$ with smooth boundary is called *regular* if there exists a finite number of subsets $S_i \subset \partial\Omega$, $i = 0, \ldots, N$, such that

1. $\varnothing = S_N \subset S_{N-1} \subset \cdots \subset S_1 \subset S_0 = \partial\Omega$.
2. If $\mathbf{z} \in S_i$, but $\mathbf{z} \notin S_{i+1}$, then there are a neighborhood $U(\mathbf{z})$ and a differentiable submanifold $M \subset U \cap \partial\Omega$ with holomorphic dimension equal to zero such that $S_i \cap U \subset M$.

If Ω is pseudoconvex and regular, then a certain *compactness estimate* holds for the Neumann problem for (p, q)-forms with $q \geq 1$ (cf. [Ca84]). Still using a result of Kohn/Nirenberg (see [KoNi65]), from the compactness estimate the existence and the compactness of the Neumann operator N follow. See also [BoStr99] for further discussion. If Ω is an arbitrary pseudoconvex domain, the Neumann operator exists as a continuous linear map $L_2^{p,q} \to \operatorname{dom}(\Box)$. This is proved in [Ca83].

A strongly pseudoconvex domain $\Omega \subset \mathbb{C}^n$ is always regular, and therefore the above theorem on the Neumann operator holds in this case. But in $\mathbb{C}^n$ the harmonic space $\mathscr{H}^{p,q}$ is always equal to zero.

Real-Analytic Boundaries. We assume again that X is a complex manifold, but we consider only a domain Ω with real-analytically smooth boundary. Here we only assume that Ω is pseudoconvex. Then the following holds:

2.4 Proposition. *The Neumann problem is subelliptic on Ω for (p, q)-forms with $q \geq 1$ if and only if all germs of local complex-analytic sets $A \subset \partial\Omega$ have dimension less than q.*

Moreover, we have the following result:

2.5 Proposition. *Assume that $\Omega \subset\subset X = \mathbb{C}^n$. Then there is no germ of a local complex-analytic set $A \subset \partial G$ of positive dimension.*

For a proof see [DiFo78] and [Ko79].

Examples. We will construct two 2-dimensional examples $\Omega \subset\subset X$, where Ω contains just one compact irreducible curve Y. From the existence of Y it will follow that the cohomology group $H^1(\Omega, \mathcal{O})$ is different from zero. The cohomology classes are given by unique harmonic forms that satisfy the boundary conditions. We can study these as if they were functions, which may be useful for our investigations. Unfortunately, the forms in $\mathscr{H}^{0,1}$ are not in general d-closed. So there is no connection between $\mathscr{H}^{0,1}$ and the cohomology with coefficients in $\mathbb{C}$ as in the case of compact Kähler manifolds. We will see this more precisely in the examples.

Example 1

We take a compact Riemann surface Y of genus $g > 1$ and a line bundle X on Y with Chern class -1. This line bundle is negative. There is a real-analytic metric on the fibers of X such that $\Omega = \{x \in X : \|x\| < 1\}$ is a strongly pseudoconvex relatively compact subdomain of X with real-analytically smooth boundary. We get a uniquely determined Kähler metric on X by using the metric of constant curvature on Y and the metric on the fibers and requiring that the surfaces $\{x \in X : \|x\| < r\}$ be orthogonal to the fibers.

Over Y we have the vector bundle $\mathcal{O}^i$ of homogeneous polynomials of degree i on the fibers of X. If i is sufficiently large, then $H^1(Y, \mathcal{O}^i)$ vanishes. So the finite-dimensional cohomology group $H^1(\Omega, \mathcal{O})$ is generated by finitely many of these groups $H^1(Y, \mathcal{O}^i)$. The cohomology classes are given by differential forms $\varphi \in \mathscr{A}^{0,1}(Y, \mathcal{O}^i) = \Gamma(Y, \bigwedge^{0,1}(Y) \otimes \mathcal{O}^i)$. They all may be viewed as differential forms on Ω whose coefficients are holomorphic polynomials of degree i on the fibers. They vanish on Y to order i. We denote the set of these forms by $\mathfrak{A}_i^{0,1}$. Examples for forms of this type are given by the derivatives $\overline{\partial} f$, where f is a differentiable function in Ω whose restriction to the fibers is a holomorphic polynomial of degree i. We take the orthogonal complement of these forms and get $\mathscr{H}^{0,1} = \bigoplus_i \mathscr{H}_i^{0,1}$, where $\mathscr{H}_i^{0,1} \subset \mathfrak{A}_i^{0,1}$. This follows by a direct computation. Thus we obtain a basis of finite length, and the restriction of these forms to $\partial\Omega$ satisfies the boundary conditions of harmonic forms.

Since $g \geq 2$, the space of harmonic forms $\mathscr{H}_1^{0,1}$ has positive dimension. We can obtain a basis by explicit computation. The elements are harmonic forms in Ω vanishing on $Y \subset \Omega$. If there were an isomorphism onto the cohomology with coefficients in $\mathbb{C}$ (as in the compact case), they would have to be zero.

Example 2

We will construct another 2-dimensional example of a completely different type. For this we use compact complex-analytic curves that are no longer regular. They are irreducible reduced 1-dimensional complex spaces (i.e., topological Hausdorff spaces that locally look like analytic sets in some complex n-space). Here we go a bit beyond the scope of this book, but we hope that the reader can nevertheless get an idea of the method.

We take the projective line $\mathbb{P}^1$ with the points 0 and ∞. We identify these two points, obtaining a compact irreducible 1-dimensional complex curve that has a "normal crossing" of 1-dimensional domains in a neighborhood of the gluing point $\widehat{0}$. This point $\widehat{0}$ is the only singular point.

The projective line $\mathbb{P}^1$ is the so-called *normalization* of Y: There is a surjective holomorphic mapping $\pi : \mathbb{P}^1 \to Y$ that maps 0 and ∞ to $\widehat{0}$ and is biholomorphic elsewhere. The first cohomology group $H^1(Y, \mathcal{O})$ of Y has dimension 1 (as can be computed).

We build a 2-dimensional complex manifold X around Y. There is a neighborhood $U = U(\widehat{0})$ in Y and a holomorphic embedding

$$j : U \hookrightarrow X_1 = \{(w, z) \in \mathbb{C}^2 \,:\, |w| < 1\}$$

with $j(U) = Y_1 := \{(w, z) \in X_1 \,:\, w^2 - z^2 = 0\}$. We choose an ε with $0 < \varepsilon < 1$ and put $P := \{(w, z) \in X_1 \,:\, |w| \le \varepsilon\}$ and $Y_1' := Y_1 - P$. We define

$$Y_2 := Y - j^{-1}(Y_1 \cap P) \quad \text{and} \quad Y_2' := j^{-1}(Y_1) \cap Y_2 = j^{-1}(Y_1 - P).$$

The manifold Y_2' consists of two disk shells

$$\begin{aligned} Y_0 &= j^{-1}(\{(w, z) \in \mathbb{C}^2 \,:\, \varepsilon < |w| < 1 \text{ and } z = w\}), \\ Y_\infty &= j^{-1}(\{(w, z) \in \mathbb{C}^2 \,:\, \varepsilon < |w| < 1 \text{ and } z = -w\}). \end{aligned}$$

On Y_1 we have the holomorphic function $w_1 := \mathrm{pr}_1|_{Y_1}$, which can serve as local (and even global) coordinate. The set Y_2 is the complement of the union of two neighborhoods of 0 and of ∞ in $\mathbb{P}^1$ and therefore isomorphic to a disk shell in $\mathbb{C}^* = \mathbb{P}^1 - \{0, \infty\}$. We have a global coordinate w_2 on Y_2, and a holomorphic function $w_1 = f(w_2)$ on Y_2' with

$$j(w_2) = \begin{cases} (f(w_2), f(w_2)) & \text{on } Y_0, \\ (f(w_2), -f(w_2)) & \text{on } Y_\infty. \end{cases}$$

Now we put $X_2 = Y_2 \times \mathbb{C}$ and use in X_2 the product coordinates (w_2, z_2). We choose suitable small open neighborhoods $W_1 = W_1(Y_1) \subset X_1$ and $W_2 = W_2(Y_2) \subset X_2$, and define the gluing map $F : (X_1 - P) \cap W_1 \to (X_2|_{Y_2'}) \cap W_2$ by

$$(w_2, z_2) = F(w, z) = \begin{cases} (f^{-1}(w), z - w) & \text{near } z = w, \\ (f^{-1}(w), z + w) & \text{near } z = -w. \end{cases}$$

So we obtain by this gluing procedure a complex manifold X around Y. The singular curve Y is a hypersurface in X.

The covariant normal bundle $N^* = N_X^*(Y)$ is given by the adjunction formula

$$N_X^*(Y) = [-Y]|_Y,$$

where $[-Y]$ is the line bundle associated to the divisor $-Y$ in X. The normal bundle $N = N_X(Y)$ is trivial over Y_2. Therefore, the holomorphic function z_2 can be regarded as a holomorphic function on N that is linear on the fibers. This defines a holomorphic cross section in N^* over Y_2 that is given by the holomorphic function $s_2(w_2) \equiv 1$ with respect to the local trivialization over Y_2. It extends to a global meromorphic section s over Y. In the local trivialization over Y_1, s is given by the meromorphic function

$$s_1(w,z) = \frac{1}{2z} = \begin{cases} 1/(z+w) & \text{on } z = w, \\ 1/(z-w) & \text{on } z = -w, \end{cases}$$

since the transition function g_{12}^* for N^* is given by

$$g_{12}^* = \frac{z_2}{z^2 - w^2} = \frac{z \mp w}{z^2 - w^2}.$$

Now we lift everything to $\mathbb{P}^1$. The lifted section $\widehat{s}$ of $\pi^*(N^*)$ has poles of first order in 0 and ∞, and no other poles and no zeros. So $\pi^*(N^*)$ has Chern class -2.

To construct a bundle with positive Chern class, we have to alter X_2. For this we take a point $p \in Y - \overline{Y}_1$ and two small disks $D' \subset\subset D$ around p such that D is still in $Y - \overline{Y}_1 \subset Y_2$. Let S be the disk shell $D - \overline{D}'$. We glue $(Y - \overline{D}') \times \mathbb{C}$ with $D \times \mathbb{C}$ over S by means of the gluing map $H : S \times \mathbb{C} \to S \times \mathbb{C}$ with

$$(w_3, z_3) = H(w_2, z_2) = (w_2, (w_2 - w_2(p)) \cdot z_2).$$

We replace our old X_2 by a new $\widetilde{X}_2$ and get a new manifold $\widetilde{X}$. Now the lifting of the covariant normal bundle $N^*_{\widetilde{X}}(Y)$ to $\mathbb{P}^1$ has Chern class 3. It follows by Riemann–Roch that $\dim_{\mathbb{C}} H^0(\mathbb{P}^1, \pi^*(N^*_{\widetilde{X}}(Y))) = 3 + 1 = 4$. By the projection $\mathbb{P}^1 \to Y$ the fibers over 0 and ∞ are identified. This gives an additional condition for the existence of global holomorphic cross sections in the covariant normal bundle of $Y \subset \widetilde{X}$. So we have 3 linearly independent holomorphic cross sections s_1, s_2, s_3 in N^*.

One can see by computation that $s_1|_{Y_1}$ may be assumed to be given by $(z-w)(z+w)$. The function $|z^2 - w^2|^2$ is an extension of $s_1\bar{s}_1$ to X_1. If we add the function $|(z-w)^2(z+w)^3|^2$, we obtain an extension that is strictly plurisubharmonic outside of Y_1. We can also extend the other $s_i|_{Y_1}$ to X_1 and the $s_i|_{Y_2}$ to X_2, form the $s_i\bar{s}_i$, sum up, and glue together after a small change in a neighborhood $U(Y) \subset\subset X$ to obtain an extension h of $s_1\bar{s}_1 + s_2\bar{s}_2 + s_3\bar{s}_3$ in U. The extension vanishes on Y and is positive outside. It is strictly plurisubharmonic in a neighborhood U' of Y. Then for very small $\varepsilon > 0$, the tube $\Omega = \{h < \varepsilon\} \subset\subset U'$ will have a smooth boundary. It is strongly pseudoconvex.

We have $\dim H^1(Y, \mathcal{O}) = 1$. The cohomology class can be given by a cocycle with coefficients in $\mathbb{C}$. It therefore extends to Ω. So there is a nonzero harmonic form $\varphi \in \mathscr{H}^{0,1}(\Omega)$. It satisfies the boundary condition. So in this example the cohomology of the curve Y, which has singularities, is given by a smooth harmonic form in a tube around the curve. This shows that the theory of harmonic forms on manifolds with boundary may be very useful.

Exercises

1. Show that the Sobolev space H_{-s} may be identified with the dual of H_s. Show that the Sobolev norm $\|\ldots\|_s$ comes from a scalar product $(\ldots,\ldots)_s$, and prove the following generalized Schwarz inequality: If $f, g \in H_0$ and $f \in H_s$ for some $s > 0$, then $(f,g)_s \le \|f\|_s \cdot \|g\|_s$.
2. It is a consequence of the famous "Rellich lemma" that the following holds: If $s > t$, then for any $\varepsilon > 0$ there is a neighborhood $V(\mathbf{0})$ such that $\|f\|_t \le \varepsilon \|f\|_s$ for all f with compact support in V. Use this to prove the same estimate for the tangential Sobolev norm.
3. Assume that the Neumann problem is solvable for (p,q)-forms. Let a form $\varphi \in L^2_{p,q}$ with $\overline{\partial}\varphi = 0$ be given that is orthogonal to $\mathscr{H}^{p,q}$. Prove that $\psi := \overline{\partial}^* N\varphi$ is the only solution of the equation $\overline{\partial}\psi = \varphi$ that is orthogonal to every $\overline{\partial}$-closed form.
4. Let $\Omega \subset\subset \mathbb{C}^n$ be a strongly pseudoconvex domain. Prove that for every positive number C there is a smooth plurisubharmonic function ϱ on $\overline{\Omega}$ with $0 \le \varrho \le 1$ such that $\mathrm{Lev}(\varrho)(\mathbf{z},\mathbf{t}) \ge C\|\mathbf{t}\|^2$ for every $\mathbf{z} \in \partial\Omega$.
5. Take Example 2 and prove that that the derivative $d(\varphi|_Y)$ is different from 0 and that this form is not harmonic.
6. Take Example 1 and prove that $\mathscr{H}^{1,0}$ has infinite dimension.
7. The following problem requires some knowledge about sheaves and complex spaces. Take integers $r < s$ such that r, s are free of common divisors. Consider at $0 \in \mathbb{C}$ the holomorphic functions that are given by convergent power series in w^r and w^s. Then replace the ring of holomorphic functions in $0 \in \mathbb{P}_1$ by these holomorphic functions and repeat the procedure of Example 2. Since now the "structure sheaf" $\mathcal{O}$ of $\mathbb{P}^1$ is smaller, a new curve Y is obtained. Construct an X around Y such that there is a 2-dimensional strongly pseudoconvex tube Ω around Y. Compute the first cohomology of Y.

3. Nebenhüllen

General Domains. Assume that $\Omega \subset \mathbb{C}^n$ is a general domain. We take the open kernel of the intersection of all domains of holomorphy $\widehat{\Omega} \supset \overline{\Omega}$ and its connected component $N(\Omega)$ that contains Ω. (More precisely, we work in the category of unbranched domains over $\mathbb{C}^n$ as in the case of the construction of the hull of holomorphy).

Definition. The domain $N(\Omega)$ is called the *Nebenhülle* of Ω.

We can see very easily:

The Nebenhülle is a domain of holomorphy.

The Nebenhülle contains the envelope of holomorphy of Ω.

If $\Omega \subset\subset \mathbb{C}^n$ is a domain of holomorphy with (smooth) real-analytic boundary, then $\partial\Omega$ does not contain any germ of an analytic set of positive dimension. Moreover, one can show that $N(\Omega) = \Omega$. That implies that one can approximate $\overline{\Omega}$ by Stein neighborhoods. See [DiFo77] for details.

The *Hartogs triangle* $\Omega := \{(w,z) \in \mathbb{C}^2 \,:\, |w| < |z| < 1\}$ is a domain of holomorphy, but $N(\Omega)$ is the unit polydisk in $\mathbb{C}^2$. In this case the envelope of holomorphy is Ω and the Nebenhülle is much bigger. We will see that this can happen even when Ω is a bounded domain with smooth boundary. The following example was constructed in [DiFo77].

A Domain with Nontrivial Nebenhülle. We define a real-analytic function $\varrho : \mathbb{C}^* \times \mathbb{C} \to \mathbb{R}$ by

$$\varrho(w,z) = |z + \exp(\mathsf{i}\ \ln(w\bar{w}))|^2 - 1$$

and put

$$\Omega := \{(w,z) \in \mathbb{C}^* \times \mathbb{C} \,:\, \varrho(w,z) < 0\}.$$

A computation of the derivatives of ϱ gives

$$\begin{aligned}
\frac{\partial\varrho}{\partial w} &= \frac{\mathsf{i}\,\bar{z}}{w}\exp(\mathsf{i}\ \ln(|w|^2)) - \frac{\mathsf{i}\,z}{w}\exp(-\mathsf{i}\ \ln(|w|^2)),\\
\frac{\partial\varrho}{\partial z} &= \bar{z} + \exp(-\mathsf{i}\ \ln(|w|^2)),\\
\frac{\partial^2\varrho}{\partial w\partial\bar{w}} &= -\frac{\bar{z}}{|w|^2}\exp(i\ \ln(|w|^2)) - \frac{z}{|w|^2}\exp(-\mathsf{i}\ \ln(|w|^2)),\\
\frac{\partial^2\varrho}{\partial w\partial\bar{z}} &= \frac{\mathsf{i}}{w}\exp(\mathsf{i}\ \ln(|w|^2)), \quad \text{and } \frac{\partial^2\varrho}{\partial z\partial\bar{z}} = 1.
\end{aligned}$$

3.1 Proposition. *$\partial\Omega$ is smooth.*

PROOF: If there were a point $(w_0, z_0) \in \partial\Omega$ where $d\varrho = 0$, then

$$\frac{\partial\varrho}{\partial z}(w_0, z_0) = 0.$$

This gives $z_0 = -\exp(\mathsf{i}\ \ln(|w_0|^2))$ and hence $\varrho(w_0, z_0) = -1$. But such a point is not on the boundary. ∎

3.2 Proposition. *The boundary of Ω is pseudoconvex. It is not strongly pseudoconvex exactly along $M = \{(w,z) : z = 0\}$.*

PROOF: The complex dimension of the complex tangent space at every point of $\partial\Omega$ is 1. A vector $\xi = a\,\partial/\partial w + b\,\partial/\partial z + \bar{a}\,\partial/\partial\bar{w} + \bar{b}\,\partial/\partial\bar{z}$ at a point of $\partial\Omega$ lies in the complex tangent space if and only if $(a\,\partial/\partial w + b\,\partial/\partial z)[\varrho] = 0$. So the vector space of complex tangents is spanned by the vector

$$\xi = -\frac{\partial\varrho}{\partial z}\frac{\partial}{\partial w} + \frac{\partial\varrho}{\partial w}\frac{\partial}{\partial z}.$$

The function ϱ is the potential of a Hermitian form H. The boundary of Ω is pseudoconvex (strongly pseudoconvex) if and only if $H \geq 0$ (respectively $H > 0$) on this vector. By a calculation we get

$$H(\xi,\xi) = \frac{|z|^2}{|w|^2} \cdot |z + \exp(\mathrm{i}\,\ln(|w|^2))|^2 \geq 0.$$

It vanishes only for $z = 0$.

We take any neighborhood U of $\overline{\Omega}$ that is a domain of holomorphy. Then U contains $\mathbb{C}^* \times \{0\}$. If $w_0 \in \mathbb{C}^*$, $z_0 \in \mathbb{C}$ and $0 < |z_0 + \exp(\mathrm{i}\,\ln(|w_0|^2))| < 1$, then $\Omega \cap (\mathbb{C}^* \times \{z_0\})$ is the disjoint union of infinitely many disk shells in $\mathbb{C}^*$. When we move z_0 on a line to 0 the shells may become larger. So by the theorem of continuity it follows that every holomorphic function in U can be extended holomorphically to the point (w_0, z_0). So U and the Nebenhülle $N(\Omega)$ contain the domain $\mathbb{C}^* \times \{z : |z| < 2\}$, and consequently the Nebenhülle is much bigger than Ω. ■

Bounded Domains. The domain Ω of the preceding paragraph lies over $\mathbb{C}^*$ and is not bounded. But one can obtain a pseudoconvex bounded domain by making Ω smaller. One takes an $r > 3\pi$, leaves Ω unchanged over the line segment $[1, r]$, and deforms the circles $\Omega \cap (\mathbb{C} \times \{w\})$ slowly to $\varnothing$ as $|w|$ goes to 0 or to ∞. The new domain is bounded and has a much bigger Nebenhülle; the boundary is $\mathscr{C}^\infty$-smooth, but not real-analytic. This is the "worm domain" constructed explicitly in [DiFo77].

Domains in $\mathbb{C}^2$. The domain Ω from above (with real-analytic boundary) lies in $\mathbb{C}^* \times \mathbb{C}$. But we also can find such a domain in $\mathbb{C}^2$. We have the universal covering $\mathbb{C}^2 \to \mathbb{C}^* \times \mathbb{C}$ by the holomorphic mapping $(w, z) = \mathbf{F}(\widetilde{w}, \widetilde{z}) := (\exp(\widetilde{w}), \widetilde{z})$. Then we lift Ω to $\mathbb{C}^2$ and obtain a domain $\widetilde{\Omega}$ that has a smooth real-analytic boundary and is not bounded.

The new domain $\widetilde{\Omega}$ is a "semitube": The translations in the imaginary $\widetilde{w}$-direction act on it. Hence, the Nebenhülle $N(\widetilde{\Omega})$ is also a semitube. So it can be written as the set $S(\widetilde{M}) = \{(\widetilde{w}, \widetilde{z}) : (\mathrm{Re}(\widetilde{w}), \widetilde{z}) \in \widetilde{M}\}$, where $\widetilde{M}$ is the smallest open set in $\{(\widetilde{u}, \widetilde{z}) : \widetilde{u} \in \mathbb{R},\ \widetilde{z} \in \mathbb{C}\}$ with the following properties:

1. $S(\widetilde{M}) \supset \widetilde{\Omega}$.
2. $S(\widetilde{M})$ is pseudoconvex.
3. $\mathbb{R} \times \{0\} \subset \widetilde{M}$.

We have $\mathbf{F}(S(\widetilde{M})) = \{(w,z) : (|w|,z) \in M\}$, where M is the image of $\widetilde{M}$ under the mapping $(u,z) \mapsto (\exp(u), z)$. The domain $S(\widetilde{M})$ is pseudoconvex if and only if $\mathbf{F}(S(\widetilde{M}))$ is pseudoconvex, since $\mathbf{F}$ is locally biholomorphic with respect to $\widetilde{w}$. So the Nebenhülle $N(\widetilde{\Omega})$ is the set $\{\widetilde{w} \in \mathbb{C} : |\widetilde{z}| < 2\}$, just as $N(\Omega)$ is $\{w \in \mathbb{C}^* : |z| < 2\}$.

Exercises

1. Let $\Omega \subset\subset \mathbb{C}^2$ be a domain of holomorphy. If there is an analytic plane E such that $\Omega \cap E$ has isolated boundary points, then Ω has a nontrivial Nebenhülle.
2. Consider domains $\Omega \subset \mathbb{C}$. When is $N(\Omega) = H(\Omega)$ (i.e., the envelope of holomorphy)?
3. A family $\Omega(t)$ of domains in $\mathbb{C}^n$ is called continuous at t_0 if for every compact set $K \subset \Omega(t_0)$ and every domain $\widehat{\Omega} \supset \overline{\Omega}$ there is an $\varepsilon > 0$ such that $K \subset \Omega(t) \subset \widehat{\Omega}$ for $|t - t_0| < \varepsilon$.
   (a) Construct a continuous family of domains $\Omega(t)$ such that the family of the envelopes of holomorphy $H(\Omega(t))$ is not continuous.
   (b) Define "upper semicontinuity" for families of domains. Find conditions such that the upper semicontinuity of $H(\Omega(t))$ follows from the continuity of $\Omega(t)$.

4. Boundary Behavior of Biholomorphic Maps

The One-Dimensional Case. Assume that $\Omega, \Omega' \subset \mathbb{C}$ are bounded domains with $\mathscr{C}^\infty$-smooth boundary and that $f : \Omega \to \Omega'$ is a biholomorphic mapping. The following theorem is well known:

4.1 Theorem. *The map f can be uniquely extended to an invertible $\mathscr{C}^\infty$ map $\widehat{f} : \overline{\Omega} \to \overline{\Omega'}$.*

In particular, if $\partial\Omega$, $\partial\Omega'$ are real-analytic, then the extension is likewise real-analytic.

The Theory of Henkin and Vormoor. Assume now that Ω is a domain in $\mathbb{C}^n$. A holomorhic map $\mathbf{F} : \Omega \to \mathbb{C}^N$ is called *Hölder continuous* of index $\varepsilon > 0$ if for all points $\mathbf{w}$, $\mathbf{z} \in \Omega$ the estimate

$$\|\mathbf{F}(\mathbf{w}) - \mathbf{F}(\mathbf{z})\| \le M \cdot \|\mathbf{w} - \mathbf{z}\|^\varepsilon$$

holds with a constant $M > 0$.

Here we state a special case of a theorem that was independently proved by Henkin and by Vormoor (see [He73] and [Vo73]).

4.2 Theorem. *Assume that $\Omega_1, \Omega_2 \in \mathbb{C}^n$ are strongly pseudoconvex domains with $\mathscr{C}^\infty$-smooth boundary and that $\mathbf{F} : \Omega_1 \to \Omega_2$ is a biholomorphic map. Then $\mathbf{F}$ has an extension to a Hölder continuous map $\widehat{\mathbf{F}} : \overline{\Omega}_1 \to \overline{\Omega}_2$ with index $\frac{1}{2}$.*

The PROOF of this theorem is very technical. The main ideas can be found in the book of Diederich/Lieb ([DL81]). An important tool is the notion of a pseudodifferential metric. Since these metrics are interesting themselves, we give the definition for general complex manifolds:

> **Definition.** Assume that X is an n-dimensional complex manifold. A *pseudodifferential metric* on X is an upper semicontinuous map $F_X : T(X) \to \mathbb{R}_0^+$ with the property
>
> $$F_X(a \cdot \xi) = |a| \cdot F_X(\xi)$$
>
> for all vectors $\xi \in T(X)$ and all $a \in \mathbb{C}$.

We do not require that $F_X(\xi)$ be positive for $\xi \neq 0$, and we do not require the triangle inequality.

As an example we consider the Kobayashi metric. It is of great importance, not only within the scope of this section. A complex manifold is called *hyperbolic* if it has a complete Kobayashi metric. The word "complete" here means that all metric balls are contained relatively compactly in X. It is known that the set of holomorphic mappings into a hyperbolic manifold is a normal family in the sense of Montel. A Riemann surface is hyperbolic in this sense if and only if it is of hyperbolic type (in the classical sense). On any other Riemann surface the Kobayashi metric degenerates to 0.

But first let us give the definition of the Kobayashi metric. As usual, we denote the unit disk by $\mathsf{D} = \{z : |z| < 1\} \subset \mathbb{C}$. If $f : \mathsf{D} \to X$ is a holomorphic map and $f(0) = x_0$, then we denote by $f'(0)$ the tangent vector $f'(0) := f_*(e)$, where $f_* : T_0(\mathsf{D}) \to T_{x_0}(X)$ is the induced mapping and e the canonical basis of $T_0(\mathsf{D}) \cong \mathbb{C}$.

> **Definition.** The *Kobayashi metric* is the pseudodifferential metric $d_X : T(X) \to \mathbb{R}_0^+$ defined by

$$d_X(\xi) := \inf\{|a|^{-1} : \exists\, f : \mathsf{D} \to X \text{ holomorphic with } f(0) = x_0,\ f'(0) = a\xi\}.$$

The following statement is a very simple consequence of the definition.

4.3 Proposition. *Assume that $F : X \to Y$ is a holomorphic map of complex manifolds. Then $d_Y(F_*(\xi)) \leq d_X(\xi)$.*

So the Kobayashi metric is invariant under biholomorphic mappings. In certain cases it is completely degenerate. For example,

> *the Kobayashi metric on $\mathbb{C}^n$ is identically* 0.

For the Kobayashi metric in the unit disk we obtain

$$d_{\mathsf{D}}(\xi) = \frac{|\xi|}{1-|z|^2}, \text{ for } z \in \mathsf{D} \text{ and } \xi \in T_z(\mathsf{D}) \cong \mathbb{C}.$$

For a bounded domain $\Omega \subset\subset \mathbb{C}^n$ one shows that

$$d_\Omega(\xi) \le a \cdot \frac{\|\xi\|}{\operatorname{dist}(\mathbf{z}, \partial\Omega)},$$

where $\xi \in T_{\mathbf{z}}(\Omega) \cong \mathbb{C}^n$, $\mathbf{z} \in \Omega$, and $a \in \mathbb{C}$ is a constant. Moreover, if Ω has a $\mathscr{C}^\infty$-smooth strongly pseudoconvex boundary, we even get

$$d_\Omega(\xi) \ge a \cdot \frac{\|\xi\|}{\operatorname{dist}(\mathbf{z}, \partial\Omega)^{1/2}}.$$

Real-Analytic Boundaries. We consider again the continuation of biholomorphic maps $\mathbf{F} : \Omega_1 \to \Omega_2$ to the boundary. But we assume now that the boundaries are real-analytically smooth. In this case we can drop the assumption of strong pseudoconvexity.

4.4 Theorem. *If $\mathbf{F} : \Omega_1 \to \Omega_2$ is a biholomorphic map of pseudoconvex domains with smooth real-analytic boundaries, then $\mathbf{F}$ extends to a (uniformly) Hölder continuous map $\widehat{\mathbf{F}} : \overline{\Omega}_1 \to \overline{\Omega}_2$.*

This was proved by Diederich and Fornæss (see [DiFo78]).

Fefferman's Result. In 1974 Charles Fefferman proved the following theorem (see [Fe74]).

4.5 Theorem. *Assume that $\Omega_1, \Omega_2 \subset\subset \mathbb{C}^n$ are strongly pseudoconvex domains with $\mathscr{C}^\infty$-smooth boundaries and that $\mathbf{F} : \Omega_1 \to \Omega_2$ is a biholomorphic mapping. Then $\mathbf{F}$ extends to a $\mathscr{C}^\infty$ diffeomorphism $\widehat{\mathbf{F}} : \overline{\Omega}_1 \cong \overline{\Omega}_2$.*

For his proof Fefferman used the Bergman metric, which is invariant under biholomorphic mappings. His proof was very difficult. There were many simplifying methods written by authors like S.M. Webster, N. Kerzman, S.R. Bell, E. Ligoçka, L. Nirenberg, P. Yang. Later, an interesting proof that uses the Bergman projection was discovered by Catlin (see [Ca84]).

We wish to indicate some ideas concerning the Bergman kernel function. We will define it in the general case of a strongly pseudoconvex domain $\Omega \subset\subset X$ with smooth boundary, where X is an n-dimensional complex manifold. If F is a complex analytic line bundle on X, one can prove as in Chapter V that the cohomology $H^1(\Omega, F)$ is finite. Then, following the ideas of Chapter V, we can construct many holomorphic cross sections in F over Ω. We take for F the line bundle of holomorphic n-forms $\chi = h(z_1, \ldots, z_n)\, dz_1 \wedge \cdots \wedge dz_n$. The norm $\|\chi\|$ is defined by

$$\|\chi\|^2 := \int_\Omega \chi \wedge \overline{*}\chi.$$

Let $\mathscr{H}$ be the space of *holomorphic* n-forms χ on Ω that are square integrable (i.e., with $\|\chi\| < \infty$). The forms $\chi \in \mathscr{H}$ with $\|\chi\| \leq 1$ are uniformly bounded (with respect to their coefficients in local coordinate systems) on every compact set $M \subset \Omega$. If A is a compact analytic subset of Ω of positive dimension, all forms χ may vanish there. This will lead to difficulties in defining the Bergman metric. The space $\mathscr{H}$ is a Hilbert space and a subspace of the Hilbert space $L^2(\Omega)$ of arbitrary square integrable complex n-forms (with the scalar product $(\varphi, \psi) = \int_\Omega \varphi \wedge \overline{*}\psi$).

Next we define the kernel form. We take an orthonormal basis $\{\chi_i : i \in \mathbb{N}\}$ of $\mathscr{H}$. Then

$$\sum_i \chi_i(\mathbf{z})\overline{\chi}_i(\mathbf{w}) = K(\mathbf{z}, \mathbf{w})\, dz_1 \wedge \cdots \wedge dz_n \wedge d\overline{w}_1 \wedge \cdots \wedge d\overline{w}_n$$

is called the *kernel form* on Ω, and the coefficient K is called the *Bergman kernel*. The theory of K is well known for domains in $\mathbb{C}^n$. We extend it here to complex manifolds, but we give only an overview and avoid going into detail. It is possible to prove that the kernel form is independent of the choice of the orthonormal basis. The convergence of the defining series is locally uniform, also for all derivatives. The kernel K is real-analytic on Ω, holomorphic in $\mathbf{z}$, and antiholomorphic in $\mathbf{w}$. As $(\mathbf{z}, \mathbf{w})$ approaches $\partial\Omega$, $K(\mathbf{z}, \mathbf{w})$ tends to ∞ to order $n+1$ (i.e., its coefficients with respect to local coordinates). When we pass over to the conjugate complex form, interchange $\mathbf{w}$ with $\mathbf{z}$, and multiply by $(-1)^n$, we get the kernel form back. Moreover, the kernel is invariant under biholomorphic mappings of Ω onto itself. In local coordinates, but independent of these coordinates, we have

$$K(\mathbf{z}, \mathbf{z}) = \sup_{\chi = f\, dz_1 \wedge \cdots \wedge dz_n,\ \|\chi\| = 1} |f(\mathbf{z})|^2.$$

We have the orthogonal projection $P : L^2(\Omega) \to \mathscr{H}$ with

$$\begin{aligned} P(a) &= \sum_i (a, \chi_i) \cdot \chi_i \\ &= \mathrm{i}^n (-1)^{n(n-1)/2} dz_1 \wedge \cdots \wedge dz_n \cdot \int_G a(\mathbf{w}) \wedge K(\mathbf{z}, \mathbf{w})\, d\overline{w}_1 \wedge \cdots \wedge d\overline{w}_n. \end{aligned}$$

It is called the *Bergman projection.*

We denote by $\mathscr{A}_0^{n,0}(\overline{\Omega})$ the set of $\mathscr{C}^\infty$ forms χ of type $(n, 0)$ in $\overline{\Omega}$ for which all derivatives of any order of the coefficient of χ vanish on $\partial\Omega$, and by $A^\infty(\Omega)$ the set of $\mathscr{C}^\infty$ forms in $\overline{\Omega}$ that are holomorphic in Ω.

4.6 Theorem. *In the case $X = \mathbb{C}^n$, for every regular pseudoconvex (and therefore for every strongly pseudoconvex) domain $\Omega \subset \mathbb{C}^n$ the Bergman projection P maps $\mathscr{A}_0^{n,0}(\overline{\Omega})$ into $A^\infty(\Omega)$.*

This was proved by Catlin in [Ca84]. The result is known as "Ω satisfies condition (R)." The *worm domain* is an example where condition (R) does not hold.

Mappings. Now we consider another strongly pseudoconvex domain $\Omega' \subset\subset \mathbb{C}^n$ with $\mathscr{C}^\infty$ smooth boundary, and a biholomorphic mapping $\mathbf{F} : \Omega \to \Omega'$. We can transform any form a of type $(n, 0)$ on Ω' to a form $a \circ \mathbf{F}$ of the same type on Ω. We state without complete proof a lemma that is trivial in the case of forms with compact support:

4.7 Lemma. *For $a \in \mathscr{A}_0^{n,0}(\overline{\Omega}')$ the transform $a \circ \mathbf{F}$ is an element of $\mathscr{A}_0^{n,0}(\overline{\Omega})$.*

We give a brief idea of the proof. We represent $a \wedge \overline{a}$ as a multiple of the kernel form with a factor $c(x) \geq 0$. The function c vanishes on $\partial\Omega'$ to arbitrarily high order. We just have to prove that $c \circ \mathbf{F}$ also has this property in Ω. To do this we take in Ω and in Ω' the metric $\delta(x, y) = \sup_f |f(x) - f(y)|$. Here f runs through the holomorphic functions in Ω (respectively Ω') with $|f(x)| \leq 1$ (this is the so-called *Carathéodory distance*). If x_0 is a fixed point of Ω (respectively Ω'), then the balls around x_0 approach $\partial\Omega$ (respectively $\partial\Omega'$) to first order for $r \to \infty$; i.e., the Euclidean distance of the ball to the boundary of the domain behaves like $\exp(-r)$. If $x_0 \in \Omega$ and $y_0 = \mathbf{F}(x_0) \in \Omega'$, then any ball of radius r around x_0 is mapped onto a ball of the same radius around y_0. This implies the vanishing of $c \circ \mathbf{F}$ with all derivatives on $\partial\Omega$.

We need the following $\mathscr{C}^\infty$ version of the Cauchy–Kovalevski theorem:

4.8 Lemma. *Assume that $h \in A^\infty(\Omega')$. Then there is a form $g \in \mathscr{A}^{n,0}(\overline{\Omega}')$ such that $\overline{\partial} g = 0$ on $\partial\Omega'$ and $h - \Delta g \in \mathscr{A}_0^{n,0}(\overline{\Omega}')$.*

We use the lemma to complete the proof. It follows by partial integration that Δg is orthogonal to all elements from $A^\infty(\Omega')$. We can define an orthonormal basis of $\mathscr{H}(\Omega')$ consisting of such functions. So Δg is orthogonal to $\mathscr{H}(\Omega')$. It follows that

$$h \circ \mathbf{F} = (P_B(h - \Delta g)) \circ \mathbf{F} = P_G((h - \Delta g) \circ \mathbf{F}) \in A^\infty(\Omega).$$

When we take for h the functions $z_1|_{\overline{\Omega}'}, \ldots, z_n|_{\overline{\Omega}'}$, we get the differentiability of $\mathbf{F}$ on $\overline{\Omega}$. The proof of Fefferman's theorem is completed.

The Bergman Metric. The Bergman metric is interesting in its own right. It is a Kähler metric and depends only on the complex structure of the manifold under consideration. Using its Riemannian curvature one can prove interesting theorems that give a deeper insight into the complex structure. The notion is well known for domains in $\mathbb{C}^n$. Here we generalize it to complex manifolds.

We will consider the situation where X is an n-dimensional complex manifold and $\Omega \subset\subset X$ a subdomain with strongly pseudoconvex boundary. Then Ω is holomorphically convex. We can construct forms χ belonging to $\mathscr{H}$. For that we use the boundary of Ω. We consider the analytic subset $E \subset \Omega$ given by

$$E \;:=\; \{x \in \Omega : \text{ all } \chi \text{ vanish at } x \textbf{ or } \text{there is a tangent vector } \xi \text{ at } x \text{ such that } \xi(b) = 0 \text{ for the coefficient } b \text{ of every } \chi\}.$$

It can be proved that E is compact and has positive dimension everywhere. So in the classical case of a domain in $\mathbb{C}^n$ it is empty. This is no longer true in complex manifolds. Therefore, we make the following additional assumption:

$$(*) \quad \text{The set } E \text{ is empty.}$$

Under this assumption we have the Bergman metric in Ω. In every local coordinate system we just define the coefficients

$$g_{ij}(\mathbf{z}) := \frac{\partial^2 \log(K(\mathbf{z}, \overline{\mathbf{z}}))}{\partial z_i \partial \overline{z}_j}$$

and the *Bergman metric*

$$ds^2 := \sum_{i,j} g_{ij}(\mathbf{z})\, dz_i\, d\overline{z}_j.$$

The definition is independent of local coordinates, and we obtain a Kähler metric in Ω that is invariant by biholomorphic transformations of Ω.

If there are no compact analytic subsets of positive dimension in Ω, then Ω is a Stein manifold. In general, Ω may contain compact analytic subsets of positive dimension. For example, we may take for X a complex analytic line bundle on $\mathbb{P}^1$ with Chern class $c \le -3$. Then a strongly pseudoconvex tube Ω around the zero section exists, and there are enough forms χ on Ω. So in this case as well we have the Bergman metric.

Near to the boundary the following estimates hold.

4.9 Proposition. *If ξ is a tangent vector in $T_{\mathbf{z}}(\Omega)$, $\mathbf{z} \in \Omega$, $\|\xi\|$ is the Euclidean norm, and* $\operatorname{dist}(\mathbf{z}, \partial\Omega)$ *is the Euclidean distance (everything with respect to local coordinates), then there are positive constants a, b such that*

$$a \cdot \frac{\|\xi\|}{\operatorname{dist}(\mathbf{z}, \partial\Omega)^{1/2}} \le \|\xi\| \le b \cdot \frac{\|\xi\|}{\operatorname{dist}(\mathbf{z}, \partial\Omega)}.$$

The distance (with respect to the Bergman metric) between a fixed point $\mathbf{z}_0 \in \Omega$ and a point $\mathbf{z}$ (near to the boundary) is approximately

$$\sqrt{n+1} \cdot (-\ln(\operatorname{dist}(\mathbf{z}, \partial\Omega))).$$

Exercises

1. Prove the properties of the Bergman kernel. Compute the Bergman kernel for the unit ball.
2. Assume that $\Omega \subset \mathbb{C}^n$ is a bounded domain. Then Ω carries the Carathéodory metric. This is a positive pseudodifferential metric on Ω and is defined in the following way: If ξ is a vector at a point $\mathbf{z}_0 \in \Omega$, we take all holomorphic functions f on Ω with $|f(\mathbf{z})| < 1$ in Ω and measure the Euclidean length of the image $f_*(\xi)$. Then we take the supremum over all these f. Prove that the Kobayashi length is at least as big as the Carathéodory length.
3. Consider the bicylinder $\mathsf{P}^n(\mathbf{0}, 2) = \{(w, z) : |w| < 1, |z| < 1\} \subset \mathbb{C}^2$ and a sequence of complex numbers w_i with $|w_i| < 1$ that converges to 1 and take two different complex numbers z_1, z_2 in the open unit disk. Does the Kobayashi distance between the points (w_i, z_1), (w_i, z_2) tend to ∞?
4. Compute the Bergman metric and the Kobayashi metric for

$$\Omega = \{z \in \mathbb{C} : 0 < |z| < 1\}.$$

Are these metrics complete?

References

[BPV84] W. Barth, C. Peters, A. Van de Ven: *Compact Complex Surfaces.* Erg. Math. 3.4, Springer, Berlin–Heidelberg (1984).

[BeSt48] H. Behnke, K. Stein: *Entwicklungen analytischer Funktionen auf Riemannschen Flächen.* Math. Ann. **120**, 430–461 (1948).

[BeTh70] H. Behnke, P. Thullen: *Theorie der Funktionen mehrerer komplexer Veränderlichen*, 2nd ed., Erg. Math. 51, Springer, Berlin–Heidelberg–New York (1970).

[Bi61] E. Bishop: *Mappings of partially analytic spaces.* Am. J. Math. **83**, 209–242 (1961).

[Bie33] L. Bieberbach: *Beispiel zweier ganzer Funktionen zweier komplexer Variablen, welche eine schlichte volumentreue Abbildung des R_4 auf einen Teil seiner selbst vermitteln.* Sitzungsb. der preuß. Akad. d. Wissensch. (1933).

[BoStr99] H.P. Boas, E.J. Straube: *Global Regularity of the $\overline{\partial}$-Neumann Problem: A Survey of the L^2-Sobolev Theory.* Several Complex Variables, MSRI Publications **37**, 79–111 (1999).

[Bou66] N. Bourbaki: *Elements of Mathematics, General Topology, part 1.* Hermann, Paris (1966).

[Br54] H. Bremermann: *Über die Äquivalenz der pseudokonvexen Gebiete und der Holomorphiegebiete im Raume von n komplexen Veränderlichen.* Math. Ann. **128**, 63–91 (1954).

[BrMo78] L. Brenton, J. Morrow: *Compactifications of* $\mathbf{C}^n$. Trans. Am. Math. Soc. **246**, 139–153 (1978).

[CS53] H. Cartan: Séminaire E.N.S. 1953/54, exposés XVI and XVII.

[Ca83] D. Catlin: *Necessary conditions for subellipticity of the $\overline{\partial}$-Neumann problem.* Ann. Math. **117**, 147–171 (1983).

[Ca84] D. Catlin: *Global regularity of the $\overline{\partial}$-Neumann problem.* Symp. Pure Appl. Math. **41**, 46–49 (1984).

[ChoKo52] W.L. Chow, K. Kodaira: *On analytic surfaces with two independent meromorphic functions.* Proc. Nat. Acad. Sci. USA **38**, 319–325 (1952).

[DL81] K. Diederich, I. Lieb: *Konvexität in der komplexen Analysis.* Birkhäuser Basel 1981, 1–150 (DMV Seminar **2**).

[DiFo77] K. Diederich, J.E. Fornæss: *Pseudoconvex Domains: An Example with Nontrivial Nebenhülle.* Math. Ann. **225**, 275–292 (1977).

[DiFo77b] K. Diederich, J.E. Fornæss: *Pseudoconvex domains. Existence of Stein neighborhoods.* Duke Math. J. **44**, 641–662 (1977).

[DiFo77c] K. Diederich, J.E. Fornæss: *Pseudoconvex domains. Bounded strictly plurisubharmonic exhaustion functions.* Invent. Math. **39**, 129–141 (1977).

[DiFo78] K. Diederich, J.E. Fornæss: *Pseudoconvex domains with real-analytic boundary.* Ann. Math. **107**, 371–384 (1978).

[DiFo79] K. Diederich, J.E. Fornæss: *Proper holomorphic maps onto pseudoconvex domains with real-analytic boundary.* Ann. Math. **110**, 575–592 (1979).

[DoGr60] F. Docquier, H. Grauert: *Levisches Problem und Rungescher Satz für Teilgebiete Steinscher Mannigfaltigkeiten.* Math. Ann. **140**, 94–123 (1960).

[Ew96] G. Ewald: *Combinatorial Convexity and Algebraic Geometry.* Springer, New York (1996).

[Fe74] C. Fefferman: *The Bergman kernel and biholomorphic mappings of pseudoconvex domains.* Invent. Math. **26**, 1–65 (1974).

[FoKo72] G.B. Folland, J.J. Kohn: *The Neumann Problem for the Cauchy–Riemann complex.* Annals of Mathematics Studies **75**, Princeton, New Jersey (1972).

[Fo70] O. Forster: *Plongements des variétés de Stein.* Comm. Math. Helv. **45**, 170–184 (1970).

[Fo81] O. Forster: *Lectures on Riemann surfaces.* Graduate texts in Math., Springer, New York (1981).

[FoSte87] J.E. Fornæss, B. Stensønes: *Lectures on counterexamples in several complex variables.* Math. Notes **33**, Princeton university press (1987).

[Fu93] W. Fulton: *Introduction to Toric Varieties.* Annals of Math. Studies **131**, Princeton (1993).

[GlSte95] J. Globevnik, B. Stensønes: *Holomorphic embeddings of planar domains into $\mathbb{C}^2$.* Math. Ann. **303**, 579–597 (1995).

[Gr55] H. Grauert: *Charakterisierung der holomorph vollständigen komplexen Räume.* Math. Ann. **129**, 233–259 (1955).

[Gr62] H. Grauert: *Über Modifikationen und exzeptionelle analytische Mengen.* Math. Ann. **146**, 331–368 (1962).

[GrLi70] H. Grauert, I. Lieb: *Das Ramirezsche Integral und die Lösung der Gleichung $\overline{\partial} f = \alpha$ im Bereich der beschränkten Formen.* Complex Analysis, Rice university studies **56**, 29–50 (1970).

[GrPeRe94] H. Grauert, Th. Peternell, R. Remmert: Encyclopaedia of Mathematical Sciences **74**, Several Complex Variables VII. Springer, Heidelberg (1994).

[GrRe55] H. Grauert, R. Remmert: *Zur Theorie der Modifikationen I. Stetige und eigentliche Modifikationen komplexer Räume.* Math. Ann. **129**, 274–296 (1955).

[GrRe56] H. Grauert, R. Remmert: *Konvexität in der komplexen Analysis (Nicht-holomorph-konvexe Holomorphiegebiete und Anwendungen auf die Abbildungstheorie).* Comm. Math. Helv. **31**, 152–183 (1956).

[GrRe57] H. Grauert, R. Remmert: *Singularitäten komplexer Mannigfaltigkeiten und Riemannsche Gebiete.* Math. Zeitschr. **67**, 103–128 (1957).

[GrRe58] H. Grauert, R. Remmert: *Komplexe Räume.* Math. Ann. **136**, 245–318 (1958).

[GrRe79] H. Grauert, R. Remmert: *Theory of Stein Spaces.* Springer, Heidelberg (1979).

[GrRe84] H. Grauert, R. Remmert: *Coherent Analytic Sheaves.* Springer, Heidelberg (1984).

[GrRie70] H. Grauert, O. Riemenschneider: *Verschwindungssätze für analytische Kohomologiegruppen auf komplexen Räumen.* Invent. Math. **11**, 263–292 (1970).

[Gre67] M.J. Greenberg: *Lectures on Algebraic Topology.* Benjamin, New York (1967).

[Gri69] Ph.A. Griffiths: *Hermitian differential geometry, Chern classes, and positive vector bundles.* Global Analysis, papers in honor of K. Kodaira, ed. by D.C. Spencer and S. Iyanaga, Princeton university press (1969).

[GriHa78] Ph.A. Griffiths, J. Harris: *Principles of Algebraic Geometry.* John Wiley & Sons, New York (1978).

[Gu90] R.C. Gunning: *Introduction to Holomorphic Functions of Several Variables I–III.* Wadsworth & Brooks/Cole, Belmont (Cal) (1990).

[GuRo65] R.C. Gunning and H. Rossi: *Analytic Functions of Several Complex Variables.* Prentice-Hall (1965).

[Hei56] E. Heinz: *Ein elementarer Beweis des Satzes von Radó–Behnke–Stein–Cartan über analytische Funktionen.* Math. Ann. **131**, 258–259 (1956).

[He73] G.M. Henkin: *An analytic polyhedron is not holomorphically equivalent to a strictly pseudoconvex domain.* Dokl. Akad. Nauk SSSR **210**, 1026–1029 (1973), and Soviet. Math. Dokl. **14**, 858–862 (1973).

[He69] G.M. Henkin: *Integral representations of functions holomorphic in strictly pseudoconvex domains and some applications.* Math. USSR Sb. **7**, 597–616 (1969).

[Hi60] H. Hironaka: *On the theory of birational blowing-up.* Thesis, Harvard (1960), unpublished.

[Hi62] H. Hironaka: *An example of a non-Kählerian complex-analytic deformation of Kählerian complex structures.* Ann. Math. **75**, 190–208 (1962).

[Ho55] H. Hopf: *Schlichte Abbildungen und lokale Modifikationen 4-dimensionaler komplexer Mannigfaltigkeiten.* Comm. Math. Helv. **29**, 132–155 (1955).

[Hoe66] L. Hörmander: *An introduction to complex analysis in several variables.* van Nostrand, New York (1966).

[KaKa83] L. Kaup, B. Kaup: *Holomorphic Functions of Several Variables.* Walter de Gruyter, Berlin (1983).

[Ko54] K. Kodaira: *On Kähler varieties of restricted type.* Ann. Math. **60**, No. 1, 28–48 (1954).

[Ko63] J.J. Kohn: *Harmonic integrals on strongly pseudoconvex manifolds.* I. Ann. Math. 78, 112–148 (1963) and II. Ann. Math. 79, 450–472 (1964).

[Ko79] J.J. Kohn: *Subellipticity of the* $\overline{\partial}$*-Neumann problem on pseudoconvex domains: sufficient conditions.* Acta math. **142**, 79–122 (1979).

[KoNi65] J.J. Kohn, L. Nirenberg: *Non-coercive boundary value problems.* Comm. Pure Appl. Math. **18**, 443–492 (1965).

[Ku60] N. Kuhlmann: *Projektive Modifikationen komplexer Räume.* Math. Ann. **139**, 217–238 (1960).

[La60] H. Langmaack: *Zur Konstruktion von Holomorphiehüllen unverzweigter Gebiete über dem* $\mathbb{C}^n$. Schriftenreihe d. Math. Inst. Münster **17**, 1–88 (1960).

[Lau71] H.B. Laufer: *Normal two-dimensional singularities.* Annals of Math. Studies **71**, Princeton (1971).

[Li70] I. Lieb: *Die Cauchy–Riemannschen DGLn auf streng pseudokonvexen Gebieten (beschränkte Lösungen).* Math. Ann. **190**, 6–44 (1970).

[Moi67] B.G. Moishezon: *On* n*-dimensional compact varieties with* n *algebraically independent meromorphic functions.* AMS Transl., II. Ser., **63**, 51–177 (1967).

[MoKo71] J. Morrow, K. Kodaira: *Complex Manifolds.* Holt, Rinehart and Winston, New York (1971).

[Nag58] M. Nagata: *Existence theorems for non-projective complete algebraic varieties.* Illinois Journal of Mathematics **2**, 490–498 (1958).

[Nar60] R. Narasimhan: *Imbedding of holomorphically complete complex spaces.* Am. J. Math. **82**, 917–934 (1960).

[Nar92] R. Narasimhan: *Compact Riemann surfaces.* Birkhäuser, Boston (1992).

[No54] F. Norguet: *Sur les domaines d'holomorphie des fonctions uniformes de plusieurs variables complexes.* Bull. Soc. France **82**, 137–159 (1954).

[Oda88] T. Oda: *Convex Bodies and Algebraic Geometry – An Introduction to the Theory of Toric Varieties.* Springer Heidelberg 1988.

[Oka53] K. Oka: *Sur les fonctions analytiques de plusieurs variables, IX Domaines finis sans point critique interieur.* Jap. J. Math. **23**, 97–155 (1953).

[Pe86] Th. Peternell: *Algebraicity criteria for compact complex manifolds.* Math. Annalen **275**, 653–672 (1986).

[Pe93] Th. Peternell: *Moishezon manifolds and rigidity theorems.* Bayreuther Mathematische Schriften **54**, 1–108 (1993).

[PS88] Th. Peternell, M. Schneider: *Compactifications of* $\mathbb{C}^3$ *– I.* Math. Ann. **280**, 129–146 (1988).

[Ra86] R.M. Range: *Holomorphic Functions and Integral Representations in Several Complex Variables.* Springer, Heidelberg (1986).

[RaSi] R.M. Range, Y.T. Siu: *Uniform estimates for the* $\overline{\partial}$*-equation on domains with piecewise smooth strictly pseudoconvex boundaries.* Math. Ann. **206**, 325–354 (1973).

[Ri68] R. Richberg: *Stetige streng pseudokonvexe Funktionen.* Math. Ann. **175**, 257–286 (1968).

[Re56] R. Remmert: *Sur les espaces analytiques holomorphiquement convexes.* C. R. Acad. Sci. Paris **243**, 118–121 (1956).

[Re57] R. Remmert: *Holomorphe und meromorphe Abbildungen komplexer Räume.* Math. Ann. **133**, 328–370 (1957).

[ReSt53] R. Remmert, K. Stein: *Über die wesentlichen Singularitäten analytischer Mengen.* Math. Ann. **126**, 263–306 (1953).

[dRh84] G. de Rham: *Differentiable Manifolds.* Grundlehren **266**, Springer, Berlin–Heidelberg–New York (1984).

[Ru74] W. Rudin: *Functional Analysis.* Tata McGraw-Hill (1974).

[Schue92] J. Schürmann: *Einbettungen Steinscher Räume in affine Räume minimaler Dimension.* Preprint, Münster (1992).

[Schue97] J. Schürmann: *Embeddings of Stein spaces into affine spaces of minimal dimension.* Math. Ann. **307**, 381–399 (1997).

[Schw50] L. Schwartz: *Théorie des distributions I, II.* Hermann, Paris (1950/51).

[Se53] J.-P. Serre: *Quelques problèmes globaux relatifs aux variétés de Stein.* Colloque sur les fonctions analytiques de plusieurs variables complexes tenu à Bruxelles, Masson, Paris (1953).

[Siu84] Y.T. Siu: *A vanishing theorem for semipositive line bundles over non-Kähler manifolds.* J. Diff. Geom. **19**, 431–452 (1984).

[Sko77] H. Skoda: *Fibrés holomorphes à base et à fibre de Stein.* Invent. Math. **43**, 97–107 (1977).

[St72] J.-L. Stehlé: *Plongements du disque dans* $\mathbb{C}^2$. Seminaire P. Lelong 1970/71. Springer Lecture Notes in Math. **275**, 119–130 (1972).

[Vo73] N. Vormoor: *Topologische Fortsetzung biholomorpher Funktionen auf dem Rande bei beschränkten streng-pseudokonvexen Gebieten im* $\mathbb{C}^n$ *mit* C^∞*-Rand.* Math. Ann. **204**, 239–261 (1973).

[vdW66] B.L. van der Waerden: *Algebra I.* Springer, Heidelberg (1966).

[War71] F.W. Warner: *Foundations of differentiable manifolds and Lie groups.* Scott, Foresman and Co, Glenview, Illinois (1971).

[Wh36] H. Whitney: *Differentiable manifolds.* Ann. Math. **37**, 645–680 (1936).

Index of Notation

$\mathbb{N}, \mathbb{Z}, \mathbb{Q}, \mathbb{R}, \mathbb{C}$	natural numbers, integers, etc.
$\mathbb{N}_0$	$= \mathbb{N} \cup \{0\}$
$\mathbb{R}_+$	$= \{x \in \mathbb{R} : x > 0\}$
$\mathbb{C}^*$	$= \mathbb{C} - \{0\}$
i	imaginary unit $\sqrt{-1}$
V^*	real dual space of a real (or complex) vector space
V'	complex dual space of a complex vector space
$F(V)$	$\mathbb{C}$-valued real linear forms on a complex vector space
$M_{p,q}(k)$	k-valued matrices with p rows and q columns
$M_n(k)$	k-valued square matrices of order n
$\mathrm{GL}_n(\mathbb{C})$	general linear group
$\mathbf{A}\,\mathbf{B}$ (or $\mathbf{A} \cdot \mathbf{B}$)	product of matrices
$\mathbf{z}$, $\mathbf{z}^{\,t}$	vector $\mathbf{z} = (z_1, \ldots, z_n)$ and transposed vector
$\left(\mathbf{z} \,\middle|\, \mathbf{w}\right)_m$	standard Euclidean scalar product in $\mathbb{R}^m$
$\lVert\mathbf{z}\rVert$, $\mathrm{dist}(\mathbf{z}, \mathbf{w})$	Euclidean norm and distance
$\left\langle \mathbf{z} \,\middle|\, \mathbf{w} \right\rangle$	standard Hermitian scalar product
$\lvert\mathbf{z}\rvert$	sup-norm (or maximum-norm)
$\mathsf{D}_r(a)$	open disk (in $\mathbb{C}$) around a with radius r
D	unit disk $\mathsf{D}_1(0)$
$\mathsf{B}_r(\mathbf{z})$	open ball with radius r around $\mathbf{z}$
$\mathsf{P}^n(\mathbf{z}, \mathbf{r})$	polydisk in $\mathbb{C}^n$ around $\mathbf{z}$ with polyradius $\mathbf{r}$
$\mathsf{P}^n_r(\mathbf{z})$	polydisk with polyradius $\mathbf{r} = (r, \ldots, r)$
P^n	unit polydisk $\mathsf{P}^n_1(\mathbf{0})$
$\tau : \mathbb{C}^n \to \mathscr{V}$	natural projection onto absolute value space
$\mathsf{P}_{\mathbf{z}}$	polydisk $\mathsf{P}^n(\mathbf{0}, \tau(\mathbf{z}))$
$\mathsf{T}^n(\mathbf{z}, \mathbf{r})$	distinguished boundary (torus)
$\mathsf{T}_{\mathbf{z}}$	distinguished boundary $\mathsf{T}^n(\mathbf{0}, \tau(\mathbf{z}))$
$U \subset\subset V$	U lies relatively compact in V
$C_f(\mathbf{z})$	Cauchy integral of $f : \mathsf{T} \to \mathbb{C}$ at $\mathbf{z} \in \mathsf{P}^n$
$\nabla f(\mathbf{z})$	holomorphic gradient $(f_{z_1}(\mathbf{z}), \ldots, f_{z_n}(\mathbf{z}))$
$\overline{\nabla} f(\mathbf{z})$	antiholomorphic gradient $(f_{\overline{z}_1}(\mathbf{z}), \ldots, f_{\overline{z}_n}(\mathbf{z}))$
$\nabla_{\mathbf{x}} f(\mathbf{z})$	$= (f_{x_1}(\mathbf{z}), \ldots, f_{x_n}(\mathbf{z}))$
$\nabla_{\mathbf{y}} f(\mathbf{z})$	$= (f_{y_1}(\mathbf{z}), \ldots, f_{y_n}(\mathbf{z}))$

$D^\nu f(\mathbf{z})$	higher partial derivative
$Df(\mathbf{z})$	total real derivative of f at $\mathbf{z}$
$(\partial f)_\mathbf{z}(\mathbf{w})$	$= f_{z_1}(\mathbf{z})w_1 + \cdots + f_{z_n}(\mathbf{z})w_n$
$(\overline{\partial} f)_\mathbf{z}(\mathbf{w})$	$= f_{\overline{z}_1}(\mathbf{z})\overline{w}_1 + \cdots + f_{\overline{z}_n}(\mathbf{z})\overline{w}_n$
$(df)_\mathbf{z}$	$= (\partial f)_\mathbf{z} + (\overline{\partial} f)_\mathbf{z}$
$J_\mathbf{f}(\mathbf{z})$	complex (or holomorphic) Jacobian matrix
$J_{\mathbb{R},\mathbf{f}}(\mathbf{z})$	real Jacobian matrix
$\mathscr{E}(G)$	space of smooth functions on G
$\mathcal{O}(G)$	space of holomorphic functions on G
$\mathsf{t} = (\mathbf{z}, \mathbf{w})$	tangent vector at $\mathbf{z}$ with direction $\mathbf{w}$
$T_\mathbf{z}$	tangent space at $\mathbf{z}$
$\dot{\alpha}(0)$	tangent vector $(\alpha(0), \alpha'(0))$
$N(f_1, \ldots, f_q)$	common zero set of $f_1, \ldots, f_q$
$\dot{A}$ (or $\operatorname{Reg}(A)$)	set of regular points of A
$\operatorname{Sing}(A)$	set of singular points of A
$\operatorname{rk}_\mathbf{z}(f_1, \ldots, f_q)$	rank of the Jacobian $J_{(f_1,\ldots,f_q)}$
$(\mathsf{P}^n, \mathsf{H})$	Euclidean Hartogs figure
$(\widetilde{P}, \widetilde{H})$	general Hartogs figure
bS_t	boundary of an analytic disk $S_t = \varphi_t(\mathsf{D})$
$C_M(\mathbf{z})$	connected component of M containing $\mathbf{z}$
δ_G	boundary distance (for a domain G)
$\operatorname{Lev}(f)(\mathbf{z}, \mathbf{w})$	Levi form $\sum_{\nu,\mu} f_{z_\nu \overline{z}_\mu}(\mathbf{z}) w_\nu \overline{w}_\mu$
$H_\mathbf{z}(\partial G)$	complex (or holomorphic) tangent space of ∂G
$\operatorname{Hess}(f)(\mathbf{z}, \mathbf{w})$	Hessian $\sum_{\nu,\mu} f_{x_\nu x_\mu}(\mathbf{z}) w_\nu w_\mu$
$\widehat{K}_G$	holomorphically convex hull of K in G
$\mathcal{G}_1 \prec \mathcal{G}_2$	the Riemann domain $\mathcal{G}_1$ is contained in $\mathcal{G}_2$
$H_\mathscr{F}(\mathcal{G})$, $H(\mathcal{G})$	$\mathscr{F}$-hull of $\mathcal{G}$ and envelope of holomorphy
$\breve{G}$	set of accessible boundary points of G
$\mathbb{C}[\![\mathbf{z}]\!]$	ring of formal power series
$B_\mathbf{t}$	$= \{f \in \mathbb{C}[\![\mathbf{z}]\!] : \Vert f \Vert_\mathbf{t} < \infty\}$
H_n	ring of convergent power series
$\mathfrak{m}$	maximal ideal in H_n
I^*, $I^\times$	set of nonzero element, set of units in an ideal I
$I^0[u]$	set of monic polynomials with coefficients in I
$f_\mathbf{z}$	germ of the holomorphic function f at $\mathbf{z}$

$\omega(u, \mathbf{z})$	pseudopolynomial
$D(\omega)$	algebraic derivative of ω
Δ_ω, D_ω	discriminant and discriminant set of ω
$\dim_{\mathbf{z}}(A)$, $\dim(A)$	local and global dimension of an analytic set
$\overline{\mathbb{C}}$	Riemann sphere $\mathbb{C} \cup \{\infty\}$
$T_a(X)$	tangent space of a manifold
$F_* = (F_*)_x$	tangential map
$X \times_Z Y$	fiber product of manifolds
K_X, $T(X)$	canonical bundle and tangent bundle on X
$\Gamma(U, V) = \mathscr{O}(U, V)$	space of holomorphic sections in a bundle V
$\mathscr{E}(U, V)$	space of smooth sections in V
$V \oplus W$, $V \otimes W$	Whitney sum and tensor product of bundles
V', F^k	complex dual bundle of V, tensor power of F
f^*V, V/W	pullback and quotient bundle
$N_X(Y)$	normal bundle of Y in X
$\mathscr{O}_X$, $\mathscr{O}_X^*$	trivial bundles $X \times \mathbb{C}$ and $X \times \mathbb{C}^*$
$C^i(\mathscr{U}, V)$	Čech cochains with values in V
$Z^i(\mathscr{U}, V)$	Čech cocycles
$B^i(\mathscr{U}, V)$	Čech coboundaries
$H^i(\mathscr{U}, V)$	Čech cohomology group with values in V
$H^i(X, V)$	cohomology group with values in V
$H_q(X), H^q(X)$	singular homology and cohomology of X
$\mathscr{O}_x$	ring of germs of holomorphic functions
$P(m)$	polar set of a meromorphic function
$\mathscr{M}(X)$	set of meromorphic functions on X
$\text{div}(m)$	divisor of a meromorphic function
$[Z]$	line bundle associated to the divisor Z
$\text{Pic}(X)$, $\mathscr{D}(X)$	Picard group and set of all divisors
$T^n = \mathbb{C}^n/\Gamma$	n-dimensional complex torus
$\mathbb{P}^n$	complex projective space
$\text{St}(k, n)$, $G_{k,n}$	Stiefel manifold and Grassmannian
$\text{pl} : G_{k,n} \to \mathbb{P}^N$	Plücker embedding
$\mathscr{O}(1)$, $\mathscr{O}(-1)$	hyperplane bundle and tautological bundle
$\mathbb{O}^m$	Osgood space $\mathbb{P}^1 \times \cdots \times \mathbb{P}^1$ (m times)

$c(h) = c(L)$	Chern class of a cocycle h (of a line bundle L)
$\bigwedge^r F(X)$	bundle of r-dimensional differential forms
Ω^p_X	bundle of holomorphic p-forms on X
$\mathscr{A}^r(U)$	r-forms
$\mathscr{A}^{p,q}(U)$	forms of type (p,q) on U
$\varphi \wedge \psi$	wedge product
$d\varphi, \partial\varphi, \overline{\partial}\varphi$	total, holomorphic and antiholomorphic differential
d^c	$= \mathrm{i}(\overline{\partial} - \partial)$ (such that $dd^c = 2\mathrm{i}\partial\overline{\partial}$)
$\mathrm{Ch}_f(z)$	$= (1/(2\pi\mathrm{i})) \int_G f(\zeta)/(\zeta - z)\, d\zeta \wedge d\overline{\zeta}$
$H^{p,q}(X)$	Dolbeault group of type (p,q) on X
$H^r(X)$	rth de Rham group of X
ω_H	fundamental form of the Hermitian metric H
$\mathrm{PU}(n+1)$	group of projective unitary transformations
dV	volume element
$*\varphi$	Hodge star operator ($\mathscr{A}^r(X) \to \mathscr{A}^{2n-r}(X)$)
(φ, ψ)	inner product $\int_X \varphi \wedge \overline{*}\psi$, with $\overline{*}\psi = \overline{*\psi}$
T_ψ	current associated to a form ψ
T_M	current associated to a submanifold M
$\mathscr{A}^r(X)'$	space of currents of degree $2n - r$
$\delta, \vartheta, \overline{\vartheta}$	adjoint operators of d, ∂, and $\overline{\partial}$
$L\varphi$	$= (1/2)\varphi \wedge \omega$, with Kähler form ω
Λ	adjoint operator of L
Δ	real Laplacian $(d\delta + \delta d)$
$\Box$	complex Laplacian $(\overline{\partial}\overline{\vartheta} + \overline{\vartheta}\overline{\partial})$
$\mathscr{H}^r(X)$	space of harmonic r-forms on X
$\mathscr{H}^{p,q}(X)$	space of harmonic (p,q)-forms
$\beta_r(X)$, $\beta_{p,q}(X)$	Betti numbers
Z_F	zero section of the bundle F
$\mathcal{H}_n$	generalized Siegel upper halfplane
$L^2_{p,q}(\Omega)$	Hilbert space closure of $\mathscr{A}^{p,q}(\overline{\Omega})$
$\mathrm{dom}(T)$	domain of the operator T
N	Neumann operator $N : L^2_{p,q}(\Omega) \to \mathrm{dom}(\Box)$

$\mathscr{S}$	rapidly decreasing functions
$H_s(\mathbb{R}^N)$	Sobolev space of index s
$\lVert\lvert f \rVert\rvert_s$	tangential Sobolev norm
$N(\Omega)$	Nebenhülle of Ω
F_X	pseudo-differential metric on X
$K(\mathbf{z}, \mathbf{w})$	Bergman kernel
$P : L^2(\Omega) \to \mathscr{H}$	Bergman projection
$A^\infty(\Omega)$	$= \mathscr{A}^{n,0}(\overline{\Omega}) \cap \mathcal{O}(\Omega)$

Index

186 Ramakrishnan/Valenza. Fourier Analysis on Number Fields.
187 Harris/Morrison. Moduli of Curves.
188 Goldblatt. Lectures on the Hyperreals: An Introduction to Nonstandard Analysis.
189 Lam. Lectures on Modules and Rings.
190 Esmonde/Murty. Problems in Algebraic Number Theory.
191 Lang. Fundamentals of Differential Geometry.
192 Hirsch/Lacombe. Elements of Functional Analysis.
193 Cohen. Advanced Topics in Computational Number Theory.
194 Engel/Nagel. One-Parameter Semigroups for Linear Evolution Equations.
195 Nathanson. Elementary Methods in Number Theory.
196 Osborne. Basic Homological Algebra.
197 Eisenbud/Harris. The Geometry of Schemes.
198 Robert. A Course in p-adic Analysis.
199 Hedenmalm/Korenblum/Zhu. Theory of Bergman Spaces.
200 Bao/Chern/Shen. An Introduction to Riemann–Finsler Geometry.
201 Hindry/Silverman. Diophantine Geometry: An Introduction.
202 Lee. Introduction to Topological Manifolds.
203 Sagan. The Symmetric Group: Representations, Combinatorial Algorithms, and Symmetric Functions.
204 Escofier. Galois Theory.
205 Félix/Halperin/Thomas. Rational Homotopy Theory. 2nd ed.
206 Murty. Problems in Analytic Number Theory.
Readings in Mathematics
207 Godsil/Royle. Algebraic Graph Theory.
208 Cheney. Analysis for Applied Mathematics.
209 Arveson. A Short Course on Spectral Theory.
210 Rosen. Number Theory in Function Fields.
211 Lang. Algebra. Revised 3rd ed.
212 Matoušek. Lectures on Discrete Geometry.
213 Fritzsche/Grauert. From Holomorphic Functions to Complex Manifolds.